Sergei M. Kopeikin (Ed.)

Frontiers in Relativistic Celestial Mechanics

De Gruyter Studies in Mathematical Physics

Volume 21

This Festschrift is dedicated to Professor *Victor A. Brumberg*,
for his enthusiasm and devotion to the science of relativistic celestial mechanics,
and to celebrate his 80th birthday.

List of contributors

Prof. Dr. Thibault Damour
Institute of Advanced Scientific Studies (IHÉS)
35 route de Chartres
Bures sur Yvette, F-91440
France
E-mail: damour@ihes.fr

Prof. Dr. Toshifumi Futamase
Tohoku University
Astronomical Institute
Sendai 980-8578
Japan
E-mail: tof@astr.tohoku.ac.jp

Prof. Dr. Sergei Kopeikin
University of Missouri
Department of Physics & Astronomy
223 Physics Bldg.
Columbia, MO 65211
USA
E-mail: kopeikins@missouri.edu

Prof. Dr. Gerhard Schäfer
Friedrich Schiller University Jena
Institute of Theoretical Physics
Max-Wien-Platz 1
D-07743 Jena
Germany
E-mail: gos@tpi.uni-jena.de

Prof. Dr. Michael Soffel
Dresden University of Technology
Lohrmann Observatory TU Dresden
Planetary Geodesy
Mommsenstr 13
01062 Dresden
Germany
E-mail: soffel@rcs.urz.tu-dresden.de

Dr. Pavel Korobkov
Solovetsky Monastery
Arkhangelsk Region 164070
Russia

Dr. Alexander Petrov
Moscow Lomonosov State University
Sternberg Astronomical Institute
Universitetskiy Prospect 13
Moscow 119992
Russia
E-mail: alex.petrov55@gmail.com

Dr. Xie Yi
Nanjing University
Department of Astronomy
22 Hankou Road
Nanjing Jiangsu 210093
P. R. China
E-mail: yixie@nju.edu.cn

Contents

List of figures

Preface

The science of relativistic celestial mechanics is an essential branch of the modern gravitational physics, a branch exploring the fundamental structure of spacetime by studying motion of massive bodies such as black holes, stars, planets, as well as elementary particles, including photons, in gravitational field. It establishes basic theoretical principles for calculation and interpretation of various relativistic effects and phenomena observed in astrophysical stellar systems and in the solar system. Relativistic celestial mechanics of massless particles like photons is more known among astronomers as relativistic astrometry. An indefeasible branch of gravitational physics, it is required to map the coordinate description of motion of celestial bodies into parameter space of observables. Theoretical progress in understanding the orbital motion of celestial bodies would be inconceivable without a corresponding improvement in mathematical description of motion of light rays in stationary and time-dependent gravitational field.

Relativistic celestial mechanics has received a special attention in the gravitational-wave astronomy. Being on its way to direct detection of gravitational waves emitted by coalescing binary stars, the gravitational-wave astronomy urgently needs highly precise templates of gravitational waves emitted by the stars at the very last stage of their orbital motion, just a few seconds before the stars collide and a catastrophic supernova explosion takes place. Therefore, development of theoretical tools of relativistic celestial mechanics has a fundamental significance for achieving further progress in gravitational-wave astronomy which is expected to become a primary experimental tool bringing much deeper understanding of the nature of gravitational field and the underlying geometric structure of the spacetime manifold.

Relativistic celestial mechanics was a subject of active research by many notable scientists, including A. Einstein, H. Lorentz, V. A. Fock, T. Levi-Civita, L. Infeld, S. Chandrasekhar, J. Ehlers, G. C. McVittie, and others who elaborated on various approaches to the equations of motion of celestial bodies and the theory of astronomical observations in general relativity. More recently, a valuable contribution to relativistic celestial mechanics was made by T. Damour, G. Schäfer, M. Soffel, C. M. Will, K. Nordtvedt, T. Futamase, K. S. Thorne, W. G. Dixon, L. Blanchet, and I. Rothstein. A key figure of relativistic celestial mechanics of the second half of twentieth century has been Victor A. Brumberg, a scholar who presently lives in Boston (USA) and who is still active in research. Victor A. Brumberg has made a significant contribution to general relativity and the science of relativistic planetary ephemerides of the solar system. He mentored and inspired many researchers around the globe (including the Editor of this book) to start working in the field of relativistic celestial mechanics. The very term "relativistic celestial mechanics" was introduced by Victor. A. Brumberg in his famous monograph "Relativistic Celestial Mechanics" published in 1972 by Nauka (Science) – the main scientific publisher of the USSR – in Moscow. For the next two decades this monograph remained the most authoritative reference and the source of invaluable information for researchers working on relativistic equations of motion and experimental testing of general relativity. Victor A. Brumberg received the 2008 Brower Award from the Division of Dynamic Astronomy of the American Astronomical Society. The Brouwer Award was established to recognize outstanding contributions to the field of dynamical astronomy, including celestial mechanics, astrometry, geophysics, stellar systems, galactic, and extragalactic dynamics.

This book is a first volume of Festschrift aimed to honor the scientific influence and achievements of V. A. Brumberg, and to celebrate his 80th birthday which took place on February 12, 2013. The book appears on the eve of another remarkable date – 100 years of Einstein's general relativity – the theory which dramatically changed the world of theoretical physics by opening new fascinating opportunities in the scientific study of fundamental laws of Nature. The volume consists of seven chapters discussing the recent theoretical advances in relativistic celestial mechanics and related areas of theoretical physics and astronomy.

Chapter 1, written by T. Damour, introduces the amazingly rich mathematics of the relativistic two-body problem. Solution of this problem within the Newtonian mechanics is cornerstone material that can be found in any textbook on celestial mechanics. On the other hand, complete solution of this problem within general relativity has not been yet obtained, even though it has been subject of numerous analytical investigations. The root of the difficulty is lying in the nonlinear character of gravitational interaction in Einstein's theory of gravity, which prevents us from finding an exact solution to the problem. Hence, the analytic solution can be ascertained only by making use of successive approximations. The method includes complicated, often diverging integrals which require development of regularization technique based on the theory of distributions. Additional difficulties arise due to the emission of gravitational waves by the two-body system, an effect generating a back reaction on the motion of the bodies – the so-called radiation-reaction force. After reviewing some of the methods used to tackle these problems, Chapter 1 focuses on a new, recently introduced approach to the motion and radiation of (comparable-mass) binary systems: the effective-onebody (EOB) formalism. The basic elements of this formalism are reviewed, and some of its recent developments are discussed. Several recent tests of EOB predictions against numerical simulations have shown the aptitude of the EOB formalism to provide accurate description of the dynamics and radiation of various binary systems (comprising black holes or neutron stars) in regimes that are inaccessible to other analytical approaches such as the last orbits and the merger of comparable mass black holes. Chapter 1 provides weighty arguments that, in synergy with numerical simulations, the post-Newtonian theory and gravitational self-force (GSF) computations, the EOB formalism is likely to provide an efficient way of accurately computing the numerous template waveforms that are needed for the purposes of gravitational wave data analysis.

Chapter 2, written by G. Schäfer, continues theoretical analysis of the two-body problem in general relativity, by making use of the advanced Hamiltonian technique introduced by Arnovitt, Deser, and Misner (ADM formalism). The Hamiltonian setting of general relativity allows a very elegant and transparent treatment of the dynamics and motion of gravitating systems. Crucial in that context is the computation of the reduced Hamiltonian which generates the dynamics of both the gravitating objects and the gravitational field. Based on the framework of post-Newtonian approximation, Chapter 2 covers the dynamics and motion of spinning compact binaries up to the fourth post-Newtonian approximation.

Chapter 3, written by Y. Xie and S. Kopeikin, presents a covariant theory of post-Newtonian equations of translational motion of extended bodies in an N-body system. It significantly extends the results obtained in 1970–80th by W. G. Dixon. The new theory is based on the combined BK-DSX theory extended to the realm of the scalar-tensor theory of gravity. It introduces one more type of multipole moments to the formalism – the scalar-type moments. The chapter explains how to build the local and global coordinates in a system of N extended bodies, and offers a procedure intended to derive the translational equations of motion of the bodies, including all internal multipoles. It is proven that any integral moment, which depends on the internal structure of the bodies in a way different from the "canonical" Blanchet–Damour moments, vanishes from the translational equations of motion. Finally, a covariant form of the post-Newtonian equations of motion of extended bodies, with all internal multipoles taken into account, is derived by applying a technique proposed by Thorne and Hartle in 1985. The translational equations of motion derived in this way represent a profound generalization of the Mathisson–Papapetrou–Dixon equations of motion.

Chapter 4, written by M. Soffel, furnishes an account of the Damour–Soffel–Xu (DSX) formalism of relativistic reference frames in N-body system. The DSX formalism is an extension of the formalism advanced in 1988 by Brumberg and Kopeikin (the BK formalism) to build the post-Newtonian theory of astronomical reference frames in the solar system. The BK-DSX theory is based on the complementary use of N local coordinate charts attached to each body, which are built to describe rotation and local dynamics of the body, and of a global coordinate chart, which is intended to describe the orbital motion of the bodies. The advantage of the DSX formalism, compared to the BK formalism, is in the systematic use of well-defined mass-type and spin-type multipole moments of the extended bodies. Chapter 4 explains the DSX formalism in a concise but mathematically rigorous form.

Chapter 5, written by P. Korobkov and S. Kopeikin, delivers theoretical tools for solving the problem of propagation of photons through multipolar gravitational field of an isolated astronomical system emitting gravitational waves. The solution is written in the first post-Minkowskian approximation of general relativity. The Chapter opens with an introduction to the linearized theory of retarded gravitational potentials of the Lienard–Wiechert type. The Chapter then deals with derivation of differential equations of light geodesics with retarded argument. Mathematical technique of integrating these equations is proposed, and a solution is found in a closed form. It is demonstrated that the leading-order observable relativistic effects depend on the value of the multipoles of the isolated system and their time derivatives taken at the retarded instant of time. This retardation is caused by finite speed of propagation of gravity, and for this reason the relativistic effects do not depend on the integrated values of the multipoles taken along the past world line of the isolated system. The integration technique reproduces the known results of integration of equations of light rays in the stationary approximation of a gravitational lens and in the approximation of a plane gravitational wave. Two limiting cases of small and large impact parameters of a light ray with respect to the isolated system are worked out in more detail. It is shown that in case of a small impact parameter the leading-order terms in the solution for light propagation depend neither on radiative nor on intermediate zone components of the gravitational field, but the main effect comes from the near-zone values of the multipole moments. This radiative-zone effacing property makes it much more difficult (but not impossible!) to directly detect gravitational waves by astronomical instruments than it was assumed by some researchers. Chapter 5 also presents analytical treatment of time-delay and light-ray bending in the case of large impact parameter corresponding to the approximation of plane gravitational wave. Explicit expressions for the time delay and the deflection angle of the light ray are obtained in terms of the transverse-traceless (TT) multipole moments of the gravitating system. This result can be directly applied to interpretation of observables in gravitational wave interferometers.

Development of the canonical theory of post-Newtonian approximations in relativistic celestial mechanics relies upon the key concept of an isolated astronomical system, under assumption that background spacetime is flat. The standard post-Newtonian theory of motion is instrumental in explanation of the existing experimental data on binary pulsars, satellite, and lunar laser ranging, and in building precise ephemerides of planets in the solar system. Recent cosmological studies indicate that the standard post-Newtonian mechanics fails to describe more subtle dynamical effects in the small-scale structure formation and in the motion of galaxy clusters comprising astronomical systems. In those settings, the curvature of the expanding universe interacts with the local gravitational field of the astronomical system and, as such, cannot be ignored. Therefore, working out theoretical foundations of relativistic celestial mechanics of isolated astronomical system residing on cosmological manifold is worthwhile. Additional motivation for this comes from the gravitational wave astronomy which will study relativistic celestial mechanics of binary systems in very distant galaxies residing at the edge of the visible universe. Dynamical evolution of the binaries on a cosmological background is primarily governed by multipolar structure of its own gravitational field, but is also intrinsically connected with the cosmological parameters of the background manifold. These parameters are determined by the content of the substance filling up the universe, whose most enigmatic components are the dark matter and dark energy. Tracking down the orbital motion of binary systems in distant galaxies at gravitational wave observatories is promising for doing precise cosmology. It is very likely that observation of binaries with gravitational wave detectors will supersede the precision of measurement of cosmological parameters by radio astronomical technique. These interesting questions are illuminated in Chapters 6 and 7 of this book.

Chapter 6, authored by T. Futamase, outlines the results of his research on the emergence of the cosmological metric in a lumpy universe, a line of study known as the averaging problem in cosmology. The Chapter also discusses the gravitational backreaction by local nonlinear inhomogeneities on the cosmic expansion, in the framework of general relativity. The problem became important after the discovery of the cosmic acceleration associated with the presence of dark energy. After a brief review of the subject, T. Futamase presents in detail his own approach to analytical calculation of the backreaction, which allows him to overview the apparent discrepancies between previous works using different approaches and gauges. Chapter 6 partially resolves these discrepancies by defining the spatially averaged energy density of matter as a conserved quantity referred to a sufficiently large volume of comoving space. It is shown that the backreaction behaves like a positive-curvature term in the averaged Friedmann–Lemître–Robertson–Walker (FLRW) universe. It neither accelerates nor decelerates the cosmic expansion in a matter-dominated universe, while the cosmological constant induces a new type of backreaction with the equation-of-state parameter being $-4/3$. However, the effective energy density remains negative, and thus it decreases the acceleration.

Chapter 7, written by A. Petrov and S. Kopeikin, extends the post-Newtonian approximation of general relativity to the realm of cosmology, by making use of a geometric theory of Lagrangian perturbations of an FLRW cosmological manifold. The Lagrangian for a perturbed cosmological model includes the dark matter, the dark energy, and the ordinary baryonic matter. The Lagrangian is decomposed in an asymptotic Taylor series around a background FLRW manifold, with the small parameter being the magnitude of the metric–tensor perturbation. Each term of the series decomposition is kept gauge invariant. The asymptotic nature of the Lagrangian decomposition does not require the post-Newtonian perturbations to be small, though computationally it works most effectively when the perturbed metric is close to the background one. The Lagrangian of dark matter is treated as an ideal fluid described by an auxiliary scalar field called the Clebsch potential. The dark energy is associated with a single scalar field of an unspecified potential energy. The scalar fields of dark matter and dark energy are taken as independent dynamical variables which play the role of generalized coordinates in the Lagrangian formalism. This allows the authors to implement the powerful methods of variational calculus, to derive gauge-invariant field equations to be used in the post-Newtonian celestial mechanics in an expanding universe. The equations generalize the field equations of the post-Newtonian theory in an asymptotically flat spacetime, by taking into account the cosmological effects without assuming a rather artificial vacuole model of an isolated system (like those proposed by Einstein and Strauss, McVittie, and Bonnor). A new cosmological gauge is proposed, which generalizes the de Donder (harmonic) gauge of the post-Newtonian theory in an asymptotically flat spacetime. The new gauge significantly simplifies the gravitational field equations and reduces them to wave equations, the latter being differential equations of Bessel's type. The new gauge also allows the authors to find out the cosmological models wherein the field equations are fully decoupled and can be solved analytically. The residual gauge freedom is explored and the residual gauge transformations are formulated in the form of wave equations for gauge functions. Chapter 7 demonstrates how cosmological effects interfere with the local distribution of matter of the isolated system and its orbital dynamics. The Chapter also offers a precise mathematical definition of the Newtonian limit for an isolated system residing on a cosmological manifold. The results of the chapter can be useful in the galactic astronomy, to study the dynamics of clusters of galaxies, and in the gravitational wave astronomy, for discussing the impact of cosmological effects on generation and propagation of gravitational waves emitted by coalescing binaries.

Over the past 30 years, relativistic celestial mechanics has experienced radical progress both in theory and in experimental testing of general relativity. The present volume cannot embrace it in its entirety. For further reading on recent developments in relativistic celestial mechanics, we recommend the following review articles and textbooks:

Asada, H., Futamase, T. and Hogan, P., "Equations of Motion in General Relativity," Oxford University Press: Oxford, 2011

Brumberg, V. A., "Celestial mechanics: past, present, future," *Solar System Research*, Vol. 47, Issue 5, pp. 347–358 (2013)

Brumberg, V. A., "Relativistic Celestial Mechanics on the verge of its 100 year anniversary" (Brouwer Award lecture), *Celestial Mechanics and Dynamical Astronomy*, Vol. 106, Issue 3, pp. 209–234 (2010)

Brumberg, V. A., "Relativistic Celestial Mechanics," *Scholarpedia*, Vol. 5, Issue 8, #10669. URL (cited on Jan 12, 2014) http://www.scholarpedia.org/article/Relativistic_Celestial_Mechanics

Brumberg, V. A., "Essential Relativistic Celestial Mechanics," Adam Hilger: Bristol, 1991

Blanchet, L., "Gravitational Radiation from Post-Newtonian Sources and Inspiralling Compact Binaries," Living Reviews in Relativity 5, 3 (2002). URL (cited on Jan. 12, 2014) http://www.livingreviews.org/lrr-2002-3

Damour, T., "The problem of motion in Newtonian and Einsteinian gravity," In: Three Hundred Years of Gravitation, Eds. Hawking, S. W. and Israel, W., Cambridge University Press: Cambridge, 1987. pp. 128–198

Goldberger, W. D. and Rothstein, I. Z., "Effective field theory of gravity for extended objects," Physical Review D, Vol. 73, Issue 10, id. 104029 (2006)

Kopeikin, S., Efroimsky, M. and G. Kaplan, "Relativistic Celestial Mechanics of the Solar System," Wiley-VCH: Berlin, 2011

Soffel, M. and Langhans, R., "Space–Time Reference Systems," Springer-Verlag: Berlin, 2013

February 12, 2014

Editor: Sergei Kopeikin
University of Missouri, USA

Thibault Damour

The general relativistic two-body problem

1 Introduction

The general relativistic problem of motion, i.e. the problem of describing the dynamics of N gravitationally interacting extended bodies, is one of the cardinal problems of Einstein's theory of gravitation. This problem has been investigated from the early days of development of general relativity, notably through the pioneering works of Einstein, Droste, and de Sitter. These authors introduced the post-Newtonian (PN) approximation method, which combines three different expansions: (i) a weak-field expansion ($g_{\mu\nu} - \eta_{\mu\nu} \equiv h_{\mu\nu} \ll 1$); (ii) a slow-motion expansion ($v/c \ll 1$); a near-zone expansion ($\frac{1}{c}\,\partial_t\, h_{\mu\nu} \ll \partial_x h_{\mu\nu}$). PN theory could be easily worked out to derive the first post-Newtonian (1PN) approximation, i.e. the leading-order general relativistic corrections to Newtonian gravity (involving one power of $1/c^2$). However, the use of the PN approximation for describing the dynamics of N extended bodies turned out to be fraught with difficulties. Most of the early derivations of the 1PN-accurate equations of motion of N bodies turned out to involve errors: this is, in particular, the case of the investigations by Droste [1], de Sitter [2], Chazy [3], and Levi-Civita [4]. These errors were linked to incorrect treatments of the internal structures of the bodies. Apart from the remarkable 1917 work of Lorentz and Droste [5] (which seems to have remained unnoticed during many years), the first correct derivations of the 1PN-accurate equations of motion date from 1938, and were obtained by Einstein et al. [6], and Eddington and Clark [7]. After these pioneering works (and the investigations they triggered, notably in Russia [8] and Poland), the general relativistic N-body problem reached a first stage of maturity and became codified in various books, notably in the books of Fock [9], Infeld and Plebanski [10], and in the second volume of the treatise of Landau and Lifshitz (starting, at least, with the 1962 second English edition).

We have started by recalling the early history of the general relativistic problem of motion both because Victor Brumberg has always shown a deep knowledge of this history, and because, as we shall discuss below, some of his research work has contributed to clarifying several of the weak points of the early PN investigations (notably those linked to the treatment of the internal structures of the N bodies).

For many years, the 1PN approximation turned out to be accurate enough for applying Einstein's theory to known N-body systems, such as the solar system, and various binary stars. It is still true today that the 1PN approximation (especially when

Thibault Damour: Institut des Hautes Études Scientifiques (IHÉS), 35 route de Chartres, F-91440, Bures sur Yvette, France

used in its multichart version, see below) is adequate for describing general relativistic effects in the solar system. However, the discovery in the 1970s of binary systems comprising strongly self-gravitating bodies (black holes or neutron stars) has obliged theorists to develop improved approaches to the N-body problem. These improved approaches are not limited (as the traditional PN method) to the case of weakly self-gravitating bodies and can be viewed as modern versions of the Einstein–Infeld–Hoffmann classic work [6].

In addition to the need of considering strongly self-gravitating bodies, the discovery of binary pulsars in the mid-1970s (starting with the Hulse–Taylor pulsar PSR 1913 + 16) obliged theorists to go beyond the 1PN ($O(v^2/c^2)$) relativistic effects in the equations of motion. More precisely, it was necessary to go to the 2.5PN approximation level, i.e. to include terms $O(v^5/c^5)$ beyond Newton in the equations of motion. This was achieved in the 1980s by several groups [11–15]. (Let us note that important progress in obtaining the N-body metric and equations of motion at the 2PN level was achieved by the Japanese school in the 1970s [16–18].)

Motivation for pushing the accuracy of the equations of motion beyond the 2.5PN level came from the prospect of detecting the gravitational wave signal emitted by inspiralling and coalescing binary systems, notably binary neutron star (BNS) and binary black hole (BBH) systems. The 3PN-level equations of motion (including terms $O(v^6/c^6)$ beyond Newton) were derived in the late 1990s and early 2000s [19–22, 80] (they have been recently rederived in [24]). Recently, the 4PN-level dynamics has been tackled in [25–28].

Separately from these purely analytical approaches to the motion and radiation of binary systems, which have been developed since the early days of Einstein's theory, numerical relativity (NR) simulations of Einstein's equations have relatively recently (2005) succeeded (after more than 30 years of developmental progress) to stably evolve binary systems made of comparable mass black holes [29–32]. This has led to an explosion of works exploring many different aspects of strong-field dynamics in general relativity, such as spin effects, recoil, relaxation of the deformed horizon formed during the coalescence of two black holes to a stationary Kerr black hole, high-velocity encounters, etc.; see [33] for a review and [34] for an impressive example of the present capability of NR codes. In addition, recently developed codes now allow one to accurately study the orbital dynamics, and the coalescence of BNSs [35]. Much physics remains to be explored in these systems, especially during and after the merger of the neutron stars (which involves a much more complex physics than the pure-gravity merger of two black holes).

Recently, a new source of information on the general relativistic two-body problem has opened: gravitational self-force (GSF) theory. This approach goes one step beyond the test-particle approximation (already used by Einstein in 1915) by taking into account self-field effects that modify the leading-order geodetic motion of a small mass m_1 moving in the background geometry generated by a large mass m_2. After some ground work (notably by DeWitt and Brehme) in the 1960s, GSF theory has recently undergone rapid developments (mixing theoretical and numerical methods) and can now yield numerical results that yield access to new information on strong-field dynamics in the extreme mass-ratio limit $m_1 \ll m_2$. See Ref. [36] for a review.

Each of the approaches to the two-body problem mentioned so far, PN theory, NR simulations, and GSF theory, has their advantages and their drawbacks. It has become recently clear that the best way to meet the challenge of accurately computing the gravitational waveforms (depending on several continuous parameters) that are needed for a successful detection and data analysis of GW signals in the upcoming LIGO/Virgo/GEO/... network of GW detectors is to combine knowledge from all the available approximation methods: PN, NR, and GSF. Several ways of doing so are a priori possible. For instance, one could try to directly combine PN-computed waveforms (approximately valid for large enough separations, say $r \gtrsim 10\,G(m_1 + m_2)/c^2$) with NR waveforms (computed with initial separations $r_0 > 10\,G(m_1 + m_2)/c^2$ and evolved up to merger and ringdown). However, this method still requires too much computational time, and is likely to lead to waveforms of rather poor accuracy, see, e.g. [37, 38].

On the other hand, 5 years before NR succeeded in simulating the late inspiral and the coalescence of BBHs, a new approach to the two-body problem was proposed: the effective one body (EOB) formalism [39–42]. The basic aim of the EOB formalism is to provide an analytical description of both the motion and the radiation of coalescing binary systems over the entire merger process, from the early inspiral, right through the plunge, merger, and final ringdown. As early as 2000 [40] this method made several quantitative and qualitative predictions concerning the dynamics of the coalescence, and the corresponding GW radiation, notably: (i) a blurred transition from inspiral to a “plunge” that is just a smooth continuation of the inspiral, (ii) a sharp transition, around the merger of the black holes, between a continued inspiral and a ring-down signal, and (iii) estimates of the radiated energy and of the spin of the final black hole. In addition, the effects of the individual spins of the black holes were investigated within the EOB [42, 43] and were shown to lead to a larger energy release for spins parallel to the orbital angular momentum, and to a dimensionless rotation parameter J/E^2 always smaller than unity at the end of the inspiral (so that a Kerr black hole can form right after the inspiral phase). All those predictions have been broadly confirmed by the results of the recent numerical simulations performed by several independent groups (for a review of numerical relativity results and references see [33]). Note that, in spite of the high computer power used in NR simulations, the calculation, checking, and processing of one sufficiently long waveform (corresponding to specific values of the many continuous parameters describing the two arbitrary masses, the initial spin vectors, and other initial data) takes on the order of 1 month. This is a very strong argument for developing analytical models of waveforms. For a recent comprehensive comparison between analytical models and numerical waveforms see [44].

In this work, we shall briefly review only a few facets of the general relativistic two-body problem (see, e.g. [45] and [46] for recent reviews dealing with other facets of, or approaches to, the general relativistic two-body problem). First, we shall recall the essential ideas of the multichart approach to the problem of motion, having especially in mind its application to the motion of compact binaries, such as BNS or BBH systems. Then we shall focus on the EOB approach to the motion and radiation of binary systems, from its conceptual framework to its comparison to NR simulations.

2 Multichart approach to the N-body problem

The traditional (text book) approach to the problem of motion of N separate bodies in GR consists of solving, by successive approximations, Einstein's field equations (we use the signature $- + ++$)

$$R_{\mu\nu} - \frac{1}{2} R\, g_{\mu\nu} = \frac{8\pi G}{c^4} T_{\mu\nu}\,, \tag{2.1}$$

together with their consequence

$$\nabla_\nu T^{\mu\nu} = 0\,. \tag{2.2}$$

To do so, one assumes some specific matter model, say a perfect fluid,

$$T^{\mu\nu} = (\varepsilon + p)\, u^\mu u^\nu + p\, g^{\mu\nu}\,. \tag{2.3}$$

One expands (say in powers of Newton's constant) the metric,

$$g_{\mu\nu}(x^\lambda) = \eta_{\mu\nu} + h^{(1)}_{\mu\nu} + h^{(2)}_{\mu\nu} + \cdots\,, \tag{2.4}$$

and use the simplifications brought by the "post-Newtonian" approximation ($\partial_0 h_{\mu\nu} = c^{-1}\partial_t h_{\mu\nu} \ll \partial_i h_{\mu\nu}$; $v/c \ll 1$, $p \ll \varepsilon$). Then one integrates the local material equation of motion (2.2) over the volume of each separate body, labelled say by $a = 1, 2, \ldots, N$. In so doing, one must define some "center of mass" z^i_a of body a, as well as some (approximately conserved) "mass" m_a of body a, together with some corresponding "spin vector" S^i_a and, possibly, higher multipole moments.

An important feature of this traditional method is to use a *unique coordinate chart* x^μ to describe the full N-body system. For instance, the center of mass, shape, and spin of each body a are all described within this common coordinate system x^μ. This use of a single chart has several inconvenient aspects, even in the case of weakly self-gravitating bodies (as in the solar system case). Indeed, it means for instance that a body which is, say, spherically symmetric in its own "rest frame" X^α will appear as deformed into some kind of ellipsoid in the common coordinate chart x^μ. Moreover, it is not clear how to construct "good definitions" of the center of mass, spin vector, and higher multipole moments of body a, when described in the common coordinate chart x^μ. In addition, as we are possibly interested in the motion of strongly self-gravitating bodies, it is not a priori justified to use a simple expansion of the type (2.4) because $h^{(1)}_{\mu\nu} \sim \sum_a Gm_a/(c^2\,|\boldsymbol{x} - \boldsymbol{z}_a|)$ will not be uniformly small in the common coordinate system x^μ. It will be small if one stays far away from each object a, but, it will become of order unity on the surface of a compact body.

These two shortcomings of the traditional "one-chart" approach to the relativistic problem of motion can be cured by using a *"multichart" approach*. The multichart approach describes the motion of N (possibly, but not necessarily, compact) bodies by using $N + 1$ separate coordinate systems: (i) one *global* coordinate chart x^μ ($\mu = 0, 1, 2, 3$) used to describe the spacetime outside N "tubes," each containing one body, and (ii) N *local* coordinate charts X^α_a ($\alpha = 0, 1, 2, 3$; $a = 1, 2, \ldots, N$) used to describe the spacetime in and around each body a. The multichart approach was first used to discuss the motion of black holes and other compact objects [47–54]. Then it was also found to be very convenient for describing, with the high-accuracy required for dealing with modern technologies such as VLBI, systems of N weakly self-gravitating bodies, such as the solar system [55, 56].

The essential idea of the multichart approach is to combine the information contained in *several expansions*. One uses both a global expansion of the type (2.4) and several local expansions of the type

$$G_{\alpha\beta}(X^\gamma_a) = G^{(0)}_{\alpha\beta}(X^\gamma_a; m_a) + H^{(1)}_{\alpha\beta}(X^\gamma_a; m_a, m_b) + \cdots\,, \tag{2.5}$$

where $G^{(0)}_{\alpha\beta}(X; m_a)$ denotes the (possibly strong-field) metric generated by an isolated body of mass m_a (possibly with the additional effect of spin).

The separate expansions (2.4) and (2.5) are then "matched" in some overlapping domain of common validity of the type $Gm_a/c^2 \lesssim R_a \ll |\boldsymbol{x} - \boldsymbol{z}_a| \ll d \sim |\boldsymbol{x}_a - \boldsymbol{x}_b|$ (with $b \neq a$), where one can relate the different coordinate systems by expansions of the form

$$x^\mu = z^\mu_a(T_a) + e^\mu_i(T_a)\, X^i_a + \tfrac{1}{2}\, f^\mu_{ij}(T_a)\, X^i_a\, X^j_a + \cdots\,. \tag{2.6}$$

The multichart approach becomes simplified if one considers *compact* bodies (of radius R_a comparable to $2\,Gm_a/c^2$). In this case, it was shown [52], by considering how the "internal expansion" (2.5) propagates into the "external" one (2.4) via the matching (2.6), that, *in general relativity*, the internal structure of each compact body was *effaced* to a very high degree, when seen in the external expansion (2.4). For instance, for nonspinning bodies, the internal structure of each body (notably the way it responds to an external tidal excitation) shows up in the external problem of motion only at the *fifth post-Newtonian* (5PN) approximation, i.e. in terms of order $(v/c)^{10}$ in the equations of motion.

This *effacement of internal structure* indicates that it should be possible to simplify the rigorous multichart approach by skeletonizing each compact body by means of some delta-function source. Mathematically, the use of distributional sources is delicate in a nonlinear theory such as GR. However, it was found that one can reproduce the results of the more rigorous matched-multichart approach by treating the divergent integrals generated by the use of delta-function sources by means of (complex) analytic continuation [52]. In particular, analytic continuation in the dimension of space d [57] is very efficient (especially at high PN orders).

Finally, the most efficient way to derive the general relativistic equations of motion of N compact bodies consists of solving the equations derived from the action (where $g \equiv -\det(g_{\mu\nu})$)

$$S = \int \frac{d^{d+1}x}{c} \sqrt{g} \frac{c^4}{16\pi G} R(g) - \sum_a m_a c \int \sqrt{-g_{\mu\nu}(z_a^\lambda)\, dz_a^\mu\, dz_a^\nu}\,, \tag{2.7}$$

formally using the standard weak-field expansion (2.4), but considering the space dimension d as an arbitrary complex number which is sent to its physical value $d = 3$ only at the end of the calculation. This "skeletonized" effective action approach to the motion of compact bodies has been extended to other theories of gravity [50, 51]. Finite-size corrections can be taken into account by adding nonminimal world line couplings to the effective action (2.7) [58, 59].

As we shall further discuss below, in the case of coalescing BNS systems, finite-size corrections (linked to tidal interactions) become relevant during late inspiral and must be included to accurately describe the dynamics of coalescing neutron stars.

Here, we shall not try to describe the results of the application of the multichart method to N-body (or two-body) systems. For applications to the solar system see the book by Brumberg [60]; see also several articles (notably by Soffel) in [61]. For applications of this method to binary pulsar systems (and to their use as tests of gravity theories) see the articles by Damour and Kramer in [62].

3 EOB description of the conservative dynamics of two-body systems

Before reviewing some of the technical aspects of the EOB method, let us indicate the historical roots of this method. First, we note that the EOB approach comprises three, rather separate, ingredients:

(1) a description of the conservative (Hamiltonian) part of the dynamics of two bodies;
(2) an expression for the radiation-reaction part of the dynamics;
(3) a description of the GW waveform emitted by a coalescing binary system.

For each one of these ingredients, the essential inputs that are used in EOB works are high-order PN expanded results which have been obtained by many years of work, by many researchers (see the review [46]). However, one of the key ideas in the EOB philosophy is to avoid using PN results in their original "Taylor-expanded" form (i.e. $c_0 + c_1 v/c + c_2 v^2/c^2 + c_3 v^3/c^3 + \cdots + c_n v^n/c^n$), but to use them instead in some *resummed* form (i.e. some nonpolynomial function of v/c, defined so as to incorporate some of the expected nonperturbative features of the exact result). The basic ideas and techniques for resumming each ingredient of the EOB are different and have different historical roots.

Concerning the first ingredient, i.e. the EOB Hamiltonian, it was inspired by an approach to electromagnetically interacting quantum two-body systems introduced by Brézin et al. [63].

The resummation of the second ingredient, i.e. the EOB radiation-reaction force $\mathcal{F}$, was initially inspired by the Padé resummation of the flux function introduced by Damour et al. [64]. More recently, a new and more sophisticated resummation technique for the (waveform and the) radiation reaction force $\mathcal{F}$ has been introduced by Damour et al. [65, 66]. It will be discussed in detail below.

As for the third ingredient, i.e. the EOB description of the waveform emitted by a coalescing black hole binary, it was mainly inspired by the work of Davis et al. [67] which discovered the transition between the plunge signal and a ringing tail when a particle falls into a black hole. Additional motivation for the EOB treatment of the transition from plunge to ring-down came from work on the, so-called close limit approximation [68].

Within the usual PN formalism, the conservative dynamics of a two-body system is currently fully known up to the 3PN level [19–24] (see below for the partial knowledge beyond the 3PN level). Going to the center of mass of the system ($\boldsymbol{p}_1 + \boldsymbol{p}_2 = 0$), the 3PN-accurate Hamiltonian (in Arnowitt–Deser–Misner-type coordinates) describing the relative motion, $\boldsymbol{q} = \boldsymbol{q}_1 - \boldsymbol{q}_2$, $\boldsymbol{p} = \boldsymbol{p}_1 = -\boldsymbol{p}_2$, has the structure

$$H_{3\mathrm{PN}}^{\mathrm{relative}}(\boldsymbol{q}, \boldsymbol{p}) = H_0(\boldsymbol{q}, \boldsymbol{p}) + \frac{1}{c^2} H_2(\boldsymbol{q}, \boldsymbol{p}) + \frac{1}{c^4} H_4(\boldsymbol{q}, \boldsymbol{p}) + \frac{1}{c^6} H_6(\boldsymbol{q}, \boldsymbol{p}) , \tag{3.1}$$

where

$$H_0(\boldsymbol{q}, \boldsymbol{p}) = \frac{1}{2\mu} \boldsymbol{p}^2 - \frac{GM\mu}{|\boldsymbol{q}|} , \tag{3.2}$$

with

$$M \equiv m_1 + m_2 \quad \text{and} \quad \mu \equiv m_1 m_2/M , \tag{3.3}$$

corresponds to the Newtonian approximation to the relative motion, while H_2 describes 1PN corrections, H_4 2PN ones and H_6 3PN ones. In terms of the rescaled variables $\boldsymbol{q}' \equiv \boldsymbol{q}/GM$, $\boldsymbol{p}' \equiv \boldsymbol{p}/\mu$, the explicit form (after dropping the primes for readability) of the 3PN-accurate rescaled Hamiltonian $\widehat{H} \equiv H/\mu$ reads [21, 70, 71]

$$\widehat{H}_N(\boldsymbol{q}, \boldsymbol{p}) = \frac{\boldsymbol{p}^2}{2} - \frac{1}{q}, \tag{3.4}$$

$$\widehat{H}_{1\mathrm{PN}}(\boldsymbol{q}, \boldsymbol{p}) = \frac{1}{8}(3\nu - 1)(\boldsymbol{p}^2)^2 - \frac{1}{2}\left[(3+\nu)\boldsymbol{p}^2 + \nu(\boldsymbol{n}\cdot\boldsymbol{p})^2\right]\frac{1}{q} + \frac{1}{2q^2}, \tag{3.5}$$

$$\begin{aligned}\widehat{H}_{2\mathrm{PN}}(\boldsymbol{q}, \boldsymbol{p}) = {} & \frac{1}{16}(1 - 5\nu + 5\nu^2)(\boldsymbol{p}^2)^3 \\ & + \frac{1}{8}\left[(5 - 20\nu - 3\nu^2)(\boldsymbol{p}^2)^2 - 2\nu^2(\boldsymbol{n}\cdot\boldsymbol{p})^2\boldsymbol{p}^2 - 3\nu^2(\boldsymbol{n}\cdot\boldsymbol{p})^4\right]\frac{1}{q} \\ & + \frac{1}{2}\left[(5+8\nu)\boldsymbol{p}^2 + 3\nu(\boldsymbol{n}\cdot\boldsymbol{p})^2\right]\frac{1}{q^2} - \frac{1}{4}(1+3\nu)\frac{1}{q^3},\end{aligned} \tag{3.6}$$

$$\begin{aligned}\widehat{H}_{3\mathrm{PN}}(\boldsymbol{q}, \boldsymbol{p}) = {} & \frac{1}{128}(-5 + 35\nu - 70\nu^2 + 35\nu^3)(\boldsymbol{p}^2)^4 \\ & + \frac{1}{16}\left[(-7 + 42\nu - 53\nu^2 - 5\nu^3)(\boldsymbol{p}^2)^3 + (2-3\nu)\nu^2(\boldsymbol{n}\cdot\boldsymbol{p})^2(\boldsymbol{p}^2)^2\right. \\ & \left. + 3(1-\nu)\nu^2(\boldsymbol{n}\cdot\boldsymbol{p})^4\boldsymbol{p}^2 - 5\nu^3(\boldsymbol{n}\cdot\boldsymbol{p})^6\right]\frac{1}{q} \\ & + \left[\frac{1}{16}(-27 + 136\nu + 109\nu^2)(\boldsymbol{p}^2)^2 + \frac{1}{16}(17 + 30\nu)\nu(\boldsymbol{n}\cdot\boldsymbol{p})^2\boldsymbol{p}^2\right. \\ & \left. + \frac{1}{12}(5 + 43\nu)\nu(\boldsymbol{n}\cdot\boldsymbol{p})^4\right]\frac{1}{q^2} \\ & + \left\{\left[-\frac{25}{8} + \left(\frac{1}{64}\pi^2 - \frac{335}{48}\right)\nu - \frac{23}{8}\nu^2\right]\boldsymbol{p}^2\right. \\ & \left. + \left(-\frac{85}{16} - \frac{3}{64}\pi^2 - \frac{7}{4}\nu\right)\nu(\boldsymbol{n}\cdot\boldsymbol{p})^2\right\}\frac{1}{q^3} \\ & + \left[\frac{1}{8} + \left(\frac{109}{12} - \frac{21}{32}\pi^2\right)\nu\right]\frac{1}{q^4}.\end{aligned} \tag{3.7}$$

In these formulas ν denotes the symmetric mass ratio:

$$\nu \equiv \frac{\mu}{M} \equiv \frac{m_1 m_2}{(m_1 + m_2)^2}. \tag{3.8}$$

The dimensionless parameter ν varies between 0 (extreme mass ratio case) and $\frac{1}{4}$ (equal mass case) and plays the rôle of a deformation parameter away from the test-mass limit.

It is well known that, at the Newtonian approximation, $H_0(\boldsymbol{q}, \boldsymbol{p})$ can be thought of as describing a "test particle" of mass μ orbiting around an "external mass" GM. The EOB approach is a *general relativistic generalization* of this fact. It consists in looking for an "effective external spacetime geometry" $g_{\mu\nu}^{\mathrm{eff}}(x^\lambda; GM, \nu)$ such that the geodesic dynamics of a "test particle" of mass μ within $g_{\mu\nu}^{\mathrm{eff}}(x^\lambda, GM, \nu)$ is *equivalent* (when expanded in powers of $1/c^2$) to the original, relative PN-expanded dynamics (3.1).

Let us explain the idea, proposed in [39], for establishing a "dictionary" between the real relative-motion dynamics, (3.1), and the dynamics of an "effective" particle of mass μ moving in $g^{\text{eff}}_{\mu\nu}(x^\lambda, GM, \nu)$. The idea consists in "thinking quantum mechanically"[1]. Instead of thinking in terms of a classical Hamiltonian, $H(\boldsymbol{q}, \boldsymbol{p})$ (such as $H^{\text{relative}}_{3\text{PN}}$, Equation (3.1)), and of its classical bound orbits, we can think in terms of the quantized energy levels $E(n, \ell)$ of the quantum bound states of the Hamiltonian operator $H(\hat{\boldsymbol{q}}, \hat{\boldsymbol{p}})$. These energy levels will depend on two (integer valued) quantum numbers n and ℓ. Here (for a spherically symmetric interaction, as appropriate to H^{relative}), ℓ parameterizes the total orbital angular momentum ($\boldsymbol{L}^2 = \ell(\ell+1)\hbar^2$), while n represents the "principal quantum number" $n = \ell + n_r + 1$, where n_r (the "radial quantum number") denotes the number of nodes in the radial wave function. The third "magnetic quantum number" m (with $-\ell \le m \le \ell$) does not enter the energy levels because of the spherical symmetry of the two-body interaction (in the center of mass frame). For instance, the nonrelativistic Newton interaction (Equation (3.2)) gives rise to the well-known result

$$E_0(n, \ell) = -\frac{1}{2}\mu\left(\frac{GM\mu}{n\hbar}\right)^2, \tag{3.9}$$

which depends only on n (this is the famous Coulomb degeneracy). When considering the PN corrections to H_0, as in Equation (3.1), one gets a more complicated expression of the form

$$\begin{aligned} E^{\text{relative}}_{3\text{PN}}(n, \ell) = -\frac{1}{2}\mu\frac{\alpha^2}{n^2}\Bigg[1 &+ \frac{\alpha^2}{c^2}\left(\frac{c_{11}}{n\ell} + \frac{c_{20}}{n^2}\right) \\ &+ \frac{\alpha^4}{c^4}\left(\frac{c_{13}}{n\ell^3} + \frac{c_{22}}{n^2\ell^2} + \frac{c_{31}}{n^3\ell} + \frac{c_{40}}{n^4}\right) + \frac{\alpha^6}{c^6}\left(\frac{c_{15}}{n\ell^5} + \cdots + \frac{c_{60}}{n^6}\right)\Bigg], \end{aligned} \tag{3.10}$$

where we have set $\alpha \equiv GM\mu/\hbar = G m_1 m_2/\hbar$, and where we consider, for simplicity, the (quasi-classical) limit where n and ℓ are large numbers. The 2PN-accurate version of Equation (3.10) had been derived by Damour and Schäfer [69] as early as 1988 while its 3PN-accurate version was derived by Damour et al. in 1999 [70]. The dimensionless coefficients c_{pq} are functions of the symmetric mass ratio $\nu \equiv \mu/M$, for instance $c_{40} = \frac{1}{8}(145 - 15\nu + \nu^2)$. In classical mechanics (i.e. for large n and ℓ), it is called the "Delaunay Hamiltonian," i.e. the Hamiltonian expressed in terms of the *action variables*[2] $J = \ell\hbar = \frac{1}{2\pi}\oint p_\varphi\, d\varphi$, and $N = n\hbar = I_r + J$, with $I_r = \frac{1}{2\pi}\oint p_r\, dr$.

The energy levels (3.10) encode, in a *gauge-invariant* way, the 3PN-accurate relative dynamics of a "real" binary. Let us now consider an auxiliary problem: the "effective" dynamics of one body, of mass μ, following (modulo the Q term discussed below)

1 This is related to an idea emphasized many times by John Archibald Wheeler: quantum mechanics can often help us in going to the essence of classical mechanics.

2 We consider, for simplicity, "equatorial" motions with $m = \ell$, i.e. classically, $\theta = \frac{\pi}{2}$.

a geodesic in some ν-dependent "effective external" (spherically symmetric) metric[3]

$$g_{\mu\nu}^{\text{eff}}\,dx^\mu\,dx^\nu = -A(R;\nu)\,c^2\,dT^2 + B(R;\nu)\,dR^2 + R^2(d\theta^2 + \sin^2\theta\,d\varphi^2)\,. \tag{3.11}$$

Here, the *a priori unknown* metric functions $A(R;\nu)$ and $B(R;\nu)$ will be constructed in the form of expansions in GM/c^2R:

$$\begin{aligned} A(R;\nu) &= 1 + \tilde{a}_1\,\frac{GM}{c^2R} + \tilde{a}_2\left(\frac{GM}{c^2R}\right)^2 + \tilde{a}_3\left(\frac{GM}{c^2R}\right)^3 + \tilde{a}_4\left(\frac{GM}{c^2R}\right)^4 + \cdots\,; \\ B(R;\nu) &= 1 + \tilde{b}_1\,\frac{GM}{c^2R} + \tilde{b}_2\left(\frac{GM}{c^2R}\right)^2 + b_3\left(\frac{GM}{c^2R}\right)^3 + \cdots\,, \end{aligned} \tag{3.12}$$

where the dimensionless coefficients $\tilde{a}_n, \tilde{b}_n$ depend on ν. From the Newtonian limit, it is clear that we should set $\tilde{a}_1 = -2$. In addition, as ν can be viewed as a deformation parameter away from the test-mass limit, we require that the effective metric (3.11) tend to the Schwarzschild metric (of mass M) as $\nu \longrightarrow 0$, that is

$$A(R;\nu=0) = 1 - 2GM/c^2R = B^{-1}(R;\nu=0)\,.$$

Let us now require that the dynamics of the "one body" μ within the effective metric $g_{\mu\nu}^{\text{eff}}$ be described by an "effective" mass-shell condition of the form

$$g_{\text{eff}}^{\mu\nu}\,p_\mu^{\text{eff}}\,p_\nu^{\text{eff}} + \mu^2\,c^2 + Q(p_\mu^{\text{eff}}) = 0\,,$$

where $Q(p)$ is (at least) *quartic* in p. Then by solving (by separation of variables) the corresponding "effective" Hamilton–Jacobi equation

$$\begin{aligned} & g_{\text{eff}}^{\mu\nu}\,\frac{\partial S_{\text{eff}}}{\partial x^\mu}\,\frac{\partial S_{\text{eff}}}{\partial x^\nu} + \mu^2c^2 + Q\left(\frac{\partial S_{\text{eff}}}{\partial x^\mu}\right) = 0\,, \\ & S_{\text{eff}} = -\mathcal{E}_{\text{eff}}\,t + J_{\text{eff}}\,\varphi + S_{\text{eff}}(R)\,, \end{aligned} \tag{3.13}$$

one can straightforwardly compute (in the quasi-classical, large quantum numbers limit) the effective Delaunay Hamiltonian $\mathcal{E}_{\text{eff}}(N_{\text{eff}}, J_{\text{eff}})$, with $N_{\text{eff}} = n_{\text{eff}}\,\hbar$, $J_{\text{eff}} = \ell_{\text{eff}}\,\hbar$ (where $N_{\text{eff}} = J_{\text{eff}} + I_R^{\text{eff}}$, with $I_R^{\text{eff}} = \frac{1}{2\pi}\oint p_R^{\text{eff}}\,dR$, $P_R^{\text{eff}} = \partial S_{\text{eff}}(R)/dR$). This yields a result of the form

$$\begin{aligned} \mathcal{E}_{\text{eff}}(n_{\text{eff}}, \ell_{\text{eff}}) = \mu c^2 - \frac{1}{2}\,\mu\,\frac{\alpha^2}{n_{\text{eff}}^2}\Bigg[1 &+ \frac{\alpha^2}{c^2}\left(\frac{c_{11}^{\text{eff}}}{n_{\text{eff}}\ell_{\text{eff}}} + \frac{c_{20}^{\text{eff}}}{n_{\text{eff}}^2}\right) \\ &+ \frac{\alpha^4}{c^4}\left(\frac{c_{13}^{\text{eff}}}{n_{\text{eff}}\ell_{\text{eff}}^3} + \frac{c_{22}^{\text{eff}}}{n_{\text{eff}}^2\ell_{\text{eff}}^2} + \frac{c_{31}^{\text{eff}}}{n_{\text{eff}}^3\ell_{\text{eff}}} + \frac{c_{40}^{\text{eff}}}{n_{\text{eff}}^4}\right) \\ &+ \frac{\alpha^6}{c^6}\left(\frac{c_{15}^{\text{eff}}}{n_{\text{eff}}\ell_{\text{eff}}^5} + \cdots + \frac{c_{60}^{\text{eff}}}{n_{\text{eff}}^6}\right)\Bigg]\,, \end{aligned} \tag{3.14}$$

---

3 It is convenient to write the "effective metric" in Schwarzschild-like coordinates. Note that the effective radial coordinate R differs from the two-body ADM-coordinate relative distance $R^{\text{ADM}} = |\boldsymbol{q}|$. The transformation between the two coordinate systems has been determined in Refs. [39, 41].

where the dimensionless coefficients c_{pq}^{eff} are now functions of the unknown coefficients $\tilde{a}_n, \tilde{b}_n$ entering the looked for "external" metric coefficients (3.12).

At this stage, one needs to define a "dictionary" between the real (relative) two-body dynamics, summarized in Equation (3.10), and the effective one-body one, summarized in Equation (3.14). As, on both sides, quantum mechanics tells us that the action variables are quantized in integers ($N_{\text{real}} = n\hbar$, $N_{\text{eff}} = n_{\text{eff}}\hbar$, etc.) it is most natural to identify $n = n_{\text{eff}}$ and $\ell = \ell_{\text{eff}}$. One then still needs a rule for relating the two different energies $E_{\text{real}}^{\text{relative}}$ and $\mathcal{E}_{\text{eff}}$. Buonanno and Damour [39] proposed to look for a general map between the real energy levels and the effective ones (which, as seen when comparing (3.10) and (3.14), cannot be directly identified because they do not include the same rest-mass contribution[4]), namely

$$\frac{\mathcal{E}_{\text{eff}}}{\mu c^2} - 1 = f\left(\frac{E_{\text{real}}^{\text{relative}}}{\mu c^2}\right) = \frac{E_{\text{real}}^{\text{relative}}}{\mu c^2}\left(1 + \alpha_1 \frac{E_{\text{real}}^{\text{relative}}}{\mu c^2} + \alpha_2 \left(\frac{E_{\text{real}}^{\text{relative}}}{\mu c^2}\right)^2 + \alpha_3 \left(\frac{E_{\text{real}}^{\text{relative}}}{\mu c^2}\right)^3 + \cdots\right). \tag{3.15}$$

The "correspondence" between the real and effective energy levels is illustrated in Figure 1.

Finally, identifying $\mathcal{E}_{\text{eff}}(n,\ell)/\mu c^2$ to $1 + f(E_{\text{real}}^{\text{relative}}(n,\ell)/\mu c^2)$ yields a system of equations for determining the unknown EOB coefficients $\tilde{a}_n, \tilde{b}_n, \alpha_n$, as well as the three coefficients z_1, z_2, z_3 parameterizing a general 3PN-level quartic mass-shell deformation:

$$Q_{3\text{PN}}(p) = \frac{1}{c^6}\frac{1}{\mu^2}\left(\frac{GM}{R}\right)^2\left[z_1\, \boldsymbol{p}^4 + z_2\, \boldsymbol{p}^2(\boldsymbol{n}\cdot\boldsymbol{p})^2 + z_3(\boldsymbol{n}\cdot\boldsymbol{p})^4\right].$$

[The need for introducing a quartic mass-shell deformation Q only arises at the 3PN level.]

The above system of equations for $\tilde{a}_n, \tilde{b}_n, \alpha_n$ (and z_i at 3PN) was studied at the 2PN level in Ref. [39], and at the 3PN level in Ref. [41]. At the 2PN level it was found that, if one further imposes the natural condition $\tilde{b}_1 = +2$ (so that the linearized effective metric coincides with the linearized Schwarzschild metric with mass $M = m_1 + m_2$), there exists a *unique* solution for the remaining five unknown coefficients $\tilde{a}_2, \tilde{a}_3, \tilde{b}_2, \alpha_1$ and α_2. This solution is very simple:

$$\tilde{a}_2 = 0\,, \quad \tilde{a}_3 = 2\nu\,, \quad \tilde{b}_2 = 4 - 6\nu\,, \quad \alpha_1 = \frac{\nu}{2}\,, \quad \alpha_2 = 0\,. \tag{3.16}$$

At the 3PN level, it was found that the system of equations is consistent, and underdetermined in that the general solution can be parameterized by the arbitrary values of

4 Indeed $E_{\text{real}}^{\text{total}} = Mc^2 + E_{\text{real}}^{\text{relative}} = Mc^2 + \text{Newtonian terms} + 1\text{PN}/c^2 + \cdots$, while $\mathcal{E}_{\text{effective}} = \mu c^2 + N + 1\text{PN}/c^2 + \cdots$.

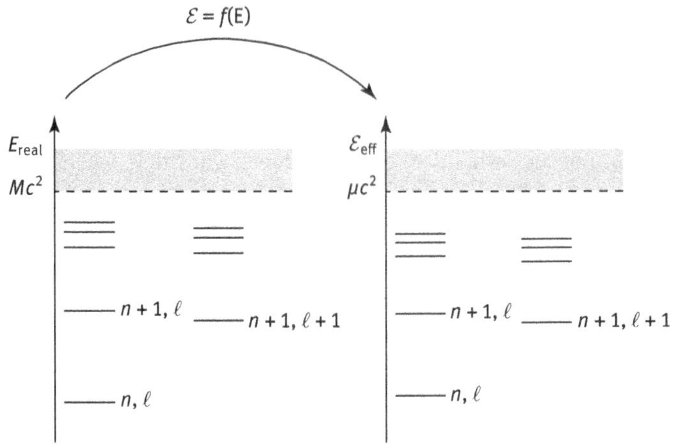

Fig. 1. Sketch of the correspondence between the quantized energy levels of the real and effective conservative dynamics. n denotes the "principal quantum number" ($n = n_r + \ell + 1$, with $n_r = 0, 1, \dots$ denoting the number of nodes in the radial function), while ℓ denotes the (relative) orbital angular momentum ($\boldsymbol{L}^2 = \ell(\ell+1)\hbar^2$). Though the EOB method is purely classical, it is conceptually useful to think in terms of the underlying (Bohr–Sommerfeld) quantization conditions of the action variables I_R and J to motivate the identification between n and ℓ in the two dynamics.

z_1 and z_2. It was then argued that it is natural to impose the simplifying requirements $z_1 = 0 = z_2$, so that Q is proportional to the fourth power of the (effective) radial momentum p_r. With these conditions, the solution is unique at the 3PN level, and is still remarkably simple, namely

$$\tilde{a}_4 = a_4\,\nu\,,\ \tilde{d}_3 = 2(3\nu - 26)\nu\,,\ \alpha_3 = 0\,,\ z_3 = 2(4 - 3\nu)\nu\,.$$

Here, a_4 denotes the number

$$a_4 = \frac{94}{3} - \frac{41}{32}\pi^2 \simeq 18.6879027 \tag{3.17}$$

while $\tilde{d}_3$ denotes the coefficient of $(GM/c^2R)^3$ in the PN expansion of the combined metric coefficient

$$D(R) \equiv A(R)\,B(R)\,.$$

Replacing $B(R)$ by $D(R)$ is convenient because (as mentioned above), in the test-mass limit $\nu \longrightarrow 0$, the effective metric must reduce to the Schwarzschild metric, namely

$$A(R; \nu = 0) = B^{-1}(R; \nu = 0) = 1 - 2\left(\frac{GM}{c^2R}\right)\,,$$

so that

$$D(R; \nu = 0) = 1\,.$$

The final result is that the three EOB potentials A, D, Q describing the 3PN two-body dynamics are given by the following very simple results. In terms of the EOB "gravitational potential"

$$u \equiv \frac{GM}{c^2 R},$$

$$A_{3PN}(R) = 1 - 2u + 2\nu u^3 + a_4 \nu u^4, \tag{3.18}$$

$$D_{3PN}(R) \equiv (A(R)B(R))_{3PN} = 1 - 6\nu u^2 + 2(3\nu - 26)\nu u^3, \tag{3.19}$$

$$Q_{3PN}(\boldsymbol{q}, \boldsymbol{p}) = \frac{1}{c^2} 2(4 - 3\nu)\nu u^2 \frac{p_r^4}{\mu^2}. \tag{3.20}$$

In addition, the map between the (real) center-of-mass energy of the binary system $E_{\text{real}}^{\text{relative}} = H^{\text{relative}} = \mathcal{E}_{\text{relative}}^{\text{tot}} - Mc^2$ and the effective one $\mathcal{E}_{\text{eff}}$ is found to have the very simple (but nontrivial) form

$$\frac{\mathcal{E}_{\text{eff}}}{\mu c^2} = 1 + \frac{E_{\text{real}}^{\text{relative}}}{\mu c^2}\left(1 + \frac{\nu}{2}\frac{E_{\text{real}}^{\text{relative}}}{\mu c^2}\right) = \frac{s - m_1^2 c^4 - m_2^2 c^4}{2 m_1 m_2 c^4} \tag{3.21}$$

where $s = (\mathcal{E}_{\text{real}}^{\text{tot}})^2 \equiv (Mc^2 + E_{\text{real}}^{\text{relative}})^2$ is Mandelstam's invariant $s = -(p_1 + p_2)^2$.

It is truly remarkable that the EOB formalism succeeds in *condensing* the complicated, original 3PN Hamiltonian, Equations (3.4)–(3.7), into the very simple potentials A, D, and Q displayed above, together with the simple energy map Equation (3.21). For instance, at the 1PN level, the already somewhat involved Lorentz–Droste–Einstein–Infeld–Hoffmann 1PN dynamics (Equations (3.4) and (3.5)) is simply described, within the EOB formalism, as a test particle of mass μ moving in an external Schwarzschild background of mass $M = m_1 + m_2$, together with the (crucial but quite simple) energy transformation (3.21). (Indeed, the ν-dependent corrections to A and D start only at the 2PN level.) At the 2PN level, the seven rather complicated ν-dependent coefficients of $\widehat{H}_{2PN}(\boldsymbol{q}, \boldsymbol{p})$, Equation (3.6), get condensed into the two very simple additional contributions $+ 2\nu u^3$ in $A(u)$, and $- 6\nu u^2$ in $D(u)$. At the 3PN level, the 11 quite complicated ν-dependent coefficients of $\widehat{H}_{3PN}$, Equation (3.7), get condensed into only three simple contributions: $+ a_4 \nu u^4$ in $A(u)$, $+ 2(3\nu - 26)\nu u^3$ in $D(u)$, and Q_{3PN} given by Equation (3.20). This simplicity of the EOB results is not only due to the reformulation of the PN-expanded Hamiltonian into an effective dynamics. Notably, the A-potential is much simpler that it could a priori have been: (i) as already noted it is not modified at the 1PN level, while one would a priori expect to have found a 1PN potential $A_{1PN}(u) = 1 - 2u + \nu a_2 u^2$ with some nonzero a_2; and (ii) there are striking cancellations taking place in the calculation of the 2PN and 3PN coefficients $\tilde{a}_2(\nu)$ and $\tilde{a}_3(\nu)$, which were a priori of the form $\tilde{a}_2(\nu) = a_2\nu + a_2'\nu^2$, and $\tilde{a}_3(\nu) = a_3\nu + a_3'\nu^2 + a_3''\nu^3$, but for which the ν-nonlinear contributions $a_2'\nu^2, a_3'\nu^2$ and $a_3''\nu^3$ precisely canceled out. Similar cancellations take place at the 4PN level (level at which it was recently possible to compute the A-potential, see below). Let us note for completeness that, starting at the 4PN level, the Taylor expansions of the A and D potentials depend on the logarithm of u. The corresponding logarithmic contributions have been computed at the 4PN level [72, 73] and even the 5PN one [74, 75]. They have been incorporated in a recent, improved implementation of the EOB formalism [76].

The fact that the 3PN coefficient a_4 in the crucial "effective radial potential" $A_{3PN}(R)$, Equation (3.18), is rather large and positive indicates that the ν-dependent nonlinear gravitational effects lead, for comparable masses ($\nu \sim \frac{1}{4}$), to a last stable (circular) orbit (LSO) which has a higher frequency and a larger binding energy than what a naive scaling from the test-particle limit ($\nu \to 0$) would suggest. Actually, the PN-expanded form (3.18) of $A_{3PN}(R)$ does not seem to be a good representation of the (unknown) exact function $A_{EOB}(R)$ when the (Schwarzschild-like) relative coordinate R becomes smaller than about $6GM/c^2$ (which is the radius of the LSO in the test-mass limit). In fact, by continuity with the test-mass case, one a priori expects that $A_{3PN}(R)$ always exhibits a simple zero defining an EOB "effective horizon" that is smoothly connected to the Schwarzschild event horizon at $R = 2GM/c^2$ when $\nu \to 0$. However, the large value of the a_4 coefficient does actually prevent A_{3PN} to have this property when ν is too large, and in particular when $\nu = 1/4$. It was therefore suggested [41] to further resum[5] $A_{3PN}(R)$ by replacing it by a suitable Padé (P) approximant. For instance, the replacement of $A_{3PN}(R)$ by[6]

$$A_3^1(R) \equiv P_3^1\left(A_{3PN}(R)\right) = \frac{1 + n_1 u}{1 + d_1 u + d_2 u^2 + d_3 u^3} \tag{3.22}$$

ensures that the $\nu = \frac{1}{4}$ case is smoothly connected with the $\nu = 0$ limit.

The same kind of ν-continuity argument, discussed so far for the A function, needs to be applied also to the $D_{3PN}(R)$ function defined in Equation (3.19). A straightforward way to ensure that the D function stays positive when R decreases (since it is $D = 1$ when $\nu \to 0$) is to replace $D_{3PN}(R)$ by $D_3^0(R) \equiv P_3^0[D_{3PN}(R)]$, where P_3^0 indicates the $(0, 3)$ Padé approximant and explicitly reads

$$D_3^0(R) = \frac{1}{1 + 6\nu u^2 - 2(3\nu - 26)\nu u^3}\,. \tag{3.23}$$

5 The PN-expanded EOB building blocks $A_{3PN}(R), B_{3PN}(R), \ldots$ already represent a *resummation* of the PN dynamics in the sense that they have "condensed" the many terms of the original PN-expanded Hamiltonian within a very concise format. But one should not refrain to further resum the EOB building blocks themselves, if this is physically motivated.

6 We recall that the coefficients n_1 and (d_1, d_2, d_3) of the $(1, 3)$ Padé approximant $P_3^1(A_{3PN}(u))$ are determined by the condition that the first four terms of the Taylor expansion of A_3^1 in powers of $u = GM/(c^2R)$ coincide with A_{3PN}.

4 EOB description of radiation reaction and of the emitted waveform during inspiral

In the previous section, we have described how the EOB method encodes the conservative part of the relative orbital dynamics into the dynamics of an "effective" particle. Let us now briefly discuss how to complete the EOB dynamics by defining some *resummed* expressions describing radiation reaction effects, and the corresponding waveform emitted at infinity. One is interested in circularized binaries, which have lost their initial eccentricity under the influence of radiation reaction. For such systems, it is enough (in the first approximation [40]; see, however, the recent results of Bini and Damour [77]) to include a radiation reaction force in the p_φ equation of motion only. More precisely, we are using phase space variables $r, p_r, \varphi, p_\varphi$ associated to polar coordinates (in the equatorial plane $\theta = \frac{\pi}{2}$). Actually it is convenient to replace the radial momentum p_r by the momentum conjugate to the "tortoise" radial coordinate $R_* = \int dR(B/A)^{1/2}$, i.e. $P_{R_*} = (A/B)^{1/2} P_R$. The real EOB Hamiltonian is obtained by first solving Equation (3.21) to get $H_{\rm real}^{\rm total} = \sqrt{s}$ in terms of $\mathcal{E}_{\rm eff}$, and then by solving the effective Hamilton–Jacobi equation to get $\mathcal{E}_{\rm eff}$ in terms of the effective phase space coordinates $\boldsymbol{q}_{\rm eff}$ and $\boldsymbol{p}_{\rm eff}$. The result is given by two nested square roots (we henceforth set $c = 1$):

$$\hat{H}_{\rm EOB}(r, p_{r_*}, \varphi) = \frac{H_{\rm EOB}^{\rm real}}{\mu} = \frac{1}{\nu}\sqrt{1 + 2\nu\,(\hat{H}_{\rm eff} - 1)}\,, \tag{4.1}$$

where

$$\hat{H}_{\rm eff} = \sqrt{p_{r_*}^2 + A(r)\left(1 + \frac{p_\varphi^2}{r^2} + z_3\,\frac{p_{r_*}^4}{r^2}\right)}\,, \tag{4.2}$$

with $z_3 = 2\nu\,(4 - 3\nu)$. Here, we are using suitably rescaled dimensionless (effective) variables: $r = R/GM$, $p_{r_*} = P_{R_*}/\mu$, $p_\varphi = P_\varphi/\mu GM$, as well as a rescaled time $t = T/GM$. This leads to equations of motion for $(r, \varphi, p_{r_*}, p_\varphi)$ of the form

$$\frac{d\varphi}{dt} = \frac{\partial \hat{H}_{\rm EOB}}{\partial p_\varphi} \equiv \Omega\,, \tag{4.3}$$

$$\frac{dr}{dt} = \left(\frac{A}{B}\right)^{1/2} \frac{\partial \hat{H}_{\rm EOB}}{\partial p_{r_*}}\,, \tag{4.4}$$

$$\frac{dp_\varphi}{dt} = \hat{\mathfrak{F}}_\varphi\,, \tag{4.5}$$

$$\frac{dp_{r_*}}{dt} = -\left(\frac{A}{B}\right)^{1/2} \frac{\partial \hat{H}_{\rm EOB}}{\partial r}\,, \tag{4.6}$$

which explicitly read

$$\frac{d\varphi}{dt} = \frac{Ap_\varphi}{\nu r^2 \hat{H}\hat{H}_{\text{eff}}} \equiv \Omega \tag{4.7}$$

$$\frac{dr}{dt} = \left(\frac{A}{B}\right)^{1/2} \frac{1}{\nu \hat{H}\hat{H}_{\text{eff}}} \left(p_{r_*} + z_3 \frac{2A}{r^2} p_{r_*}^3\right) \tag{4.8}$$

$$\frac{dp_\varphi}{dt} = \hat{\mathcal{F}}_\varphi \tag{4.9}$$

$$\frac{dp_{r_*}}{dt} = -\left(\frac{A}{B}\right)^{1/2} \frac{1}{2\nu \hat{H}\hat{H}_{\text{eff}}} \left\{A' + \frac{p_\varphi^2}{r^2}\left(A' - \frac{2A}{r}\right) + z_3\left(\frac{A'}{r^2} - \frac{2A}{r^3}\right) p_{r_*}^4\right\} , \tag{4.10}$$

where $A' = dA/dr$. As explained above the EOB metric function $A(r)$ is defined by Padé resumming the Taylor-expanded result (3.12) obtained from the matching between the real and effective energy levels (as we were mentioning, one uses a similar Padé resumming for $D(r) \equiv A(r)\,B(r)$). One similarly needs to resum $\hat{\mathcal{F}}_\varphi$, i.e. the φ component of the radiation reaction which has been introduced on the right-hand side of Equation (4.5).

Several methods have been tried during the development of the EOB formalism to resum the radiation reaction $\hat{\mathcal{F}}_\varphi$ (starting from the high-order PN-expanded results that have been obtained in the literature). Here, we shall briefly explain the new, *parameter-free* resummation technique for the multipolar waveform (and thus for the energy flux) introduced in Refs. [78, 79] and perfected in [65]. To be precise, the new results discussed in Ref. [65] are twofold: on the one hand, that work generalized the $\ell = m = 2$ resummed factorized waveform of [78, 79] to higher multipoles by using the most accurate currently known PN-expanded results [80–83] as well as the higher PN terms which are known in the test-mass limit [84, 85]; on the other hand, it introduced a *new resummation procedure* which consists in considering a new theoretical quantity, denoted as $\rho_{\ell m}(x)$, which enters the (ℓ, m) waveform (together with other building blocks, see below) only through its ℓth power: $h_{\ell m} \propto (\rho_{\ell m}(x))^\ell$. Here, and below, x denotes the invariant PN-ordering parameter given during inspiral by $x \equiv (GM\Omega/c^3)^{2/3}$.

The main novelty introduced by Ref. [65] is to write the (ℓ, m) multipolar waveform emitted by a circular nonspinning compact binary as the *product* of several factors, namely

$$h_{\ell m}^{(\epsilon)} = \frac{GM\nu}{c^2 R} n_{\ell m}^{(\epsilon)} c_{\lambda+\epsilon}(\nu) x^{(\ell+\epsilon)/2} Y^{\ell-\epsilon,-m}\left(\frac{\pi}{2}, \Phi\right) \hat{S}_{\text{eff}}^{(\epsilon)} T_{\ell m} e^{i\delta_{\ell m}} \rho_{\ell m}^\ell . \tag{4.11}$$

Here ϵ denotes the parity of $\ell + m$ ($\epsilon = \pi(\ell + m)$), i.e. $\epsilon = 0$ for “even-parity” (mass-generated) multipoles ($\ell + m$ even), and $\epsilon = 1$ for “odd-parity” (current-generated) ones ($\ell + m$ odd); $n_{\ell m}^{(\epsilon)}$ and $c_{\lambda+\epsilon}(\nu)$ are numerical coefficients; $\hat{S}_{\rm eff}^{(\epsilon)}$ is a μ-normalized effective source (whose definition comes from the EOB formalism); $T_{\ell m}$ is a resummed version [78, 79] of an infinite number of "leading logarithms" entering the *tail effects* [86, 87]; $\delta_{\ell m}$ is a supplementary phase (which corrects the phase effects not included in the *complex* tail factor $T_{\ell m}$), and, finally, $(\rho_{\ell m})^\ell$ denotes the ℓth power of the quantity $\rho_{\ell m}$ which is the new building block introduced in [65]. Note that in previous papers [78, 79] the quantity $(\rho_{\ell m})^\ell$ was denoted as $f_{\ell m}$ and we will often use this notation below. Before introducing explicitly the various elements entering the waveform (4.11) it is convenient to decompose $h_{\ell m}$ as

$$h_{\ell m}^{(\epsilon)} = h_{\ell m}^{(N,\epsilon)} \hat{h}_{\ell m}^{(\epsilon)}, \tag{4.12}$$

where $h_{\ell m}^{(N,\epsilon)}$ is the Newtonian contribution (i.e. the product of the first five factors in Equation (4.11)) and

$$\hat{h}_{\ell m}^{(\epsilon)} \equiv \hat{S}_{\rm eff}^{(\epsilon)} T_{\ell m} e^{i\delta_{\ell m}} f_{\ell m} \tag{4.13}$$

represents a resummed version of all the PN corrections. The PN correcting factor $\hat{h}_{\ell m}^{(\epsilon)}$, as well as all its building blocks, has the structure $\hat{h}_{\ell m}^{(\epsilon)} = 1 + \mathcal{O}(x)$.

The reader will find in Ref. [65] the definitions of the quantities entering the "Newtonian" waveform $h_{\ell m}^{(N,\epsilon)}$, as well as the precise definition of the effective source factor $\widehat{S}_{\rm eff}^{(\epsilon)}$, which constitutes the first factor in the PN-correcting factor $\widehat{h}_{\ell m}^{(\epsilon)}$. Let us only note here that the definition of $\widehat{S}_{\rm eff}^{(\epsilon)}$ makes use of EOB-defined quantities. For instance, for even-parity waves ($\epsilon = 0$) $\widehat{S}_{\rm eff}^{(0)}$ is defined as the μ-scaled *effective* energy $\mathcal{E}_{\rm eff}/\mu c^2$. (We use the "$J$-factorization" definition of $\widehat{S}_{\rm eff}^{(\epsilon)}$ when $\epsilon = 1$, i.e. for odd parity waves.)

The second building block in the factorized decomposition is the "tail factor" $T_{\ell m}$ (introduced in Refs. [78, 79]). As mentioned above, $T_{\ell m}$ is a resummed version of an infinite number of "leading logarithms" entering the transfer function between the near-zone multipolar wave and the far-zone one, due to *tail effects* linked to its propagation in a Schwarzschild background of mass $M_{\rm ADM} = H_{\rm EOB}^{\rm real}$. Its explicit expression reads

$$T_{\ell m} = \frac{\Gamma(\ell + 1 - 2i\hat{\hat{k}})}{\Gamma(\ell + 1)} e^{\pi\hat{\hat{k}}} e^{2i\hat{\hat{k}} \log(2kr_0)}, \tag{4.14}$$

where $r_0 = 2GM/\sqrt{e}$ and $\hat{\hat{k}} \equiv GH_{\rm EOB}^{\rm real} m\Omega$ and $k \equiv m\Omega$. Note that $\hat{\hat{k}}$ differs from k by a rescaling involving the *real* (rather than the *effective*) EOB Hamiltonian, computed at this stage along the sequence of circular orbits.

The tail factor $T_{\ell m}$ is a complex number which already takes into account some of the dephasing of the partial waves as they propagate out from the near zone to infinity. However, as the tail factor only takes into account the leading logarithms, one needs to correct it by a complementary dephasing term, $e^{i\delta_{\ell m}}$, linked to subleading logarithms and other effects. This subleading phase correction can be computed as being the phase $\delta_{\ell m}$ of the complex ratio between the PN-expanded $\hat{h}_{\ell m}^{(\epsilon)}$ and the above defined source and tail factors. In the comparable-mass case ($\nu \neq 0$), the 3PN δ_{22} phase correction to the leading quadrupolar wave was originally computed in Ref. [79] (see also Ref. [78] for the $\nu = 0$ limit). Full results for the subleading partial waves to the highest possible PN-accuracy by starting from the currently known 3PN-accurate ν-dependent waveform [83] have been obtained in [65]. For higher order test-mass ($\nu \to 0$) contributions, see [88, 89]. For extensions of the (nonspinning) factorized waveform of [65] see [90–92].

The last factor in the multiplicative decomposition of the multipolar waveform can be computed as being the modulus $f_{\ell m}$ of the complex ratio between the PN-expanded $\hat{h}^{(\epsilon)}_{\ell m}$ and the above defined source and tail factors. In the comparable mass case ($\nu \neq 0$), the f_{22} modulus correction to the leading quadrupolar wave was computed in Ref. [79] (see also Ref. [78] for the $\nu = 0$ limit). For the subleading partial waves, Ref. [65] explicitly computed the other $f_{\ell m}$'s to the highest possible PN-accuracy by starting from the currently known 3PN-accurate ν-dependent waveform [83]. In addition, as originally proposed in Ref. [79], to reach greater accuracy the $f_{\ell m}(x;\nu)$'s extracted from the 3PN-accurate $\nu \neq 0$ results are completed by adding higher order contributions coming from the $\nu = 0$ results [84, 85]. In the particular f_{22} case discussed in [79], this amounted to adding 4PN and 5PN $\nu = 0$ terms. This "hybridization" procedure was then systematically pursued for all the other multipoles, using the 5.5PN accurate calculation of the multipolar decomposition of the gravitational wave energy flux of Refs. [84, 85].

The decomposition of the total PN-correction factor $\hat{h}^{(\epsilon)}_{\ell m}$ into several factors is in itself a resummation procedure which already improves the convergence of the PN series one has to deal with: indeed, one can see that the coefficients entering increasing powers of x in the PN expansion of the $f_{\ell m}$'s tend to be systematically smaller than the coefficients appearing in the usual PN expansion of $\hat{h}^{(\epsilon)}_{\ell m}$. The reason for this is essentially twofold: (i) the factorization of $T_{\ell m}$ has absorbed powers of $m\pi$ which contributed to make large coefficients in $\hat{h}^{(\epsilon)}_{\ell m}$, and (ii) the factorization of either $\hat{H}_{\rm eff}$ or $\hat{j}$ has (in the $\nu = 0$ case) removed the presence of an inverse square-root singularity located at $x = 1/3$ which caused the coefficient of x^n in any PN-expanded quantity to grow as 3^n as $n \to \infty$.

To further improve the convergence of the waveform several resummations of the factor $f_{\ell m}(x) = 1 + c_1^{\ell m} x + c_2^{\ell m} x^2 + \cdots$ have been suggested. First, Refs. [78, 79] proposed to further resum the $f_{22}(x)$ function via a Padé (3,2) approximant, $P^3_2\{f_{22}(x;\nu)\}$, so as to improve its behavior in the strong-field-fast-motion regime. Such a resummation gave an excellent agreement with numerically computed waveforms, near the end of the inspiral and during the beginning of the plunge, for different mass ratios [78, 93, 94]. As we were mentioning above, a new route for resumming $f_{\ell m}$ was explored in Ref. [65]. It is based on replacing $f_{\ell m}$ by its ℓth root, say

$$\rho_{\ell m}(x;\nu) = [f_{\ell m}(x;\nu)]^{1/\ell}. \tag{4.15}$$

The basic motivation for replacing $f_{\ell m}$ by $\rho_{\ell m}$ is the following: the leading "Newtonian-level" contribution to the waveform $h_{\ell m}^{(\epsilon)}$ contains a factor $\omega^{\ell} r_{\text{harm}}^{\ell} v^{\epsilon}$, where r_{harm} is the harmonic radial coordinate used in the MPM formalism [95, 96]. When computing the PN expansion of this factor one has to insert the PN expansion of the (dimensionless) harmonic radial coordinate r_{harm}, $r_{\text{harm}} = x^{-1}(1 + c_1 x + \mathcal{O}(x^2))$, as a function of the gauge-independent frequency parameter x. The PN re-expansion of $[r_{\text{harm}}(x)]^{\ell}$ then generates terms of the type $x^{-\ell}(1 + \ell c_1 x + cdots)$. This is one (though not the only one) of the origins of 1PN corrections in $h_{\ell m}$ and $f_{\ell m}$ whose coefficients grow linearly with ℓ. The study of [65] has pointed out that these ℓ-growing terms are problematic for the accuracy of the PN-expansions. The replacement of $f_{\ell m}$ by $\rho_{\ell m}$ is a cure for this problem.

Several studies, both in the test-mass limit, $\nu \to 0$ (see Figure 1 in [65]) and in the comparable-mass case (see notably Figure 4 in [66]), have shown that the resummed factorized (inspiral) EOB waveforms defined above provided remarkably accurate analytical approximations to the "exact" inspiral waveforms computed by numerical simulations. These resummed multipolar EOB waveforms are much closer (especially during late inspiral) to the exact ones than the standard PN-expanded waveforms given by Equation (4.12) with a PN-correction factor of the usual "Taylor-expanded" form

$$\hat{h}_{\ell m}^{(\epsilon)\text{PN}} = 1 + c_1^{\ell m} x + c_{3/2}^{\ell m} x^{3/2} + c_2^{\ell m} x^2 + \cdots$$

See Figure 1 in [65].

Finally, one uses the newly resummed multipolar waveforms (4.11) to define a resummation of the *radiation reaction force* $\mathcal{F}_{\varphi}$ defined as

$$\mathcal{F}_{\varphi} = -\frac{1}{\Omega} F^{(\ell_{\max})}, \tag{4.16}$$

where the (instantaneous, circular) GW flux $F^{(\ell_{\max})}$ is defined as

$$F^{(\ell_{\max})} = \frac{2}{16\pi G} \sum_{\ell=2}^{\ell_{\max}} \sum_{m=1}^{\ell} (m\Omega)^2 |Rh_{\ell m}|^2. \tag{4.17}$$

Summarizing: Equations (4.11) and (4.16), (4.17) define resummed EOB versions of the waveform $h_{\ell m}$, and of the radiation reaction $\hat{\mathcal{F}}_{\varphi}$, during inspiral. A crucial point is that these resummed expressions are *parameter free*. Given some current approximation to the conservative EOB dynamics (i.e. some expressions for the A, D, Q potentials) they *complete* the EOB formalism by giving explicit predictions for the radiation reaction (thereby completing the dynamics, see Equations (4.3)–(4.6)), and for the emitted inspiral waveform.

5 EOB description of the merger of binary black holes and of the ringdown of the final black hole

Up to now we have reviewed how the EOB formalism, starting only from *analytical* information obtained from PN theory, and adding extra resummation requirements (both for the EOB conservative potentials A, Equation (3.22), and D, Equation (3.23), and for the waveform, Equation (4.11), and its associated radiation reaction force, Equations (4.16), (4.17)) make specific predictions, both for the motion and the radiation of BBHs. The analytical calculations underlying such an EOB description are essentially based on skeletonizing the two black holes as two, sufficiently separated point masses, and therefore seem unable to describe the merger of the two black holes, and the subsequent ringdown of the final, single black hole formed during the merger. However, as early as 2000 [40], the EOB formalism went one step further and proposed a specific strategy for describing the *complete* waveform emitted during the entire coalescence process, covering inspiral, merger, and ringdown. This EOB proposal is somewhat crude. However, the predictions it has made (years before NR simulations could accurately describe the late inspiral and merger of BBHs) have been broadly confirmed by subsequent NR simulations. (See Section 1 for a list of EOB predictions.) Essentially, the EOB proposal (which was motivated partly by the closeness between the 2PN-accurate effective metric $g_{\mu\nu}^{\rm eff}$ [39] and the Schwarzschild metric, and by the results of Refs. [67] and [68]) consists of:

(i) defining, within EOB theory, the instant of (effective) "*merger*" of the two black holes as the (dynamical) EOB time t_m where the orbital frequency $\Omega(t)$ reaches its *maximum*;
(ii) describing (for $t \leq t_m$) the inspiral-plus-plunge (or simply *insplunge*) waveform, $h^{\rm insplunge}(t)$, by using the inspiral EOB dynamics and waveform reviewed in the previous section; and
(iii) describing (for $t \geq t_m$) the merger-plus-ringdown waveform as a superposition of several quasi-normal-mode (QNM) complex frequencies of a final Kerr black hole (of mass M_f and spin parameter a_f, self-consistency estimated within the EOB formalism), say
$$\left(\frac{Rc^2}{GM}\right) h_{\ell m}^{\rm ringdown}(t) = \sum_N C_N^+ \, e^{-\sigma_N^+(t-t_m)} \,, \qquad (5.1)$$
with $\sigma_N^+ = \alpha_N + i\,\omega_N$, and where the label N refers to indices (ℓ, ℓ', m, n), with (ℓ, m) being the Schwarzschild-background multipolarity of the considered (metric) waveform $h_{\ell m}$, with $n = 0, 1, 2\ldots$, being the "overtone number" of the considered Kerr-background Quasi-Normal-Mode, and ℓ' the degree of its associated spheroidal harmonics $S_{\ell' m}(a\sigma, \theta)$;
(iv) determining the excitation coefficients C_N^+ of the QNM's in Equation (5.1) by using a simplified representation of the transition between plunge and ring-down ob-

tained by smoothly *matching* (following Ref. [78]), on a $(2p+1)$-toothed "comb" $(t_m - p\delta, \dots, t_m - \delta, t_m, t_m + \delta, \dots, t_m + p\delta)$ centered around the merger (and matching) time t_m, the inspiral-plus-plunge waveform to the above ring-down waveform.

Finally, one defines a complete, quasi-analytical EOB waveform (covering the full process from inspiral to ring-down) as

$$h_{\ell m}^{\mathrm{EOB}}(t) = \theta(t_m - t)\, h_{\ell m}^{\mathrm{insplunge}}(t) + \theta(t - t_m)\, h_{\ell m}^{\mathrm{ringdown}}(t)\,, \tag{5.2}$$

where $\theta(t)$ denotes Heaviside's step function. The final result is a waveform that essentially depends only on the choice of a resummed EOB $A(u)$ potential, and, less importantly, on the choice of resummation of the main waveform amplitude factor $f_{22} = (\rho_{22})^2$.

We have emphasized here that the EOB formalism is able, in principle, starting only from the best currently known analytical information, to predict the full waveform emitted by coalescing BBHs. The early comparisons between 3PN-accurate EOB predicted waveforms[7] and NR-computed waveforms showed a satisfactory agreement between the two, within the (then relatively large) NR uncertainties [97, 98]. Moreover, as we shall discuss below, it has been recently shown that the currently known Padé-resummed 3PN-accurate $A(u)$ potential is able, as is, to describe with remarkable accuracy several aspects of the dynamics of coalescing BBHs [99, 100].

On the other hand, when NR started delivering high-accuracy waveforms, it became clear that the 3PN-level analytical knowledge incorporated in EOB theory was not accurate enough for providing waveforms agreeing with NR ones within the high-accuracy needed for detection, and data analysis of upcoming GW signals. (See, e.g. the discussion in Section II of Ref. [91].) At that point, one made use of the *natural flexibility* of the EOB formalism. Indeed, as already emphasized in early EOB work [42, 101], we know from the analytical point of view that there are (yet uncalculated) further terms in the u-expansions of the EOB potentials $A(u), D(u), \dots$ (and in the x-expansion of the waveform), so that these terms can be introduced either as "free parameter(s) in constructing a bank of templates, and (one should) wait until" GW observations determine their value(s) [42], or as "*fitting parameters* and adjusted so as to reproduce other information one has about the exact results" (to quote Ref. [101]). For instance, modulo logarithmic corrections that will be further discussed below, the Taylor expansion in powers of u of the main EOB potential $A(u)$ reads

$$A^{\mathrm{Taylor}}(u;\nu) = 1 - 2u + \tilde{a}_3(\nu)u^3 + \tilde{a}_4(\nu)u^4 + \tilde{a}_5(\nu)u^5 + \tilde{a}_6(\nu)u^6 + \cdots\,,$$

where the 2PN and 3PN coefficients $\tilde{a}_3(\nu) = 2\nu$ and $\tilde{a}_4(\nu) = a_4\nu$ have been known since 2001, but where the 4PN, 5PN, ... coefficients, $\tilde{a}_5(\nu), \tilde{a}_6(\nu), \dots$ were not known

7 The new, resummed EOB waveform discussed above was not available at the time, so that these comparisons employed the coarser "Newtonian-level" EOB waveform $h_{22}^{(N,\epsilon)}(x)$.

at the time (see below for the recent determination of $\tilde{a}_5(\nu)$). A first attempt was made in [101] to use numerical data (on circular orbits of corotating black holes) to fit for the value of a (single, effective) 4PN parameter of the simple form $\tilde{a}_5(\nu) = a_5\nu$ entering a Padé-resummed 4PN-level A potential, i.e.

$$A_4^1(u; a_5, \nu) = P_4^1\left[A_{3\text{PN}}(u) + \nu a_5 u^5\right] . \tag{5.3}$$

This strategy was pursued in Refs. [79, 102] and many subsequent works. It was pointed out in Ref. [66] that the introduction of a further 5PN coefficient $\tilde{a}_6(\nu) = a_6\nu$, entering a Padé-resummed 5PN-level A potential, i.e.

$$A_5^1(u; a_5, a_6, \nu) = P_5^1\left[A_{3\text{PN}}(u) + \nu a_5 u^5 + \nu a_6 u^6\right] , \tag{5.4}$$

helped in having a closer agreement with accurate NR waveforms.

In addition, Refs. [78, 79] introduced another type of flexibility parameters of the EOB formalism: the non-quasi-circular (NQC) parameters accounting for uncalculated modifications of the quasi-circular inspiral waveform presented above, linked to deviations from an adiabatic quasi-circular motion. These NQC parameters are of various types, and subsequent works [66, 91, 93, 94, 103, 104] have explored several ways of introducing them. They enter the EOB waveform in two separate ways. First, through an explicit, additional complex factor multiplying $h_{\ell m}$, e.g.

$$f_{\ell m}^{\text{NQC}} = (1 + a_1^{\ell m} n_1 + a_2^{\ell m} n_2)\exp[i(a_3^{\ell m} n_3 + a_4^{\ell m} n_4)] ,$$

where the n_i's are dynamical functions that vanish in the quasi-circular limit (with n_1, n_2 being time-even, and n_3, n_4 time-odd). For instance, one usually takes $n_1 = (p_{r_*}/r\Omega)^2$. Second, through the (discrete) choice of the argument used during the plunge to replace the variable x of the quasi-circular inspiral argument: e.g. either $x_\Omega \equiv (GM\Omega)^{2/3}$, or (following [106]) $x_\varphi \equiv v_\varphi^2 = (r_\omega\Omega)^2$ where $v_\varphi \equiv \Omega r_\omega$, and $r_\omega \equiv r[\psi(r, p_\varphi)]^{1/3}$ is a modified EOB radius, with ψ being defined as

$$\psi(r, p_\varphi) = \frac{2}{r^2}\left(\frac{dA(r)}{dr}\right)^{-1}\left[1 + 2\nu\left(\sqrt{A(r)\left(1 + \frac{p_\varphi^2}{r^2}\right)} - 1\right)\right] . \tag{5.5}$$

For a given value of the symmetric mass ratio, and given values of the A-flexibility parameters $\tilde{a}_5(\nu), \tilde{a}_6(\nu)$ one can determine the values of the NQC parameters $a_i^{\ell m}$'s from accurate NR simulations of BBH coalescence (with mass ratio ν) by imposing, say, that the complex EOB waveform $h_{\ell m}^{\text{EOB}}(t^{\text{EOB}}; \tilde{a}_5, \tilde{a}_6; a_i^{\ell m})$ *osculates* the corresponding NR one $h_{\ell m}^{\text{NR}}(t^{\text{NR}})$ at their respective instants of "merger", where $t_{\text{merger}}^{\text{EOB}} \equiv t_m^{\text{EOB}}$ was defined above (maximum of $\Omega^{\text{EOB}}(t)$), while $t_{\text{merger}}^{\text{NR}}$ is defined as the (retarded) NR time where the modulus $|h_{22}^{\text{NR}}(t)|$ of the quadrupolar waveform reaches its maximum. The order of osculation that one requires between $h_{\ell m}^{\text{EOB}}(t)$ and $h_{\ell m}^{\text{NR}}(t)$ (or, separately, between their moduli and their phases or frequencies) depends on the number of NQC parameters $a_i^{\ell m}$. For instance, $a_1^{\ell m}$ and $a_2^{\ell m}$ affect only the modulus of $h_{\ell m}^{\mathrm{EOB}}$ and allow one to match both $|h_{\ell m}^{\mathrm{EOB}}|$ and its first time derivative, at merger, to their NR counterparts, while $a_3^{\ell m}$, $a_4^{\ell m}$ affect only the phase of the EOB waveform, and allow one to match the GW frequency $\omega_{\ell m}^{\mathrm{EOB}}(t)$ and its first time derivative, at merger, to their NR counterparts. The above EOB/NR matching scheme has been developed and varied in various versions in Refs. [66, 76, 91, 93, 94, 103–105]. One has also extracted the needed matching data from accurate NR simulations, and provided explicit, analytical ν-dependent fitting formulas for them [66, 76, 91].

Having so "calibrated" the values of the NQC parameters by extracting nonperturbative information from a sample of NR simulations, one can then, for any choice of the A-flexibility parameters, compute a full EOB waveform (from early inspiral to late ringdown). The comparison of the latter EOB waveform to the results of NR simulations is discussed in the next section.

6 EOB vs NR

There have been several different types of comparison between EOB and NR. For instance, the early work [97] pioneered the comparison between a purely analytical EOB waveform (uncalibrated to any NR information) and a NR wavform, while the early work [107] compared the predictions for the final spin of a coalescing black hole binary made by EOB, *completed* by the knowledge of the energy and angular momentum lost during ringdown by an extreme mass ratio binary (computed by the test-mass NR code of [108]), to comparable-mass NR simulations [109]. Since then, many other EOB/NR comparisons have been performed, both in the comparable-mass case [66, 79, 93, 94, 98, 102, 103] and in the small-mass-ratio case [78, 104, 110, 111]. Note in this respect that the numerical simulations of the GW emission by extreme mass-ratio binaries have provided (and still provide) a very useful "laboratory" for learning about the motion and radiation of binary systems, and their description within the EOB formalism.

Here we shall discuss only two recent examples of EOB/NR comparisons, which illustrate different facets of this comparison.

6.1 EOB[NR] waveforms vs NR ones

We explained above how one could complete the EOB formalism by calibrating some of the natural EOB flexibility parameters against NR data. First, for any given mass ratio ν and any given values of the A-flexibility parameters $\tilde{a}_5(\nu)$, $\tilde{a}_6(\nu)$, one can use NR data to uniquely determine the NQC flexibility parameters a_i's. In other words, we have (for a given ν)

$$a_i = a_i[\text{NR data}; a_5, a_6]\,,$$

where we defined a_5 and a_6 so that $\tilde{a}_5(\nu) = a_5\nu, \tilde{a}_6(\nu) = a_6\nu$. (We allow for some residual ν-dependence in a_5 and a_6.) Inserting these values in the (analytical) EOB waveform then defines an NR-completed EOB waveform which still depends on the two unknown flexibility parameters a_5 and a_6.

In Ref. [66] the (a_5, a_6)-dependent predictions made by such a NR-completed EOB formalism were compared to the high-accuracy waveform from an equal-mass BBH ($\nu = 1/4$) computed by the Caltech–Cornell–CITA group [112], (and then made available on the web). It was found that there is a strong degeneracy between a_5 and a_6 in the sense that there is an excellent EOB-NR agreement for an extended region in the (a_5, a_6)-plane. More precisely, the phase difference between the EOB (metric) waveform and the Caltech–Cornell–CITA one, considered between GW frequencies $M\omega_L = 0.047$ and $M\omega_R = 0.31$ (i.e. the last 16 GW cycles before merger), stays smaller than 0.02 radians within a long and thin banana-like region in the (a_5, a_6)-plane. This "good region" approximately extends between the points $(a_5, a_6) = (0, -20)$ and $(a_5, a_6) = (-36, +520)$. As an example (which actually lies on the boundary of the "good region"), we shall consider here (following Ref. [113]) the specific values $a_5 = 0, a_6 = -20$ (to which correspond, when $\nu = 1/4$, $a_1 = -0.036347, a_2 = 1.2468$). (Ref. [66] did not make use of the NQC phase flexibility; i.e. it took $a_3 = a_4 = 0$. In addition, it introduced a (real) modulus NQC factor $f_{\ell m}^{\text{NQC}}$ only for the dominant quadrupolar wave $\ell = 2 = m$.) We henceforth use M as time unit. This result relies on the proper comparison between NR and EOB time series, which is a delicate subject. In fact, to compare the NR and EOB phase time-series $\phi_{22}^{\text{NR}}(t_{\text{NR}})$ and $\phi_{22}^{\text{EOB}}(t_{\text{EOB}})$ one needs to shift, by additive constants, both one of the time variables, and one of the phases. In other words, we need to determine τ and α such that the "shifted" EOB quantities

$$t'_{\text{EOB}} = t_{\text{EOB}} + \tau \qquad \phi_{22}^{\prime\text{EOB}} = \phi_{22}^{\text{EOB}} + \alpha \tag{6.1}$$

"best fit" the NR ones. One convenient way to do so is first to "pinch" (i.e. constrain to vanish) the EOB/NR phase difference at two different instants (corresponding to two different frequencies ω_1 and ω_2). Having so related the EOB time and phase variables to the NR ones we can straigthforwardly compare the EOB time series to its NR correspondant. In particular, we can compute the (shifted) EOB–NR phase difference

$$\Delta^{\omega_1,\omega_2}\phi_{22}^{\text{EOBNR}}(t_{\text{NR}}) \equiv \phi_{22}^{\prime\text{EOB}}(t'EOB) - \phi_{22}^{\text{NR}}(t^{\text{NR}}). \tag{6.2}$$

Figure 2 compares[8] (the real part of) the analytical EOB *metric* quadrupolar waveform $\Psi_{22}^{\text{EOB}}/\nu$ to the corresponding (Caltech–Cornell–CITA) NR *metric* waveform $\Psi_{22}^{\text{NR}}/\nu$. (Here, Ψ_{22} denotes the Zerilli-normalized asymptotic quadrupolar waveform, i.e.

8 The two "pinching" frequencies used for this comparison are $M\omega_1 = 0.047$ and $M\omega_2 = 0.31$.

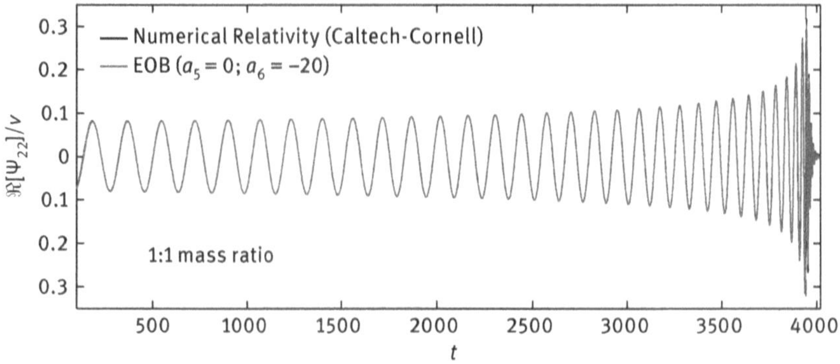

Fig. 2. This figure illustrates the comparison (made in Refs. [66, 113]) between the (NR-completed) EOB waveform (Zerilli-normalized quadrupolar ($\ell = m = 2$) metric waveform (5.2) with parameter-free radiation reaction (4.16) and with $a_5 = 0, a_6 = -20$) and one of the most accurate numerical relativity waveform (equal-mass case) nowadays available [112]. The phase difference between the two is $\Delta\phi \leq \pm 0.01$ radians during the entire inspiral and plunge, which is at the level of the numerical error.

$\Psi_{22} \equiv \hat{R}h_{22}/\sqrt{24}$ with $\hat{R} = Rc^2/GM$.) This NR metric waveform has been obtained by a double time-integration (following the procedure of Ref. [94]) from the original, publicly available, *curvature* waveform ψ_4^{22} [112]. Such a curvature waveform has been extrapolated *both* in resolution and in extraction radius. The agreement between the analytical prediction and the NR result is striking, even around the merger. See Figure 3 which closes up on the merger. The vertical line indicates the location of the EOB-merger time, i.e. the location of the maximum of the orbital frequency.

The phasing agreement between the waveforms is excellent over the full time span of the simulation (which covers 32 cycles of inspiral and about 6 cycles of ringdown), while the modulus agreement is excellent over the full span, apart from two cycles after merger where one can notice a difference. More precisely, the phase difference, $\Delta\phi = \phi^{\rm EOB}_{\rm metric} - \phi^{\rm NR}_{\rm metric}$, remains remarkably small ($\sim \pm 0.02$ radians) during the entire inspiral and plunge ($\omega_2 = 0.31$ being quite near the merger). By comparison, the root-sum of the various numerical errors on the phase (numerical truncation, outer boundary, extrapolation to infinity) is about 0.023 radians during the inspiral [112]. At the merger, and during the ringdown, $\Delta\phi$ takes somewhat larger values ($\sim \pm 0.1$ radians), but it oscillates around zero, so that, on average, it stays very well in phase with the NR waveform whose error rises to ± 0.05 radians during ringdown. In addition, Ref. [66] compared the EOB waveform to accurate numerical relativity data (obtained by the Jena group [94]) on the coalescence of *unequal mass-ratio* black-hole binaries. Again, the agreement was good, and within the numerical error bars.

This type of high-accuracy comparison between NR waveforms and EOB[NR] ones (where EOB[NR] denotes a EOB formalism which has been completed by fitting some EOB-flexibility parameters to NR data) has been pursued and extended in Ref. [91]. The

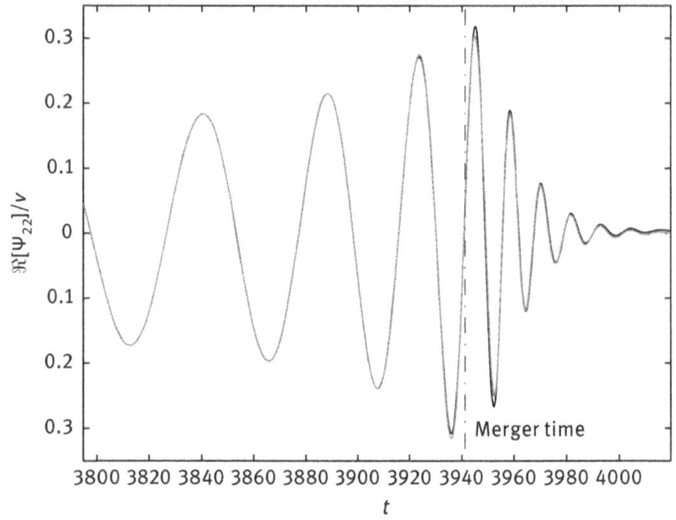

Fig. 3. Close up around merger of the waveforms of Figure 2. Note the excellent agreement between *both* modulus and phasing also during the ringdown phase.

latter reference used the "improved" EOB formalism of Ref. [66] with some variations (e.g. a third modulus NQC coefficient a_i, two phase NQC coefficients, the argument x_Ω in $(\rho_{\ell m}^{\text{Taylor}}(x))^\ell$, eight QNM modes) and calibrated it to NR simulations of mass ratios $q = m_2/m_1 = 1, 2, 3, 4$, and 6. They considered not only the leading $(\ell, m) = (2, 2)$ GW mode, but also the subleading ones $(2, 1), (3, 3), (4, 4)$, and $(5, 5)$. They found that, for this large range of mass ratios, EOB[NR] (with suitably fitted, ν-dependent values of a_5 and a_6) was able to describe the NR waveforms essentially within the NR errors. See also the recent Ref. [76] which incorporated several analytical advances in the two-body problem. This confirms the usefulness of the EOB formalism in helping the detection and analysis of upcoming GW signals.

Here, having in view GW observations from ground-based interferometric detectors we focussed on comparable-mass systems. The EOB formalism has also been compared to NR results in the extreme mass-ratio limit $\nu \ll 1$. In particular, Ref. [104] found an excellent agreement between the analytical and numerical results.

6.2 EOB[3PN] dynamics vs NR one

Let us also mention other types of EOB/NR comparisons. Several examples of EOB/NR comparisons have been performed directly at the level of the *dynamics* of a BBH, rather than at the level of the waveform. Moreover, contrary to the waveform comparisons of the previous subsection which involved an NR-completed EOB formalism

("EOB[NR]"), several of the dynamical comparisons we are going to discuss involve the purely analytical 3PN-accurate EOB formalism ("EOB[3PN]"), without any NR-based improvement.

First, Le Tiec et al. [99] have extracted from accurate NR simulations of slightly eccentric BBH systems (for several mass ratios $q = m_1/m_2$ between 1/8 and 1) the function relating the periastron-advance parameter

$$K = 1 + \frac{\Delta\Phi}{2\pi} ,$$

(where $\Delta\Phi$ is the periastron advance per radial period) to the dimensionless averaged angular frequency $M\Omega_\varphi$ (with $M = m_1 + m_2$ as above). Then they compared the NR-estimate of the mass-ratio-dependent functional relation

$$K = K(M\Omega_\varphi; \nu) ,$$

where $\nu = q/(1+q)^2$, to the predictions of various analytic approximation schemes: PN theory, EOB theory and two different ways of using GSF theory. Let us only mention here that the prediction from the purely analytical EOB[3PN] formalism for $K(M\Omega_\varphi; \nu)$ [72] agreed remarkably well (essentially within numerical errors) with its NR estimate for all mass ratios, while, by contrast, the PN-expanded prediction for $K(M\Omega_\varphi; \nu)$ [70] showed a much poorer agreement, especially as q moved away from 1.

Second, Damour et al. [100] have extracted from accurate NR simulations of black-hole binaries (with mass ratios $q = m_2/m_1 = 1, 2$, and 3) the gauge-invariant relation between the (reduced) binding energy $E = (\mathcal{E}^{\text{tot}} - M)/\mu$ and the (reduced) angular momentum $j = J/(G\mu M)$ of the system. Then they compared the NR-estimate of the mass-ratio-dependent functional relation

$$E = E(j; \nu)$$

to the predictions of various analytic approximation schemes: PN theory and various versions of EOB theory (some of these versions were NR-completed). Let us only mention here that the prediction from the purely analytical, 3PN-accurate EOB [3PN] for $E(j; \nu)$ agreed remarkably well with its NR estimate (for all mass ratios) essentially *down to the merger*. This is illustrated in Figure 4 for the $q = 1$ case. By contrast, the 3PN expansion in (powers of $1/c^2$) of the function $E(j; \nu)$ showed a much poorer agreement (for all mass ratios).

Recently, several other works have (successfully) compared EOB dynamical predictions to NR results. Ref. [114] compared the EOB[NR] predictions for the dynamical state of a nonspinning, coalescing BBH *at merger* to NR results and found agreement at the per mil level. Ref. [115] compared the predictions of an analytical (3.5PN-accurate) spinning EOB model to NR simulations and found a very good agreement.

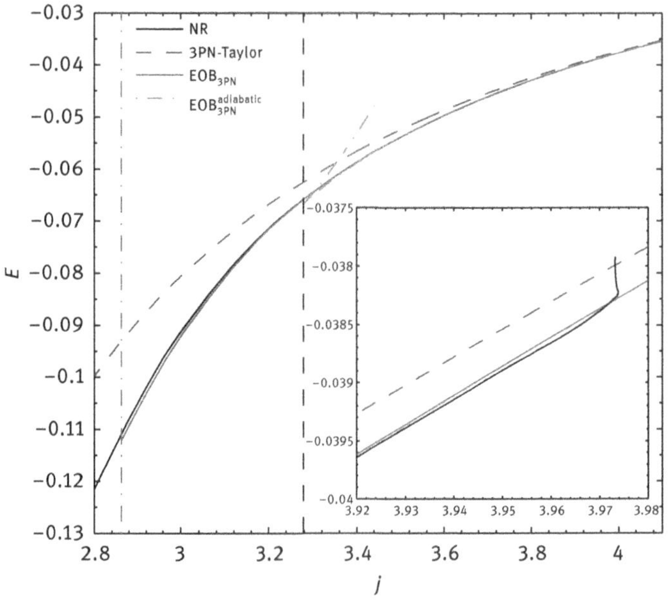

Fig. 4. Comparison (made in [100]) between various analytical estimates of the energy-angular momentum functional relation and its numerical-relativity estimate (equal-mass case). The standard "Taylor-expanded" 3PN $E(j)$ curve shows the largest deviation from NR results, especially at low j's, while the two (adiabatic and nonadiabatic) 3PN-accurate, *non-NR-calibrated* EOB $E(j)$ curves agree remarkably well with the NR one.

7 Other developments

7.1 EOB with spinning bodies

We do not wish to enter into a detailed discussion of the extension of the EOB formalism to binary systems made of spinning bodies. Let us only mention that the spin-extension of the EOB formalism was initiated in Ref. [42], that the first EOB-based analytical calculation of a complete waveform from a spinning binary was performed in Ref. [43], and that the first attempt at calibrating a spinning EOB model to accurate NR simulations of spinning (nonprecessing) black-hole binaries was presented in [116]. In addition, several formal aspects related to the inclusion of spins in the EOB formalism have been discussed in Refs. [117–121] (see references within these papers for PN works dealing with spin effects) and a generalization of the factorized multipolar waveform of Ref. [65] to spinning, nonprecessing binaries has been constructed in Refs. [90, 92]. Comparisons between spinning-EOB models and NR simulations have been obtained in [122, 123] and, recently, in the spinning, precessing case, in [124].

7.2 EOB with tidally deformed bodies

In binary systems comprising *neutron stars*, rather than black holes, the tidal deformation of the neutron star(s) will significantly modify the phasing of the emitted gravitational waveform during the late inspiral, thereby offering the possibility to measure the tidal polarizability of neutron stars [125–128]. As GWs from BNSs are expected sources for upcoming ground-based GW detectors, it is important to extend the EOB formalism by including tidal effects. This extension has been defined in Refs. [133, 134]. The comparison between this tidal-extended EOB and state-of-the-art NR simulations of neutron–star binaries has been discussed in Refs. [129–132]. It appears from these comparisons that the tidal-extended EOB formalism is able to describe the motion and radiation of neutron-star binaries within NR errors. More accurate simulations will be needed to ascertain whether one needs to calibrate some higher order flexibility parameters of the tidal-EOB formalism, or whether the currently known analytic accuracy is sufficient [35, 131].

7.3 EOB and GSF

We mentioned in Section 1 that GSF theory has recently opened a new source of information on the general relativistic two-body problem. Let us briefly mention here that there has been a quite useful transfer of information from GSF theory to EOB theory. The program of using GSF-theory to improve EOB-theory was first highlighted in Ref. [72]. That work pointed to several concrete gauge-invariant calculations (within GSF theory) that would provide accurate information about the $O(\nu)$ contributions to several EOB potentials. More precisely, let us define the functions $a(u)$ and $\bar{d}(u)$ as the ν-linear contributions to the EOB potentials $A(u;\nu)$ and $\overline{D}(u;\nu) \equiv D^{-1}(u;\nu)$:

$$A(u;\nu) = 1 - 2u + \nu\, a(u) + O(\nu^2)\,,$$
$$\overline{D}(u;\nu) = (AB)^{-1} = 1 + \nu\, \bar{d}(u) + O(\nu^2)\,.$$

Ref. [72] has shown that a computation of the GSF-induced correction to the periastron advance of slightly eccentric orbits would allow one to compute the following combination of EOB functions:

$$\bar{\rho}(u) = a(u) + u\, a'(u) + \tfrac{1}{2}\, u(1-2u)\, a''(u) + (1-6u)\, \bar{d}(u)\,.$$

The GSF-calculation of the EOB function $\bar{\rho}(u)$ was then performed in Ref. [135] (in the range $0 \le u \le \frac{1}{6}$).

Later, a series of works by Le Tiec and collaborators [75, 136, 137] have (through an indirect route) shown how GSF calculations could be used to compute the EOB ν-linear $a(u)$ function separately from the $\bar{d}(u)$ one. Barausse et al. [75] then gave a fitting formula for $a(u)$ over the interval $0 \leq u \leq \frac{1}{5}$ as well as accurate estimates of the coefficients of the Taylor expansion of $a(u)$ around $u = 0$ (corresponding to the knowledge of the PN expansion of $a(u)$ to a very high PN order). More recently, Ackay et al. [138] succeeded in accurately computing (through GSF theory) the EOB $a(u)$ function over the larger interval $0 \leq u \leq \frac{1}{3}$. It was (surprisingly) found that $a(u)$ *diverges* like $a(u) \approx 0.25(1-3u)^{-1/2}$ at the light-ring limit $u \to (\frac{1}{3})^-$. The meaning for EOB theory of this singular behavior of $a(u)$ at the light ring is discussed in detail in Ref. [138].

Let us finally mention that Ref. [28] has recently showed how to combine analytical GSF theory with the partial 4PN-level results of Ref. [27] so as to obtain the complete analytical expression of the 4PN-level contribution to the A potential. Specifically, Ref. [28] found that the coefficient $\tilde{a}_5(\nu; \ln a)$ of u^5 in the PN expansion, of $A(u;\nu)$,

$$A^{\text{Taylor}}(u;\nu) = 1 - 2u + \tilde{a}_3(\nu)u^3 + \tilde{a}_4(\nu)u^4 + \tilde{a}_5(\nu;\ln u)u^5 + \tilde{a}_6(\nu;\ln u)u^6 + \cdots$$

was equal to

$$\tilde{a}_5(\nu;\ln u) = \left(a_5 + \frac{64}{5}\ln u\right)\nu + a_5'\nu^2 ,$$

with

$$a_5 = -\frac{4237}{60} + \frac{2275}{512}\pi^2 + \frac{256}{5}\ln 2 + \frac{128}{5}\gamma ,$$

$$a_5' = -\frac{221}{6} + \frac{41}{32}\pi^2 .$$

Note that $\tilde{a}_5(\nu)$ is *no more than quadratic* in ν, i.e. *without* contributions of degree ν^3 and ν^4. (Contributions of degree ν^3 and ν^4 would a priori be expected in a 4PN level quantity; see, e.g. $e_{4\text{PN}}(\nu; \ln x)$ below.) We recall that similar cancellations of higher ν^n terms were found at lower PN orders in the EOB $A(u;\nu)$ function. Namely, they were found to contain only terms *linear* in ν, while $\tilde{a}_3(\nu)$ could a priori have been quadratic in ν, and $\tilde{a}_4(\nu)$ could a priori have been cubic in ν. The fact that similar remarkable cancellations still hold, at the 4PN level, is a clear indication that the EOB packaging of information of the dynamics in the $A(u;\nu)$ potential is quite compact. By contrast, the PN expansions of other dynamical functions do not exhibit such cancellations. For instance, the coefficients entering the PN expansion of the (gauge-invariant) function $E(x;\nu)$ relating the total energy to the frequency parameter $x \equiv (M\,\Omega_\varphi)^{2/3}$, namely

$$\begin{aligned} E(x;\nu) = -\frac{1}{2}\mu c^2 x(1 &+ e_{1PN}(\nu)x + e_{2PN}(\nu)x^2 + e_{3PN}(\nu)x^3 \\ &+ e_{4PN}(\nu;\ln x)x^4 + O(x^5\ln x)) , \end{aligned}$$

contain all the a priori possible powers of ν. In particular, at the 4PN level $e_{4\text{PN}}(\nu; \ln x)$ is a polynomial of fourth degree in ν.

8 Conclusions

Though this work did not attempt to expound the many different approaches to the general relativistic two-body problem but focussed only on a few approaches, we hope to have made it clear that there is a *complementarity* between the various current ways of tackling this problem: post-Newtonian,[9] effective one body, gravitational self-force, and numerical relativity simulations. Among these approaches, the effective one body formalism plays a special role in that it allows one to combine, in a synergetic manner, information coming from the other approaches. As we are approaching the 100th anniversary of the discovery of general relativity, it is striking to see how this theory has not only passed with flying colors many stringent tests, but has established itself as an essential tool for describing many aspects of the Universe from, say, the Big Bang to an accurate description of planets and satellites. Though the two-body (and, more generally, the N-body) problem is one of the oldest problems in general relativity, it is more lively than ever. Indeed, several domains of (astro-)physics and astronomy are providing new incentives for improving the various ways of describing general relativistic N-body systems: the developement of (ground-based and space-based) detectors of gravitational waves, the development of improved techniques for observing binary pulsars, the prospect of observing soon (with Gaia) a billion stars with $\sim 10^{-5}$ arcsec accuracy, ... Together with our esteemed friend and colleague Victor Brumberg, who pioneered important developments in relativistic celestial mechanics, we are all looking forward to witnessing new applications of Einstein's vision of gravity to the description and understanding of physical reality.

References

[1] J. Droste, *Versl. K. Akad. Wet. Amsterdam* **19**, 447–455 (1916).

[2] W. de Sitter, *Mon. Not. R. Astr. Soc.* **76**, 699–728 (1916); **77**, 155–184 (1916).

[3] J. Chazy, *La théorie de la Relativité et la Mécanique Céleste*, vols. 1 and 2 (Gauthier-Villars, Paris, 1928 and 1930).

[4] T. Levi-Civita, *Am. J. Math.* **59**, 9–22 (1937); *Am. J. Math.* **59**, 225–234 (1937); "Le problème des n corps en relativité générale", *Mémorial des Sciences Mathématiques* **116** (Gauthier-Villars, Paris, 1950).

[5] H. A. Lorentz and J. Droste, *Versl. K. Akad. Wet. Amsterdam* **26**, 392 (1917); and **26**, 649 (1917).

[6] A. Einstein, L. Infeld and B. Hoffmann, *Ann. Math.* **39**, 65–100 (1938).

[7] A. Eddington and G. L. Clark, *Proc. Roy. Soc. (Lond.)* **A166**, 465–475 (1938).

[8] V. A. Fock, *Zh. Eksp. i. Teor. Fiz.* **9**, 375 (1939).

---

9 including the effective-field-theory reformulation of the computation of the PN-expanded Fokker-action [45, 139].

[9] V. A. Fock, *The Theory of Space, Time and Gravitation* (Russian edition, Moscow, State Technical Publications, 1955).
[10] L. Infeld and J. Plebanski, *Motion and Relativity* (Pergamon Press, Oxford, 1960).
[11] T. Damour and N. Deruelle, Radiation Reaction and Angular Momentum Loss in Small Angle Gravitational Scattering, *Phys. Lett. A* **87**, 81 (1981).
[12] T. Damour, Problème des deux corps et freinage de rayonnement en relativité générale, *C.R. Acad. Sc. Paris, Série II* **294**, 1355–1357, 1982.
[13] T. Damour, Gravitational radiation and the motion of compact bodies, in *Gravitational Radiation*, N. Deruelle and T. Piran (eds.), North-Holland, Amsterdam, 1983, 59–144.
[14] S. M. Kopeikin, *Astron. Zh.* **62**, 889 (1985).
[15] G. Schaefer, The Gravitational Quadrupole Radiation Reaction Force and the Canonical Formalism of ADM, *Annals Phys.* **161**, 81 (1985).
[16] T. Ohta, H. Okamura, T. Kimura and K. Hiida, Physically acceptable solution of einstein's equation for many-body system, *Prog. Theor. Phys.* **50**, 492 (1973).
[17] H. Okamura, T. Ohta, T. Kimura and K. Hiida, Perturbation calculation of gravitational potentials, *Prog. Theor. Phys.* **50**, 2066 (1973).
[18] T. Ohta, H. Okamura, T. Kimura and K. Hiida, Coordinate Condition and Higher Order Gravitational Potential in Canonical Formalism, *Prog. Theor. Phys.* **51**, 1598 (1974).
[19] P. Jaranowski and G. Schäfer, 3rd post-Newtonian higher order Hamilton dynamics for two-body point-mass systems ([Erratum *Phys. Rev. D* **63**, 029902 (2001)]), *Phys. Rev. D* **57**, 72–74, (1998) [gr-qc/9712075].
[20] L. Blanchet and G. Faye, General relativistic dynamics of compact binaries at the third post-Newtonian order, *Phys. Rev. D* **63**, 062005 (2001) [gr-qc/0007051].
[21] T. Damour, P. Jaranowski and G. Schäfer, Dimensional regularization of the gravitational interaction of point masses, *Phys. Lett. B* **513**, 147–155 (2001) [gr-qc/0105038].
[22] Y. Itoh and T. Futamase, New derivation of a third post-Newtonian equation of motion for relativistic compact binaries without ambiguity, *Phys. Rev. D* **68**, 121501 (2003) [gr-qc/0310028].
[23] L. Blanchet, T. Damour and G. Esposito-Farèse, Dimensional regularization of the third post-Newtonian dynamics of point particles in harmonic coordinates, *Phys. Rev. D* **69**, 124007 (2004) [gr-qc/0311052].
[24] S. Foffa and R. Sturani, Effective field theory calculation of conservative binary dynamics at third post-Newtonian order, *Phys. Rev. D* **84**, 044031 (2011) [1104.1122 [gr-qc]].
[25] P. Jaranowski and G. Schafer, Towards the 4th post-Newtonian Hamiltonian for two-point-mass systems, *Phys. Rev. D* **86**, 061503 (2012) [arXiv:1207.5448 [gr-qc]].
[26] S. Foffa and R. Sturani, Dynamics of the gravitational two-body problem at fourth post-Newtonian order and at quadratic order in the Newton constant, *Phys. Rev. D* **87** (6), 064011 (2013) [arXiv:1206.7087 [gr-qc]].
[27] P. Jaranowski and G. Schäfer, Dimensional regularization of local singularities in the 4th post-Newtonian two-point-mass Hamiltonian, *Phys. Rev. D* **87**, 081503 (2013) [arXiv:1303.3225 [gr-qc]].
[28] D. Bini and T. Damour, Analytical determination of the two-body gravitational interaction potential at the 4th post-Newtonian approximation, *Phys. Rev. D* **87**, 121501 (2013) [arXiv:1305.4884 [gr-qc]].
[29] F. Pretorius, Evolution of binary black hole spacetimes, *Phys. Rev. Lett.* **95**, 121101 (2005) [gr-qc/0507014].
[30] M. Campanelli, C. O. Lousto, P. Marronetti and Y. Zlochower, Accurate Evolutions of Orbiting Black-Hole Binaries Without Excision, *Phys. Rev. Lett.* **96**, 111101 (2006) [gr-qc/0511048].

[31] J. G. Baker, J. Centrella, D. I. Choi, M. Koppitz and J. van Meter, Gravitational wave extraction from an inspiraling configuration of merging black holes, *Phys. Rev. Lett.* **96**, 111102 (2006) [gr-qc/0511103].
[32] M. Boyle et al., High-accuracy comparison of numerical relativity simulations with post-Newtonian expansions, *Phys. Rev. D* **76**, 124038 (2007) [0710.0158 [gr-qc]].
[33] F. Pretorius, Binary black hole coalescence, in *Relativistic Objects in Compact Binaries: From Birth to Coalescense*, M. Colpi et al. (eds.), Springer Verlags, Canopus Publishing Limited (2007) [arXiv:0710.1338 [gr-qc]].
[34] A. H. Mroue, M. A. Scheel, B. Szilagyi, H. P. Pfeiffer, M. Boyle, D. A. Hemberger, L. E. Kidder, G. Lovelace et al., A catalog of 171 high-quality binary black-hole simulations for gravitational-wave astronomy, *Phys. Rev. Lett.* **111**, 241104 (2013) [arXiv:1304.6077 [gr-qc]].
[35] D. Radice, L. Rezzolla and F. Galeazzi, Beyond second-order convergence in simulations of binary neutron stars in full general-relativity, *Mon. Not. Roy. Astron. Soc.* **437**, L46–L50 (2014) [arXiv:1306.6052 [gr-qc]].
[36] L. Barack, Gravitational self force in extreme mass-ratio inspirals, *Class. Quant. Grav.* **26**, 213001 (2009) [0908.1664 [gr-qc]].
[37] M. Hannam, S. Husa, B. Bruegmann and A. Gopakumar, Comparison between numerical-relativity and post-Newtonian waveforms from spinning binaries: the orbital hang-up case, *Phys. Rev. D* **78**, 104007 (2008) [0712.3787 [gr-qc]].
[38] M. Hannam, S. Husa, B. Bruegmann and A. Gopakumar, Comparison between numerical-relativity and post-Newtonian waveforms from spinning binaries: the orbital hang-up case, *Phys. Rev. D* **78**, 104007 (2008) [0712.3787 [gr-qc]].
[39] A. Buonanno and T. Damour, Effective one-body approach to general relativistic two-body dynamics, *Phys. Rev. D* **59**, 084006, (1999) [gr-qc/9811091].
[40] A. Buonanno and T. Damour, Transition from inspiral to plunge in binary black hole coalescences, *Phys. Rev. D* **62**, 064015 (2000). [gr-qc/0001013].
[41] T. Damour, P. Jaranowski and G. Schäfer, On the determination of the last stable orbit for circular general relativistic binaries at the third post-Newtonian approximation, *Phys. Rev. D* **62**, 084011 (2000) [gr-qc/0005034].
[42] T. Damour, Coalescence of two spinning black holes: An effective one-body approach, *Phys. Rev. D* **64**, 124013 (2001) [gr-qc/0103018].
[43] A. Buonanno, Y. Chen and T. Damour, Transition from inspiral to plunge in precessing binaries of spinning black holes, *Phys. Rev. D* **74**, 104005 (2006) [gr-qc/0508067].
[44] I. Hinder, A. Buonanno, M. Boyle, Z. B. Etienne, J. Healy, N. K. Johnson-McDaniel, A. Nagar, H. Nakano et al., Error-analysis and comparison to analytical models of numerical waveforms produced by the NRAR Collaboration, *Class. Quant. Grav.* **31**, 025012 (2014) [arXiv:1307.5307 [gr-qc]].
[45] S. Foffa and R. Sturani, Effective field theory methods to model compact binaries, arXiv:1309.3474 [gr-qc].
[46] L. Blanchet, Gravitational Radiation from Post-Newtonian Sources and Inspiralling Compact Binaries, *Living Rev. Rel.* **17**, 2 (2014) [arXiv:1310.1528 [gr-qc]].
[47] F. K. Manasse, J. Math. Phys. **4**, 746 (1963).
[48] P. D. D'Eath, *Phys. Rev. D* **11**, 1387 (1975).
[49] R. E. Kates, *Phys. Rev. D* **22**, 1853 (1980).
[50] D. M. Eardley, *Astrophys. J.* **196**, L59 (1975).
[51] C. M. Will, D. M. Eardley, *Astrophys. J.* **212**, L91 (1977).
[52] T. Damour, *Gravitational radiation and the motion of compact bodies*, in Gravitational Radiation, N. Deruelle and T. Piran (eds.), North-Holland, Amsterdam, 1983, 59–144.

[53] K. S. Thorne and J. B. Hartle, Laws of motion and precession for black holes and other bodies, *Phys. Rev. D* **31**, 1815 (1984).

[54] C. M. Will, *Theory and experiment in gravitational physics*, Cambridge University Press, 1993.

[55] S. M. Kopeikin, *Celestial Mechanics* **44**, 87 (1988); V. A. Brumberg and S. M. Kopejkin, *Nuovo Cimento B* **103**, 63 (1988); S. A. Klioner and A. V. Voinov, *Phys. Rev. D* **48**, 1451 (1993).

[56] T. Damour, M. Soffel and C. M. Xu, General relativistic celestial mechanics. 1. Method and definition of reference system, *Phys. Rev. D* **43**, 3273 (1991); General relativistic celestial mechanics. 2. Translational equations of motion, *Phys. Rev. D* **45**, 1017 (1992); General relativistic celestial mechanics. 3. Rotational equations of motion, *Phys. Rev. D* **47**, 3124 (1993); General relativistic celestial mechanics. 4. Theory of satellite motion, *Phys. Rev. D* **49**, 618 (1994).

[57] G. 't Hooft and M. J. G. Veltman, Regularization and renormalization of gauge fields, *Nucl. Phys. B* **44**, 189 (1972).

[58] T. Damour and G. Esposito-Farèse, Gravitational-wave versus binary-pulsar tests of strong-field gravity, *Phys. Rev. D* **58**, 042001 (1998).

[59] W. D. Goldberger and I. Z. Rothstein, An effective field theory of gravity for extended objects, *Phys. Rev. D* **73**, 104029 (2006) [DOI], [arXiv:hep-th/0409156 [hep-th]].

[60] V. A. Brumberg, *Essential Relativistic Celestial Mechanics*, Adam Hilger (ed.), Bristol, Philadelphia and New York 1991.

[61] S. A. Klioner, P. K. Seidelman and M. H. Soffel (eds.), *Relativity in Fundamental Astronomy: Dynamics, Reference Frames and Data Analysis*, Proceedings of the IAU Symposium 261, Cambridge University Press, Cambridge, 2010.

[62] M. Colpi et al. (eds.), *Physics of Relativistic Objects in Compact Binaries: From Birth to Coalescence*, Astrophysics and Space Science Library 359, Springer, Dordrecht, 2009.

[63] E. Brézin, C. Itzykson and J. Zinn-Justin, Relativistic Balmer formula including recoil effects, *Phys. Rev. D* **1**, 2349, (1970).

[64] T. Damour, B. R. Iyer and B. S. Sathyaprakash, Improved filters for gravitational waves from inspiralling compact binaries, *Phys. Rev. D* **57**, 885, (1998) [gr-qc/9708034].

[65] T. Damour, B. R. Iyer and A. Nagar, Improved resummation of post-Newtonian multipolar waveforms from circularized compact binaries, *Phys. Rev. D* **79**, 064004 (2009) [0811.2069 [gr-qc]]

[66] T. Damour and A. Nagar, An improved analytical description of inspiralling and coalescing black-hole binaries, *Phys. Rev. D* **79**, 081503 (2009) [0902.0136 [gr-qc]].

[67] M. Davis, R. Ruffini and J. Tiomno, Pulses of gravitational radiation of a particle falling radially into a Schwarzschild black hole, *Phys. Rev. D* **5**, 2932, (1972).

[68] R. H. Price and J. Pullin, Colliding black holes: The Close limit, *Phys. Rev. Lett.* **72**, 3297, (1994) [gr-qc/9402039].

[69] T. Damour and G. Schäfer, Higher order relativistic periastron advances and binary pulsars, *Nuovo Cim. B* **101**, 127, (1988).

[70] T. Damour, P. Jaranowski and G. Schäfer, Dynamical invariants for general relativistic two-body systems at the third post-Newtonian approximation, *Phys. Rev. D*, **62**, 044024 (2000) [gr-qc/9912092].

[71] T. Damour, P. Jaranowski and G. Schäfer, Poincaré invariance in the ADM Hamiltonian approach to the general relativistic two-body problem, [Erratum *Phys. Rev. D* **63**, 029903 (2001)], *Phys. Rev. D* **62**, 021501 (2000) [gr-qc/0003051].

[72] T. Damour, Gravitational self force in a Schwarzschild background and the effective one body formalism, *Phys. Rev. D* **81**, 024017 (2010) [0910.5533 [gr-qc]].

[73] L. Blanchet, S. L. Detweiler, A. Le Tiec and B. F. Whiting, High-order post-Newtonian fit of the gravitational self-force for circular orbits in the Schwarzschild geometry, *Phys. Rev. D* **81**, 084033 (2010) [1002.0726 [gr-qc]].

[74] T. Damour (unpublished), cited in Ref. [135], which quoted and used some combinations of the logarithmic contributions to $a(u)$ and $\bar{d}(u)$, 2010.

[75] E. Barausse, A. Buonanno and A. Le Tiec, The complete non-spinning effective-one-body metric at linear order in the mass ratio, *Phys. Rev. D* **85**, 064010 (2012) [1111.5610 [gr-qc]].

[76] T. Damour, A. Nagar and S. Bernuzzi, Improved effective-one-body description of coalescing nonspinning black-hole binaries and its numerical-relativity completion, *Phys. Rev. D* **87**, 084035 (2013) [arXiv:1212.4357 [gr-qc]].

[77] D. Bini and T. Damour, Gravitational radiation reaction along general orbits in the effective one-body formalism, *Phys. Rev. D* **86**, 124012 (2012) [1210.2834[gr-qc]].

[78] T. Damour and A. Nagar, Faithful effective-one-body waveforms of small-mass-ratio coalescing black-hole binaries, *Phys. Rev. D* **76**, 064028 (2007) [0705.2519 [gr-qc]].

[79] T. Damour and A. Nagar, Comparing Effective-One-Body gravitational waveforms to accurate numerical data, *Phys. Rev. D* **77**, 024043 (2008) [0711.2628 [gr-qc]].

[80] L. Blanchet, T. Damour, G. Esposito-Farèse and B. R. Iyer, Gravitational radiation from inspiralling compact binaries completed at the third post-Newtonian order, *Phys. Rev. Lett.* **93**, 091101 (2004) [gr-qc/0406012].

[81] L. E. Kidder, Using full information when computing modes of post-Newtonian waveforms from inspiralling compact binaries in circular orbit, *Phys. Rev. D* **77**, 044016 (2008) [0710.0614 [gr-qc]].

[82] E. Berti, V. Cardoso, J. A. Gonzalez, U. Sperhak, M. Hannam, S. Husa and B. Bruegmann, Inspiral, merger and ringdown of unequal mass black hole binaries: A multipolar analysis, *Phys. Rev. D* **76**, 064034 (2007) [gr-qc/0703053].

[83] L. Blanchet, G. Faye, B. R. Iyer and S. Sinha, The third post-Newtonian gravitational wave polarisations and associated spherical harmonic modes for inspiralling compact binaries in quasi-circular orbits, *Class. Quant. Grav.* **25**, 165003 (2008) [0802.1249 [gr-qc]].

[84] H. Tagoshi and M. Sasaki, Post-Newtonian expansion of gravitational waves from a particle in circular orbit around a Schwarzschild black hole, *Prog. Theor. Phys*, **92**, 745–771, (1994) [gr-qc/9405062].

[85] T. Tanaka, H. Tagoshi and M. Sasaki, Gravitational Waves by a Particle in Circular Orbit around a Schwarzschild Black Hole, *Prog. Theor. Phys* **96**, 1087–1101, (1996) [gr-qc/9701050].

[86] L. Blanchet and T. Damour, Hereditary effects in gravitational radiation, *Phys. Rev. D* **46**, 4304, (1992).

[87] L. Blanchet, Gravitational-wave tails of tails, [Erratum *Class. Quant. Grav.* **22**, 3381 (2005)], *Class. Quant. Grav.* **15**, 113, (1998) [gr-qc/9710038].

[88] R. Fujita and B. R. Iyer, Spherical harmonic modes of 5.5 post-Newtonian gravitational wave polarisations and associated factorised resummed waveforms for a particle in circular orbit around a Schwarzschild black hole, *Phys. Rev. D* **82**, 044051 (2010) [1005.2266 [gr-qc]].

[89] R. Fujita, Gravitational radiation for extreme mass ratio inspirals to the 14th post-Newtonian order, *Prog. Theor. Phys.* **127**, 583 (2012) [1104.5615 [gr-qc]].

[90] Y. Pan, A. Buonanno, R. Fujita, E. Racine and H. Tagoshi, Post-Newtonian factorized multipolar waveforms for spinning, non-precessing black-hole binaries,.*Phys. Rev. D* **83**, 064003 (2011) [1006.0431 [gr-qc]].

[91] Y. Pan, A. Buonanno, M. Boyle, L. T. Buchman, L. E. Kidder, H. P. Pfeiffer and M. A. Scheel, Inspiral-merger-ringdown multipolar waveforms of nonspinning black-hole binaries using the effective-one-body formalism, *Phys. Rev. D* **84**, 124052 (2011) [1106.1021 [gr-qc]].

[92] A. Taracchini, Y. Pan, A. Buonanno, E. Barausse, M. Boyle, T. Chu, G. Lovelace, H. P. Pfeiffer et al., Prototype effective-one-body model for nonprecessing spinning inspiral-merger-ringdown waveforms, *Phys. Rev. D* **86**, 024011 (2012) [1202.0790 [gr-qc]].

[93] T. Damour, A. Nagar, E. N. Dorband, D. Pollney and L. Rezzolla, Faithful Effective-One-Body waveforms of equal-mass coalescing black-hole binaries, *Phys. Rev. D*, **77**, 084017 (2008) [0712.3003 [gr-qc]].

[94] T. Damour, A. Nagar, M. Hannam, S. Husa and B. Bruegmann, Accurate Effective-One-Body waveforms of inspiralling and coalescing black-hole binaries, *Phys. Rev. D* **78**, 044039 (2008) [0803.3162 [gr-qc]].

[95] L. Blanchet and T. Damour, Post-Newtonian generation of gravitational waves, *Annales Poincaré Phys. Theor.* **50**, 377, (1989).

[96] T. Damour and B. R. Iyer, Post-Newtonian generation of gravitational waves. 2. The Spin moments, *Annales Poincaré Phys. Theor.* **54**, 115, (1991).

[97] A. Buonanno, G. B. Cook and F. Pretorius, Inspiral, merger and ring-down of equal-mass black-hole binaries, *Phys. Rev. D* **75**, 124018 (2007) [gr-qc/0610122].

[98] Y. Pan et al., A data-analysis driven comparison of analytic and numerical coalescing binary waveforms: Nonspinning case, *Phys. Rev. D* **77**, 024014 (2008) [0704.1964 [gr-qc]].

[99] A. Le Tiec, A. H. Mroue, L. Barack, A. Buonanno, H. P. Pfeiffer, N. Sago and A. Taracchini, Periastron advance in black hole binaries, *Phys. Rev. Lett.* **107**, 141101 (2011) [1106.3278 [gr-qc]].

[100] T. Damour, A. Nagar, D. Pollney and C. Reisswig, Energy versus angular momentum in black hole binaries, *Phys. Rev. Lett.* **108**, 131101 (2012) [1110.2938 [gr-qc]].

[101] T. Damour, E. Gourgoulhon and P. Grandclément, Circular orbits of corotating binary black holes: Comparison between analytical and numerical results, *Phys. Rev. D*, **66**, 024007 (2002) [gr-qc/0204011].

[102] A. Buonanno, Y. Pan, J. G. Baker, J. Centrella, B. J. Kelly, S. T. McWilliams, and J. R. van Meter, Toward faithful templates for non-spinning binary black holes using the effective-one-body approach, *Phys. Rev. D* **76**, 104049 (2007) [0706.3732 [gr-qc]].

[103] A. Buonanno, Y. Pan, H. P. Pfeiffer, M. A. Scheel, L. T. Buchman and L. E. Kidder, Effective-one-body waveforms calibrated to numerical relativity simulations: coalescence of non-spinning, equal-mass black holes, *Phys. Rev. D* **79**, 124028 (2009) [0902.0790 [gr-qc]].

[104] S. Bernuzzi, A. Nagar and A. Zenginoglu, Binary black hole coalescence in the extreme-mass-ratio limit: testing and improving the effective-one-body multipolar waveform, *Phys. Rev. D* **83**, 064010 (2011) [1012.2456 [gr-qc]].

[105] E. Barausse, A. Buonanno, S. A. Hughes, G. Khanna, S. O'Sullivan, et al., Modeling multipolar gravitational-wave emission from small mass-ratio mergers, *Phys. Rev. D* **85**, 024046 (2012) [DOI], [arXiv:1110.3081 [gr-qc]].

[106] T. Damour and A. Gopakumar, Gravitational recoil during binary black hole coalescence using the effective one body approach, *Phys. Rev. D* **73**, 124006 (2006) [gr-qc/0602117].

[107] T. Damour and A. Nagar, Final spin of a coalescing black-hole binary: an effective-one-body approach, *Phys. Rev. D* **76**, 044003 (2007) [0704.3550 [gr-qc]].

[108] T. Damour, A. Nagar and A. Tartaglia, Binary black hole merger in the extreme mass ratio limit, *Class. Quant. Grav.* **24**, S109 (2007) [gr-qc/0612096].

[109] J. A. Gonzalez, U. Sperhake, B. Bruegmann, M. Hannam and S. Husa, Total recoil: the maximum kick from nonspinning black-hole binary inspiral, *Phys. Rev. Lett.*, **98**, 091101 (2007) [gr-qc/0610154].

[110] N. Yunes, A. Buonanno, S. A. Hughes, M. Coleman Miller and Y. Pan, Modeling extreme mass ratio inspirals within the effective-one-body approach, *Phys. Rev. Lett.*, **104**, 091102 (2010) [0909.4263 [gr-qc]].

[111] N. Yunes, A. Buonanno, S. A. Hughes, Y. Pan, E. Barausse, M. C. Miller and W. Throwe, Extreme mass-ratio inspirals in the effective-one-body approach: Quasi-circular, equatorial orbits around a spinning black hole, *Phys. Rev. D* **83**, 044044 (2011) [1009.6013 [gr-qc]].
[112] M. A. Scheel, M. Boyle, T. Chu, L. E. Kidder, K. D. Matthews and H. P. Pfeiffer, High-accuracy waveforms for binary black hole inspiral, merger, and ringdown, *Phys. Rev. D* **79**, 024003 (2009) [0810.1767 [gr-qc]].
[113] T. Damour and A. Nagar, The effective one body description of the two-body problem, *Fundam. Theor. Phys.* **162**, 211 (2011) [0906.1769 [gr-qc]].
[114] T. Damour, A. Nagar and L. Villain, Merger states and final states of black hole coalescences: A numerical-relativity-assisted effective-one-body approach, *Phys. Rev. D* **89**, 024031 (2014) [arXiv:1307.2868 [gr-qc]].
[115] T. Hinderer, A. Buonanno, A. H. Mroué, D. A. Hemberger, G. Lovelace and H. P. Pfeiffer, Periastron advance in spinning black hole binaries: comparing effective-one-body and numerical relativity, *Phys. Rev. D* **88**, 084005 (2013) [arXiv:1309.0544 [gr-qc]].
[116] Y. Pan, A. Buonanno, L. T. Buchman, T. Chu, L. E. Kidder, H. P. Pfeiffer and M. A. Scheel, Effective-one-body waveforms calibrated to numerical relativity simulations: coalescence of non-precessing, spinning, equal-mass black holes, *Phys. Rev. D*, **81**, 084041 (2010) [0912.3466 [gr-qc]].
[117] T. Damour, P. Jaranowski and G. Schäfer, Effective one body approach to the dynamics of two spinning black holes with next-to-leading order spin–orbit coupling, *Phys. Rev. D* **78**, 024009 (2008) [0803.0915 [gr-qc]].
[118] E. Barausse, E. Racine and A. Buonanno, Hamiltonian of a spinning test-particle in curved spacetime [Erratum *Phys. Rev. D* **85**, 069904 (2012)], *Phys. Rev. D* **80**, 104025 (2009) [0907.4745 [gr-qc]].
[119] E. Barausse and A. Buonanno, An Improved effective-one-body Hamiltonian for spinning black-hole binaries, *Phys. Rev. D* **81**, 084024 (2010) [0912.3517 [gr-qc]].
[120] A. Nagar, Effective one body Hamiltonian of two spinning black-holes with next-to-next-to-leading order spin–orbit coupling, *Phys. Rev. D* **84**, 084028 (2011) [1106.4349 [gr-qc]].
[121] E. Barausse and A. Buonanno, Extending the effective-one-body Hamiltonian of black-hole binaries to include next-to-next-to-leading spin–orbit couplings, *Phys. Rev. D*, **84**, 104027 (2011) [1107.2904 [gr-qc]].
[122] Y. Pan, A. Buonanno, L. T. Buchman, T. Chu, L. E. Kidder, H. P. Pfeiffer and M. A. Scheel, Effective-one-body waveforms calibrated to numerical relativity simulations: coalescence of non-precessing, spinning, equal-mass black holes, *Phys. Rev. D* **81**, 084041 (2010) [arXiv:0912.3466 [gr-qc]].
[123] A. Taracchini, Y. Pan, A. Buonanno, E. Barausse, M. Boyle, T. Chu, G. Lovelace, H. P. Pfeiffer et al., Prototype effective-one-body model for nonprecessing spinning inspiral-merger-ringdown waveforms, *Phys. Rev. D* **86**, 024011 (2012) [arXiv:1202.0790 [gr-qc]].
[124] Y. Pan, A. Buonanno, A. Taracchini, L. E. Kidder, A. H. Mroue, H. P. Pfeiffer, M. A. Scheel and B. Szilagyi, Inspiral-merger-ringdown waveforms of spinning, precessing black-hole binaries in the effective-one-body formalism, *Phys. Rev. D* **89**, 084006 (2014) [arXiv:1307.6232 [gr-qc]].
[125] E. E. Flanagan and T. Hinderer, Constraining neutron star tidal Love numbers with gravitational wave detectors, *Phys. Rev. D* **77**, 021502 (2008) [arXiv:0709.1915 [astro-ph]].
[126] T. Hinderer, B. D. Lackey, R. N. Lang and J. S. Read, Tidal deformability of neutron stars with realistic equations of state and their gravitational wave signatures in binary inspiral, *Phys. Rev. D* **81**, 123016 (2010) [arXiv:0911.3535 [astro-ph.HE]].

[127] T. Damour, A. Nagar and L. Villain, Measurability of the tidal polarizability of neutron stars in late-inspiral gravitational-wave signals, *Phys. Rev. D* **85**, 123007 (2012) [arXiv:1203.4352 [gr-qc]].
[128] W. Del Pozzo, T. G. F. Li, M. Agathos, C. V. D. Broeck and S. Vitale, Demonstrating the feasibility of probing the neutron star equation of state with second-generation gravitational wave detectors, *Phys. Rev. Lett.* **111**, 071101 (2013) [arXiv:1307.8338 [gr-qc]].
[129] L. Baiotti, T. Damour, B. Giacomazzo, A. Nagar and L. Rezzolla, Analytic modelling of tidal effects in the relativistic inspiral of binary neutron stars, *Phys. Rev. Lett.* **105**, 261101 (2010) [1009.0521 [gr-qc]].
[130] L. Baiotti, T. Damour, B. Giacomazzo, A. Nagar and L. Rezzolla, Accurate numerical simulations of inspiralling binary neutron stars and their comparison with effective-one-body analytical models, *Phys. Rev. D* **84**, 024017 (2011) [1103.3874 [gr-qc]].
[131] S. Bernuzzi, A. Nagar, M. Thierfelder and B. Brugmann, Tidal effects in binary neutron star coalescence, *Phys. Rev. D* **86**, 044030 (2012) [arXiv:1205.3403 [gr-qc]].
[132] K. Hotokezaka, K. Kyutoku and M. Shibata, Exploring tidal effects of coalescing binary neutron stars in numerical relativity, *Phys. Rev. D* **87** (4), 044001 (2013) [arXiv:1301.3555 [gr-qc]].
[133] T. Damour and A. Nagar, Effective one body description of tidal effects in inspiralling compact binaries, *Phys. Rev. D* **81**, 084016 (2010) [0911.5041 [gr-qc]].
[134] D. Bini, T. Damour and G. Faye, Effective action approach to higher-order relativistic tidal interactions in binary systems and their effective one body description, *Phys. Rev. D* **85**, 124034 (2012) [1202.3565 [gr-qc]].
[135] L. Barack, T. Damour and N. Sago, Precession effect of the gravitational self-force in a Schwarzschild spacetime and the effective one-body formalism, *Phys. Rev. D* **82**, 084036 (2010) [1008.0935 [gr-qc]].
[136] A. Le Tiec, L. Blanchet and B. F. Whiting, The first law of binary black hole mechanics in general relativity and post-Newtonian theory, *Phys. Rev. D* **85**, 064039 (2012) [1111.5378 [gr-qc]].
[137] A. Le Tiec, E. Barausse and A. Buonanno, Gravitational Self-Force Correction to the Binding Energy of Compact Binary Systems, *Phys. Rev. Lett.* **108**, 131103 (2012) [1111.5609 [gr-qc]].
[138] S. Akcay, L. Barack, T. Damour and N. Sago, Gravitational self-force and the effective-one-body formalism between the innermost stable circular orbit and the light ring, *Phys. Rev. D* **86**, 104041 (2012) [DOI], [arXiv:1209.0964 [gr-qc]].
[139] T. Damour and G. Esposito-Farese, Testing gravity to second post-Newtonian order: A Field theory approach, *Phys. Rev. D* **53**, 5541 (1996) [gr-qc/9506063].

Gerhard Schäfer

Hamiltonian dynamics of spinning compact binaries through high post-Newtonian approximations

1 Introduction

The book by Professor Brumberg entitled *Essential Relativistic Celestial Mechanics*, published in 1991 [1], was the long-awaited successor of his much earlier book "Relativistic Celestial Mechanics" of 1972 [2], published in Russian only. In these accounts, the celestial mechanics has been presented with unprecedented details based on relativistic physics. Fully in their tradition, the recent book by Kopeikin, Efroimsky, and Kaplan entitled *Relativistic Celestial Mechanics of the Solar System* [3] is the most advanced account on the dynamics of the solar system. The mentioned books mainly rely on the first post-Newtonian (1pN) approximation of Einstein's general relativity theory but in part and to some extent also results are presented from the 2pN and 2.5pN approximations and the leading order gravitational spin coupling. Focusing on compact binary systems, black holes, and neutron stars, this chapter covers the dynamics and motion up to the 4pN approximation including proper rotation (spin) of the bodies.

The Einstein theory of gravitation delivers an excellent and well tested description of the dynamics of gravitating systems despite its very intricate and complex structure from both mathematical and conceptual points of view [4]. The very reason for this complication is the fact that the gravitational potential functions, which are identical with the metric coefficients of spacetime, together with their first derivatives, the so-called Christoffel symbols, are no observables but depend on the chosen coordinate system. Selecting a specific coordinate system thus means choosing a specific gauge. The freedom of choice of the coordinate system renders the highly nonlinear Einstein field equations degenerate, e.g. [5]. A most elegant way to cope with these difficulties is the Hamiltonian approach where finally one functional, the nondegenerate physical Hamiltonian, determines the dynamics of the whole gravitating system. It is a Hamiltonian setting of general relativity which is used in the present contribution. If at various places in the following the Newton's gravitational constant, G, and the speed of light, c, do not show up explicitly, the units $16\pi G = 1$ and $c = 1$ are employed. Notice should also be made to the fact that the chapter is devoted to the

Gerhard Schäfer: Friedrich-Schiller-Universität Jena, Theoretisch-Physikalisches Institut, Max-Wien-Platz 1, D-07743 Jena, Germany

Hamiltonian treatment of spinning and nonspinning binaries; for related results from other formalisms, the reader is advised to consult [6–8].

2 Hamiltonian formulation of general relativity

The most often used Hamiltonian formulation of general relativity is the one developed by Arnowitt, Deser, and Misner (ADM) around 1960 [9], which is nearly half a century after Einstein's publication of his theory in 1915. Their primary interest was the quantization of the gravitational field which is still an unsolved problem. On the other side, the application of the ADM approach in classical general relativity has turned out very efficient as can be clearly seen in the following.

In the ADM formalism, the full Einstein field equations

$$G^{\mu\nu} = \frac{1}{2} T^{\mu\nu}, \tag{2.1}$$

where $G^{\mu\nu}$ denotes the Einstein tensor and $T^{\mu\nu}$ the energy–momentum tensor ($\mu, \nu = 0, 1, 2, 3; i, j = 1, 2, 3$), are obtained by applying functional derivatives to the formal Hamiltonian (symbolic writing in part)

$$H = \int d^3x [N\mathcal{H}(\gamma, \partial\gamma, \partial\partial\gamma, \pi, MV) - N^i \mathcal{H}_i(\gamma, \partial\gamma, \pi, \partial\pi, MV)] \tag{2.2}$$

with compact-support perturbations of the field functions N and N^i, coined lapse and shift functions by Wheeler [10], respectively, as well as of the three-dimensional metric coefficients $\gamma_{ij} \equiv g_{ij}$ and their canonical conjugates π^{ij}. The field functions γ_{ij} and π^{ij} are dynamical variables whereas the field functions N and N^i are Lagrange multipliers which do not show up in the functions $\mathcal{H}$, called super Hamiltonian, and $\mathcal{H}_i$, called super momentum (the minus sign in Equation (2.2) results from the definition of $\mathcal{H}_i$ to contain the momentum density of the matter and not the negative of it). The canonical conjugates of N and N^i do not appear in the Hamiltonian. Variation of the formal Hamiltonian with respect to N and N^i results in the constraint equations which are of fundamental importance in the ADM formalism. MV is an abbreviation for the canonical matter variables. In terms of the metric coefficients $g_{\mu\nu}$, or their inverse $g^{\mu\nu}$, the lapse and shift functions can easily be deduced from the following equations, see e.g. [11]:

$$ds^2 \equiv g_{\mu\nu} dx^\mu dx^\nu = -N^2 dx^0 dx^0 + \gamma_{ij}(dx^i + N^i dx^0)(dx^j + N^j dx^0), \tag{2.3}$$

$$g^{\mu\nu} \frac{\partial S}{\partial x^\mu} \frac{\partial S}{\partial x^\nu} = -\frac{1}{N^2} \left(\frac{\partial S}{\partial x^0} - N^i \frac{\partial S}{\partial x^i} \right)^2 + \gamma^{ij} \frac{\partial S}{\partial x^i} \frac{\partial S}{\partial x^j}, \tag{2.4}$$

where γ^{ij} denotes the inverse three-dimensional metric, $\gamma^{ik}\gamma_{kj} = \delta_{ij}$. ds^2 is the line element and S is a scalar function; if applied to a particle, their meanings are particle's proper time τ ($ds^2 = -c^2 d\tau^2$) and action, respectively. The motion orthogonal to x^0 = const reads $dx^i + N^i dx^0 = 0$ and the derivative orthogonal to x^0 = const takes the form $\partial_0 - N^i\partial_i$, where $\partial_\mu = \frac{\partial}{\partial x^\mu}$.

The following identities hold:

$$\mathcal{H} \equiv \sqrt{-g}N(T^{00} - 2G^{00}), \quad \mathcal{H}_i \equiv \sqrt{-g}(T_i^0 - 2G_i^0) \tag{2.5}$$

and

$$\pi^{ij} \equiv -\sqrt{\gamma}(\gamma^{ik}\gamma^{jl} - \gamma^{ij}\gamma^{kl})K_{kl}\,, \tag{2.6}$$

where $K_{kl}dx^k dx^l$ is the second fundamental form or extrinsic curvature of a hypersurface x^0 = const whereas the first fundamental form or intrinsic metric is given by $\gamma_{kl}dx^k dx^l$. The determinants of $g_{\mu\nu}$ and γ_{ij} are denoted by g and γ, respectively. In terms of Christoffel symbols, $K_{ij} = -N\Gamma^0_{ij}$ holds showing that the extrinsic curvature depends on the first partial spacetime derivatives of the metric coefficients. In contrast herewith, the intrinsic curvature form $R_{kl}dx^k dx^l$ of a hypersurface x^0 = const depends on partial space derivatives of the metric coefficients through second order.

If the field functions are assumed to decay at spacelike flat infinity (denoted by i^0) in the way physical gravitational fields have to decay, as being inferred from physical solutions of the Einstein field equations for isolated systems, namely

$$N = 1 + O(1/r), \quad N^i = O(1/r), \quad \gamma_{ij} = \delta_{ij} + O(1/r), \quad \pi^{ij} = O(1/r^2)\,, \tag{2.7}$$

a surface term has to be added to the Hamiltonian for not spoiling the Hamilton principle, reading, see [9, 12, 13],

$$H = \int d^3x(N\mathcal{H} - N^i\mathcal{H}_i) + \oint_{i^0} d^2s_i\partial_j(\gamma_{ij} - \delta_{ij}\gamma_{kk})\,, \tag{2.8}$$

where d^2s_i denotes the two-dimensional surface area element with outward-pointing unit normal vector.

Splitting naturally $\mathcal{H} = \mathcal{H}^{\text{matter}} + \mathcal{H}^{\text{field}}$ and $\mathcal{H}_i = \mathcal{H}_i^{\text{matter}} + \mathcal{H}_i^{\text{field}}$, it is interesting to point out that the Hamiltonian for test matter (TM) in an external gravitational field (passive interaction) reads

$$H_{\text{TM}} = \int d^3x(N\mathcal{H}^{\text{matter}} - N^i\mathcal{H}_i^{\text{matter}})\,. \tag{2.9}$$

For the treatment of gravitationally passively and actively interacting matter the free-field (FF) Hamiltonian

$$H_{\text{FF}} = \int d^3x(N\mathcal{H}^{\text{field}} - N^i\mathcal{H}_i^{\text{field}}) \tag{2.10}$$

has to be added as well as the above surface term because gravitational interaction implies $H_{\text{TM}} + H_{\text{FF}} = 0$ so otherwise the Hamiltonian of the total system would vanish, e.g. [12]. Indeed, on the submanifold where the constraint equations hold,

$$\mathcal{H} = 0, \qquad \mathcal{H}_i = 0\,, \tag{2.11}$$

the total Hamiltonian reduces to a surface term only,

$$H = \oint_{i^0} d^2 s_i \partial_j (\gamma_{ij} - \delta_{ij}\gamma_{kk}) \,. \tag{2.12}$$

The present considerations clearly show that gravitational energy is not localizable. The rewriting of the energy surface integral as volume integral, using the Gaussian theorem, i.e.

$$H = \int d^3x \, \partial_i \partial_j (\gamma_{ij} - \delta_{ij}\gamma_{kk}) \,, \tag{2.13}$$

neither gives a unique integrand nor a one which transforms as some tensor-density component under general coordinate transformations. The Hamiltonian within the Dirac approach reads

$$H_{\mathrm{D}} = - \int d^3x \, \partial_i (\gamma^{-1/2} \partial_j (\gamma \gamma^{ij})) \tag{2.14}$$

and the one in the Schwinger formalism takes the form

$$H_{\mathrm{S}} = - \int d^3x \, \partial_i \partial_j (\gamma \gamma^{ij}) \,. \tag{2.15}$$

The three Hamiltonians evidently fulfil $H = H_{\mathrm{D}} = H_{\mathrm{S}}$, taking into account the asymptotic behavior of the metric coefficients.

The surface-integral Hamiltonian is still not able to generate the remaining Einstein field equations and, thus, is not a Hamiltonian at all; it only delivers the total energy of the system and coincides with, e.g. the energy expression from the Landau–Lifshitz pseudo-tensor complex for the gravitational field, [14]. However, by the additional choice of four appropriate coordinate conditions, e.g.

$$\pi^{ii} = 0 \quad \text{and} \quad 3\partial_j \gamma_{ij} - \partial_i \gamma_{jj} = 0 \,, \tag{2.16}$$

resulting in

$$\pi^{ij} = \tilde{\pi}^{ij} + \pi^{ij\mathrm{TT}} \quad \text{and} \quad \gamma_{ij} = \psi \delta_{ij} + h_{ij}^{\mathrm{TT}} \tag{2.17}$$

with

$$\tilde{\pi}^{ij} = \partial_j \pi^i + \partial_i \pi^j - \frac{2}{3}\delta_{ij}\partial_k \pi^k , \tag{2.18}$$

the energy expression in question turns into a Hamitonian for the matter variables and the "true" degrees of freedom of the gravitational field h_{ij}^{TT} and their canonical conjugate $\pi^{ij\mathrm{TT}}$ which both are transverse and traceless, i.e. $h_{ii}^{\mathrm{TT}} = \pi^{ii\mathrm{TT}} = \partial_j h_{ij}^{\mathrm{TT}} = \partial_j \pi^{ij\mathrm{TT}} = 0$. The outcome is the ADM-Hamiltonian. Importantly, the derivation of the ADM-Hamiltonian needs the constraint equations and the coordinate conditions only (the functions ψ and π^i are determined by the constraint equations). The lapse and shift functions do not enter; only the asymptotic value 1 of the lapse function entered into the obtention of the ADM-Hamiltonian. For the calculation of the functions N and N^i, the coordinate conditions are needed as well as the field equations for g_{ij}

and π^{ij}. Importantly too, the resulting equations for N and N^i are of instantaneous type without any time derivatives. Only the variables h_{ij}^{TT} and $\pi^{ij\mathrm{TT}}$ fulfil evolution equations and result from functional derivatives of the ADM-Hamiltonian which can be put into the form

$$H_{\mathrm{ADM}} = -2\oint_{i^0} d^2 s_i \partial_i \psi[h_{kl}^{\mathrm{TT}}, \pi^{kl\mathrm{TT}}, MV] = -2\int d^3x \partial_i\partial_i\psi \,. \tag{2.19}$$

The Poisson-bracket commutation relations read

$$\{h_{ij}^{\mathrm{TT}}(\mathbf{x},t), \pi^{kl\mathrm{TT}}(\mathbf{x}',t)\} = \delta_{ij}^{\mathrm{TT}kl}\delta(\mathbf{x}-\mathbf{x}')\,, \tag{2.20}$$

and zero otherwise, where $\delta_{ij}^{\mathrm{TT}kl}$, the transverse-traceless projection operator, is defined by

$$\delta_{ij}^{\mathrm{TT}kl} = \frac{1}{2}(P_{il}P_{jk} + P_{ik}P_{jl} - P_{kl}P_{ij}), \quad P_{ij} = \delta_{ij} - \Delta^{-1}\partial_i\partial_j\,, \tag{2.21}$$

where Δ^{-1} denotes the usual inverse Laplacian in flat three-dimensional space.

2.1 Point particles

In regard to the matter variables MV, the canonical structure is simple for point masses and for perfect fluids, e.g. [9, 15]. Also the inclusion of the electromagnetic field is straightforward, [9]. Quite involved, however, is the canonical setting of spinning particles. For point particles, the total super Hamiltonian and super momentum read

$$\mathcal{H} = \sum_a \left(m_a^2c^2 + \gamma^{ij}p_{ai}p_{aj}\right)^{1/2}\delta_a + \frac{1}{\gamma^{1/2}}\left(\pi^i_j\pi^j_i - \frac{1}{2}\pi^i_i\pi^j_j\right) - \gamma^{1/2}R\,, \tag{2.22}$$

$$\mathcal{H}_i = \sum_a p_{ai}\delta_a - \pi^{kl}\gamma_{kl,i} + 2\pi^j_{i,j}\,, \tag{2.23}$$

where $\pi^i_j = \gamma_{jk}\pi^{ik}$, $R = \gamma^{ij}R_{ij}$, $A_{,i} = \partial_i A$, and $\delta_a = \delta(x^i - x^i_a)$ with $\int d^3x\delta_a = 1$. The canonical particle variables, positions, and momenta, are simply x^i_a and p_{ai}, with the particle label $a = 1, 2, \ldots$, fulfilling the standard Poisson-bracket relations,

$$\{x^i_a, p_{aj}\} = \delta_{ij}\,, \tag{2.24}$$

and zero otherwise.

The reader may wonder how the use of Dirac delta functions can make sense in a nonlinear theory such as Einstein's general relativity theory. Here it is to be said that with dimensional regularization, e.g. [16, 17], i.e. performimg calculations in d space-dimensions and making the limit $d \to 3$ at the very end of the calculations, all local singularities could be uniquely controlled through 4pN order including spin, [16,

18, 19]. The problem with the badly defined outer near zone, starting at the 4pN approximation, is not connected with the application of Dirac delta functions [19]. Interestingly, Brill–Lindquist black holes can be represented by Dirac delta functions with support in conformally related three-dimensional space, i.e. fictitious point masses in fictitious space generate black holes (extended objects!) in physical space [20], similarly to the image charges in the electrodynamics which are also not located in the region where the physical fields operate.

2.2 Spinning particles

In the case of spinning particles, the matter super Hamiltonian and super momentum are given by [21, 22]

$$\mathcal{H}^{\text{matter}} = \sum_a \Bigg[\Bigg(-np_a - \frac{1}{2}\left(\frac{S_{ali}p_{aj}}{np_a} + \gamma^{mn}\frac{S_{ali}p_{aj}p_{an}p_{al}}{(np_a)^2(m_a - np_a)} \right)\gamma^{kl}\gamma_{ij,k} + \frac{p_{aj}\gamma^{ji}}{np_a}\pi_a^{kl}\gamma_{kl,i} \Bigg)\delta_a - \left(\frac{p_{al}}{m_a - np_a}\gamma^{ij}\gamma^{kl}S_{ajk}\delta_a \right)_{,i} \Bigg], \tag{2.25}$$

$$\mathcal{H}_i^{\text{matter}} = \sum_a \left[\left(p_{ai} - \pi_a^{kl}\gamma_{kl,i} \right)\delta_a + \frac{1}{2}\left(s_a^{ij}\delta_a \right)_{,j} \right], \tag{2.26}$$

where

$$\pi_a^{ij} = \gamma^{ik}\gamma^{jl}\frac{m_a p_{a(k}nS_{al)}}{2np_a(m_a - np_a)} + \frac{1}{2}B_{kl}^{ij}\gamma^{km}\gamma^{ln}S_{amn}, \tag{2.27}$$

$$np_a = -\left(m_a^2c^2 + \gamma^{ij}p_{ai}p_{aj} \right)^{1/2}, \quad nS_{ai} = -\frac{p_{ak}\gamma^{kj}S_{aji}}{m}, \tag{2.28}$$

$$s_a^{ij} = \gamma^{jk}S_{aik} + \gamma^{jk}\gamma^{lp}\frac{2p_{al}p_{a(i}S_{ak)p}}{np_a(m_a - np_a)}. \tag{2.29}$$

If the symmetric root of the symmetric matrix γ_{ij} is denoted by e_{ij}, i.e. $e_{ik}e_{kj} = \gamma_{ij}$, B_{kl}^{ij} reads

$$2B_{kl}^{ij} = e_{km}\frac{\partial e_{ml}}{\partial \gamma_{ij}} - e_{lm}\frac{\partial e_{mk}}{\partial \gamma_{ij}}. \tag{2.30}$$

The canonical field momentum is now given by

$$\pi_{\text{can}}^{ij} = \pi^{ij} + \sum_a \pi_a^{ij}\delta_a. \tag{2.31}$$

Imposing coordinate conditions of the form

$$\pi_{\text{can}}^{ii} = 0 \quad \text{and} \quad 3\partial_j\gamma_{ij} - \partial_i\gamma_{jj} = 0, \tag{2.32}$$

a new decomposition

$$\pi_{\text{can}}^{ij} = \tilde{\pi}_{\text{can}}^{ij} + \pi_{\text{can}}^{ij\text{TT}} \tag{2.33}$$

arises. Also applying the constraint equations $\mathcal{H} = 0$ and $\mathcal{H}_i = 0$, wherein π^{ij} has to be substituted through π^{ij}_{can}, finally, the reduced total action W for spinning particles turns out to be

$$W = \int dt \sum_a \left(p_{ai}\dot{x}^i_a + \frac{1}{2} S_{a(i)(j)} \dot{\theta}^{(i)(j)}_a \right) + \int d^4x\, \pi^{ij\text{TT}}_{\text{can}} h^{\text{TT}}_{ij,0}$$
$$- \int dt\, H_{\text{ADM}} \left[x^i_a, p_{ai}, S_{a(i)(j)}, h^{\text{TT}}_{ij}, \pi^{ij\text{TT}}_{\text{can}} \right] , \qquad (2.34)$$

where $S_{aij} = e_{ik} e_{jl} S_{a(k)(l)}$ holds. The Poisson-bracket relations read

$$\left\{ x^i_a, p_{aj} \right\} = \delta_{ij}, \quad \left\{ S_{a(i)}, S_{a(j)} \right\} = \epsilon_{ijk} S_{a(k)} , \qquad (2.35)$$
$$\left\{ h^{\text{TT}}_{ij}(\mathbf{x}, t), \pi^{kl\text{TT}}_{\text{can}}(\mathbf{x}', t) \right\} = \delta^{\text{TT}kl}_{ij} \delta(\mathbf{x} - \mathbf{x}') , \qquad (2.36)$$

and zero otherwise, with $S_{a(k)} = \frac{1}{2} \epsilon_{kij} S_{a(i)(j)}$. The evolution equations for the canonical variables have the forms

$$\frac{d}{dt}(x^i_a, p_{ai}, S_{a(i)}) = \left\{ (x^i_a, p_{ai}, S_{a(i)}), H_{\text{ADM}} \right\} , \qquad (2.37)$$
$$\frac{\partial}{\partial t}(h^{\text{TT}}_{ij}, \pi^{ij\text{TT}}_{\text{can}}) = \left\{ (h^{\text{TT}}_{ij}, \pi^{ij\text{TT}}_{\text{can}}), H_{\text{ADM}} \right\} . \qquad (2.38)$$

2.3 Introducing the Routhian

A priori for self-gravitating objects, the total Hamiltonian results in the form $H = H[p, q, S, h^{\text{TT}}, \pi^{\text{TT}}]$. The transition to a Routhian description of the type $R = R[p, q, S, h^{\text{TT}}, \dot{h}^{\text{TT}}]$ allows the derivation of an autonomous matter Hamiltonian for the conservative dynamics,

$$H_c \equiv R[p, q, S, h^{\text{TT}}(x; p, q, S), \dot{h}^{\text{TT}}(x; p, q, S)] = H_c(p, q, S) \qquad (2.39)$$

as well as a nonautonomous or nonconservative one resulting in

$$H_{nc}(p, q, S; t) = R[p, q, S, h^{\text{TT}}(x; p', q', S'), \dot{h}^{\text{TT}}(x; p', q', S')]$$
$$= H_{nc}(p, q, S; p', q', S') . \qquad (2.40)$$

In more detail, the definition of the Routhian in question is given by, in the pN context first applied in [23],

$$R\left[x^i_a, p_{ai}, S_{a(i)}, h^{\text{TT}}_{ij}, \partial_t h^{\text{TT}}_{ij} \right] = H - \int d^3\mathbf{x}\, \pi^{ij\text{TT}}_{\text{can}} \partial_t h^{\text{TT}}_{ij} . \qquad (2.41)$$

The equations of motion for the matter read

$$\dot{p}_{ai} = -\frac{\partial R}{\partial x^i_a}, \quad \dot{x}^i_a = \frac{\partial R}{\partial p_{ai}}, \quad \dot{S}_{a(i)} = \{S_{a(i)}, R\} , \qquad (2.42)$$

and the field equations take the form

$$\frac{\delta \int R(t')dt'}{\delta h_{ij}^{\mathrm{TT}}(x^k, t)} = 0 . \tag{2.43}$$

The insertion of the solution of the field equations into the Routhian results in the mentioned autonomous and nonautonomous Hamiltonians for the matter degrees of freedom. The use of equations of motion is allowed on the action or Hamiltonian level. However, one should keep in mind that a change of variables is getting performed implicitly [24, 25]. The final equations of motion are of the form

$$\dot{p}_{ai} = -\frac{\partial H_{c;nc}}{\partial x_a^i} , \quad \dot{x}_a^i = \frac{\partial H_{c;nc}}{\partial p_{ai}} , \tag{2.44}$$

$$\dot{S}_{a(i)} = -\epsilon_{ijk} S_{a(j)} \Omega_{a(k)} \quad \text{with } \Omega_{a(i)} = \frac{\partial H_{c,nc}}{\partial S_{a(i)}} . \tag{2.45}$$

3 The Poincaré algebra

In asymptotically flat spacetimes the Poincaré or inhomogeneous Lorentz group is a global symmetry group. Its generators P^μ and $J^{\mu\nu}$ are conserved in time and fulfil the Poincaré algebra, see e.g. [13],

$$\{P^\mu, P^\nu\} = 0, \tag{3.1}$$

$$\{P^\mu, J^{\rho\sigma}\} = -\eta^{\mu\rho} P^\sigma + \eta^{\mu\sigma} P^\rho, \tag{3.2}$$

$$\{J^{\mu\nu}, P^{\rho\sigma}\} = -\eta^{\nu\rho} J^{\mu\sigma} + \eta^{\mu\rho} J^{\nu\sigma} + \eta^{\sigma\mu} J^{\rho\nu} - \eta^{\sigma\nu} J^{\rho\mu} . \tag{3.3}$$

The meaning of the components are energy $P^0 = H/c$, linear momentum $P^i = P_i$, angular momentum $J^{ij} = J_{ij}$, and Lorentz boost $J^{i0}/c \equiv K^i = G^i - t P^i$. A center-of-energy vector is defined by $X^i = c^2 G^i / H$. This vector, however, is not a canonical position vector, see e.g. [26]. In terms of three-dimensional quantities, the Poincaré algebra reads, see e.g. [27], with $J_{ij} = \epsilon_{ijk} J_k$,

$$\{P_i, H\} = \{J_i, H\} = 0 , \tag{3.4}$$

$$\{J_i, P_j\} = \varepsilon_{ijk} P_k , \; \{J_i, J_j\} = \varepsilon_{ijk} J_k , \tag{3.5}$$

$$\{J_i, G_j\} = \varepsilon_{ijk} G_k , \tag{3.6}$$

$$\{G_i, H\} = P_i , \tag{3.7}$$

$$\{G_i, P_j\} = \frac{1}{c^2} H \delta_{ij} , \tag{3.8}$$

$$\{G_i, G_j\} = -\varepsilon_{ijk} \frac{1}{c^2} J_k . \tag{3.9}$$

In terms of the field function ϕ, with $\psi = (1+\phi/8)^4$, the energy or Hamiltonian H and the center-of-energy vector $G^i = G_i$ have the representations

$$H = -\int d^3\mathbf{x}\,\Delta\phi = -\oint_{i^0} r^2 d\Omega n^i\,\partial_i\phi, \tag{3.10}$$

$$G_i = -\int d^3\mathbf{x}\,x^i\Delta\phi = -\oint_{i^0} r^2 d\Omega n^j\left(x^i\partial_j - \delta_{ij}\right)\phi\,, \tag{3.11}$$

whereas the total linear and angular momentum are given by

$$P_i = -2\int d^3\mathbf{x}\partial_j\tilde{\pi}^{ij} = -2\oint_{i^0} r^2 d\Omega n^j\,\tilde{\pi}^{ij}, \tag{3.12}$$

$$J_i = -2\int d^3\mathbf{x}\epsilon_{ijk}x^j\partial_l\tilde{\pi}^{kl} = -2\oint_{i^0} r^2 d\Omega n^l\,\epsilon_{ijk}x^j\tilde{\pi}^{kl}\,, \tag{3.13}$$

where $r^2 d\Omega n^i$ is the two-dimensional surface-area element and n^i is the outward-pointing radial unit vector.

The Poincaré algebra has been extensively used in the calculations and checking of higher order pN Hamiltonians [18, 19, 27–33]. Hereby the most important equation was (3.7) which tells that the total linear momentum is a total time derivative. Once in previous calculations, this equation has finally fixed the kinetic ambiguity in the nondimensional regularization calculations in [27]. The reader might perhaps also be interested in the construction of canonical center-of-energy and relative coordinates in asymptotically flat spacetimes [34].

4 Post-Newtonian binary Hamiltonians

4.1 Spinless binaries

For nonspinning compact objects, the binary dynamics is fully known up to the 3.5pN order and in part even at 4pN order. On reasons of readability, the Hamiltonians for spinless particles are given in the center-of-energy frame only and the 3.5pN part is not shown. The known but not shown Hamiltonians can easily be found in the quoted literature below.

Introducing the reduced quantities, $\hat{H} = (H - M)/\mu$, $\mu = m_1 m_2/M$, $M = m_1 + m_2$, $\nu = \mu/M$ with $0 \leq \nu \leq 1/4$ (test particle case $\nu = 0$, equal mass case $\nu = 1/4$), $\mathbf{p} = \mathbf{p}_1/\mu$, $r = r_{12} = |\mathbf{x}_1 - \mathbf{x}_2|$, $p_r = (\mathbf{n}\cdot\mathbf{p})$, $\mathbf{q} = (\mathbf{x}_1 - \mathbf{x}_2)/GM$, and $\mathbf{n} = \mathbf{n}_{12} = \mathbf{q}/|\mathbf{q}|$, in the center-of-energy frame, $\mathbf{p}_1 + \mathbf{p}_2 = 0$, the Hamiltonian

$$\begin{aligned}\hat{H}(t) = {} & \hat{H}_{\mathrm{N}} + \hat{H}_{1\mathrm{pN}} + \hat{H}_{2\mathrm{pN}} + \hat{H}_{3\mathrm{pN}} + \hat{H}_{4\mathrm{pN}} + \cdots \\ & + \hat{H}_{2.5\mathrm{pN}}(t) + \hat{H}_{3.5\mathrm{pN}}(t) + \cdots\,,\end{aligned} \tag{4.1}$$

where the 2.5pN and 3.5pN Hamiltonians are nonconservative (dissipative) ones [35–38], contains the following conservative terms [16, 19, 39, 40]:

$$\hat{H}_{\mathrm{N}} = \frac{p^2}{2} - \frac{1}{q}, \tag{4.2}$$

$$\hat{H}_{1\mathrm{pN}} = \frac{1}{8}(3\nu - 1)p^4 - \frac{1}{2}\left[(3+\nu)p^2 + \nu p_r^2\right]\frac{1}{q} + \frac{1}{2q^2}, \tag{4.3}$$

$$\begin{aligned}\hat{H}_{2\mathrm{pN}} = {} & \frac{1}{16}(1 - 5\nu + 5\nu^2)p^6 \\ & + \frac{1}{8}\left[(5 - 20\nu - 3\nu^2)p^4 - 2\nu^2 p_r^2 p^2 - 3\nu^2 p_r^4\right]\frac{1}{q} \\ & + \frac{1}{2}\left[(5 + 8\nu)p^2 + 3\nu p_r^2\right]\frac{1}{q^2} - \frac{1}{4}(1 + 3\nu)\frac{1}{q^3},\end{aligned} \tag{4.4}$$

$$\begin{aligned}\hat{H}_{3\mathrm{pN}} = {} & \frac{1}{128}(-5 + 35\nu - 70\nu^2 + 35\nu^3)p^8 \\ & + \frac{1}{16}\left[(-7 + 42\nu - 53\nu^2 - 5\nu^3)p^6 + (2 - 3\nu)\nu^2 p_r^2 p^4 \right. \\ & \left. + 3(1-\nu)\nu^2 p_r^4 p^2 - 5\nu^3 p_r^6\right]\frac{1}{q} \\ & + \left[\frac{1}{16}(-27 + 136\nu + 109\nu^2)p^4 + \frac{1}{16}(17 + 30\nu)\nu p_r^2 p^2 \right. \\ & \left. + \frac{1}{12}(5 + 43\nu)\nu p_r^4\right]\frac{1}{q^2} \\ & + \left[\left(-\frac{25}{8} + \left(\frac{1}{64}\pi^2 - \frac{335}{48}\right)\nu - \frac{23}{8}\nu^2\right)p^2 \right. \\ & \left. + \left(-\frac{85}{16} - \frac{3}{64}\pi^2 - \frac{7}{4}\nu\right)\nu p_r^2\right]\frac{1}{q^3} \\ & + \left[\frac{1}{8} + \left(\frac{109}{12} - \frac{21}{32}\pi^2\right)\nu\right]\frac{1}{q^4},\end{aligned} \tag{4.5}$$

$$\begin{aligned}\hat{H}_{4\mathrm{pN}} = {} & \left(\frac{7}{256} - \frac{63}{256}\nu + \frac{189}{256}\nu^2 - \frac{105}{128}\nu^3 + \frac{63}{256}\nu^4\right)p^{10} \\ & + \left\{\frac{45}{128}p^8 - \frac{45}{16}p^8\nu + \left(\frac{423}{64}p^8 - \frac{3}{32}p_r^2 p^6 - \frac{9}{64}p_r^4 p^4\right)\nu^2 \right. \\ & + \left(-\frac{1013}{256}p^8 + \frac{23}{64}p_r^2 p^6 + \frac{69}{128}p_r^4 p^4 - \frac{5}{64}p_r^6 p^2 + \frac{35}{256}p_r^8\right)\nu^3 \\ & \left. + \left(-\frac{35}{128}p^8 - \frac{5}{32}p_r^2 p^6 - \frac{9}{64}p_r^4 p^4 - \frac{5}{32}p_r^6 p^2 - \frac{35}{128}p_r^8\right)\nu^4\right\}\frac{1}{q}\end{aligned}$$

$$+\left\{\frac{13}{8}p^6+\left(-\frac{791}{64}p^6+\frac{49}{16}p_r^2p^4-\frac{889}{192}p_r^4p^2+\frac{369}{160}p_r^6\right)\nu\right.$$
$$+\left(\frac{4857}{256}p^6-\frac{545}{64}p_r^2p^4+\frac{9475}{768}p_r^4p^2-\frac{1151}{128}p_r^6\right)\nu^2$$
$$\left.+\left(\frac{2335}{256}p^6+\frac{1135}{256}p_r^2p^4-\frac{1649}{768}p_r^4p^2+\frac{10\,353}{1280}p_r^6\right)\nu^3\right\}\frac{1}{q^2}$$
$$+\left\{\frac{105}{32}p^4+\left[C_{41}(\mathbf{n},\mathbf{p})+\left(\frac{237}{40}p^4-\frac{1293}{40}p_r^2p^2+\frac{97}{4}p_r^4\right)\ln\frac{q}{\hat{s}}\right]\nu\right.$$
$$\left.+C_{42}(\mathbf{n},\mathbf{p})\,\nu^2+\left(-\frac{553}{128}p^4-\frac{225}{64}p_r^2p^2-\frac{381}{128}p_r^4\right)\nu^3\right\}\frac{1}{q^3}$$
$$+\left\{\frac{105}{32}p^2+\left[C_{21}(\mathbf{n},\mathbf{p})+\left(\frac{233}{40}p^2-\frac{29}{6}p_r^2\right)\ln\frac{q}{\hat{s}}\right]\nu+C_{22}(\mathbf{n},\mathbf{p})\,\nu^2\right\}\frac{1}{q^4}$$
$$+\left\{-\frac{1}{16}+\left[c_{01}+\frac{21}{20}\ln\frac{q}{\hat{s}}\right]\nu+c_{02}\,\nu^2\right\}\frac{1}{q^5}\,, \tag{4.6}$$

$$C_{42}(\mathbf{n},\mathbf{p})=\left(-\frac{1189\,789}{28\,800}+\frac{18\,491}{16\,384}\pi^2\right)p^4+\left(-\frac{127}{3}-\frac{4035}{2048}\pi^2\right)p_r^2p^2$$
$$+\left(\frac{57\,563}{1920}-\frac{38\,655}{16\,384}\pi^2\right)p_r^4\,, \tag{4.7}$$

$$C_{22}(\mathbf{n},\mathbf{p})=\left(\frac{672\,811}{19\,200}-\frac{158\,177}{49\,152}\pi^2\right)p^2$$
$$+\left(-\frac{21\,827}{3840}+\frac{110\,099}{49\,152}\pi^2\right)p_r^2\,, \tag{4.8}$$

$$C_{02}=-\frac{1256}{45}+\frac{7403}{3072}\pi^2\,, \tag{4.9}$$

and the terms $C_{41}(\mathbf{n},\mathbf{p})$ and $C_{21}(\mathbf{n},\mathbf{p})$ have the structure

$$C_{41}(\mathbf{n},\mathbf{p})=c_{411}p^4+c_{412}p_r^2p^2+c_{413}p_r^4\,, \tag{4.10}$$
$$C_{21}(\mathbf{n},\mathbf{p})=c_{211}p^2+c_{212}p_r^2\,. \tag{4.11}$$

The six constants c_{411}, c_{412}, c_{413}, c_{211}, c_{212}, c_{01} are still unknown. Their calculations will need an intricate treatment of spacelike infinity in a post-Newtonian setting. The both articles [41, 42] clearly indicate this intricate problem.

To complete, the center-of-energy vectors through 3pN order can be found in [27]: they, of course, do fulfil the Poincaré algebra through this order.

Finally, also the n-body Hamiltonian through linear order in Newton's gravitational constant G, coined first post-Minkowskian approximation, is known [46].

Radiation reaction

The dissipative 2.5pN Hamiltonian from the gravitational radiation reaction is given by [35, 36]

$$\hat{H}_{2.5\text{pN}}(\hat{t}) = \frac{2}{5}\left[p_i p_j - \frac{n^i n^j}{q}\right] \frac{d^3 \hat{Q}_{ij}(\hat{t})}{d\hat{t}^3}, \tag{4.12}$$

where $\hat{Q}_{ij}(\hat{t}) = \nu(q'^i q'^j - \delta_{ij} q'^2/3)$ holds. The time derivatives ($\hat{t} = t/GM$) in the Hamiltonian are allowed to be eliminated using the equations of motion. Only after the performance of the phase-space derivatives, the primed variables are allowed to be identified with the unprimed ones. The resulting reaction force can be found in [43, 44] (in [43], there is a factor 1/2 misprinted in the definition of the center-of-energy reaction force and in the 2pN expressions the factor $7m_1m_2$ has to be replaced through $5m_1m_2$ [39, 45]). In Section 5.1 the orbital period change resulting from the 2.5 Hamiltonian dynamics is presented.

4.2 Spinning binaries

Often in the literature, the spin is counted of 0.5pN order using typical black hole dimensions, $S \sim m\,v\,r \sim m\,c\,Gm/c^2 \sim m^2G/c$. This ordering will also be applied in the present contribution.

The leading order spin–orbit Hamiltonian reads, e.g. [47, 48],

$$H_{\text{SO}}^{1.5\text{pN}} = \frac{G}{c^2}\sum_a \sum_{b\neq a} \frac{1}{r_{ab}^2} (\mathbf{S}_a \times \mathbf{n}_{ab}) \cdot \left[\frac{3m_b}{2m_a}\mathbf{p}_a - 2\mathbf{p}_b\right]. \tag{4.13}$$

The leading order spin(1)–spin(2) Hamiltonian takes the form, e.g. [47, 48],

$$H_{\text{S}_1\text{S}_2}^{2\text{pN}} = G\sum_a \sum_{b\neq a} \frac{1}{2r_{ab}^3} \left[3(\mathbf{S}_a \cdot \mathbf{n}_{ab})(\mathbf{S}_b \cdot \mathbf{n}_{ab}) - (\mathbf{S}_a \cdot \mathbf{S}_b)\right] \tag{4.14}$$

and the leading order spin(1)–spin(1) dynamics is given by, going beyond the linear order in spin, e.g. [47, 48], $r = r_{12}$, $\mathbf{n} = \mathbf{n}_{12}$,

$$H_{\text{S}_1\text{S}_1}^{2\text{pN}} = G\frac{m_2}{2m_1 r^3} \left[3(\mathbf{S}_1 \cdot \mathbf{n})(\mathbf{S}_1 \cdot \mathbf{n}) - (\mathbf{S}_1 \cdot \mathbf{S}_1)\right]. \tag{4.15}$$

The next-to-leading order spin–orbit Hamiltonian reads [29],

$$\begin{aligned}
H_{\mathrm{SO}}^{2.5\mathrm{pN}} = \frac{G}{r^2}\Bigg[&-((\mathbf{p}_1 \times \mathbf{S}_1)\cdot\mathbf{n})\left[\frac{5m_2\mathbf{p}_1^2}{8m_1^3} + \frac{3(\mathbf{p}_1\cdot\mathbf{p}_2)}{4m_1^2}\right. \\
&\left. - \frac{3\mathbf{p}_2^2}{4m_1m_2} + \frac{3(\mathbf{p}_1\cdot\mathbf{n})(\mathbf{p}_2\cdot\mathbf{n})}{4m_1^2} + \frac{3(\mathbf{p}_2\cdot\mathbf{n})^2}{2m_1m_2}\right] \\
&+ ((\mathbf{p}_2 \times \mathbf{S}_1)\cdot\mathbf{n})\left[\frac{(\mathbf{p}_1\cdot\mathbf{p}_2)}{m_1m_2} + \frac{3(\mathbf{p}_1\cdot\mathbf{n})(\mathbf{p}_2\cdot\mathbf{n})}{m_1m_2}\right] \\
&+ ((\mathbf{p}_1 \times \mathbf{S}_1)\cdot\mathbf{p}_2)\left[\frac{2(\mathbf{p}_2\cdot\mathbf{n})}{m_1m_2} - \frac{3(\mathbf{p}_1\cdot\mathbf{n})}{4m_1^2}\right]\Bigg] \\
+ \frac{G^2}{c^4r^3}\Bigg[&-((\mathbf{p}_1 \times \mathbf{S}_1)\cdot\mathbf{n})\left[\frac{11m_2}{2} + \frac{5m_2^2}{m_1}\right] \\
&+ ((\mathbf{p}_2 \times \mathbf{S}_1)\cdot\mathbf{n})\left[6m_1 + \frac{15m_2}{2}\right]\Bigg] + (1 \leftrightarrow 2)
\end{aligned} \tag{4.16}$$

and the next-to-leading order spin(1)–spin(2) Hamiltonian is given by [30]

$$\begin{aligned}
H_{\mathrm{S_1S_2}}^{3\mathrm{pN}} = \frac{G}{2m_1m_2r^3}\big[&3((\mathbf{p}_1 \times \mathbf{S}_1)\cdot\mathbf{n})((\mathbf{p}_2 \times \mathbf{S}_2)\cdot\mathbf{n})/2 \\
&+ 6((\mathbf{p}_2 \times \mathbf{S}_1)\cdot\mathbf{n})((\mathbf{p}_1 \times \mathbf{S}_2)\cdot\mathbf{n}) \\
&- 15(\mathbf{S}_1\cdot\mathbf{n})(\mathbf{S}_2\cdot\mathbf{n})(\mathbf{p}_1\cdot\mathbf{n})(\mathbf{p}_2\cdot\mathbf{n}) \\
&- 3(\mathbf{S}_1\cdot\mathbf{n})(\mathbf{S}_2\cdot\mathbf{n})(\mathbf{p}_1\cdot\mathbf{p}_2) + 3(\mathbf{S}_1\cdot\mathbf{p}_2)(\mathbf{S}_2\cdot\mathbf{n})(\mathbf{p}_1\cdot\mathbf{n}) \\
&+ 3(\mathbf{S}_2\cdot\mathbf{p}_1)(\mathbf{S}_1\cdot\mathbf{n})(\mathbf{p}_2\cdot\mathbf{n}) + 3(\mathbf{S}_1\cdot\mathbf{p}_1)(\mathbf{S}_2\cdot\mathbf{n})(\mathbf{p}_2\cdot\mathbf{n}) \\
&+ 3(\mathbf{S}_2\cdot\mathbf{p}_2)(\mathbf{S}_1\cdot\mathbf{n})(\mathbf{p}_1\cdot\mathbf{n}) - 3(\mathbf{S}_1\cdot\mathbf{S}_2)(\mathbf{p}_1\cdot\mathbf{n})(\mathbf{p}_2\cdot\mathbf{n}) \\
&+ (\mathbf{S}_1\cdot\mathbf{p}_1)(\mathbf{S}_2\cdot\mathbf{p}_2) - (\mathbf{S}_1\cdot\mathbf{p}_2)(\mathbf{S}_2\cdot\mathbf{p}_1)/2 + (\mathbf{S}_1\cdot\mathbf{S}_2)(\mathbf{p}_1\cdot\mathbf{p}_2)/2\big] \\
+ \frac{3}{2m_1^2r^3}\big[&-((\mathbf{p}_1 \times \mathbf{S}_1)\cdot\mathbf{n})((\mathbf{p}_1 \times \mathbf{S}_2)\cdot\mathbf{n}) \\
&+ (\mathbf{S}_1\cdot\mathbf{S}_2)(\mathbf{p}_1\cdot\mathbf{n})^2 - (\mathbf{S}_1\cdot\mathbf{n})(\mathbf{S}_2\cdot\mathbf{p}_1)(\mathbf{p}_1\cdot\mathbf{n})\big] \\
+ \frac{3}{2m_2^2r^3}\big[&-((\mathbf{p}_2 \times \mathbf{S}_2)\cdot\mathbf{n})((\mathbf{p}_2 \times \mathbf{S}_1)\cdot\mathbf{n}) \\
&+ (\mathbf{S}_1\cdot\mathbf{S}_2)(\mathbf{p}_2\cdot\mathbf{n})^2 - (\mathbf{S}_2\cdot\mathbf{n})(\mathbf{S}_1\cdot\mathbf{p}_2)(\mathbf{p}_2\cdot\mathbf{n})\big] \\
+ \frac{6G^2(m_1+m_2)}{c^4r^4}\big[&(\mathbf{S}_1\cdot\mathbf{S}_2) - 2(\mathbf{S}_1\cdot\mathbf{n})(\mathbf{S}_2\cdot\mathbf{n})\big]\,.
\end{aligned} \tag{4.17}$$

Finally, the next-to-leading order spin(1)–spin(1) dynamics reads [31]

$$
\begin{aligned}
H^{3\mathrm{pN}}_{\mathbf{S}_1\mathbf{S}_1} = \frac{G}{r^3}\Bigg[& \frac{m_2}{4m_1^3}(\mathbf{p}_1\cdot\mathbf{S}_1)^2 + \frac{3m_2}{8m_1^3}(\mathbf{p}_1\cdot\mathbf{n})^2\,\mathbf{S}_1^2 \\
& - \frac{3m_2}{8m_1^3}\mathbf{p}_1^2(\mathbf{S}_1\cdot\mathbf{n})^2 - \frac{3m_2}{4m_1^3}(\mathbf{p}_1\cdot\mathbf{n})(\mathbf{S}_1\cdot\mathbf{n})(\mathbf{p}_1\cdot\mathbf{S}_1) - \frac{3}{4m_1m_2}\mathbf{p}_2^2\mathbf{S}_1^2 \\
& + \frac{9}{4m_1m_2}\mathbf{p}_2^2(\mathbf{S}_1\cdot\mathbf{n})^2 + \frac{3}{4m_1^2}(\mathbf{p}_1\cdot\mathbf{p}_2)\,\mathbf{S}_1^2 - \frac{9}{4m_1^2}(\mathbf{p}_1\cdot\mathbf{p}_2)(\mathbf{S}_1\cdot\mathbf{n})^2 \\
& - \frac{3}{2m_1^2}(\mathbf{p}_1\cdot\mathbf{n})(\mathbf{p}_2\cdot\mathbf{S}_1)(\mathbf{S}_1\cdot\mathbf{n}) + \frac{3}{m_1^2}(\mathbf{p}_2\cdot\mathbf{n})(\mathbf{p}_1\cdot\mathbf{S}_1)(\mathbf{S}_1\cdot\mathbf{n}) \\
& + \frac{3}{4m_1^2}(\mathbf{p}_1\cdot\mathbf{n})(\mathbf{p}_2\cdot\mathbf{n})\,\mathbf{S}_1^2 - \frac{15}{4m_1^2}(\mathbf{p}_1\cdot\mathbf{n})(\mathbf{p}_2\cdot\mathbf{n})(\mathbf{S}_1\cdot\mathbf{n})^2\Bigg] \\
& - \frac{G^2m_2}{2r^4}\left[5\left(1+\frac{6m_2}{5m_1}\right)\left((\mathbf{S}_1\cdot\mathbf{n})^2 - \mathbf{S}_1^2\right) + 4\left(1+\frac{2m_2}{m_1}\right)(\mathbf{S}_1\cdot\mathbf{n})^2\right].
\end{aligned}
\tag{4.18}
$$

Also this Hamiltonian goes beyond linear order in spin. For its derivation, the Tulczyjew energy–momentum tensor was needed one step beyond pole–dipole-particle approximation. All the previous Hamiltonians, spinless and spinning, were valid for both neutron stars and black holes with the exception of the spin(1)–spin(1) ones which are valid for black holes only. Their generalizations to neutron stars are presented in [49, 50]. The next-to-leading order Hamiltonians found confirmations through [49, 51–54], see [55].

As examples, the up to the next-to-leading order spin–orbit center-of-energy vector is given by [29, 56, 57]

$$
\begin{aligned}
\mathbf{G}^{1.5\mathrm{pN}}_{\mathrm{SO}} + \mathbf{G}^{2.5\mathrm{pN}}_{\mathrm{SO}} = & \sum_a \frac{1}{2m_a}(\mathbf{p}_a\times\mathbf{S}_a) - \sum_a \frac{\mathbf{p}_a^2}{8m_a^3}(\mathbf{p}_a\times\mathbf{S}_a) \\
& + \sum_a\sum_{b\neq a}\frac{m_b}{4m_a r_{ab}}\left[((\mathbf{p}_a\times\mathbf{S}_a)\cdot\mathbf{n}_{ab})\frac{5\mathbf{x}_a+\mathbf{x}_b}{r_{ab}} - 5(\mathbf{p}_a\times\mathbf{S}_a)\right] \\
& + \sum_a\sum_{b\neq a}\frac{1}{r_{ab}}\Bigg[\frac{3}{2}(\mathbf{p}_b\times\mathbf{S}_a) - \frac{1}{2}(\mathbf{n}_{ab}\times\mathbf{S}_a)(\mathbf{p}_b\cdot\mathbf{n}_{ab}) \\
& - ((\mathbf{p}_a\times\mathbf{S}_a)\cdot\mathbf{n}_{ab})\frac{\mathbf{x}_a+\mathbf{x}_b}{r_{ab}}\Bigg],
\end{aligned}
\tag{4.19}
$$

and the spin(1)–spin(2) one reads [30, 56, 57]

$$
\mathbf{G}^{3\mathrm{pN}}_{\mathbf{S}_1\mathbf{S}_2} = \frac{1}{2}\sum_a\sum_{b\neq a}\left\{[3(\mathbf{S}_a\cdot\mathbf{n}_{ab})(\mathbf{S}_b\cdot\mathbf{n}_{ab}) - (\mathbf{S}_a\cdot\mathbf{S}_b)]\,\frac{\mathbf{x}_a}{r_{ab}^3} + (\mathbf{S}_b\cdot\mathbf{n}_{ab})\frac{\mathbf{S}_a}{r_{ab}^2}\right\}. \tag{4.20}
$$

The leading order spin(1)–spin(2) expression of the spin–orbit center-of-energy vector renders zero. The contribution to the center-of-energy vector from the spin(a)-spin(a) interaction can be found in [57].

To summarize, the following pN-Hamiltonians with spin are known fully explicitly,

$$\begin{aligned} H(t) = {} & H_{\mathrm{SO}}^{1.5\mathrm{pN}} + H_{\mathrm{SO}}^{2.5\mathrm{pN}} + H_{\mathrm{SO}}^{3.5\mathrm{pN}} + H_{\mathrm{SO}}^{4\mathrm{pN}}(t) \\ & + H_{S_1S_2}^{2\mathrm{pN}} + H_{S_1S_2}^{3\mathrm{pN}} + H_{S_1S_2}^{4\mathrm{pN}} + H_{S_1S_2}^{4.5\mathrm{PN}}(t) \\ & + H_{S_1S_1}^{2\mathrm{pN}} + H_{S_2S_2}^{2\mathrm{pN}} + H_{S_1S_1}^{3\mathrm{pN}} + H_{S_2S_2}^{3\mathrm{pN}} . \end{aligned} \tag{4.21}$$

The conservative Hamiltonians not given above can be found in [18, 32, 33], the nonconservative ones in [58, 59]. Related results for the conservative parts were achieved in [60, 61] and for the nonconservative ones in [62, 63]. Also the test-spin Hamiltonian in the Kerr metric is known [64].

5 Binary motion

It is a remarkable fact that the high pN motion of spinless binaries can be given in full-analytic form if restricted to the conservative part of the motion. Taking the radiation reaction into account, the secular aspects of the motion can be analytically treated too, apart from trivial numerical integrations. Of particular interest is the circular motion and its innermost stable circular orbit (ISCO) where plunge motion happens beyond. During the inspiral phase an initial eccentricity is damped out so that the orbits become more and more circular. Only through external actions, e.g. third-body collision or asymmetric body explosions, tight orbits can have high eccentricities.

5.1 Spinless two-body systems

Circular motion

For circular motion, the energy of a binary depends on the orbital frequency ω only. Introducing the dimensionless variable $x = (GM\omega/c^3)^{2/3}$ and identifying $\hat{s} = cP/GM$, where $P = 2\pi/\omega$ is the orbital period, i.e. $\hat{s} = 2\pi c^{-2} x^{-3/2}$, the 4pN-accurate conservative binding energy, Equation (4.1), can be put into the form

$$\begin{aligned} \hat{E}_{4\mathrm{PN}}(x;\nu) = {} & -\frac{c^2 x}{2}\Big(1 + e_1(\nu)\,x + e_2(\nu)\,x^2 \\ & + e_3(\nu)\,x^3 + \left(e_4(\nu) + \frac{448}{15}\nu\ln x\right) x^4 + \mathcal{O}\left(x^5\right)\Big) , \end{aligned} \tag{5.1}$$

where the fractional corrections to the Newtonian energy at various pN orders read

$$e_1(\nu) = -\frac{3}{4} - \frac{1}{12}\nu\,, \tag{5.2}$$

$$e_2(\nu) = -\frac{27}{8} + \frac{19}{8}\nu - \frac{1}{24}\nu^2\,, \tag{5.3}$$

$$e_3(\nu) = -\frac{675}{64} + \left(\frac{34\,445}{576} - \frac{205}{96}\pi^2\right)\nu - \frac{155}{96}\nu^2 - \frac{35}{5184}\nu^3\,, \tag{5.4}$$

$$e_4(\nu) = -\frac{3969}{128} + c_1\,\nu + \left(-\frac{498\,449}{3456} + \frac{3157}{576}\pi^2\right)\nu^2 + \frac{301}{1728}\nu^3 + \frac{77}{31\,104}\nu^4 \tag{5.5}$$

with

$$c_1 = -\frac{123\,671}{5760} + \frac{9037}{1536}\pi^2 + \frac{1792}{15}\ln 2 + \frac{896}{15}\gamma\,, \tag{5.6}$$

where γ is the Euler constant. The $\ln x$-term has been calculated for the first time in [65, 66] and later verified in [28, 67]. The terms in e_4 proportional to ν^4 and ν^3 have been obtained for the first time in [28] and were later confirmed by [68], and the term proportional to ν^3 has been obtained only quite recently [19]. The same holds with c_1 which has been derived in [42]. A numerical value of c_1 was known before [41], coinciding, however, only in the first five digits, 153.88, with the analytic result.

In the test-mass limit $\nu = 0$, the exact ISCO occurs for $x_{\text{ISCO}} = 1/6$ corresponding to the minimum of the function

$$\frac{\hat{E}(x)}{c^2} = \frac{1-x}{(1-3x)^{1/2}} - 1\,. \tag{5.7}$$

The 4pN prediction for the location of the ISCO in the test-mass limit is 0.179... which deviates by ~7.7% from the exact value. In the equal-mass case ($\nu = 1/4$), the corresponding values are 0.648... (1pN), 0.265... (2pN), 0.254... (3pN), 0.236... (4pN), [19].

Eccentric orbital motion

Dynamical invariants related to our previous dynamics are easily calculated within a Hamiltonian framework [69]. Let us denote the radial action by $i_r(\hat{E}, j)$ with $\hat{E} = \hat{H}$ and $p^2 = p_r^2 + j^2/r^2$ ($\mathbf{p} = p_r\mathbf{e}_r + p_\varphi\mathbf{e}_\varphi$ with orthonormal basis $\mathbf{e}_r$, $\mathbf{e}_\varphi$ in the orbital plane). Then it holds

$$i_r(\hat{E}, j) = \frac{1}{2\pi}\oint dr\, p_r\,, \tag{5.8}$$

where the integration is defined from minimum to minimum of the radial distance. Thus, all expressions derived hereof relate to orbits completed in this sense. From analytical mechanics it is known that the phase of the completed orbit revolution Φ is given by

$$\frac{\Phi}{2\pi} = 1 + k = -\frac{\partial}{\partial j} i_r(\hat{E}, j) \tag{5.9}$$

and the orbital period P reads

$$\frac{P}{2\pi Gm} = \frac{\partial}{\partial \hat{E}} i_r(\hat{E}, j) \,. \tag{5.10}$$

Explicitly, up to 3pN order, we get for the periastron advance parameter k [16, 69],

$$\begin{aligned} k = \frac{1}{c^2}\frac{3}{j^2}\left\{1 + \frac{1}{c^2}\left[\frac{5}{4}(7-2\nu)\frac{1}{j^2} + \frac{1}{2}(5-2\nu)\,\hat{E}\right]\right. \\ \left. + \frac{1}{c^4}\left[a_1(\nu)\frac{1}{j^4} + a_2(\nu)\frac{\hat{E}}{j^2} + a_3(\nu)\,\hat{E}^2\right]\right\}\,, \end{aligned} \tag{5.11}$$

and for the orbital period [16, 69],

$$\begin{aligned} \frac{P}{2\pi Gm} = \frac{1}{(-2\hat{E})^{3/2}}\left\{1 - \frac{1}{c^2}\frac{1}{4}(15-\nu)\hat{E}\right. \\ + \frac{1}{c^4}\left[\frac{3}{2}(5-2\nu)\frac{(-2\hat{E})^{3/2}}{j} - \frac{3}{32}(35+30\nu+3\nu^2)\,\hat{E}^2\right] \\ \left. + \frac{1}{c^6}\left[a_2(\nu)\frac{(-2\hat{E})^{3/2}}{j^3} - 3a_3(\nu)\frac{(-2\hat{E})^{5/2}}{j} + a_4(\nu)\,\hat{E}^3\right]\right\}\,, \end{aligned} \tag{5.12}$$

where

$$a_1(\nu) = \frac{5}{2}\left(\frac{77}{2} + \left(\frac{41}{64}\pi^2 - \frac{125}{3}\right)\nu + \frac{7}{4}\nu^2\right), \tag{5.13}$$

$$a_2(\nu) = \frac{105}{2} + \left(\frac{41}{64}\pi^2 - \frac{218}{3}\right)\nu + \frac{45}{6}\nu^2, \tag{5.14}$$

$$a_3(\nu) = \frac{1}{4}(5 - 5\nu + 4\nu^2), \tag{5.15}$$

$$a_4(\nu) = \frac{5}{128}(21 - 105\nu + 15\nu^2 + 5\nu^3)\,. \tag{5.16}$$

Through 2pN order, applications to binary pulsar systems have been worked out, particularly for the binary pulsar PSR 1913+16 [39].

Explicit analytical orbit solutions of the conservative dynamics through 3pN order are given in [70]. Up to 2pN order, these expressions read [39, 71],

$$r = a_r(1 - e_r \cos u), \tag{5.17}$$

$$\frac{2\pi}{P}(t - t_0) = u - e_t \sin u + (v - u)F + F_v \sin v\,, \tag{5.18}$$

$$\frac{2\pi}{\Phi}(\phi - \phi_0) = v + G_{2v}\sin(2v) + G_{3v}\sin(3v)\,, \tag{5.19}$$

$$v = 2\arctan\left[\sqrt{\frac{1+e_\phi}{1-e_\phi}}\tan\frac{u}{2}\right]\,, \tag{5.20}$$

where the coefficients on the right-hand sides depend on energy and angular momentum in the form

$$a_r = \frac{1}{(-2\hat{E})}\left(1 + \frac{(-2\hat{E})}{4c^2}(-7+\nu) + \frac{(-2\hat{E})^2}{16c^4}\left((1+10\nu+\nu^2) + \frac{1}{(-2\hat{E}j^2)}(-68+44\nu)\right)\right), \tag{5.21}$$

$$e_r^2 = 1 + 2\hat{E}j^2 + \frac{(-2\hat{E})}{4c^2}\left(24 - 4\nu + 5(-3+\nu)(-2\hat{E}j^2)\right) + \frac{(-2\hat{E})^2}{8c^4}\left(52 + 2\nu + 2\nu^2 - (80 - 55\nu + 4\nu^2)(-2\hat{E}j^2) - \frac{8}{(-2\hat{E}j^2)}(-17+11\nu)\right), \tag{5.22}$$

$$e_t^2 = 1 + 2\hat{E}j^2 + \frac{(-2\hat{E})}{4c^2}\left(-8 + 8\nu - (-17+7\nu)(-2\hat{E}j^2)\right) + \frac{(-2\hat{E})^2}{8c^4}\left(8 + 4\nu + 20\nu^2 - (112 - 47\nu + 16\nu^2)(-2\hat{E}j^2) - 24\sqrt{(-2\hat{E}j^2)}(-5+2\nu) - \frac{4}{(-2\hat{E}j^2)}(-17+11\nu) - \frac{24}{\sqrt{(-2\hat{E}j^2)}}(5-2\nu)\right), \tag{5.23}$$

$$e_\phi^2 = 1 + 2\hat{E}j^2 + \frac{(-2\hat{E})}{4c^2}\left(24 + (-15+\nu)(-2\hat{E}j^2)\right) + \frac{(-2\hat{E})^2}{16c^4}\left(-32 + 176\nu + 18\nu^2 - (160 - 30\nu + 3\nu^2)(-2\hat{E}j^2) + \frac{1}{(-2\hat{E}j^2)}(408 - 232\nu - 15\nu^2)\right), \tag{5.24}$$

$$F = \frac{3(-2\hat{E})^2}{2c^4} \frac{5 - 2\nu}{\sqrt{(-2\hat{E} j^2)}} , \tag{5.25}$$

$$F_\upsilon = -\frac{1}{8c^4} \frac{(-2\hat{E})^2}{\sqrt{(-2\hat{E} j^2)}} (4 + \nu)\nu \sqrt{1 + 2\hat{E} j^2} , \tag{5.26}$$

$$G_{2\upsilon} = \frac{(-2\hat{E})^2}{8c^4} \frac{1 + 2\hat{E} j^2}{(-2\hat{E} j^2)^2} \nu(1 - 3\nu) , \tag{5.27}$$

$$G_{3\upsilon} = -\frac{3(-2\hat{E})^2}{32c^4} \frac{(1 + 2\hat{E} j^2)^{3/2}}{(-2\hat{E} j^2)^2} \nu^2 . \tag{5.28}$$

The given parametrization of the orbits is based on the 1pN parametrization by [72]. The latter parametrization is different from the one in [1] but identical to the one in [73] in the test-body limit. Through 1pN order the parametrization given above plays an important role in current observations of binary pulsar systems, e.g. [74].

Radiation damping

The 2.5pN gravitational radiation damping of the orbital period reads, e.g. [75, 76], where also the 3.5pN damping is treated via balance equations,

$$\frac{\dot{P}}{P} = -\frac{96G^3}{5c^5} \frac{\mu M^2}{a^4(1 - e^2)^{7/2}} \left(1 + \frac{73}{24} e^2 + \frac{37}{96} e^4\right) , \tag{5.29}$$

where for the orbital elements the Newtonian approximation is sufficient, $a = a_r$ and $e = e_r = e_t = e_\phi$. Current confirmations of this damping formula at 2.5pN order were obtained from the binary pulsar PSR B1913+16 with uncertainty 2×10^{-3} [74], and from the double pulsar PSR J0737-3039A/B within an error bar of 3×10^{-4}, M. Kramer and N. Wex (personal communication). The 3.5pN Hamiltonian dynamics has been worked out in [37, 38] but it is far too small for current observations.

5.2 Particle motion in Kerr geometry

The binding energy of a test particle in the Kerr geometry can be put into the form, taking into account linear and quadratic spin terms, each up to some pN order, of the central black hole [77],

$$E = \frac{p_r^2}{2m} + \frac{\mathfrak{C}^2}{2r^2m} - \frac{GMm}{r} + \frac{1}{c^2}\left[G\left(-\frac{\mathfrak{C}^2M}{2r^3m} - \frac{3Mp_r^2}{2rm}\right) - \frac{p_r^4}{8m^3} - \frac{\mathfrak{C}^4}{8r^4m^3} - \frac{\mathfrak{C}^2p_r^2}{4r^2m^3} - \frac{G^2M^2m}{2r^2}\right] + \frac{1}{c^3}\frac{2aG^2L_zM^2}{r^3}$$
$$+ \frac{1}{c^4}\left[G\left(\frac{\mathfrak{C}^4M}{8r^5m^3} + \frac{3\mathfrak{C}^2Mp_r^2}{4r^3m^3} + \frac{5Mp_r^4}{8rm^3}\right) - \frac{G^3M^3m}{2r^3} + G^2\left(-\frac{\mathfrak{C}^2M^2}{4r^4m} + \frac{3M^2p_r^2}{4r^2m}\right) + \frac{\mathfrak{C}^6}{16r^6m^5} + \frac{3\mathfrak{C}^4p_r^2}{16r^4m^5} + \frac{p_r^6}{16m^5}\right.$$
$$\left. + \frac{3\mathfrak{C}^2p_r^4}{16r^2m^5} + a^2G^2\left(-\frac{L_z^2M^2}{2r^4m} + \frac{M^2p_r^2}{2r^2m}\right)\right] + O\left(\frac{1}{c^6}\right), \tag{5.30}$$

with the Carter-like constant, $\mathfrak{C}$,

$$\mathfrak{C}^2 = L^2 + \alpha^2a^2\cos^2\theta \quad \text{with}$$
$$\alpha^2 = -\frac{2E}{mc^2}\left(\frac{GMm}{c}\right)^2 + O\left(\frac{1}{c^6}\right), \tag{5.31}$$

where the dimensionless spin parameter $a = Sc/GM^2$ has the range $a \in [-1, 1]$. The canonical variables are $(p_r, L, L_z; r, v, \Omega)$, where r and p_r denote the radial coordinate and its canonical conjugate momentum, respectively, v being the true anomaly, i.e. the angle between the position vector $\mathbf{r} = r\mathbf{n}$ and the direction of the ascending node $\mathbf{N} = (\mathbf{e}_z \times \mathbf{L})/|\mathbf{e}_z \times \mathbf{L}|$: $\cos v = \mathbf{n}\cdot\mathbf{N}$ and $\sin v = \mathbf{n}\cdot\mathbf{W}$ with $\mathbf{W} = \mathbf{L}\times\mathbf{N}/L$. Ω is the angle of the ascending node as measured from the x-axis of the nonrotating orthonormal basis $(\mathbf{e}_x, \mathbf{e}_y, \mathbf{e}_z)$: $\cos\Omega = \mathbf{N}\cdot\mathbf{e}_x$ and $\sin\Omega = \mathbf{N}\cdot\mathbf{e}_y$, measuring the precession of the orbital plane about L_z which is tilted from the equatorial plane with the inclination angle i, the angle between the background z-axis and the orbital angular momentum vector $\mathbf{L}$, so $L_z = L\cos i$.

The canonical Poisson-bracket relations read

$$\{r, p_r\} = \{v, L\} = \{\Omega, L_z\} = 1 \tag{5.32}$$

with all other brackets being zero.

The derivation of observables is straightforward with the knowledge of the action of a completed revolution, from periastron to periastron,

$$S = S(E, L_z, \mathfrak{C}, P, \Phi, U)$$
$$= -E_bP + L_z\Phi + \oint p_r dr + \oint L dv\,. \tag{5.33}$$

The result depends on the three constants of motion, $E, L_z, \mathfrak{C}$, and three orbit-completed variables

$$P = \oint dt\,, \quad \Phi = \oint d\Omega\,, \quad U = \oint dv\,, \tag{5.34}$$

P being the orbital period, Φ the precession of the orbital plane per revolution, and U the intrinsic periastron advance. The relation between the variable v and the space-fixed Boyer–Lindquist coordinate θ reads

$$\sin v \sin i = \cos\theta \,. \tag{5.35}$$

Solving Equation (5.30) for p_r and Equation (5.31) for L allows the calculation of the action, Equation (5.33). Within leading order calculations, the variable L in the expression for $\cos i$ can be replaced by $\mathcal{C}$. The next step is to calculate the remaining integrals in the action (5.33), resulting in

$$\begin{aligned}\frac{1}{2\pi}\oint p_r dr = {} & -\mathcal{C} + \frac{GMm}{\sqrt{-\frac{2E}{m}}} + \frac{1}{c^2}\left(\frac{3G^2M^2m^2}{\mathcal{C}}\right.\\ & \left. + \frac{15GMm}{4}\sqrt{\frac{-E}{2m}}\right) - \frac{1}{c^3}\frac{2aG^3L_zM^3m^3}{\mathcal{C}^3}\\ & + \frac{1}{c^4}\left[\frac{15EG^2M^2m^4}{2\mathcal{C}} + \frac{35G^4M^4m^4}{4\mathcal{C}^3}\right.\\ & + \frac{35EGM}{32}\sqrt{-\frac{E_b}{2m}} + a^2\left(EG^2\left(-\frac{M^2m}{2\mathcal{C}} + \frac{L_z^2M^2m}{2\mathcal{C}^3}\right)\right.\\ & \left.\left. - \frac{G^4M^4m^4}{4\mathcal{C}^3} + \frac{3G^4L_z^2M^4m^4}{4\mathcal{C}^5}\right)\right]\\ & - \frac{a}{c^5}\left(\frac{12EG^3M^3L_zm^2}{\mathcal{C}^3} + \frac{21G^5M^5L_zm^5}{\mathcal{C}^5}\right)\end{aligned} \tag{5.36}$$

and

$$\begin{aligned}\oint L dv &= \oint \sqrt{\mathcal{C}^2 - \alpha^2 a^2 \sin^2 i \,\sin^2 v}\, dv\\ &= \mathcal{C}U - \frac{\pi}{2\mathcal{C}}a^2\alpha^2\left(1 - \frac{L_z^2}{\mathcal{C}^2}\right) + \mathcal{O}(a^2/c^6)\,,\end{aligned} \tag{5.37}$$

where integration over a closed Newtonian orbit has been performed for the α^2-term.

The action principle states that

$$\frac{\partial S}{\partial \mathcal{C}} = \frac{\partial S}{\partial L_z} = \frac{\partial S}{\partial E} = 0\,. \tag{5.38}$$

Hereof the two periastron shifts, $\Delta\bar{U}$ and $\Delta\bar{\Phi}$, the intrinsic and the one related with the precession of the orbital plane, respectively, result in the form

$$\begin{aligned}\Delta\bar{U} &= \frac{1}{2\pi}(U-2\pi) = \frac{3G^2M^2m^2}{c^2\mathcal{C}^2} - \frac{6aG^3M^3m^3}{c^3\mathcal{C}^3}\cos i\\&+\frac{1}{c^4}\left[\frac{15EG^2M^2m}{2\mathcal{C}^2}+\frac{105G^4M^4m^4}{4\mathcal{C}^4}\right.\\&\left.+a^2\frac{3G^4M^4m^4}{4\mathcal{C}^4}\left(5\cos^2 i-1\right)\right]\\&-\frac{aG^3M^3m^2}{c^5\mathcal{C}^3}\left(36E+105\frac{G^2M^2m^3}{\mathcal{C}^2}\right)\cos i\end{aligned}\tag{5.39}$$

and

$$\begin{aligned}\Delta\bar{\Phi} &= \frac{1}{2\pi}\Phi = \frac{2aG^3M^3m^3}{c^3\mathcal{C}^3}+\frac{a^2}{c^4}\cos i\left(-\frac{3G^4M^4m^4}{2\mathcal{C}^4}\right)\\&+\frac{a}{c^5}\left(\frac{12EG^3M^3m^2}{\mathcal{C}^3}+\frac{21G^5M^5m^5}{\mathcal{C}^5}\right).\end{aligned}\tag{5.40}$$

These periastron advances agree with corresponding results in [39, 47] in the test-mass limit, apart from the higher pN-order term linear in a which does not appear therein. For the orbital period we get

$$\begin{aligned}P &= \frac{2\pi GM}{(-2E/m)^{3/2}}\left[1-\frac{1}{c^2}\frac{15}{4}\frac{E}{m}\right.\\&+\frac{1}{c^4}\left(\frac{15}{2}\frac{GMm}{\mathcal{C}}\left(\frac{-2E}{m}\right)^{3/2}-\frac{105}{32}\frac{E^2}{m^2}\right)\\&\left.-\frac{12aG^2M^2m^2}{c^5\mathcal{C}^2}\left(\frac{-2E}{m}\right)^{3/2}\cos i\right].\end{aligned}\tag{5.41}$$

In the case of equatorial motion the resulting shift, say $\Delta\tilde{\Phi}$, is given by

$$\begin{aligned}\Delta\tilde{\Phi} &= \Delta\bar{U}+\Delta\bar{\Phi} = -1-\frac{1}{2\pi}\left(\frac{\partial}{\partial\mathcal{C}}+\frac{\partial}{\partial L_z}\right)\oint p_r dr\\&=\frac{3G^2M^2m^2}{c^2\mathcal{C}^2}-\frac{4aG^3M^3m^3}{c^3\mathcal{C}^3}\\&+\frac{1}{c^4}\left[\frac{15EG^2M^2m}{2\mathcal{C}^2}+\frac{105G^4M^4m^4}{4\mathcal{C}^4}+a^2\frac{3G^4M^4m^4}{2\mathcal{C}^4}\right]\\&+\frac{a}{c^5}\left(-\frac{24EG^3M^3m^2}{\mathcal{C}^3}-\frac{84G^5M^5m^5}{\mathcal{C}^5}\right).\end{aligned}\tag{5.42}$$

This quantity can also be deduced from [78] apart from the higher pN-order term linear in a which has not been treated therein. As $\mathcal{C}$ deviates from L only at the order a^2/c^4, in the obtained results, $\mathcal{C}$ may always be replaced with L.

5.3 Two-body systems with spinning components

The motion of spins and precession of orbits of spinning binaries are important indicators for the gravitational interaction. From an observational point of view, only the leading pN-order interactions are important, so in the following we will restrict ourselves to those interactions, particularly to the leading order spin–orbit and spin(1)–spin(2) interactions. Also, we shall restrict ourselves to the center-of-energy systems.

Introducing the Runge–Lenz–Laplace–Lagrange vector (often only called Laplace–Runge–Lenz vector or even only Runge–Lenz vector as coined by Pauli in the quantum mechanics; but Lagrange introduced it first, cf. [39, 79]),

$$\mathbf{A} \equiv \mathbf{p} \times \mathbf{L} - \frac{GM\mu^2}{r}\mathbf{r}\,, \tag{5.43}$$

the time averaged or secular precession of the binary orbit takes the form

$$\left\langle \left(\frac{d\mathbf{L}}{dt}\right)^{\mathrm{SO}} \right\rangle_t = \Omega_{\mathrm{SO}} \times \mathbf{L}\,, \tag{5.44}$$

$$\left\langle \left(\frac{d\mathbf{A}}{dt}\right)^{\mathrm{SO}} \right\rangle_t = \Omega_{\mathrm{SO}} \times \mathbf{A}\,, \tag{5.45}$$

where $\mathbf{L} = \mathbf{r} \times \mathbf{p}$ denotes the orbital angular momentum and $\langle \cdots \rangle_t$ orbital averaging over time. As above, $M = M_1 + M_2$ is the total mass and μ the reduced one; only the single masses are now put with capital letters. The precessional frequency vector reads

$$\Omega_{\mathrm{SO}} = \frac{2G}{c^2}\frac{1}{a^3(1-e^2)^{3/2}}\left(\mathbf{S}_{\mathrm{eff}} - 3\frac{(\mathbf{L}\cdot\mathbf{S}_{\mathrm{eff}})\mathbf{L}}{L^2}\right), \tag{5.46}$$

$$\mathbf{S}_{\mathrm{eff}} = \mathbf{S} + \frac{3}{4}\left(\frac{M_2}{M_1}\mathbf{S}_1 + \frac{M_1}{M_2}\mathbf{S}_2\right), \tag{5.47}$$

where for the orbital elements, in Newtonian approximation, $a = a_r$ and $e = e_r = e_t = e_\phi$ hold. This is the famous Lense–Thirring effect. With the satellites LAGEOS(1&2) the predicted effect of 31 mas/yr has been measured with an accuracy of 1×10^{-1} [80].

The Schiff effect, also called Lense–Thirring effect for spin or frame-dragging effect, is given by

$$\left(\frac{d\mathbf{S}_1}{dt}\right)^{\mathrm{S}_1\mathrm{S}_2} = \Omega_{\mathrm{S}_1\mathrm{S}_2} \times \mathbf{S}_1\,, \tag{5.48}$$

where

$$\Omega_{\mathrm{S}_1\mathrm{S}_2} = \frac{G}{c^2 r^3}\left(\frac{3(\mathbf{r}\cdot\mathbf{S}_2)\mathbf{r}}{r^2} - \mathbf{S}_2\right). \tag{5.49}$$

The space mission GP-B succeeded in measuring the predicted effect of 39.2 mas/yr with an accuracy of 2×10^{-1} [81].

A further effect is the de Sitter effect, also called Fokker effect or geodetic precession, which results from the spin(1)–spin(2) coupling in the form

$$\left(\frac{d\mathbf{S}_1}{dt}\right)^{\mathrm{SO}} = \boldsymbol{\Omega}^{\mathrm{s}}_{\mathrm{SO}} \times \mathbf{S}_1 \,, \tag{5.50}$$

$$\boldsymbol{\Omega}^{\mathrm{s}}_{\mathrm{SO}} = \frac{2G}{c^2 r^3}\left(1 + \frac{3M_2}{4M_1}\right)\mathbf{L} \,. \tag{5.51}$$

The precession of the "spin" (orbital angular momentum) of the Earth–Moon system in the gravitational field of the Sun has the value 19 mas/yr and could be verified with precision of 2×10^{-2} [82]. The precession rate in the GP-B mission of 6606 mas/yr could be seen with a precision of 3×10^{-3} by [81]. Also in the case of binary pulsar systems, the spin precession has been observed, e.g. [83].

References

[1] V. A. Brumberg, *Essential Relativistic Celestial Mechanics*, (Adam Hilger, Bristol, 1991).

[2] V. A. Brumberg, *Relativistic Celestial Mechanics (in Russian)*, (Nauka, Moscow, 1972).

[3] S. Kopeikin, M. Efroimsky and G. Kaplan, *Relativistic Celestial Mecahnics in the Solar System*, (Wiley-VCH, Weinheim, 2011).

[4] C. M. Will, Living Rev. Relativity 9:3, 2006 (www.livingreviews.org/lrr-2006-3).

[5] B. S. DeWitt, *Dynamical Theory of Groups and Fields*, (Gordon and Breach, New York, 1965).

[6] L. Blanchet, Living Rev. Relativity 9:4, 2006 (www.livingreviews.org/lrr-2006-4); update: arXiv:1310.1528.

[7] T. Futamase and Y. Itoh, Living Rev. Relativity 10:2, 2007 (www.livingreviews.org/lrr-2007-2).

[8] L. Blanchet, in *Mass and Motion in General Relativity*, Fundamental Theories of Physics 162, ed. by L. Blanchet, A. Spallicci and B. Whiting (Springer, Dordrecht, 2011), pp. 125–166.

[9] R. Arnowitt, S. Deser and C. W. Misner, in *Gravitation: An Introduction to Current Research*, ed. by L. Witten (John Wiley, New York, 1962), pp. 227–265 [arXiv:gr-qc/0405109].

[10] J. A. Wheeler, in *Relativity, Groups, and Topology*, ed. by C. DeWitt and B. S. DeWitt (Gordon and Breach, New York, 1964), pp. 315–520.

[11] C. W. Misner, K. S. Thorne and J. A. Wheeler, *Gravitation*, (W. H. Freeman and Co., San Francisco, 1973).

[12] B. S. DeWitt, Phys. Rev. 160:1113, 1967.

[13] T. Regge and T. Teitelboim, Ann. Phys. (N.Y.) 88:286, 1974.

[14] L. D. Landau and E. M. Lifshitz, *The Classical Theory of Fields*, (Pergamon Press, Oxford, 1985).

[15] D. D. Holm, Physica D 17:1, 1985.

[16] T. Damour, P. Jaranowski and G. Schäfer, Phys. Lett. B 513:147, 2001.

[17] T. Damour, P. Jaranowski and G. Schäfer, in *Proceedings of the 11th Marcel Grossmann Meeting on General Relativity*, ed. by H. Kleinert, R. T. Jantzen and R. Ruffini, (World Scientific, Singapore, 2008), pp. 2490–2492.

[18] J. Hartung, J. Steinhoff and G. Schäfer, Ann. Phys. (Berlin) 525:359, 2013.

[19] P. Jaranowski and G. Schäfer, Phys. Rev. D 87:081503, 2013.

[20] G. Schäfer, in *Mass and Motion in General Relativity*, Fundamental Theories of Physics 162, ed. by L. Blanchet, A. Spallicci and B. Whiting (Springer, Dordrecht, 2011), pp. 167–210.

[21] J. Steinhoff and G. Schäfer, Europhys. Lett. 87:50004, 2009.
[22] J. Steinhoff, Ann. Phys. (Berlin) 523:296, 2011.
[23] P. Jaranowski and G. Schäfer, Phys. Rev. D 57:7274, 1998.
[24] G. Schäfer, Phys. Lett. A 100:128, 1984.
[25] T. Damour and G. Schäfer, J. Math. Phys. 32:127, 1991.
[26] A. Hanson and T. Regge, Ann. Phys. (N.Y.) 87:498, 1974.
[27] T. Damour, P. Jaranowski and G. Schäfer, Phys. Rev. D 62:021501, 2000; Erratum Phys. Rev. D 63:029903, 2001.
[28] P. Jaranowski and G. Schäfer, Phys. Rev. D 86:061503, 2012.
[29] T. Damour, P. Jaranowski and G. Schäfer, Phys. Rev. D 77:064032, 2008.
[30] J. Steinhoff, S. Hergt and G. Schäfer, Phys. Rev. D 77:081501, 2008.
[31] J. Steinhoff, S. Hergt and G. Schäfer, Phys. Rev. D 78:101503, 2008.
[32] J. Hartung and J. Steinhoff, Ann. Phys. (Berlin) 523:783, 2011.
[33] J. Hartung and J. Steinhoff, Ann. Phys. (Berlin) 523:919, 2011.
[34] T. J. Rothe and G. Schäfer, J. Math. Phys. 51:082501, 2010.
[35] G. Schäfer, Astron. Nachr. 311:4, 1990.
[36] G. Schäfer, in *Symposia Gaussiana, Conf. A: Mathematiocal and Theoretical Physics*, ed. by M. Behara, R. Fritsch and R. G. Lintz (Walter de Gruyter, Berlin, 1995), pp. 667–679.
[37] P. Jaranowski and G. Schäfer, Phys. Rev. D 55:4712, 1997.
[38] C. Königsdörffer, G. Faye and G. Schäfer, Phys. Rev. D 68:044004, 2003.
[39] T. Damour and G. Schäfer, Nuovo Cimento B 101:127, 1988.
[40] T. Damour, P. Jaranowski and G. Schäfer, Phys. Rev. D 62:084011, 2000.
[41] A. Le Tiec, L. Blanchet and B. F. Whiting, Phys. Rev. D 85:064039, 2012.
[42] D. Bini and T. Damour, Phys. Rev. D 87:121501, 2013.
[43] G. Schäfer, Prog. Theor. Phys. 68:2191, 1982.
[44] G. Schäfer, Ann. Phys. (N.Y.) 161:81, 1985.
[45] T. Damour and G. Schäfer, Gen. Relativ. Gravit. 17:879, 1985.
[46] T. Ledvinka, G. Schäfer and J. Bičák, Phys. Rev. Lett. 100:251101, 2008.
[47] B. M. Barker and R. F. O'Connell, Gen. Relativ. Gravit. 11:149, 1979.
[48] T. Damour, Phys. Rev. D 64:124013, 2001.
[49] R. A. Porto and I. Z. Rothstein, Phys. Rev. D 78:044013, 2008; Erratum Phys. Rev. D 81:029905, 2010.
[50] S. Hergt, J. Steinhoff and G. Schäfer, Class. Quantum Grav. 27:135007, 2010.
[51] R. A. Porto and I. Z. Rothstein, Phys. Rev. D 78:044012, 2008.
[52] R. A. Porto, Class. Quant. Grav. 27:205001, 2010.
[53] M. Levi, Phys. Rev. D 82:064029, 2010.
[54] M. Levi, Phys. Rev. D 82:104004, 2010.
[55] S. Hergt, J. Steinhoff and G. Schäfer, Ann. Phys. (N.Y.) 327:1494, 2012.
[56] S. Hergt and G. Schäfer, Phys. Rev. D 77:104001, 2006.
[57] S. Hergt and G. Schäfer, Phys. Rev. D 78:124004, 2008.
[58] J. Steinhoff and H. Wang, Phys. Rev. D 81:024022, 2010.
[59] H. Wang, J. Steinhoff, J. Zeng and G. Schäfer, Phys. Rev. D 84:124005, 2011.
[60] M. Levi, Phys. Rev. D 85:064043, 2012.
[61] S. Marsat, A. Bohé, G. Faye and L. Blanchet, Class. Quantum Grav. 30:055007, 2013.
[62] H. Wang and C. M. Will, Phys. Rev. D 75:064017, 2007.
[63] J. Zeng and C. M. Will, Gen. Relativ. Gravit. 39:1661, 2007.
[64] E. Barausse, E. Racine and A. Buonanno, Phys. Rev. D 80:104025, 2009.
[65] T. Damour, Phys. Rev. D 81:024017, 2010.
[66] L. Blanchet, S. L. Detweiler, A. Le Tiec and B. F. Whiting, Phys. Rev. D 81:084033, 2010.

[67] S. Foffa and R. Sturani, Phys. Rev. D 87:044056, 2013.
[68] S. Foffa and R. Sturani, Phys. Rev. D 87:064011, 2013.
[69] T. Damour, P. Jaranowski and G. Schäfer, Phys. Rev. D 62:044024, 2000.
[70] R.-M. Memmesheimer, A. Gopakumar and G. Schäfer, Phys. Rev. D 70:104011, 2004.
[71] G. Schäfer and N. Wex, Phys. Lett. A 174:196, 1993; Erratum Phys. Lett. A 177:461, 1993.
[72] T. Damour and N. Deruelle, Ann. Inst. H. Poincaré 43:107, 1985.
[73] D. V. Gal'tsov, A. A. Matiukhin and V. I. Petukhov, Phys. Lett A 77:387, 1980.
[74] J. M. Weisberg, D. J. Nice and J. H. Taylor, Astrophys. J. 722:1030, 2010.
[75] L. Blanchet and G. Schäfer, Mon. Not. R. Astr. Soc. 239:845, 1989.
[76] W. Junker and G. Schäfer, Mon. Not. R. Astr. Soc. 254:146, 1992.
[77] S. Hergt, A. Shah and G. Schäfer, Phys. Rev. Lett. 111:021101, 2013.
[78] M. Tessmer, J. Hartung and G. Schäfer, Class. Quant. Grav. 30:015007, 2013.
[79] G. Schäfer, Gen. Relativ. Gravit. 36:2223, 2004.
[80] I. Ciufolini, Nature 449:41, 2007.
[81] C. W. F. Everitt, D. B. DeBra, B. W. Parkinson et al., Phys. Rev. Lett. 106:221101, 2011.
[82] I. I. Shapiro, R. D. Reasenberg, J. F. Chandler and R. W. Babcock, Phys. Rev. Lett. 61:2643, 1988.
[83] R. P. Breton, V. M. Kaspi, M. Kramer et al., Science 321:104, 2008.

Yi Xie and Sergei Kopeikin

Covariant theory of the post-Newtonian equations of motion of extended bodies

1 Introduction

Theoretical formulation of relativistic equations of motion of massive particles and extended bodies have been an important research problem since the discovery of general theory of relativity by Albert Einstein. It continues to be of paramount importance for further theoretical development of general theory of relativity and its experimental testing in the solar system [93, 154] and in the binary pulsars [37, 40, 106, 151]. Rapidly growing new branch of relativistic astrophysics – gravitational wave astronomy – urgently demands significantly better understanding of the theoretical principles underlying the derivation of relativistic equations of motion of compact and extended bodies comprising the astrophysical N-body system [4, 10, 36, 134].

During the last three decades the most theoretical efforts in solving the problem of motion in general relativity have been focused on the astronomical systems consisting of point-like particles with the goal to advance the analytic derivation of the higher order post-Newtonian corrections beyond the famous quadrupole formula of Landau and Lifshitz [107] describing the loss of energy and angular momentum from the system due to the emission of gravitational waves. Theoretical difficulties in solving this problem are mainly caused by the nonlinear nature of the gravitational field leading to its self-interaction and appearance of divergent integrals which have to be regularized [13, 15, 16]. Additional problem is to take into account the internal structure of the bodies comprising the astronomical N-body system because the point-particle limit is not sufficient to correctly predict the templates of gravitational waves emitted by coalescing binaries consisting of neutron stars or black holes – the multipole moments of the moving bodies – spin, quadrupole, etc. – are a matter of importance and are to be taken into account especially at the very last stage of the coalescing process [14, 145, 159].

Calculations of equations of motion of binary systems for gravitational wave astronomy are primarily pursued either in the Arnovitt–Deser–Misner (ADM) or in har-

Yi Xie: Department of Astronomy, Nanjing University, 22 Hankou Road, Nanjing Jiangsu 210093, China and
Key Laboratory of Modern Astronomy and Astrophysics, Nanjing University, Nanjing Jiangsu 210093, China
Sergei Kopeikin: Department of Physics & Astronomy, University of Missouri, 322 Physics Bldg., Columbia, Missouri 65211, USA

monic coordinates [84, 114]. Hence, the resulting equations of motion are not covariant that requires careful consideration not only the process of generation of gravitational waves but also their propagation and eventual detection by gravitational wave observatories. Under assumption that background spacetime of a binary system emitting gravitational waves is asymptotically flat the gauge invariant quantities are orbital period and periastron advance of the binary system that can be expressed in terms of the energy and angular momentum of the system representing the integrals of motion in nonradiative post-Newtonian approximations [41, 45, 76]. These quantities are adiabatic invariants under a secular change of the orbital parameters of the binary system caused by the emission of gravitational waves.

In this chapter, we do not discuss the radiative celestial mechanics of astronomical N-body systems and restrict ourselves with a conservative, post-Newtonian approximation of the first order but with a full account of internal structure of the bodies. The goal is to build the covariant post-Newtonian theory of motion of extended bodies and to find out relativistic corrections to the point-like particle limit which accounts for *all* multipole moments characterizing the interior structure of the extended bodies. This programmatic task was put forward by Mathisson [111, 112, 133] and originally explored by Fock [73], Papapetrou [123–125], Tulczyjew [149]. Significant progress in advancing solution of this problem was achieved by Dixon [57, 58, 60, 62] who proposed to derive an *exact* covariant equations of motion of extended bodies from microscopic conservation law[1]

$$\nabla_\alpha T^{\alpha\beta} = 0\,, \tag{1.1}$$

where ∇_α denotes a covariant derivative and $T^{\alpha\beta}$ is the stress–energy–momentum tensor of matter, by making use of a novel method of integration of the linear connection in general relativity (Christoffel symbols). The generic mathematical techniques used by Dixon to achieve this goal were the formalism of two-point world function σ of Synge [144] and its derivatives (called sometimes bi-tensors), the horizontal and vertical covariant derivatives defined on the tangent bundle of spacetime manifold (also known as Ehresmann's connection [91]) and the distributional theory of multipole moments.

An extended body in Dixon's approach is idealized as a time-like world tube filled with matter which stress–energy–momentum tensor $T^{\alpha\beta}$ vanishes outside the tube. By making use of the bi-tensor propagators, $K^{\alpha}{}_{\mu'}$ and $H^{\alpha}{}_{\mu'}$, composed out of the second-order derivatives of the world function, Dixon defined the total linear momentum p^α and spin $S^{\alpha\beta}$ of the extended body by integrals over a spacelike hyper-

1 Greek indices take values from 0 to 3 and numerate spacetime coordinates. Roman indices take values 1, 2, 3 and numerate spatial coordinates only. Repeated indices indicate application of Einstein's summation rule.

surface Σ, [62, Equations 66, 67]

$$p^{\alpha}(z,\Sigma) \equiv \int_{\Sigma} K^{\alpha}{}_{\mu'} T^{\mu'\nu'} \sqrt{-g(x')} d\Sigma_{\nu'} , \tag{1.2}$$

$$S^{\alpha\beta}(z,\Sigma) \equiv -2 \int_{\Sigma} X^{[\alpha} H^{\beta]}{}_{\mu'} T^{\mu'\nu'} \sqrt{-g(x')} d\Sigma_{\nu'} , \tag{1.3}$$

where the primed indices belong to the point of integration x' on Σ, $z = z^{\alpha}(s)$ is a reference world line $\mathcal{Z}$ of a representative point that is associated with the center of mass of the body, s is a parameter on this world line, X^{α} is a partial derivative of the world function that can be viewed as a vector in tangent space attached to each point z of the world line $\mathcal{Z}$. The cross section $\Sigma = \Sigma(s)$ of the world tube of the body consists of all spacelike geodesics passing through $z(s)$ orthogonal to the *dynamical* 4-velocity, n^{α} of $z(s)$.

Definition of mass $M = M(z,\Sigma)$, and the mass dipole moment $m^{\alpha} = m^{\alpha}(z,\Sigma)$,of the body are given by

$$p^{\alpha} \equiv M n^{\alpha} , \tag{1.4}$$

$$m^{\alpha} \equiv S^{\alpha\beta} n_{\beta} , \tag{1.5}$$

where n^{α} is the *dynamical* velocity of the body moving along the world line $\mathcal{Z}$ related to its *kinematic* velocity $u^{\alpha} = dz^{\alpha}/ds$ by $n_{\alpha}u^{\alpha} = -1$, and the normalization condition is $u_{\alpha}u^{\alpha} = -1$. The world line $z = z(s)$ of the center of mass of the body is, then, defined by choosing $m^{\alpha} = 0$. Due to (1.4) it is equivalent to

$$S^{\alpha\beta} p_{\beta} = 0 , \tag{1.6}$$

which is known as Dixon's supplementary condition [62, Equation 81].

High-order multipoles of the body are defined by means of 2^l-pole moment tensor [62, Equations 140–145],

$$I^{\alpha_1\ldots\alpha_l\mu\nu}(z) = \int X^{\alpha_1} \cdots X^{\alpha_l} \hat{T}^{\mu\nu}(z,X) DX , \tag{1.7}$$

where X^{α} is the same vector as in (1.3), $\hat{T}^{\mu\nu}(z,X)$ is a *skeleton* of the stress–energy–momentum tensor $T^{\mu\nu}$ of the body, $DX \equiv \sqrt{-g(z)} d^4X$ is the invariant volume of integration on the tangent spacetime at point z on $\mathcal{Z}$, and the integration is performed over the tangent spacetime. According to Dixon, the skeleton $\hat{T}^{\mu\nu}(z,X)$ is a distribution (called a generalized function in Russian [136]) defined on tangent spacetime in such a way that it contains only information about the body but is entirely independent of the geometry of the surrounding spacetime which contains the bodies. Definition (1.7) assumes that

$$I^{\alpha_1\ldots\alpha_l\mu\nu} = I^{(\alpha_1\ldots\alpha_l)(\mu\nu)} , \tag{1.8}$$

where the round parentheses around the tensor indices denote a full symmetrization. Furthermore, the stress–energy conservation law imposes one more constrain [62]

$$I^{(\alpha_1 \ldots \alpha_l \mu)\nu} = 0 \, . \tag{1.9}$$

The 2^l-pole moments are coupled to the Riemann tensor $R^{\alpha}{}_{\mu\beta\nu}$ characterizing the curvature of spacetime. Therefore, they can be replaced with a more suitable set of *reduced* moments $J^{\alpha_1 \ldots \alpha_l \lambda\mu\nu\rho}$ which are defined by the following formulas:

$$J^{\alpha_1 \ldots \alpha_l \lambda\mu\nu\rho} = I^{(\alpha_1 \ldots \alpha_l)[\lambda[\nu\mu]\rho]} \, , \qquad (l \geq 0) \tag{1.10a}$$

$$J^{\alpha_1 \ldots \alpha_l \lambda[\mu\nu\rho]} = 0 \, , \qquad (l \geq 0) \tag{1.10b}$$

$$J^{\alpha_1 \ldots [\alpha_l \lambda\mu]\nu\rho} = 0 \, , \qquad (l \geq 1) \tag{1.10c}$$

where the square parentheses around the tensor indices denote a full antisymmetrization, and the nested square brackets in (1.10a) denote anti-simmetrization on pairs of indices λ, μ and ν, ρ independently. Inverting (1.10) yields

$$I^{\alpha_1 \ldots \alpha_l \mu\nu} = \frac{4(l-1)}{l+1} J^{(\alpha_1 \ldots \alpha_{l-1}/\mu/\alpha_l)\nu} \, , \tag{1.11}$$

where the forward slashes embracing the index μ mean that it is excluded from the symmetrization denoted with the round parentheses. The moments satisfy the orthogonality condition

$$J^{\alpha_1 \ldots \alpha_l \lambda\mu\nu\rho} p_{\alpha_1} = 0 \, , \tag{1.12}$$

being valid for any index from the set $\alpha_1, \alpha_2, \ldots, \alpha_l$ due to the property of symmetry (1.7).

Dixon presented a number of arguments suggesting that the covariant equations of motion of the extended body must have the following *exact* form [59, Equations 4.9, 4.10]:

$$\frac{dp_\alpha}{ds} = \frac{1}{2} v^\beta S^{\mu\nu} \bar{R}_{\alpha\beta\mu\nu} - \frac{1}{2} \sum_{l=2}^{\infty} \frac{1}{l!} \nabla_{(\alpha} A_{\beta_1 \ldots \beta_l)\mu\nu} I^{\beta_1 \ldots \beta_l \mu\nu} \tag{1.13}$$

$$\frac{dS^{\alpha\beta}}{ds} = 2p^{[\alpha} v^{\beta]} - \sum_{l=1}^{\infty} \frac{1}{l!} B_{\gamma_1 \ldots \gamma_l \mu\nu}{}^{[\alpha} I^{\beta]\gamma_1 \ldots \gamma_l \mu\nu} \, , \tag{1.14}$$

where $d/ds \equiv v^\alpha \nabla_\alpha$ is the (horizontal) covariant derivative along the reference line $z = z(s)$, $A_{\beta_1 \ldots \beta_l \mu\nu}$ and $B_{\beta_1 \ldots \beta_l \mu\nu\sigma}$ are the l-fold symmetrized tensors composed from the covariant derivatives of the metric tensor and the world function at point x taken, then, in the limit $x \to z$ and the bar above the Riemann tensor indicates that it is calculated on the background spacetime manifold from which the body under consideration is excluded. These tensors are expressed in terms of the Riemann tensor, $\bar{R}^{\alpha}{}_{\mu\beta\nu}$, of the background spacetime and its covariant derivatives. We follow [147] and call tensors, $A_{\beta_1 \ldots \beta_l \mu\nu}$ and $B_{\beta_1 \ldots \beta_l \mu\nu\sigma}$, the *external* multipole moments of the background spacetime. The multipole moments characterizing the interior structure and gravitational field of the body itself are called the *internal* multipole moments. Analytic calculations of tensors $A_{\beta_1\ldots\beta_l\mu\nu}$ and $B_{\beta_1\ldots\beta_l\mu\nu\sigma}$ in the framework of Dixon's theory are lengthy and were never performed by Dixon or anybody else except the quadrupole approximation in which equations of motion (1.13) and (1.14) read [62, Equations 171, 172], [59, Equations 4.11, 4.12]

$$\frac{dp^\alpha}{ds} = \frac{1}{2}v^\beta S^{\mu\nu}\bar{R}^\alpha{}_{\beta\mu\nu} + \frac{1}{6}\nabla^\alpha \bar{R}_{\sigma\rho\mu\nu}J^{\sigma\rho\mu\nu} + \cdots, \tag{1.15}$$

$$\frac{dS^{\alpha\beta}}{ds} = 2p^{[\alpha}v^{\beta]} + \frac{4}{3}J^{\mu\nu\sigma[\alpha}\bar{R}_{\mu\nu\sigma}{}^{\beta]} + \cdots, \tag{1.16}$$

where the ellipsis indicate the presence of higher order multipole terms.

The mathematical elegance and the apparently covariant nature of the Dixon theory of equations of motion is striking and attracted a peer attention of a number of experts in general relativity after it was proposed [6, 7, 66, 67, 135]. However, it was soon realized [147] that there are several issues which make the theory look rather abstract and not directly suitable for astrophysical applications or in relativistic celestial mechanics of the solar system. It reduced the number of publications pursuing a la Dixon's study of equations of motion. The main problem is that Dixon's theory does not deal with the gravitational field equations. It tacitly assumes that they are solved and the metric tensor is known. However, the Einstein equations and the equations of motion of the bodies creating the field are bounded in a closed tie – matter generates gravity while gravity governs the motion of matter. Dixon's definitions of the multipole moments of the bodies and other auxiliary geometric structures depend on the metric tensor which, in its own turn and in accordance with Einstein's equations, depends on the multipole moments and those structures. Their general relativistic interaction is nonlinear which assumes that the field equations cannot be mapped to the linear multipolar structure of equations of motion (1.13), (1.14) that seems are missing the nonlinear contributions of the body's moments. This is indeed the case as we shall see later on in Section 8. Moreover, the Dixon's equations are incomplete even at a linear level as they neither include the spin-type (internal) multipole moments of extended bodies besides their spins nor the time derivatives of the internal multipoles (see equations (8.47)–(8.59) in this chapter). Because of these drawbacks the Dixon theory is imperfect and can hardly be called "standard" as some authors have been recently claiming [9].

Dixon's approach is missing several critical ingredients which has to be added in order to make the theory complete and capable to determine the motion of extended bodies in practical situations. More precisely, the following additional ingredients of the theory are required:

(1) the procedure of finding the metric tensor from the field equations;
(2) the procedure of selection of the center-of-mass world line $\mathcal{Z}$ within each body;
(3) the momentum–velocity relation;

(4) the procedure of unambiguous characterization and determination of the gravitational self-field and self-force inside the body;
(5) the precise algorithm for calculation the body's internal multipoles (1.7) as the stress–energy–momentum tensor's skeleton, $\hat{T}^{\alpha\beta}$, introduced by Dixon, is lacking this algorithm;
(6) the precise algorithm for calculation the external multipoles $A_{\beta_1 \dots \beta_l \mu\nu}$ and $B_{\beta_1 \dots \beta_l \mu\nu\sigma}$ of the background spacetime manifold in the right-hand side of (1.13) and (1.14) in terms of the internal multipoles of the bodies.

The goal of this chapter is to discuss the missing elements of Dixon's approach and to present a comprehensive covariant theory of post-Newtonian equations of motion. Herein, we focus on translational equations of the bodies. Covariant rotational equations of motion will be considered somewhere else. We shall step beyond general theory of relativity and work with the field equations of a scalar-tensor theory of gravity. Therefore, in addition to the metric tensor, a scalar field will be also a carrier of a long-range gravitational interaction. This introduces some additional complications compared with Dixon's general-relativistic theory of equations of motion. In particular, instead of two sets of general-relativistic multipole moments we shall have to define additional set associated with the scalar field. The introduction of the scalar field has certain advantages for experimental tests of general relativity as it allows us to parameterize the field equations and equations of motion with two parameters, β and γ, of the parameterized post-Newtonian (PPN) theory which is a covariant generalization of a purely phenomenological parametrization of general-relativistic metric of the "canonical" Will–Nordtvedt PPN formalism [153]. We have started the development of this theory in [93, 94, 158]. Here, we present new results completing Dixon's program of derivation of translational equations of motion with accounting for all internal and external multipoles of extended bodies in N-body system. We also give a covariant form of the equations of motion which generalize Dixon's equations (1.13) and (1.14).

The brief content of our study is as follows. In the next Section 2 we discuss a scalar-tensor theory of gravity for post-Newtonian celestial mechanics of N-body system. Parametrization of the field equations, small parameters, and post-Newtonian approximations and gauges are introduced in Section 3. Parameterized post-Newtonian coordinate charts covering the spacetime manifold globally and in a local neighborhood of each body are set up in Section 4. They make up an atlas of spacetime manifold. The coordinates in relativity are characterized in terms of the metric tensor and its parameters – multipole moments which are also explained in Section 4. The differential structure of spacetime manifold presupposes that the parametric descriptions of the metric tensor given in different coordinates must smoothly match in the regions where the coordinate charts overlap. The matching procedure is described in Section 5 and gives relations between the multipole moments and the metric functionals from the distribution of matter density, its current, pressure, etc. It also establishes parametrization of the world line $\mathcal{W}$ of the origin of each local coordinates with respect to the global coordinate chart and defines the effective background metric, $\bar{g}_{\alpha\beta}$, for each extended body that is used later on for derivation of covariant equations of motion. The local coordinate chart introduced around each extended body is used for detailed description of the body's gravitational field and for definition of the body's mass and the center of mass in Section 6. It is also used to derive the local equations of motion of the body's center of mass with respect to the origin of the local coordinates. Because the world line $\mathcal{W}$ of the origin of the local coordinates has been derived by the method of asymptotic matching in Section 5, the equations of motion of the body follow immediately after substitution of the local equations of motion to the parametric description of the world line $\mathcal{W}$. This calculation is performed in Section 7 which derives equations of translational motion (7.20) of each body from N-body astronomical system in terms of their internal multipoles as well as global coordinates and velocities of their centers of mass. Finally, Section 8 establishes covariant form (8.47) of the translational equations of motion of extended bodies which completes the Dixon's program.

2 A theory of gravity for post-Newtonian celestial mechanics

We consider an isolated N-body system comprised of N extended bodies with non-singular interior described by the stress–energy–momentum tensor $T^{\alpha\beta}$. The bodies have a localized matter support and are supposed to be well isolated one from another in space. Accretion and other fluxes of matter outside of the bodies are absent or neglected.

Post-Newtonian celestial mechanics describes orbital and rotational motions of the bodies on a curved spacetime manifold described by the metric tensor, $g_{\alpha\beta}$ obtained as a solution of the field equations of a metric-based theory of gravitation in the slow-motion and weak-gravitational field approximation. Class of the viable metric theories of gravity, which can be employed for developing relativistic celestial mechanics, ranges from general theory of relativity [21, 107] to a scalar-vector-tensor theory of gravity recently proposed by [8] for description the motion of galaxies at the cosmological scale. It is inconceivable to review all these theories in this chapter and we refer the reader to [155] for further details on those alternative theories. As for our goal, we shall build a PPN theory of celestial mechanics in the framework of a scalar-tensor theory of gravity introduced by Jordan and Fierz [71, 86, 87] and rediscovered later by Brans and Dicke [17, 52, 53]. The Jordan–Fierz–Brans–Dicke (JFBD) theory extends the Lagrangian of general relativity by introducing a long range, nonlinear scalar field (or fields [38]) minimally coupled to gravity. The presence of the scalar field causes deviation of the metric-based gravity theory from pure geometry. The scalar field highlights the geometric role of the metric tensor and makes the physical content of the gravitational theory richer. Recent discovery of the scalar Higgs boson at LHC [55] and its possible connection to the JFBD scalar field in gravitation and cosmology [50] reinforces the theoretical basis and value of the scalar-tensor theory in astrophysics and relativistic celestial mechanics.

2.1 The field equations

Gravitational field in the scalar-tensor theory of gravity is described by the metric tensor $g_{\alpha\beta}$ and a long-range scalar field ϕ with a self-interaction described by means of a coupling function $\theta(\phi)$. The field equations in the scalar-tensor theory are derived from the action [153]

$$S = \frac{c^3}{16\pi}\int\left(\phi R - \theta(\phi)\frac{\phi^{,\alpha}\phi_{,\alpha}}{\phi} - \frac{16\pi}{c^4}\mathcal{L}(g_{\mu\nu}, \Psi)\right)\sqrt{-g}\,d^4x\,, \tag{2.1}$$

where the first, second, and third terms in the integrand of (2.1) are the Lagrangian densities of the gravitational field, scalar field, and matter of the N-body system respectively, $g = \det[g_{\alpha\beta}] < 0$ is the determinant of the metric tensor $g_{\alpha\beta}$, $R = g^{\alpha\beta}R_{\alpha\beta}$ is the Ricci scalar, $R_{\alpha\beta}$ is the Ricci tensor, Ψ indicates the set of matter variables, and $\theta(\phi)$ is the coupling function of the scalar field which is kept unspecified for the purpose of further parametrization of the possible deviation from general relativity. The action (2.1) is written in the Jordan–Fierz frame in which the metric tensor $g_{\alpha\beta}$ has a standard physical meaning of observable quantity in definition of the proper time and proper length [153].

For the sake of simplicity, we postulate that the potential $V(\phi)$ of the scalar field ϕ is nil so that the scalar field propagates freely. Discarding the potential $V(\phi)$ is justified in the first post-Newtonian approximation as it does not yield any measurable relativistic effects within the boundaries of the solar system [155]. However, if the potential is nonlinear it can be important in strong gravitational fields of neutron stars and black holes, and its inclusion to the theory leads to interesting physical consequences [38, 39]. We shall not analyze in this chapter the nonlinear effects of the scalar field ϕ.

Field equations for the metric tensor are obtained by variation of action (2.1) with respect to $g_{\alpha\beta}$. It yields [153]

$$R_{\mu\nu} = \frac{8\pi}{\phi c^2}\left(T_{\mu\nu} - \frac{1}{2}g_{\mu\nu}T\right) + \theta(\phi)\frac{\phi_{,\mu}\phi_{,\nu}}{\phi^2} + \frac{1}{\phi}\left(\phi_{;\mu\nu} + \frac{1}{2}g_{\mu\nu}\Box_g\phi\right), \tag{2.2}$$

where

$$\Box_g \equiv g^{\mu\nu}\frac{\partial^2}{\partial x^\mu \partial x^\nu} - g^{\mu\nu}\Gamma^\alpha_{\mu\nu}\frac{\partial}{\partial x^\alpha} \tag{2.3}$$

is the differential Laplace–Beltrami operator for the scalar field [69, 115], and $T_{\mu\nu}$ is the stress–energy–momentum tensor of matter comprising the N-body system defined by the variational derivative from the Lagrangian $\mathcal{L}$ of matter [107]

$$\frac{c^2}{2}\sqrt{-g}\,T_{\mu\nu} \equiv \frac{\partial(\sqrt{-g}\mathcal{L})}{\partial g^{\mu\nu}} - \frac{\partial}{\partial x^\alpha}\frac{\partial(\sqrt{-g}\mathcal{L})}{\partial g^{\mu\nu}{}_{,\alpha}}\,. \tag{2.4}$$

Equation for the scalar field ϕ is obtained by variation of action (2.1) with respect to ϕ. After making use of the contracted form of (2.2) it yields [153]

$$\Box_g\phi = \frac{1}{3+2\theta(\phi)}\left(\frac{8\pi}{c^2}\,T - \phi_{,\alpha}\,\phi^{,\alpha}\,\frac{d\theta}{d\phi}\right), \tag{2.5}$$

where $T = g^{\alpha\beta}T_{\alpha\beta}$ is the trace of the stress–energy–momentum tensor of matter which serves as a source of the scalar field. The last term in (2.5) shows that $\theta(\phi)$ is responsible for self-interaction of the field ϕ.

2.2 The energy–momentum tensor

Gravitational field and matter are tightly connected via the Bianchi identity of the field equations for the metric tensor [107, 115]. The Bianchi identity makes four out of ten components of the metric tensor fully independent so that they can be chosen arbitrary. This freedom is usually fixed by picking up a specific gauge condition, which imposes four restrictions on four components of the metric tensor and/or its first derivatives but no restriction on the scalar field. On the other hand, the Bianchi identity imposes four differential constraints (1.1) on the stress–energy–momentum tensor of matter which constitutes microscopic equations of motion of matter [107].

We have assumed that the astronomical N-body system is isolated, which means that we neglect any influence from the gravitational environment outside the system and ignore cosmological effects. This makes the spacetime asymptotically flat so that the center of mass of the system can be set at rest with respect to an inertial coordinate chart at infinity. We postulate that the matter of the system is described by the stress energy tensor

$$T^{\alpha\beta} = \rho\left(1 + c^{-2}\Pi\right)u^\alpha u^\beta + c^{-2}\pi^{\alpha\beta}\,, \tag{2.6}$$

where ρ and Π are the density and the specific internal energy of matter, respectively, $u^\alpha = dx^\alpha/cd\tau$ is 4-velocity of the matter with τ being the proper time along the world line of matter's volume element, and $\pi^{\alpha\beta}$ is a symmetric tensor of spatial stresses being orthogonal to the 4-velocity of matter

$$u^\alpha\pi_{\alpha\beta} = 0\,. \tag{2.7}$$

Equation (2.7) means that the stress tensor has only spatial components in the frame co-moving with matter.

Due to the Bianchi identity the energy–momentum tensor is conserved, that is satisfies the microscopic equation of motion (1.1) or

$$T^{\alpha\beta}{}_{;\beta} = 0\,, \tag{2.8}$$

where here and everywhere else the semicolon denotes a covariant derivative on the spacetime manifold with respect to the Christoffel symbols calculated from the metric $g_{\alpha\beta}$. The conservation of the energy–momentum tensor leads to the equation of continuity

$$(\rho u^{\alpha})_{;\alpha} = \frac{1}{\sqrt{-g}}(\rho\sqrt{-g}u^{\alpha})_{,\alpha} = 0\,, \tag{2.9}$$

and to the second law of thermodynamics that is expressed as a differential relation between the specific internal energy and the tensor of stresses

$$\rho u^{\alpha}\Pi_{,\alpha} + \pi^{\alpha\beta}u_{\alpha;\beta} = 0\,. \tag{2.10}$$

These equations will be employed later for solving the field equations and for derivation of the equations of motion of the bodies.

3 Parameterized post-Newtonian celestial mechanics

Post-newtonian celestial mechanics is an extension of Newtonian mechanics on a curved spacetime manifold [19, 21]. Mathematical description of the manifold requires a particular theory of gravity going beyond the Newtonian theory. General-relativistic celestial mechanics operates with Einstein's general relativity having a minimal number of free fundamental parameters characterizing geometry of spacetime. This chapter deals with JFBD scalar-tensor theory of gravity which can be considered as a covariant extension of general relativity parameterizing possible deviations from a pure geometric theory of spacetime. The basic principles of the PPN celestial mechanics remains the same as in general relativity and in the Newtonian gravitational theory though the mathematical derivation of equations of motion of extended bodies become more complicated. PPN formalism described in [153] contains more parameters than the scalar-tensor theory. However, it is not covariant (though Lorentz invariant) and is limited to consideration of gravitational field of massive point-like particles only. We do not apply it anywhere in this chapter to avoid noncovariant conclusions which can easily become misleading and/or controversial. We explain the basic principle of PPN celestial mechanics in this section. More details are given in textbook [93].

3.1 External and internal problems of motion

Celestial mechanics subdivides equations of motion of N-body system in three classes corresponding to the number of degrees of freedom of matter of the bodies [73]. The main degrees of freedom are characterized by motion of linear and angular momenta of the bodies. Corresponding equations are

(1) translational equations of motion consisting of
   - the equation of motion of the total linear momentum of the entire system,
   - the equations of motion of the linear momentum of each body,

(2) rotational equations of motion consisting of
   - the equation of motion of the total angular momentum of the entire system;
   - the equations of motion of the angular momentum (spin) of each body.

(3) equations of temporal evolution of multipole moments of the system as a whole and each body separately.

The translational and rotational equations are typically refer to the translational motion of rigidly rotating bodies. If bodies have the time-dependent internal structure, the equations of motion of higher order internal multipoles must be included.

Derivation of the translational and rotational equations of motion naturally suggest separation of the problem of motion in two parts: external and internal. The external problem refers to the derivation of the translational equations, and the internal problem deals with the derivation of the rotational equations of motion as well as equations of motion of the body's internal multipoles. Though, the final form of the equations of motion must be covariant and is not expected to depend on the choice of a particular (privileged) coordinate chart on spacetime manifold, the coordinate approach is the most effective for unambiguous separation of the internal and external degrees of motion and for building the background effective manifold with metric, $\bar{g}_{\alpha\beta}$, admitting the covariant formulation of the equations of motion for the linear momentum, spin, and other internal multipoles.

The absence of a privileged coordinate chart is ensured by the covariant nature of the field equations of the scalar-tensor theory of gravity used in this chapter. In particular, the theory is Lorentz invariant which means that in case of an isolated astronomical system embedded to asymptotically flat spacetime we can always introduce a global coordinate chart with the origin located at the center of mass of the system. In what follows, we shall systematically neglect the cosmological effects in the motion of the bodies and assume that (1) the metric tensor, $g_{\alpha\beta}$, in the global coordinates approaches the Minkowski metric, $\eta_{\alpha\beta}$, at infinity, and (2) the global coordinates smoothly match the inertial coordinates of the Minkowski spacetime at infinity. However, the global coordinate chart is not sufficient for solving the problem of motion of extended bodies as it is not well suited for description of the internal structure of each body from the astronomical N-body system.

Indeed, the motion of matter is naturally split in two components – the orbital motion of the center of mass of each body and the internal motion of matter with respect to the body's center of mass. The global chart is adequate for describing the orbital dynamics. On the other hand, description of the internal motion of matter demands introduction of a local coordinate chart attached to each gravitating body as it excludes spurious effects (like Lorentz contraction, etc.) which have no relation to internal structure of the body [97]. Construction of the local coordinates should be reconciled with the principle of equivalence as long as it is allowed by the scalar-tensor theory of gravity. The body-related coordinates replicate the inertial coordinates only locally and cover a limited domain of spacetime manifold around the body under consideration. Thus, a full solution of the external and internal problems of celestial mechanics is based on consideration of $N + 1$ coordinate charts – one global and N local ones. It agrees with the topological structure of manifold defined by a set of overlapping coordinate charts making the atlas of the manifold [3, 63]. The equations of motion in the post-Newtonian celestial mechanics are intimately connected to the differential structure of the spacetime manifold characterized by the metric tensor and affine connection. It means that the mathematical presentations of the metric tensor in the local and global coordinates must be diffeomorphically equivalent that is the transition functions defining the transformation from local to global coordinates must map the components of the metric tensor of the internal problem of motion to the external one and vice versa.

Newtonian mechanics of N-body system describes translational motion of the bodies in global coordinates which origin is placed at the Newtonian center of mass of the system. Local coordinates for each body are constructed by a spatial translation of the origin of the global coordinates to the Newtonian center of mass of the body. Time in the Newtonian theory is absolute, and, hence, does not change when one transforms it from global to local coordinates. Newtonian space is also absolute, which makes the difference between the global and local coordinates physically insignificant. The theory changes dramatically as one switches from the Newtonian concepts to a consistent relativistic theory of gravity. There is no longer the absolute time nor the absolute space which are replaced with a pseudo-Riemannian spacetime manifold endowed with a rather complicated set of differential equations for gravitational variables and matter fields accounting for various relativistic effects. Construction of the post-Newtonian global and local coordinates depends on the boundary conditions imposed on the field equations [93]. The principle of relativity should be satisfied when the law of transformation from the global to local coordinates is derived. Not only should it be consistent with the Lorentz transformation of special relativity but must account for the full gauge freedom of the relativistic theory of gravity as well [96]. Time and spatial coordinates are transformed simultaneously making up a class of nonlinear coordinate transformations establishing mutual functional relations between various geometric objects and world lines of the bodies [21, 101, 137].

At the first glance the coordinate-related approach to the problem of motion looks quite disentangled from the covariant approach advocated by Dixon. In fact, the two approaches complement each other in a constructive way. In order to prove its consistency the Dixon's theory must be linked to the differential structure of spacetime manifold which is given in terms of the transition functions between coordinate charts. Moreover, it must be connected to the solutions of the field equations in order to identify the background spacetime manifold used to build skeleton of the stress–energy–momentum tensor, and to link it to observational values of multipole moments in astrophysics. It is impossible to identify the center of mass of an extended body and derive the equation of its world line without solving the internal problem of motion in the local coordinates with its subsequent matching to the external solution. Dixon's covariant formulation merely states that the center of mass inside of each body in N-body system exists but it does not provide the criterion for making the unique choice of the center of mass. This problem cannot be solved without resorting to the method of asymptotic matching of the solutions of the field equations. For these reasons we tackle the problem of covariant formulation of equations of motion, first, from the coordinate-related approach which allows us to find out the unique definition of the center of mass of the body at least in the post-Newtonian approximation. Covariant description of the equations of motion is achieved in our theoretical scheme later on, at last stage of calculations, by mapping back the locally defined quantities and equations to arbitrary coordinates on the effective background manifold with metric $\bar{g}_{\alpha\beta}$ (see Section 8 for more details). This procedure has been proposed by Landau and Lifshitz [107] and applied to the problem of motion by Thorne and Hartle [147]. It works perfect on torsionless manifolds with the affine connection being fully determined by the metric tensor. Its extension to the manifolds with torsion and/or nonminimal coupling of matter with gravity requires further theoretical study. Some steps forward in this direction have been made, for example, in [72, 83, 109, 110, 160] and [128, 129]. We do not discuss these extensions over here.

3.2 Solving the field equations by post-Newtonian approximations

Small parameters

Field equations (2.2) and (2.5) of the scalar-tensor theory of gravity represent a system of eleventh nonlinear differential equations in partial derivatives. It is challenging to find their solution in the case of N-body system made of self-gravitating extended bodies which back reaction on the geometry of spacetime manifold cannot be neglected. Like in general relativity, an exact solution of this problem is not known and may not be available in analytic form. Hence, one has to resort to approximations to apply the analytic methods. Two basic methods are known in asymptotically flat spacetime, namely, the post-Minkowskian (PMA) and the post-Newtonian (PNA) approximations [35]. Post-Newtonian approximations are applicable in case when matter moves slowly and gravitational field is weak everywhere – the conditions, which are satisfied, e.g. within the solar system. Post-Minkowskian approximations relax the requirement of the slow motion but the weak-field limitation remains. We use the post-Newtonian approximations in this chapter.

Post-Newtonian approximations assume that the metric tensor can be expanded in the near zone of N-body system with respect to the ultimate speed of gravity c called the speed of light for historical reasons [93]. This expansion may be not analytic in higher post-Newtonian approximations in a certain class of coordinates including the harmonic coordinates [11, 88]. Exact formulation of basic axioms underlying the post-Newtonian expansion was given by Rendall [132]. Practically, it requires to have several small parameters characterizing N-body system and the interior structure of the bodies. They are: $\epsilon_{\rm i} \sim v_{\rm i}/c$, $\epsilon_{\rm e} \sim v_{\rm e}/c$, and $\eta_{\rm i} \sim U_{\rm i}/c^2$, $\eta_{\rm e} \sim U_{\rm e}/c^2$, where $v_{\rm i}$ is the characteristic internal velocity of motion of matter inside an extended body, $v_{\rm e}$ is the characteristic velocity of the relative motion of the bodies with respect to each other, $U_{\rm i}$ is the internal gravitational potential inside each body, and $U_{\rm e}$ is the external gravitational potential in the space between the bodies. If we denote a characteristic radius of body as L and a characteristic distance between the bodies as R, the internal and external gravitational potentials will be $U_{\rm i} \simeq GM/L$ and $U_{\rm e} \simeq GM/R$, where M is the characteristic mass of the body.

Due to the virial theorem of the Newtonian gravity [107] the small parameters are not fully independent. Specifically, one has $\epsilon_{\rm i}^2 \sim \eta_{\rm i}$ and $\epsilon_e^2 \sim \eta_e$. Hence, two parameters $\epsilon_{\rm i}$ and $\epsilon_{\rm e}$ are sufficient in doing the post-Newtonian expansion. In what follows, we shall use notation $\epsilon \equiv 1/c$ to mark the presence of the post-Newtonian parameter and the fundamental speed c in the post-Newtonian series.

Besides the small relativistic parameters ϵ and η, the post-Newtonian approximations utilize one more small parameter which, in fact, is not relativistic parameter as it presents already in the Newtonian gravity. This parameter is denoted by $\delta \sim L/R$, and it characterizes the influence of the finite size of the body on the gravitational field outside of it. It is well-known that in the Newtonian mechanics gravitational field of a spherically symmetric body is the same as the field of a point-like particle having the same mass [28]. It suggests that for spherically symmetric bodies parameter $\delta = L/R$ does not play any role in the Newtonian approximation. However, it may appear directly in PNA expansion even if the body is spherically symmetric. Such appearance of the parameter δ is not supported in general relativity which satisfy the, so-called, *effacing principle* [34, 35, 97], but is allowed in scalar-tensor theory of gravity where terms of the order of $(\beta - 1)\epsilon^2\delta^2$ appear in the translational equations of motion of spherically symmetric bodies [95].

If the bodies are not spherically symmetric, parameter δ appears in both Newtonian and post-Newtonian approximations as a result of the multipolar expansion of gravitational field with respect to body's internal multipoles. A multipole of order l

depends on the parameter of ellipticity of the body, J_l that is related to the elastic properties of matter which are characterized by Love's numbers κ_{nl} (n = 1, 2, 3). Generally, they are different for each multipole [29, 78, 162]. This chapter will account for all gravitational multipoles of the massive bodies without truncation of the multipolar series at some finite order l.

The post-Newtonian series

The post-Newtonian series are expansions of the geometric, scalar field, and matter variables around their background values with respect to the small parameters introduced above. We denote ϕ_0 the background value of the scalar field ϕ and assume that the dimensionless perturbation of the field, ζ, is small. Then, we can write an *exact* decomposition

$$\phi = \phi_0(1 + \zeta) . \tag{3.1}$$

In principle, the background value ϕ_0 of the scalar field depends on time due to the Hubble expansion of the universe. Because ϕ_0 defines the current value of the universal gravitational constant, G, the time dependence of ϕ_0 causes a secular drift of $G = G_0 + \dot{G}(t - t_0)$ (see (3.29)). However, in our model of N-body system, the spacetime is assumed to be asymptotically flat which excludes cosmological scenario and makes ϕ_0 = const.

According to theoretical expectations [43] and experimental limitation on PPN parameters [155], the post-Newtonian perturbation ζ of the scalar field have a very small magnitude, so that we can expand all quantities depending on the scalar field in the Taylor series with respect to ζ using it as a small parameter in the expansion. In particular, the post-Newtonian decomposition of the coupling function $\theta(\phi)$ can be written as

$$\theta(\phi) = \omega + \omega' \zeta + \mathcal{O}\left(\zeta^2\right) , \tag{3.2}$$

where $\omega \equiv \theta(\phi_0)$, $\omega' \equiv (d\theta/d\zeta)_{\phi=\phi_0}$, and we impose the boundary condition on the scalar field such that ζ approaches zero as the distance from the N-body system approaches spatial infinity. The post-Newtonian expansion of the perturbation ζ is given in the form

$$\zeta = \epsilon \overset{(1)}{\zeta} + \epsilon^2 \overset{(2)}{\zeta} + \mathcal{O}\left(\epsilon^3\right) , \tag{3.3}$$

where the post-Newtonian corrections $\overset{(1)}{\zeta}$, $\overset{(2)}{\zeta}$, etc. will be defined below.

The background value of the metric tensor $g_{\alpha\beta}$ in our model of asymptotically flat spacetime is the Minkowski metric $\eta_{\alpha\beta}$. Cosmological post-Newtonian approximations with the background Friedmann–Lemaître–Roberston–Walker metric is considered in Chapter 7 of this book. The metric tensor is expanded in the post-Newtonian series with respect to parameter $\epsilon \equiv 1/c$ as follows:

$$g_{\alpha\beta} = \eta_{\alpha\beta} + \epsilon \overset{(1)}{h}_{\alpha\beta} + \epsilon^2 \overset{(2)}{h}_{\alpha\beta} + \epsilon^3 \overset{(3)}{h}_{\alpha\beta} + \epsilon^4 \overset{(4)}{h}_{\alpha\beta} + \mathcal{O}(\epsilon^5) . \tag{3.4}$$

The generic post-Newtonian expansion of the metric tensor is not analytic [11, 35, 88]. However, the nonanalytic terms emerge only in higher post-Newtonian approximations and do not affect results of this chapter since we restrict ourselves with the first post-Newtonian approximation. We also notice that the linear, with respect to ϵ, terms in the metric tensor expansion (3.4) can be eliminated by making coordinate adjustments [147]. These terms do not originate from the field equations and are pure coordinate-dependent effect. If we kept them, they would make the coordinate grid nonorthogonal and rotating at the classic level. Reference frames with such properties are rarely used in astronomy and astrophysics. Therefore, we assume that the linear term in expansion (3.4) is absent.

After eliminating the linear terms in the post-Newtonian expansion of the metric tensor and substituting the expansion to the field equations (2.2) we can check by inspection that various components of the metric tensor and the scalar field have in the first post-Newtonian approximation the following form:

$$g_{00} = -1 + \epsilon^2 \overset{(2)}{h}_{00} + \epsilon^4 \overset{(4)}{h}_{00} + \mathcal{O}\left(\epsilon^5\right) , \tag{3.5}$$

$$g_{0i} = \epsilon^3 \overset{(3)}{h}_{0i} + \mathcal{O}\left(\epsilon^5\right) , \tag{3.6}$$

$$g_{ij} = \delta_{ij} + \epsilon^2 \overset{(2)}{h}_{ij} + \mathcal{O}\left(\epsilon^4\right) , \tag{3.7}$$

where each term of the expansions will be explained below. Scalar field equation (2.5) shows that there is no linear term in (3.3) which reduces it in the first post-Newtonian approximation to

$$\zeta = \epsilon^2 \overset{(2)}{\zeta} + \mathcal{O}\left(\epsilon^4\right) . \tag{3.8}$$

In order to simplify notations, we shall use the following abbreviations:

$$h_{00} \equiv \overset{(2)}{h}_{00}, \qquad l_{00} \equiv \overset{(4)}{h}_{00}, \qquad h_{0i} \equiv \overset{(3)}{h}_{0i}, \qquad h_{ij} \equiv \overset{(2)}{h}_{ij}, \qquad h \equiv \overset{(2)}{h}_{kk} , \tag{3.9}$$

and

$$\varphi \equiv (\omega + 2) \overset{(2)}{\zeta} . \tag{3.10}$$

Post-Newtonian expansion of the metric tensor and the scalar field introduces a corresponding expansion of the stress–energy–momentum tensor of matter

$$T_{00} = \overset{(0)}{T}_{00} + \epsilon^2 \overset{(2)}{T}_{00} + \mathcal{O}\left(\epsilon^4\right) , \tag{3.11}$$

$$T_{0i} = \epsilon \overset{(1)}{T}_{0i} + \mathcal{O}\left(\epsilon^3\right) , \tag{3.12}$$

$$T_{ij} = \epsilon^2 \overset{(2)}{T}_{ij} + \mathcal{O}\left(\epsilon^4\right) , \tag{3.13}$$

where $\overset{(n)}{T}_{\alpha\beta}$ $(n = 0, 1, 2, 3\ldots)$ denotes terms of the order ϵ^n. In the first post-Newtonian approximation the components of the energy–momentum tensor were derived by Fock [73]

$$\overset{(0)}{T}_{00} = \rho^* \,, \tag{3.14}$$

$$\overset{(1)}{T}_{0i} = -\rho^* v^i \,, \tag{3.15}$$

$$\overset{(2)}{T}_{ij} = \rho^* v^i v^j + \pi^{ij} \,, \tag{3.16}$$

$$\overset{(2)}{T}_{00} = \rho^* \left(\frac{v^2}{2} + \Pi - h_{00} - \frac{h}{2} \right) , \tag{3.17}$$

where $v^i = cu^i/u^0$ is the three-dimensional velocity of matter.

Fock also introduced the invariant density of matter [73]

$$\rho^* \equiv \sqrt{-g}u^0 \rho = \rho + \frac{\epsilon^2}{2}\rho(v^2 + h) + \mathcal{O}\left(\epsilon^4\right) , \tag{3.18}$$

which is a useful mathematical tool in relativistic hydrodynamics [115, 153]. The reason is that the invariant density, unlike density ρ, obeys an exact equation of continuity (2.9) that can be recast to a Newtonian-like form [73]

$$c\rho^*_{,0} + (\rho^* v^i)_{,i} = 0 \,, \tag{3.19}$$

where $f_{,0} \equiv \epsilon \partial f / \partial t$, and comma with an index behind it denotes a partial derivative with respect to a corresponding coordinate. Equation (3.19) is valid in any post-Newtonian approximation and it makes calculation of time derivative of a volume integral of any function $f(t, \boldsymbol{x})$ simple

$$\frac{d}{dt} \int\limits_{V_A} \rho^*(t, \boldsymbol{x}) f(t, \boldsymbol{x}) d^3x = \int\limits_{V_A} \rho^*(t, \boldsymbol{x}) \frac{df(t, \boldsymbol{x})}{dt} d^3x \,, \tag{3.20}$$

where V_A is the volume of body A, the total time derivative

$$\frac{d}{dt} = \frac{\partial}{\partial t} + v^i \frac{\partial}{\partial x^i} \,, \tag{3.21}$$

and we have taken into account in the derivation of (3.20) that body A moves, and its shape and internal structure depend on time [93]. Equation (3.20) is exact.

3.3 The post-Newtonian field equations

The post-Newtonian field equations for the post-Newtonian components of the metric tensor and scalar field variables can be derived after substituting the post-Newtonian series of the previous section to the covariant equations (2.2) and (2.5), and arranging the terms in the expansion in the order of smallness with respect to parameter $\epsilon \equiv 1/c$. The post-Newtonian equations are covariant like the original field equations that is they are valid in arbitrary coordinates. Hence, their solution depends on four arbitrary functions reflecting the gauge freedom. It is a common practice to limit this arbitrariness by imposing a gauge condition which is equivalent to a choice of a specific set of coordinates on spacetime manifold. The gauge condition does not fix the freedom in choosing coordinates completely – a restricted class of coordinate transformations within the imposed gauge is still allowed. This class of transformations is called the residual gauge freedom which plays an essential role in theoretical formulation of relativistic celestial mechanics of a N-body system.

The most convenient gauge condition in the scalar-tensor theory of gravity were proposed by Nutku [121, 122] as a generalization of the harmonic gauge

$$\left(\phi\sqrt{-g}\, g^{\mu\nu}\right)_{,\nu} = 0\,, \tag{3.22}$$

where the comma denotes a partial derivative with respect to a corresponding coordinate, $f_{,\nu} = \partial f/\partial x^{\nu}$. By making use of the conformal metric tensor, $\tilde{g}_{\alpha\beta} \equiv \phi g_{\alpha\beta}$, equation (3.22) can be recast to the harmonic gauge condition for the conformal metric tensor [73, 123]

$$\left(\sqrt{-\tilde{g}}\tilde{g}^{\mu\nu}\right)_{,\nu} = 0\,. \tag{3.23}$$

Post-Newtonian expansion of gauge condition (3.22) yields

$$\left(\frac{2\phi}{\omega+2} + h_{00} + h\right)_{,0} = 2\epsilon h_{0k,k}\,, \tag{3.24}$$

$$\left(\frac{2\phi}{\omega+2} - h_{00} + h\right)_{,i} = 2h_{ik,k}\,. \tag{3.25}$$

It is worth noting that in the first post-Newtonian approximation, equations (3.24) and (3.25) do not involved the metric tensor component $\overset{(4)}{h}_{00} \equiv l_{00}$, which is directly obtained from the field equation and is fixed.

The post-Newtonian field equations for the scalar field and the metric tensor are obtained from equations (2.5) and (2.2) after making use of the post-Newtonian expansions, given by equations (3.5)–(3.13), and the gauge conditions (3.24) and (3.25). The scalar-tensor theory of gravity with variable coupling function $\theta(\phi)$ has two constant parameters, ω and ω', characterizing deviation from general relativity. They are related to the standard PPN parameters γ and β as follows [153]:

$$\gamma = \gamma(\omega) = \frac{\omega+1}{\omega+2}\,, \tag{3.26}$$

$$\beta = \beta(\omega) = 1 + \frac{\omega'}{(2\omega+3)(2\omega+4)^2}\,. \tag{3.27}$$

General relativity is obtained as a limiting case of the scalar-tensor theory when parameters $\gamma = \beta = 1$ or $\omega = \infty$. Notice that in order to get this limit convergent, parameter ω' must grow slower than ω^3 as ω approaches infinity. Currently, there are no experimental data restricting the functional behavior of $\omega' \sim \omega^3\beta(\omega)$ which could help us to understand better the nature of the coupling function $\theta(\phi)$. This makes parameter β a primary target for experimental study in the near-future gravitational experiments [1, 31, 56] including the advanced lunar laser ranging (LLR) [100].

The scalar field perturbation (3.10) is expressed in terms of γ as

$$\overset{(2)}{\zeta} = (1-\gamma)\varphi\,. \tag{3.28}$$

The background scalar field ϕ_0 and the parameter of coupling ω determine the observed numerical value of the universal gravitational constant

$$G = \frac{2\omega+4}{2\omega+3}\frac{1}{\phi_0}\,. \tag{3.29}$$

Had the background value ϕ_0 of the scalar field been driven by cosmological evolution, the measured value of the universal gravitational constant G would depend on time, $G = G_0 + \dot{G}(t-t_0)$, and might be detected experimentally. Currently, the best upper limit on time variability of G is imposed by LLR as $|\dot{G}/G| < (4 \pm 9) \times 10^{-13}\,\mathrm{yr}^{-1}$ [156].

After making use of definition of the tensor of energy–momentum (3.14)–(3.17), and PPN parameters (3.26)–(3.29), one obtains the final form of the post-Newtonian field equations:

$$\Box\varphi = -4\pi G\rho^*\,, \tag{3.30}$$

$$\begin{aligned}\Box\left\{h_{00} + \epsilon^2\left[l_{00} + \frac{h_{00}^2}{2} + 2(\beta-1)\varphi^2\right]\right\} &= \\ -\,8\pi G\rho^*\left\{1 + \epsilon^2\left[\left(\gamma+\frac{1}{2}\right)v^2 + \Pi + \gamma\frac{\pi^{kk}}{\rho^*} - \frac{h}{6} - (2\beta-\gamma-1)\varphi\right]\right\}& \\ +\,\epsilon^2 h_{\langle ij\rangle}h_{00,ij}\,,&\end{aligned} \tag{3.31}$$

$$\Box h_{0i} = 8\pi G(1+\gamma)\rho^* v^i\,, \tag{3.32}$$

$$\Box h_{ij} = -8\pi G\gamma\rho^*\delta_{ij}\,, \tag{3.33}$$

where $\Box \equiv \eta^{\mu\nu}\partial_\mu\partial_\nu$ is the D'Alembert (wave) operator of the Minkowski spacetime, $h \equiv h_{ii} = \delta^{ij}h_{ij}$, and the angular brackets around tensor indices in $h_{\langle ij\rangle} \equiv h_{ij} - \delta_{ij}h/3$, denote the symmetric trace-free (STF) part of the spatial components of the metric tensor (the STF tensors which are explained in Chapter 5 of this book). They are also thoroughly discussed by Thorne [146], Blanchet and Damour [11] and in a review article by Poisson et al. [127]. Equations (3.30)–(3.33) are valid in the class of coordinates defined by the gauge condition (3.22).

3.4 Conformal harmonic gauge

Let us rewrite the Nutku gauge condition (3.22) as follows:

$$g^{\mu\nu}\Gamma^{\alpha}_{\mu\nu} = \left(\ln\frac{\phi}{\phi_0}\right)^{,\alpha} . \tag{3.34}$$

It reveals that the Laplace–Beltrami operator (2.3) depends on the scalar field,

$$\Box_g \equiv g^{\mu\nu}\left(\frac{\partial^2}{\partial x^{\mu}\partial x^{\nu}} - \frac{1}{\phi}\frac{\partial\phi}{\partial x^{\mu}}\frac{\partial}{\partial x^{\nu}}\right) . \tag{3.35}$$

Any function $F(x^{\alpha})$ satisfying the homogeneous Laplace–Beltrami equation, $\Box_g F(x^{\alpha}) = 0$, is called harmonic. The gauge condition (3.22) assumes that $\Box_g x^{\alpha} = -(\ln\phi)^{,\alpha} \neq 0$, that is the coordinate x^{α} is not a harmonic function on spacetime manifold in the Jordan–Fierz frame metric. Nonetheless, such nonharmonic coordinates are more convenient in the scalar-tensor theory of gravity because they allow us to eliminate more spurious terms from the field equations than the harmonic gauge condition does. We call the class of coordinates singled out by the Nutku gauge conditions (3.22), conformal harmonic coordinates [94]. This is because these coordinates are harmonic functions in the conformal Einstein frame metric $\tilde{g}_{\alpha\beta}$.

The conformal harmonic coordinates have many properties similar to the harmonic coordinates in general relativity. The choice of the conformal harmonic coordinates for constructing the theory of motion of extended bodies is justified by the following three factors: (1) the conformal harmonic coordinates approach harmonic coordinates in general relativity when the scalar field $\phi \to \phi_0$ so that $\beta = \gamma = 1$, (2) the conformal harmonic coordinates are natural for scalar-tensor parametrization of equations used in resolutions of the IAU 2000 General Assembly [137] on relativistic reference frames, and (3) the gauge condition (3.22) significantly simplifies the post-Newtonian field equations, thus, facilitating their solution. Harmonic coordinates were used in [90] for construction of post-Newtonian reference frames in PPN formalism. Pitfalls associated with this choice have been analyzed in [93, 94].

Gauge condition (3.34) does not fix the conformal harmonic coordinates uniquely. Let us change the coordinates

$$x^{\alpha} \mapsto w^{\alpha} = w^{\alpha}(x^{\alpha}) , \tag{3.36}$$

but keep the Nutku gauge condition (3.34) the same. Simple calculation shows that under such condition the new coordinates w^{α} must satisfy a homogeneous wave equation

$$g^{\mu\nu}(x^{\beta})\frac{\partial^2 w^{\alpha}}{\partial x^{\mu}\partial x^{\nu}} = 0 , \tag{3.37}$$

which means that the new coordinates, w^{α}, are harmonic functions. Equation (3.37) describes the residual gauge freedom and has an infinite number of nontrivial solutions defining the entire set of the local coordinates w^{α} on spacetime manifold of the

N-body system. This residual gauge freedom in the scalar-tensor theory of gravity is the same as in the harmonic gauge in general relativity. We specify the set of the conformal harmonic coordinates for N-body system in the next section.

4 Parameterized post-Newtonian coordinates

Our task in this chapter is to derive fully *covariant* post-Newtonian equations of translational motion of extended bodies with accounting for all their multipole moments. Standard textbooks on the post-Newtonian celestial mechanics [19, 21, 32, 73, 138] derives the post-Newtonian equations of motion in a particular, usually harmonic, gauge which freezes the gauge-dependent modes and brings the equations to a form which is suitable for practical applications. As we explained above, the post-Newtonian equations admit a large freedom in making the gauge (coordinate) transformations on spacetime manifold [21, 61, 96, 124] as well as in the configuration space of orbital and other parameters characterizing the motion of bodies [64, 65]. Therefore, single terms taken in such post-Newtonian equations separately from the others, make no physical sense – they can be always changed (or even eliminated) by making coordinate transformations. Only after the equations are solved and their solutions are substituted to observables, we can unambiguously discuss gravitational physics because the observables are invariantly defined [20, 22, 93]. This subtle difference between coordinate and observable effects is not so explicit in the Newtonian theory but should be carefully taken into account in relativistic celestial mechanics to avoid misinterpretation of gravitational physics on spacetime manifold [92, 99]. Coordinate-independent form of equations of motion is important to facilitate this task.

Unfortunately, the road to success is paved with failure, and Dixon's straightforward approach to build covariant theory of motion of celestial bodies faces with obstacles and, in fact, is inapplicable for derivation of equations of motion of compact astrophysical objects like black holes where the only way to proceed is the method of asymptotic matching of vacuum solutions of the field equations in global and local coordinates [48, 49, 75, 79, 80]. The case of Dixon's extended bodies is manageable but the price to pay is to replace the interior structure of the bodies with a skeleton of the stress–energy–momentum tensor of matter which definition admits a large freedom in interpretation and is not unique. The only definite statement which could be made so far about the stress–energy–momentum tensor skeleton, is that it may exist (see [62, Appendix]). It is clear that this situation is unsatisfactory and the only remedy is to make use of the local coordinates to identify Dixon's multipole moments with the multipole moments of gravitational field which emerge unambiguously in the process of solution of the field equations.

One more problem that was not solved satisfactory in Dixon's covariant approach is the identification of the center of mass of the extended body and its world line.

Its solution is easily achieved in the coordinate-based approached amended with the method of asymptotic matching as will be demonstrated below. Nonetheless, the reader should understand that introduction of coordinates is just an auxiliary intermediate step in building the covariant theory of motion of celestial bodies as they are required for unique solution of the internal and external problems of celestial mechanics. The coordinates play a role of scaffolding which is necessary to single out the world line of the center of mass of each body, say B, and to separate a self-force from the external gravitational force exerted on the body B by other $N-1$ bodies of the N-body system. The coordinate scaffolding is removed as soon as the theory is completed in Section 8.

As we have learned above, N-body problem requires introduction of one global and N local coordinate charts – one for each body. Geometric attributes of the coordinate charts as well as their kinematic and dynamic properties are distinguished in scalar-tensor theory of gravity by specification of the metric tensor and the scalar field.

4.1 The global post-Newtonian coordinates

Boundary conditions and kinematic properties

We have assumed that N-body system is isolated and there is no matter outside of it. We have to decide which bodies in the system should be considered as the sources of the gravitational field. It is clear that the number N of such bodies depends on the accuracy of astronomical observations and on the precision of calculation of their celestial ephemeris (position and velocities). Since there are no external astronomical bodies outside the system, the spacetime can be considered as flat at infinity with the metric tensor, $g_{\alpha\beta}$, asymptotically approaching the Minkowski metric $\eta_{\alpha\beta} = \mathrm{diag}(-1, +1, +1, +1)$. We further assume that there are no singularities on the manifold like black holes, wormholes, etc., the bodies move slowly and gravitational field is weak everywhere.

These limitations allow us to cover the whole spacetime manifold with a single coordinate chart denoted as $x^\alpha = (x^0, x^i)$, where $x^0 = ct$ is the coordinate time and $x^i \equiv \boldsymbol{x}$ are spatial coordinates. The global coordinates are used for the description of orbital dynamics of the bodies with respect to the center of mass of the N-body system. The coordinate time and spatial coordinates have no immediate physical meaning in those domains of spacetime where gravitational field is not negligible. However, when one approaches to infinity the global coordinates approximates the inertial coordinates of observer in the Minkowski space. For this reason, one can think about the coordinate time t and the spatial coordinates x^i as the proper time and the proper distance measured by fictitious observers located at rest at infinity with respect to the center of mass of the system [73]. These coordinates have been used some researchers for invariant description of observable effects in astronomy [19, 85]. They are also useful for formulation of conformal infinity in isolated astronomical systems that provides a very powerful method within numerical relativity to study global problems such as gravitational wave propagation and detection [74].

Precise mathematical definition of the global coordinates can be given in terms of the metric tensor, which is a solution of the field equations (3.31)–(3.33) with boundary conditions imposed at infinity. To formulate the boundary conditions, we introduce the metric perturbation

$$h_{\alpha\beta}(t,\boldsymbol{x}) \equiv g_{\alpha\beta}(t,\boldsymbol{x}) - \eta_{\alpha\beta}\,, \tag{4.1}$$

where $h_{\alpha\beta}$ is the full post-Newtonian series defined in (3.4). The very existence of the global coordinates matching asymptotically with the inertial coordinates of the Minkowski spacetime, presumes that the products, $rh_{\alpha\beta}$ and $r^2h_{\alpha\beta,\gamma}$, where $r = |\boldsymbol{x}|$, are bounded at spatial infinity while at null past infinity

$$\lim_{\substack{r\to\infty\\ t+r/c=\text{const.}}} h_{\alpha\beta}(t,\boldsymbol{x}) = 0\,. \tag{4.2}$$

Additional boundary condition must be imposed on the first derivatives of the metric tensor to exclude nonphysical radiative solutions associated with gravitational waves incoming to N-body system [73]. This condition is imposed because there are no sources of gravitational field outside of the N-body system. It is formulated as follows [34, 73]:

$$\lim_{\substack{r\to\infty\\ t+r/c=\text{const.}}} \left[(rh_{\alpha\beta})_{,r} + (rh_{\alpha\beta})_{,0}\right] = 0\,, \tag{4.3}$$

where the comma denotes a partial derivative with respect to the radial, r, and time, $x^0 = ct$, coordinates. Though, the first post-Newtonian approximation does not include gravitational waves, the boundary condition (4.3) tells us to choose the retarded solution of the field equation (3.31)–(3.33).

Similar “no-incoming-radiation” conditions are imposed on the perturbation, φ, of the scalar field defined in (3.28),

$$\lim_{\substack{r\to\infty\\ t+r/c=\text{const.}}} \varphi(t,\boldsymbol{x}) = 0\,, \tag{4.4}$$

$$\lim_{\substack{r\to\infty\\ t+r/c=\text{const.}}} \left[(r\varphi)_{,r} + (r\varphi)_{,0}\right] = 0\,. \tag{4.5}$$

The origin of the global coordinates coincides with the center of mass of N-body system at any instant of time. This condition can be satisfied after choosing a suitable definition of the post-Newtonian dipole moment, $\mathbb{D}^i$, of N-body system and equating its numerical value to zero along with its first and second time derivatives. This requirement can be fulfilled, at least in the first post-Newtonian approximation, because of the law of conservation of the linear momentum, $\mathbb{P}^i$, of the system [19, 21, 93]. The law of conservation of the angular momentum of N-body system allows us to make the spatial axes of the global coordinates nonrotating in space either in kinematic or dynamic sense [24, 25]. Coordinates are called kinematically nonrotating if their spatial orientation does not change with respect to the Minkowskian coordinates at infinity as time goes on [104, 105]. Dynamically nonrotating coordinates are defined by the condition that equations of motion of test particles moving with respect to these coordinates do not have terms that can be interpreted as the Coriolis or centripetal forces [104]. This definition operates only with the local properties of the spacetime manifold. Because of the assumption that N-body system is isolated the global coordinate grid do not rotate in any sense about the coordinate origin.

Metric tensor and scalar field

The metric tensor $g_{\alpha\beta}(t, \boldsymbol{x})$ and the scalar field $\varphi(t, \boldsymbol{x})$ are obtained in the global coordinates by solving the field equations (3.30)–(3.33) with the boundary conditions (4.2)–(4.4). It yields [93, 94]

$$\varphi(t, \boldsymbol{x}) = U(t, \boldsymbol{x}) , \tag{4.6}$$

$$h_{00}(t, \boldsymbol{x}) = 2U(t, \boldsymbol{x}) , \tag{4.7}$$

$$l_{00}(t, \boldsymbol{x}) = 2\Psi(t, \boldsymbol{x}) - 2(\beta - 1)\varphi^2(t, \boldsymbol{x}) - 2U^2(t, \boldsymbol{x}) - \chi_{,tt}(t, \boldsymbol{x}) , \tag{4.8}$$

$$h_{0i}(t, \boldsymbol{x}) = -2(1 + \gamma)\, U_i(t, \boldsymbol{x}) , \tag{4.9}$$

$$h_{ij}(t, \boldsymbol{x}) = 2\gamma\delta_{ij}U(t, \boldsymbol{x}) , \tag{4.10}$$

where $\chi_{,tt} \equiv \partial^2\chi/\partial t^2$, the post-Newtonian potential

$$\Psi(t, \boldsymbol{x}) \equiv \left(\gamma + \frac{1}{2}\right)\Psi_1(t, \boldsymbol{x}) - \frac{1}{6}\Psi_2(t, \boldsymbol{x}) + (1 + \gamma - 2\beta)\Psi_3(t, \boldsymbol{x}) + \Psi_4(t, \boldsymbol{x}) + \gamma\Psi_5(t, \boldsymbol{x}) , \tag{4.11}$$

and parameters γ and β have been defined in (3.26) and (3.27), respectively.

Gravitational potentials U, U^i, χ, $\Psi = \sum_{k=1}^{5}\Psi_k$ are linear combinations of the respective gravitational potentials of the bodies of N-body system,

$$U = \sum_A U_A, \qquad U_i = \sum_A U_A^i, \qquad \Psi_k = \sum_A \Psi_{Ak}, \qquad \chi = \sum_A \chi_A , \tag{4.12}$$

where the summation index $A = 1, 2, \ldots, N$ numerates the bodies of the astronomical system.

Each gravitational potential of body A is defined as an integral over the spatial volume V_A occupied by matter of this body,

$$U_A(t, \boldsymbol{x}) = G\int\limits_{V_A} \frac{\rho^*(t, \boldsymbol{x}')}{|\boldsymbol{x} - \boldsymbol{x}'|} d^3x' , \tag{4.13}$$

$$U_A^i(t, \boldsymbol{x}) = G\int\limits_{V_A} \frac{\rho^*(t, \boldsymbol{x}')v^i(t, \boldsymbol{x}')}{|\boldsymbol{x} - \boldsymbol{x}'|} d^3x' , \tag{4.14}$$

$$\chi_A(t, \boldsymbol{x}) = -G \int\limits_{V_A} \rho^*(t, \boldsymbol{x}')|\boldsymbol{x} - \boldsymbol{x}'| d^3x' , \tag{4.15}$$

$$\Psi_{A1}(t, \boldsymbol{x}) = G \int\limits_{V_A} \frac{\rho^*(t, \boldsymbol{x}')v^2(t, \boldsymbol{x}')}{|\boldsymbol{x} - \boldsymbol{x}'|} d^3x' , \tag{4.16}$$

$$\Psi_{A2}(t, \boldsymbol{x}) = G \int\limits_{V_A} \frac{\rho^*(t, \boldsymbol{x}')h(t, \boldsymbol{x}')}{|\boldsymbol{x} - \boldsymbol{x}'|} d^3x' , \tag{4.17}$$

$$\Psi_{A3}(t, \boldsymbol{x}) = G \int\limits_{V_A} \frac{\rho^*(t, \boldsymbol{x}')\varphi(t, \boldsymbol{x}')}{|\boldsymbol{x} - \boldsymbol{x}'|} d^3x' , \tag{4.18}$$

$$\Psi_{A4}(t, \boldsymbol{x}) = G \int\limits_{V_A} \frac{\rho^*(t, \boldsymbol{x}')\Pi(t, \boldsymbol{x}')}{|\boldsymbol{x} - \boldsymbol{x}'|} d^3x' , \tag{4.19}$$

$$\Psi_{A5}(t, \boldsymbol{x}) = G \int\limits_{V_A} \frac{\pi^{kk}(t, \boldsymbol{x}')}{|\boldsymbol{x} - \boldsymbol{x}'|} d^3x' , \tag{4.20}$$

where

$$h(t, \boldsymbol{x}) \equiv h_{ii}(t, \boldsymbol{x}) = 6U(t, \boldsymbol{x}) . \tag{4.21}$$

Potential χ is determined as a particular solution of the inhomogeneous Poisson equation

$$\triangle\chi(t, \boldsymbol{x}) = -2U(t, \boldsymbol{x}) \tag{4.22}$$

with the right-hand side defined in a whole space, $\triangle \equiv \delta^{ij}\partial^2/\partial x^i \partial x^j$. Nevertheless, its solution given by equation (4.15), has a compact support inside the volumes of the bodies of N-body system [73, 153]. It is worthwhile to emphasize that all integrals defining the metric tensor in the global coordinates are taken over the hypersurface of constant coordinate time t. Post-Newtonian spacetime transformations change the time hypersurface, hence, transforming the corresponding integral.

Notice that the Newtonian gravitational potential $U(t, \boldsymbol{x})$ has a double camouflage in scalar-tensor theory of gravity. It appears in the solution of the field equations, first, in (4.6) for the scalar field perturbation ϕ, and second, in (4.7) and (4.10) for the perturbation of the metric tensor components h_{00} and h_{ij}. It would be wrong, however, to identify the metric tensor components with the scalar field [like $h_{00} = 2\phi$] in the scalar-tensor theory of gravity – they are equal only in the Newtonian approximation, and only in the global coordinates x^α. In all other circumstances the metric tensor must be carefully distinguished from the scalar field to prevent incorrect formulation of post-Newtonian transformation of potential Ψ which is a functional of both, the metric tensor and the scalar field. This is because the scalar field and the time–time component of the metric tensor transform differently and cannot be used to substitute one another. This important point is not emphasized in PPN formalism [153, 155] and has been overlooked, for example, in [90]. We discuss this issue in more detail in [94, Appendix C].

4.2 The local post-Newtonian coordinates

Boundary conditions and dynamic properties

We denote the local coordinates by $w^\alpha = (w^0, w^i) = (cu, w^i)$ where u stands for the local coordinate time. A local coordinate chart is constructed in a close proximity to the world line $\mathcal{Z}$ of the center of mass of a self-gravitating body in the N-body system. There are N local coordinates – one for each body. The local coordinates are used to describe internal motion of matter inside the body, to define its center of mass, linear momentum, spin, as well as multipolar fine structure of the body's gravitational field. In practical applications, the local coordinates are used, for example, in geodynamics and satellite geodesy to determine precession, nutation and Earth Orientation Parameters (EOP) as well as orbital motion of Moon and satellites, and to build the global positioning system (GPS).

The metric in the local coordinates is approximately Minkowskian inside the body. This is because we have assumed the gravitational field is weak everywhere. In case when the body is a compact relativistic object, the local metric in the close proximity to the body must be approximated by an exact solution of the scalar-tensor theory corresponding to the strong gravitational field like the Schwarzschild solution in general relativity. This case was treated, for example, in [34, 48, 49, 147] but we shall not dwell here upon further details as it will take us too far aside.

The local coordinates do not cover the entire spacetime manifold. Therefore, the metric in the local coordinates is not asymptotically approaching the Minkowski metric as the radial distance from the body is increasing. The local metric diverges at infinity. This is because the local metric is a solution of the field equations (3.31)–(3.33) in the local coordinates, w^α, which must smoothly match the global solution of the same equations in the global coordinates x^α. The global solution includes the contribution of gravitational field of other (external) bodies of the N-body system which reveals locally as a tidal force. Newtonian gravitational potential of the tidal force is represented by a harmonic polynomial with respect to the local spatial coordinates with time-dependent, STF tensor coefficients $Q_L \equiv Q_{i_1 \ldots i_l}(u)$ which are called the external (or tidal) multipoles [46, 101, 146, 147] in contrast to the internal multipoles, $I^L \equiv I^{i_1 \ldots i_l}(u)$ which characterize the structure of the gravitational field of the body under consideration. In the Newtonian approximation this polynomial is a solution of a homogeneous Laplace equation, $\Delta U^{(\mathrm{ext})} = 0$, which has a general polynomial solution

$$U^{(\mathrm{ext})} = Q + Q_i w^i + \sum_{l=2}^{\infty} \frac{1}{l!} Q_{\langle i_2 \ldots i_l \rangle} w^{\langle i_2 \ldots i_l \rangle} , \tag{4.23}$$

where the external multipoles $Q, Q_i, Q_{ij}, \ldots$, are functions of time, and the angular brackets around spatial indices indicate the symmetric and trace-free (STF) Cartesian tensor [146].

Since the potential $U^{(\text{ext})}$ enters the metric tensor the monopole, Q, and dipole, Q_i, external multipoles can be eliminated by rescaling of units of measurement and making a transformation to a freely falling frame. This effacing of the monopole and dipole moments from the external gravitational field in the post-Newtonian approximation is a consequence of Einstein's equivalence principle (EEP) [107, 115, 153]. In particular, EEP suggests that it is always possible to choose the local coordinates in such a way that the first derivatives of the metric tensor (the Christoffel symbols) vanish along a geodesic world line of a freely falling particle [115, 117]. In general relativity EEP is also valid for self-gravitating bodies [18, 47, 101, 116]. In the latter case it is called the strong principle of equivalence (SEP).

Besides the metric tensor the scalar-tensor theory of gravity has additional long-range gravitational field caused by the presence of scalar field that can not be eliminated by a coordinate transformation to a freely falling frame. This is because the scalar field is a single function which does not change its numerical value under coordinate transformations and cannot vanish in any coordinates had it been present, at least, in one. The scalar field couples with the intrinsic gravitational field of an extended body and affects its internal multipoles like mass, dipole moment, etc. This explains the mathematical reason behind the mechanism of violation of SEP in scalar-tensor theory of gravity discussed by Dicke [52, 54] and Nordtvedt [118, 153]. SEP is also violated in any alternative theory of gravity which violates the local Lorentz invariance [103, 113]. The SEP violation is called Nordtvedt's effect [153]. Its experimental testing is conducted with Lunar Laser Ranging (LLR) [51, 156, 157] and pulsar timing observations [44, 108, 152].

The origin of the local coordinates moves along some, yet unspecified, world line, $\mathcal{W}$, which will be determined later on as a result of matching of the mathematical solutions of the field equations obtained in the local and global coordinates. We demand that the origin of the local coordinates coincides with the center of mass of the body under consideration at any instant of time. This requires a precise post-Newtonian definition of the internal dipole moment, I^i, and the center of mass of the body beyond the Newtonian approximation. Mathematically inadequate or insufficiently precise definition of body's center of mass introduces fictitious inertial forces that will cause it to move with respect to the origin of the local coordinates. Because the scalar-tensor theory of gravity does not violate the law of conservation of the linear momentum [122, 153] this motion is spurious and must be removed. This problem was studied and resolved in [93, 94].

In order to keep the center of mass of the body at the origin of the local coordinates we have to take into account that the linear momentum of the body is subject to the force caused by the coupling of the internal multipole moments $I^{i_1 \ldots i_l}$ $(l \geq 2)$ of the body with the external tidal gravitational field of other bodies in the N-body system characterized by the external multipoles $Q_{i_1\dots i_l}$. This coupling exists already in the Newtonian approximation [62, 101, 147] and is shown on the right-hand side of Dixon's equation (1.13). If one assumes that the origin of the local coordinates moves along a world line $\mathcal{W}$ which does not coincide with world line $\mathcal{Z}$ of the body's center of mass, the coupling force makes the time derivative of the linear momentum of the body different from zero which means that the center of mass of the body moves with acceleration. In order to make the center of mass of the body be always located at the origin of the local coordinates we have to make the time derivative of the linear momentum nil. This can be achieved by choosing the external dipole moment Q_i in the homogeneous solution of the field equations in such a way that it compensates the dynamic force stemming from the coupling of the internal and external multipoles with $l \geq 2$ (see equation (5.26) for more details).

We postulate that the spatial axes of the local coordinates are not rotating dynamically about the origin. It means that the geodesic equations of motion of test particles in the local coordinates do not contain the Coriolis and centrifugal forces. However, the post-Newtonian nature of the gravitational interaction suggests that the spatial axes of the dynamically nonrotating local coordinates must slowly rotate (precess) in the kinematic sense with respect to the spatial axes of the global coordinates. This precessional motion of the spatial axes of the local coordinates has a pure geometric origin and follows from the condition of a smooth matching of the local and global coordinates on spacetime manifold [93]. The relativistic precession includes three terms that are called, respectively, de-Sitter (geodetic), Lense–Thirring (gravitomagnetic), and Thomas precession [115]. Exact formulation of this precession is given below in equation (5.25).

Metric tensor and scalar field

We are looking for solution of the field equations (3.30)–(3.33) inside a world tube covered by the local coordinates. This world tube spreads out from the body under consideration to another nearest body from the N-body system. Thus, the right-hand side of inhomogeneous equations (3.30)–(3.33) includes only the matter of the body. It does not mean that the other bodies are not included into these equations. They participate implicitly through the solution of the homogeneous equation. Indeed, solution of the field equations in the local coordinates is a linear combination of a particular solution of the inhomogeneous equation and a general solution of a homogeneous equation. In order to distinguish solutions in the local coordinates from the corresponding solutions of the field equations in the global coordinates, we put a hat over any function of the local coordinates. This is because one and the same mathematical function has different forms in different coordinates. For example, when we apply a coordinate transformation $x = x(w)$ to a scalar function $F(x)$, it becomes $\hat{F}(w) = F[x(w)]$. It is erroneous to write $F(w)$ instead of $\hat{F}(w)$ because $F(w) \neq F[x(w)]$ [3, 63, 69].

Accounting for this remark, the post-Newtonian solution of the scalar field equation (3.30) in the local coordinates is written as a sum of two terms

$$\hat{\varphi}(u, \boldsymbol{w}) = \hat{\varphi}^{(\mathrm{int})}(u, \boldsymbol{w}) + \hat{\varphi}^{(\mathrm{ext})}(u, \boldsymbol{w}) , \tag{4.24}$$

respectively, describing the contributions of the internal matter and external bodies. Perturbation of the metric tensor, $\hat{g}_{\mu\nu}(u, \boldsymbol{w})$, in the local coordinates is denoted by

$$\hat{h}_{\mu\nu}(u, \boldsymbol{w}) = \hat{g}_{\mu\nu}(u, \boldsymbol{w}) - \eta_{\mu\nu} . \tag{4.25}$$

Post-Newtonian solution of the field equations (3.31)–(3.33) in the local coordinates is given as a sum of three terms

$$\hat{h}_{\mu\nu}(u, \boldsymbol{w}) = \hat{h}^{(\mathrm{int})}_{\mu\nu}(u, \boldsymbol{w}) + \hat{h}^{(\mathrm{ext})}_{\mu\nu}(u, \boldsymbol{w}) + \hat{h}^{(\mathrm{mix})}_{\mu\nu}(u, \boldsymbol{w}) , \tag{4.26}$$

where $\hat{h}^{(\mathrm{int})}_{\mu\nu}$ describes the solution of the inhomogeneous equation generated by internal matter, $\hat{h}^{(\mathrm{ext})}_{\mu\nu}$ describes the solution of the homogeneous equation produced by external bodies, and $\hat{h}^{(\mathrm{mix})}_{\mu\nu}$ arises due to a nonlinearity of the field equation (3.31) for the metric tensor perturbation. In the first post-Newtonian approximation the mixed terms $\hat{h}^{(\mathrm{mix})}_{\mu\nu}$ appear only in $\hat{g}_{00}(u, \boldsymbol{w})$ component of the metric tensor.

Scalar Field: internal and external solutions

The internal, $\hat{\varphi}^{(\mathrm{int})}(u, \boldsymbol{w})$, and external, $\hat{\varphi}^{(\mathrm{ext})}(u, \boldsymbol{w})$, solutions for the scalar field have the following form:

$$\hat{\varphi}^{(\mathrm{int})}(u, \boldsymbol{w}) = \hat{U}_B(u, \boldsymbol{w}) , \tag{4.27}$$

$$\hat{\varphi}^{(\mathrm{ext})}(u, \boldsymbol{w}) = \sum_{l=0}^{\infty} \frac{1}{l!} P_L w^L . \tag{4.28}$$

Here, the internal solution $\hat{\varphi}^{(\mathrm{int})}(u, \boldsymbol{w})$ describes the scalar field which is generated by the matter of the body (index B) covered by the local coordinates. It is expressed in terms of the Newtonian gravitational potential $\hat{U}_B(u, \boldsymbol{w})$ that is defined in the next section by equation (4.34). The external solution, $\hat{\varphi}^{(\mathrm{ext})}(u, \boldsymbol{w})$, is given in the form of the multipolar expansion with respect to scalar external multipoles $P_L \equiv P_{\langle i_1 \dots i_l \rangle}(u)$ which are STF Cartesian tensors. These multipoles are functions of the local time u only. Finally, $w^L \equiv w^{\langle i_1 \dots i_l \rangle}$ is the STF harmonic polynomial made out of spatial local coordinates.

A subtle point of interpretation of relation between the gravitational potential U and perturbation of the scalar field φ should be discussed here. By definition, the scalar field φ is invariant under a coordinate transformation. On the other hand, the Newtonian potential U is not an independent scalar field because it appears in time–time component of the metric tensor, g_{00}, which transforms as a tensor. This remark indicates that the scalar field φ cannot be equal to the Newtonian potential U in *all* post-Newtonian approximations. However, the exact relation between φ and U is not required in this chapter as the scalar field directly perturbs the metric tensor only in the post-Newtonian terms and, thus, the approximate equations (4.6) and (4.27) are sufficient in our calculations. Nonetheless, we should not replace φ with U when doing coordinate transformations.

Metric tensor: internal solution

The boundary conditions imposed on the internal solution for the metric tensor in the local coordinates are identical with those given in equations (4.2) and (4.3). For this reason, the internal solution for the metric tensor has the same form as in the global coordinates, but all functions refers now only to the body B covered by the local coordinates. We obtain

$$\hat{h}_{00}^{(\text{int})}(u, \boldsymbol{w}) = 2\hat{U}_B(u, \boldsymbol{w}) \,, \tag{4.29}$$

$$\hat{l}_{00}^{(\text{int})}(u, \boldsymbol{w}) = 2\hat{\Psi}_B(u, \boldsymbol{w}) - 2(\beta - 1)\left[\hat{\varphi}^{(\text{int})}(u, \boldsymbol{w})\right]^2 - 2\hat{U}_B^2(u, \boldsymbol{w}) - \frac{\partial^2 \hat{\chi}_B(u, \boldsymbol{w})}{\partial u^2} \,, \tag{4.30}$$

$$\hat{h}_{0i}^{(\text{int})}(u, \boldsymbol{w}) = -2(1 + \gamma)\hat{U}_B^i(u, \boldsymbol{w}) \,, \tag{4.31}$$

$$\hat{h}_{ij}^{(\text{int})}(u, \boldsymbol{w}) = 2\gamma\delta_{ij}\hat{U}_B(u, \boldsymbol{w}) \,, \tag{4.32}$$

where

$$\hat{\Psi}_B(u, \boldsymbol{w}) = \left(\gamma + \frac{1}{2}\right)\hat{\Psi}_{B1}(u, \boldsymbol{w}) - \frac{1}{6}\hat{\Psi}_{B2}(u, \boldsymbol{w}) + (1 + \gamma - 2\beta)\hat{\Psi}_{B3}(u, \boldsymbol{w}) + \hat{\Psi}_{B4}(u, \boldsymbol{w}) + \gamma\hat{\Psi}_{B5}(u, \boldsymbol{w}) \,, \tag{4.33}$$

and the index B indicates the local body. Notice that we have not replaced the scalar field $\hat{\varphi}^{(\text{int})}(u, \boldsymbol{w})$ with the Newtonian potential $\hat{U}_B(u, \boldsymbol{w})$ in (4.30) to keep track of the scalar field contribution.

All these functions are defined as integrals over a volume V_B occupied by matter of the body B:

$$\hat{U}_B(u, \boldsymbol{w}) = G\int_{V_B} \frac{\rho^*(u, \boldsymbol{w}')}{|\boldsymbol{w} - \boldsymbol{w}'|} d^3w' \,, \tag{4.34}$$

$$\hat{U}_B^i(u, \boldsymbol{w}) = G\int_{V_B} \frac{\rho^*(u, \boldsymbol{w}')\nu^i(u, \boldsymbol{w}')}{|\boldsymbol{w} - \boldsymbol{w}'|} d^3w' \,, \tag{4.35}$$

$$\hat{\Psi}_{B1}(u, \boldsymbol{w}) = G\int_{V_B} \frac{\rho^*(u, \boldsymbol{w}')\nu^2(u, \boldsymbol{w}')}{|\boldsymbol{w} - \boldsymbol{w}'|} d^3w' \,, \tag{4.36}$$

$$\hat{\Psi}_{B2}(u,\boldsymbol{w}) = G\int\limits_{V_B} \frac{\rho^*(u,\boldsymbol{w}')\hat{h}^{(\mathrm{int})}(u,\boldsymbol{w}')}{|\boldsymbol{w}-\boldsymbol{w}'|}d^3w' \,, \tag{4.37}$$

$$\hat{\Psi}_{B3}(u,\boldsymbol{w}) = G\int\limits_{V_B} \frac{\rho^*(u,\boldsymbol{w}')\hat{\varphi}^{(\mathrm{int})}(u,\boldsymbol{w}')}{|\boldsymbol{w}-\boldsymbol{w}'|}d^3w' \,, \tag{4.38}$$

$$\hat{\Psi}_{B4}(u,\boldsymbol{w}) = G\int\limits_{V_B} \frac{\rho^*(u,\boldsymbol{w}')\Pi(u,\boldsymbol{w}')}{|\boldsymbol{w}-\boldsymbol{w}'|}d^3w' \,, \tag{4.39}$$

$$\hat{\Psi}_{B5}(u,\boldsymbol{w}) = G\int\limits_{V_B} \frac{\pi^{kk}(u,\boldsymbol{w}')}{|\boldsymbol{w}-\boldsymbol{w}'|}d^3w' \,, \tag{4.40}$$

$$\hat{\chi}_B(u,\boldsymbol{w}) = -G\int\limits_{V_B} \rho^*(u,\boldsymbol{w}')|\boldsymbol{w}-\boldsymbol{w}'|d^3w' \,, \tag{4.41}$$

where $\hat{h}^{(\mathrm{int})} = \hat{h}^{(\mathrm{int})}_{ii}$, $v^i = dw^i/du$ is a coordinate velocity of the body's matter element with respect to the origin of the local coordinates w^α. It is worth to emphasize that the integrals given in this section are taken over a hypersurface of the coordinate time u. It does not coincide with the hypersurface of the coordinate time t, which is used in the integrals (4.13)–(4.20) defining gravitational potentials in the global coordinates x^α. This remark is important for making the post-Newtonian transformations of the potentials from one frame to another as it requires to use the Lie transport of the integrand function between the hypersurfaces [93, Section 5.2.3].

The internal terms of the metric tensor in the local coordinates given by equations (4.29)–(4.32) are subject to the gauge condition (3.22). We shall see later that the external part of the local metric satisfy this condition by construction. The internal part of the metric must satisfy the gauge condition as well. It yields

$$\frac{\partial \hat{U}_B(u,\boldsymbol{w})}{\partial u} + \frac{\partial \hat{U}^i_B(u,\boldsymbol{w})}{\partial w^i} = \mathcal{O}\left(\epsilon^2\right) \,, \tag{4.42}$$

which relates the potentials of the internal solution of the inhomogeneous field equations in the first post-Newtonian approximations. We note that equation (4.42) is validated by the equation of continuity (3.19).

Metric tensor: external solution

Solution of the homogeneous field equations for the metric tensor in the local coordinates yields tidal gravitational field of the bodies of the N-body system in terms of the external STF multipoles [93, 94]. It also includes the inertial force exerted on the center of mass of the body B covered by the local coordinates, caused by acceleration of the body with respect to a time-like geodesic world line of a freely falling particle. The external solution must converge to a finite value at the origin of the local coordinates, that is at the point $\boldsymbol{w} = 0$. The external solution should also match the tidal gravitational field of other bodies far away from the origin of the local coordinates. These boundary conditions are typical for constructing local coordinates in curved spacetime [24, 46, 93, 94, 117, 143, 147].

Explicit form of the most general external solution for the linearized metric tensor perturbation in the local coordinates is given by [93, 94]

$$\hat{h}^{(\text{ext})}_{00}(u,\boldsymbol{w}) = 2\sum_{l=0}^{\infty}\frac{1}{l!}Q_L w^{\langle L\rangle} + C_pC_q\left(\delta^{pq}w^2 - w^p w^q\right), \tag{4.43}$$

$$\begin{aligned}\hat{h}^{(\text{ext})}_{0i}(u,\boldsymbol{w}) &= \varepsilon_{ipq}C_p w^q + \sum_{l=2}^{\infty}\frac{1}{l!}\varepsilon_{ipq}C_{pL-1}w^{\langle qL-1\rangle} + \sum_{l=0}^{\infty}\frac{1}{l!}Z_{iL}w^{\langle L\rangle}\\ &+\sum_{l=0}^{\infty}\frac{1}{l!}S_L w^{\langle iL\rangle},\end{aligned} \tag{4.44}$$

$$\begin{aligned}\hat{h}^{(\text{ext})}_{ij}(u,\boldsymbol{w}) &= 2\delta_{ij}\sum_{l=0}^{\infty}\frac{1}{l!}A_L w^{\langle L\rangle} + \sum_{l=0}^{\infty}\frac{1}{l!}B_L w^{\langle ijL\rangle} + \frac{1}{3}(\delta_{ij}C_pC^p - C_iC_j)w^2\\ &+\sum_{l=1}^{\infty}\frac{1}{l!}\left(D_{iL-1}w^{\langle jL-1\rangle} + \varepsilon_{ipq}E_{pL-1}w^{\langle jqL-1\rangle}\right)^{\text{sym}(ij)}\\ &+\sum_{l=2}^{\infty}\frac{1}{l!}\left(F_{ijL-2}w^{\langle L-2\rangle} + \varepsilon_{pq(i}G_{j)pL-2}w^{\langle qL-2\rangle}\right),\end{aligned} \tag{4.45}$$

where the external multipoles $Q_L = Q_L(u), C_L = C_L(u), S_L = S_L(u)$, etc. are functions of time u, the multi-index notation $L \equiv i_1i_2\ldots i_l$, $L-1 \equiv i_1i_2\ldots i_{l-1}$, $L-2 \equiv i_1i_2\ldots i_{l-2}$, the symbol sym$(ij)$ and the round brackets around tensor indices denote symmetrization with respect to the indices, for instance, $[T_{ijL}]^{\text{sym}(ij)} \equiv T_{(ij)L} = (1/2)[T_{ijL} + T_{jiL}]$, and $T_{(ij)} = (1/2)(T_{ij} + T_{ji})$, repeated indices imply Einstein's summation rule, spatial indices are raised and lowered with the help of the Kronecker symbol δ_{ij}, the angular brackets around indices denote symmetric and trace-free (STF) Cartesian tensors which vanish if any two indices are contracted [11, 146]. Notice that because the spatial indices are raised and lowered with δ^{ij} the position of the spatial index (subscript or superscript) does not make any difference in the local coordinates.

Vector C_i in (4.43)–(4.45) is the angular velocity of the kinematic rotation of the spatial axes of the local coordinates with respect to the global coordinates. We keep its contribution only up to quadratic terms which is sufficient for our goal. We also assume that C^i has the post-Newtonian order of magnitude being comparable with the rate of the geodetic precession. The external solution contains monopole terms, $Q = Q(u)$ and $A = A(u)$, entering $\hat{h}^{(\text{ext})}_{00}$ and $\hat{h}^{(\text{ext})}_{ij}$, respectively. Function Q defines the unit of measurement of the coordinate time u at the origin of the local coordinates, and function A defines the unit of measurement of spatial distances on the hypersuface of constant time u. Physical meaning of the external multipoles Q_L can be understood if one writes down the Newtonian equation of motion of freely falling test particle [93, 94]. It turns out that Q_i is an acceleration of the local frame with respect to the geodesic world line of the particle, and Q_L ($l \geq 2$) describes the tidal force with multipolarity l (quadrupole, octupole, etc.).

A set of 11 external STF multipole moments A_L, B_L, C_L, D_L, E_L, F_L, G_L, P_L, Q_L, S_L, Z_L is defined on the world line of the origin of the local coordinates so that these multipoles are functions of the local coordinate time u only. Furthermore, the external multipole moments are symmetric and trace-free (STF) Cartesian tensors with respect to any pair of free indices [93, 94]. Imposing four gauge conditions (3.24) and (3.25) on the external solution of metric tensor (4.43)–(4.45) reveals that only 7 out of 11 external multipole moments are algebraically independent. Indeed, the gauge conditions can be satisfied if, and only if, the external multipole moments B_L, E_L, S_L, D_L are eliminated from the local metric [93, 94]. The remaining multipoles: A_L, C_L, F_L, G_L, P_L, Q_L, Z_L can be constrained by making use of the residual gauge freedom allowed by differential equation (3.37), which excludes four other multipoles A_L, F_L, G_L, Z_L [93, 94]. We conclude that only three families of external moments P_L, Q_L, and C_L have real physical meaning reflecting one degree of freedom for the scalar field and two degrees of freedom for tidal gravitational field associated with the metric tensor. It is convenient to keep some gauge freedom by not fixing the external multipoles Z_L with $l \geq 2$. They can be chosen later to simplify equations of motion.

After fixing the gauge freedom as indicated above, the external metric tensor assumes in the local coordinates the following form:

$$\hat{h}^{(\text{ext})}_{00}(u, \boldsymbol{w}) = 2Q + 2\sum_{l=1}^{\infty} \frac{1}{l!} Q_L w^L , \tag{4.46}$$

$$\begin{aligned}\hat{h}^{(\text{ext})}_{0i}(u, \boldsymbol{w}) = &\left(\dot{A} + \frac{\dot{Q}}{3} + \frac{1-\gamma}{3}\dot{P} \right) w^i + \sum_{l=1}^{\infty} \frac{1}{l!} \varepsilon_{ipq} C_{pL-1} w^{\langle qL-1 \rangle} \\ &+ 2\sum_{l=1}^{\infty} \frac{2l+1}{(2l+3)(l+1)!} \left[2\dot{Q}_L + (\gamma - 1)\dot{P}_L \right] w^{\langle iL \rangle} \\ &+ \sum_{l=1}^{\infty} \frac{1}{l!} Z_{iL} w^L ,\end{aligned} \tag{4.47}$$

$$\hat{h}^{(\text{ext})}_{ij}(u, \boldsymbol{w}) = 2\delta_{ij} \left\{ A + \sum_{l=1}^{\infty} \frac{1}{l!} [Q_L + (\gamma - 1)P_L] w^L \right\} , \tag{4.48}$$

where the dot above the external multipoles denotes a derivative with respect to time u.

The nonlinear part $\hat{l}_{00}$ of the perturbation of the external metric tensor is determined from (3.31) up to a solution of the homogeneous Laplace equation which is absorbed to the definition of the external multipoles Q^L and is not shown for this reason.

The particular solution of (3.31) is [94]

$$\hat{l}^{(\text{ext})}_{00}(u,\boldsymbol{w}) = -2\left(\sum_{l=1}^{\infty}\frac{1}{l!}Q_L w^L\right)^2 - 2(\beta-1)\left(\sum_{l=1}^{\infty}\frac{1}{l!}P_L w^L\right)^2 + \sum_{l=0}^{\infty}\frac{1}{(2l+3)l!}\ddot{Q}_L w^L w^2, \tag{4.49}$$

where a double dot above Q_L denotes a second derivative with respect to time u. We have excluded the monopole, Q, and dipole, Q^i terms from the nonlinear part of the local metric tensor because they are absorbed to (yet unknown) Q and Q^i in the linear part of the metric (4.46). We could also decompose the product of the two sums in (4.49) into irreducible parts and absorb the STF piece of this decomposition to multipoles Q_L ($l \geq 2$). This procedure is done later in the process of matching the internal and external solutions.

The mixed terms

The mixed terms entering the metric tensor in the local coordinates are given as a particular solution of the inhomogeneous field equation (3.31) with the right-hand side taken as a product of the internal and external solutions found on the previous step of the post-Newtonian iterations. Solving (3.31) yields

$$\begin{aligned}\hat{l}^{(\text{mix})}_{00}(u,\boldsymbol{w}) &= -2\hat{U}_B(u,\boldsymbol{w})\left[A + (2\beta-\gamma-1)P\right] \\ &- 4\hat{U}_B(u,\boldsymbol{w})\sum_{l=0}^{\infty}\frac{1}{l!}\left[Q_L + (\beta-1)P_L\right]w^L \\ &- 2G\sum_{l=1}^{\infty}\frac{1}{l!}\left[Q_L + 2(\beta-1)P_L\right]\int_{V_B}\frac{\rho^*(u,\boldsymbol{w}')w'^L}{|\boldsymbol{w}-\boldsymbol{w}'|}d^3w',\end{aligned} \tag{4.50}$$

where V_B denotes a volume of the body B. This completes derivation of the metric tensor in the local coordinates.

Body's internal multipoles in the local coordinates

Had we ignored the tidal gravitational field of all bodies of the N-body system but the body B it would be considered as an isolated object and its gravitational field would be characterized by the sum of the internal gravitational potentials defined in (4.34)–(4.41). Multipolar decomposition of the metric tensor of an isolated gravitating system residing in asymptotically flat spacetime has been well understood and can be found in papers [12, 42, 146]. This technique has been extended to the case of a self-gravitating system embedded to a curved background spacetime in [46, 147] and [94].

The body B interacts gravitationally with other bodies of the N-body system and this interaction cannot be ignored. It brings about the mixed terms (4.50) to the metric tensor in the local coordinates which contribute to the numerical values of the body B internal moments in the multipolar decomposition of the local metric tensor. The presence of the mixed terms raised a question about precise definition of the internal multipole moments of the body B that is to be used in deriving translational and rotational equations of motion [147]. There are two options – either to include the contribution of the mixed terms to the internal multipole moments or to exclude it. Both options look theoretically possible but one of them is actually more preferred. Straightforward calculations [93, 94] proved that equations of motion of extended bodies can be significantly simplified if the mixed terms are included to the definition of the internal multipole moments. In this way, the mixed terms do not appear explicitly in the equations of motion confirming the principle of effacing of internal structure of the bodies in scalar-tensor theory of gravity.

There are three families of the internal mass-type multipole moments in the scalar-tensor theory of gravity – active, conformal, and scalar multipoles [153]. The *active* STF mass-type multipoles are defined by equation [93, 94]

$$
\begin{aligned}
\mathcal{I}^{\langle L\rangle} = & \int\limits_{V_B} \sigma(u,\boldsymbol{w})w^{\langle L\rangle}d^3w + \frac{\epsilon^2}{2(2l+3)}\left[\frac{d^2}{du^2}\int\limits_{V_B}\sigma(u,\boldsymbol{w})w^{\langle L\rangle}w^2 d^3w \right.\\
& \left. - 4(1+\gamma)\frac{2l+1}{l+1}\frac{d}{du}\int\limits_{V_B}\sigma^i(u,\boldsymbol{w})w^{\langle iL\rangle}d^3w\right]\\
& - \epsilon^2\int\limits_{V_B} d^3w\,\sigma(u,\boldsymbol{w})\left\{A + (2\beta-\gamma-1)P\right.\\
& \left. + \sum_{k=1}^{\infty}\frac{1}{k!}\left[Q_K + 2(\beta-1)P_K\right]w^{\langle K\rangle}\right\}w^{\langle L\rangle}\,,
\end{aligned}
\tag{4.51}
$$

where V_B is the volume occupied by the matter of body B, the matter current density

$$\sigma^i(u,\boldsymbol{w}) = \rho^*(u,\boldsymbol{w})v^i(u,\boldsymbol{w})\,, \tag{4.52}$$

and the *active* mass density

$$
\begin{aligned}
\sigma(u,\boldsymbol{w}) = \rho^*(u,\boldsymbol{w})\left\{1 + \epsilon^2\left[\left(\gamma+\tfrac{1}{2}\right)v^2(u,\boldsymbol{w}) + \Pi(u,\boldsymbol{w})(2\beta-1)\hat{U}_B(u,\boldsymbol{w})\right]\right\}\\
+ \epsilon^2\gamma\pi^{kk}(u,\boldsymbol{w})\,.
\end{aligned}
\tag{4.53}
$$

The *conformal* STF mass-type multipoles of the local system are defined as follows [93, 94]:

$$I^{\langle L\rangle} = \int_{V_B} \varrho(u, \boldsymbol{w}) \left\{1 - \epsilon^2 \left[A + (1-\gamma)P + \sum_{k=1}^{\infty} \frac{1}{k!} Q_K w^{\langle K\rangle}\right]\right\} w^{\langle L\rangle}\, d^3w$$
$$+ \frac{\epsilon^2}{2(2l+3)} \left[\frac{d^2}{du^2} \int_{V_B} \varrho(u, \boldsymbol{w}) w^{\langle L\rangle} w^2\, d^3w\right.$$
$$\left. - \frac{8(2l+1)}{l+1} \frac{d}{du} \int_{V_B} \sigma^i(u, \boldsymbol{w}) w^{\langle iL\rangle}\, d^3w\right] , \tag{4.54}$$

with the *conformal* mass density of matter

$$\varrho = \rho^*(u, \boldsymbol{w}) \left[1 + \epsilon^2 \left(\frac{3}{2} v^2(u, \boldsymbol{w}) + \Pi(u, \boldsymbol{w}) - \hat{U}_B(u, \boldsymbol{w})\right)\right] + \epsilon^2 \pi^{kk}(u, \boldsymbol{w}) . \tag{4.55}$$

The *conformal* mass density does not depend on the PPN parameters β and γ as contrasted to the definition of the *active* mass density. The *scalar* field multipoles, $\bar{I}^L$, are not independent and are related to the *active* and *conformal* multipoles [93, 94]

$$\bar{I}^{\langle L\rangle} = 2\mathcal{I}^{\langle L\rangle} - (1+\gamma) I^{\langle L\rangle} . \tag{4.56}$$

Integration in (4.51) and (4.54) is performed over a hypersurface of constant coordinate time u.

In addition to the gravitational mass-type multipoles, $\mathcal{I}^L$ and I^L, there is a set of internal spin-type multipoles. In the post-Newtonian approximation they are defined by equation [12, 93, 94, 146]

$$\mathcal{S}^{\langle L\rangle} = \int_{V_B} \varepsilon^{pq<i_l} w^{L-1>p} \sigma^q(u, \boldsymbol{w}) d^3w , \tag{4.57}$$

where the matter current density σ^q has been defined in (4.52). All multipole moments are functions of time u only. They can be considered in the local coordinates as the STF Cartesian tensors attached to the world line $\mathcal{W}$ of the origin of the local coordinates.

5 Post-Newtonian coordinate transformations by asymptotic matching

5.1 General structure of the transformation

Post-Newtonian transformations between the global and local coordinate charts are derived by making use of the mathematical technique known as asymptotic matching of the post-Newtonian expansions of the scalar field and the metric tensor. This technique was originally proposed in general relativity by D'Eath [48, 49] as a tool for derivation of equations of motion of black holes [75, 79, 80]. It also proved out to be efficient in the post-Newtonian theory of reference frames in the solar system [5, 24, 46, 47, 94, 97, 98, 101]. Perturbations of the metric tensor $g_{\alpha\beta}$ and the scalar field φ are given on spacetime manifold as mathematical solutions of the field equations expressed in terms of various functions which exact form depends on the choice of coordinates. However, these solutions describe the same physical situation irrespectively of the choice of the coordinates which means that the solutions must match smoothly in the spacetime domain where the coordinate charts overlap. The matching relies upon the tensor transformation law applied to the post-Newtonian metric tensor and the scalar field. The matching domain is bounded by the radius of convergence of the post-Newtonian series in the local coordinates. After the matching is finished and the post-Newtonian transition functions establishing the mapping between coordinates, are found, the domain of validity of the local coordinates can be analytically extended to a larger radius of convergence if it is required by practical applications [93].

As we have imposed the conformal harmonic gauge condition (3.34), the transition functions of the post-Newtonian coordinate transformation are constrained and must comply with the homogeneous differential equation (3.37) describing the residual gauge freedom. Solution of this equation must be convergent at the origin of the local coordinates and consistent with the tidal gravitational field of the external solution of the local metric. Therefore, we are looking for the post-Newtonian coordinate transformation in the form of a series of harmonic polynomials expanded in powers of the spatial coordinates of the local coordinates. Because we are solving the homogeneous Laplace equation, the coefficients of the polynomials are STF Cartesian tensors defined on the world line $\mathcal{W}$ of the origin of the local coordinates [47, 101]. The transition functions are substituted to the equations matching the internal and external solutions of the field equations expressed in the global and local coordinates. Matching of the post-Newtonian expansions of the scalar field and the metric tensor allows us to fix all degrees of the gauge freedom in the final form of the post-Newtonian coordinate transformation except for the multipoles Q and A defining the units of measurement, and the external dipole moment Q_i which is not constrained by the matching conditions. Notice that we have partially used this gauge freedom in Section 4.2 to remove the nonphysical multipoles in the external solution for the local metric tensor.

The post-Newtonian transformation between coordinate times, t and u, describes the Lorentz (velocity-dependent) and Einstein (gravitational-field-dependent) time dilations associated with the different simultaneity of events in the two coordinate charts [5, 101]. It also includes an infinite series of polynomial terms [23, 24]. The post-Newtonian transformation between spatial coordinates, x^i and w^i, is a quadratic function of spatial coordinates. The linear part of the transformation includes the Lorentz and Einstein contractions of length as well as a matrix of rotation describing the post-Newtonian precession of the spatial axes of the local coordinates with respect to the global coordinates due to the translational and rotational motion of the bodies [35, 97]. The Lorentz length contraction takes into account the kinematic aspects of the post-Newtonian transformation and depends on the relative velocity of motion of the local coordinates with respect to the global coordinates. The Einstein (gravitational) length contraction accounts for static effects of the scalar field and the metric tensor [93, 94]. The quadratic part of the spatial transformation depends on the orbital acceleration of the local coordinates and accounts for the effects of the affine connection (the Christoffel symbols) of spacetime manifold.

Let us now discuss the mathematical structure of the post-Newtonian transformation between the local coordinates, $w^\alpha = (w^0, w^i) = (cu, \boldsymbol{w})$, and the global coordinates, $x^\alpha = (x^0, x^i) = (ct, \boldsymbol{x})$ in more detail. This coordinate transformation must be compatible with the weak-field and slow-motion approximation used in the post-Newtonian iterations. Hence, the coordinate transformation is given as a post-Newtonian series for time and spatial coordinates:

$$u = t + \epsilon^2 \xi^0(t, \boldsymbol{x}) \,, \tag{5.1}$$

$$w^i = R^i_B + \epsilon^2 \xi^i(t, \boldsymbol{x}) \,, \tag{5.2}$$

where ξ^0 and ξ^i are the post-Newtonian corrections to the Galilean transformation, $u = t$, $R^i_B = x^i - x^i_B(t)$, and $x^i_B(t)$ is the spatial position of the origin of the local coordinates expressed in terms of the global coordinates. We denote velocity and acceleration of the origin of the local coordinates as $v^i_B \equiv \dot{x}^i_B$ and $a^i_B \equiv \ddot{x}^i_B$, respectively, where here and everywhere else the dot above function is understood as a derivative with respect to time t. The world line $\mathcal{W}$ of the origin of the local coordinates is decoupled from the world line $\mathcal{Z}$ of the center of mass of the body. Nonetheless, the two world lines can be superposed and tied up by demanding conservation of the linear momentum of the body [93, 94]. It can be always achieved by introducing the local acceleration Q_i which compensates the tidal forces acting on the body. It makes the local coordinates noninertial. It is relevant to point out that Dixon's translational equations of motion (1.13) implicitly assume that the world line $\mathcal{W}$ of the origin of local coordinates is a time-like geodesic of an effective background spacetime manifold because the external dipole moment $Q_i \equiv 0$ so that only quadrupole, octupole and higher order external multipoles are present. Dixon [62] defined the body's center of mass in terms of the mass internal dipole m^α given in (1.5), and defined the world line $\mathcal{Z}$ of this center of mass by condition (1.6) that is equivalent to $m^\alpha = 0$. This condition makes the two world lines, $\mathcal{W}$ and $\mathcal{Z}$, be identical but it means that Dixon's definition of the center of mass does not coincide with the center of mass of the body defined by the condition that the dipole moment, I^α, in the multipolar expansion of gravitational field of the body vanishes, $I^\alpha = 0$. We discuss this issue in more detail in Section 6.3.

Matching equations for the scalar field, the metric tensor, and the Christoffel symbols are given by the general law of coordinate transformations of these geometric objects [63] [63]

$$\varphi(t, \boldsymbol{x}) = \hat{\varphi}(u, \boldsymbol{w}) , \tag{5.3}$$

$$g_{\mu\nu}(t, \boldsymbol{x}) = \hat{g}_{\alpha\beta}(u, \boldsymbol{w}) \frac{\partial w^\alpha}{\partial x^\mu} \frac{\partial w^\beta}{\partial x^\nu} , \tag{5.4}$$

$$\Gamma^\mu_{\alpha\beta}(t, \boldsymbol{x}) = \hat{\Gamma}^\nu_{\rho\sigma}(u, \boldsymbol{w}) \frac{\partial x^\mu}{\partial w^\nu} \frac{\partial w^\rho}{\partial x^\alpha} \frac{\partial w^\sigma}{\partial x^\beta} + \frac{\partial x^\mu}{\partial w^\nu} \frac{\partial^2 w^\nu}{\partial x^\alpha \partial x^\beta} , \tag{5.5}$$

where

$$\Gamma^\mu_{\alpha\beta}(t, \boldsymbol{x}) = \frac{1}{2} g^{\mu\nu} \left(\frac{\partial g_{\nu\alpha}}{\partial x^\beta} + \frac{\partial g_{\nu\beta}}{\partial x^\alpha} - \frac{\partial g_{\alpha\beta}}{\partial x^\nu} \right) , \tag{5.6}$$

$$\hat{\Gamma}^\mu_{\alpha\beta}(u, \boldsymbol{w}) = \frac{1}{2} \hat{g}^{\mu\nu} \left(\frac{\partial \hat{g}_{\nu\alpha}}{\partial w^\beta} + \frac{\partial \hat{g}_{\nu\beta}}{\partial w^\alpha} - \frac{\partial \hat{g}_{\alpha\beta}}{\partial w^\nu} \right) , \tag{5.7}$$

are the Christoffel symbols expressed in the global and local coordinates, respectively.

Matching equations (5.3)–(5.5) are valid in the domain covered simultaneously by the local and global coordinates both inside and outside of the extended body. The scalar field, the metric tensor, and their first derivatives are continuously differentiated functions in this domain. The matching equations are not identities which are automatically satisfied. We accept that functions on the left-hand side of these equations are known and given in Section 4.1 as integrals from matter variables (density, pressure, etc.) performed over volumes of the bodies of the N-body system on hypersurface of constant time t. The right-hand side of the matching equations contains the known integrals from matter variables of the body B explained in Section 4.2 and yet unknown functions which are the external multipoles, P^L, Q^L, C^L, etc., of the metric tensor in the local coordinate along with function ξ^α entering the post-Newtonian transformation (5.1) and (5.2). All these functions are determined by solving the matching equations. Matching also allows us to derive equation of motion of the world line of the origin of the local coordinates in terms of the global coordinates. Matching is necessary but not sufficient for derivation of equations of motion of the center of mass of the extended bodies of the N-body system. Additional procedure is required that will be explained later.

The starting point for solving matching equations is $\hat{g}_{0i}$ component of the metric tensor in the local coordinates. It does not contain the linear terms of the order of $O(\epsilon)$ because the local coordinates are not dynamically rotating and their time axis is orthogonal to space axes. It eliminates the angular and linear velocity terms in $\hat{g}_{0i}$ [46, 94]. This information is used in (5.4) which implies that function $\xi^0(t, \boldsymbol{x})$ in (5.1) must satisfy the following restriction:

$$\xi^0_{,i} = -v^i_B + \mathcal{O}\left(\epsilon^2\right) . \tag{5.8}$$

This is a partial differential equation which can be solved. It yields

$$\xi^0(t, \boldsymbol{x}) = -\mathcal{A}(t) - v^k_B R^k_B + \epsilon^2 \kappa(t, \boldsymbol{x}) + \mathcal{O}\left(\epsilon^4\right) , \tag{5.9}$$

where $\mathcal{A}(t)$ and $\kappa(t, \boldsymbol{x})$ are analytic but yet arbitrary functions. Notice that function $\mathcal{A}(t)$ depends only on time t while $\kappa(t, \boldsymbol{x})$ depends on both time and spatial coordinates.

We now use the differential equation (3.37) in order to impose further restrictions of the transition functions. We substitute ξ^0 from (5.9) to $w^0 = cu$ in (5.1) and use it along with w^i from (5.2) in equation (3.37) which is reduced to the inhomogeneous Poisson equations

$$\Delta\kappa(t, \boldsymbol{x}) = 3v_B^k a_B^k - \ddot{\mathcal{A}} - \dot{a}_B^k R_B^k + \mathcal{O}\left(\epsilon^2\right) , \tag{5.10}$$

$$\Delta\xi^i(t, \boldsymbol{x}) = -a_B^i + \mathcal{O}\left(\epsilon^2\right) , \tag{5.11}$$

where $\Delta \equiv \delta^{ij}\partial^2/\partial x^i \partial x^j$. General solution of these elliptic-type equations can be written in the form of the Taylor series in which coefficients are the irreducible STF Cartesian tensors. Furthermore, solution for functions $\kappa(t, \boldsymbol{x})$ and $\xi^i(t, \boldsymbol{x})$ in (5.10) and (5.11) consists of two parts – a fundamental solution of the homogeneous Laplace equation and a particular solution of the inhomogeneous Poisson equation. We discard the fundamental solution of the Laplace equation that diverges at the origin of the local coordinates, where $w^i = 0$. This is because we work with a continuous distribution of matter which has no singular points inside the bodies.

Integrating equations (5.10) and (5.11) results in

$$\kappa = \left(\frac{1}{2}v_B^k a_B^k - \frac{1}{6}\ddot{\mathcal{A}}\right) R_B^2 - \frac{1}{10}\dot{a}_B^k R_B^k R_B^2 + \Xi(t, \boldsymbol{x}) , \tag{5.12}$$

$$\xi^i = -\frac{1}{6}a_B^i R_B^2 + \Xi^i(t, \boldsymbol{x}) , \tag{5.13}$$

where functions Ξ and Ξ^i are the fundamental solutions of the homogeneous Laplace equation which are convergent at the origin of the local coordinates. These solutions can be written in the form of scalar and vector harmonic polynomials

$$\Xi(t, \boldsymbol{x}) = \sum_{l=0}^{\infty} \frac{1}{l!}\mathcal{B}^L R_B^{\langle L\rangle} , \tag{5.14}$$

$$\Xi^i(t, \boldsymbol{x}) = \sum_{l=1}^{\infty} \frac{1}{l!}\mathcal{D}^{iL} R_B^{\langle L\rangle} + \sum_{l=0}^{\infty} \frac{\varepsilon_{ipq}}{(l+1)!}\mathcal{F}^{pL} R_B^{\langle qL\rangle} + \sum_{l=0}^{\infty} \frac{1}{l!}\mathcal{E}^L R_B^{\langle iL\rangle} , \tag{5.15}$$

where coefficients $\mathcal{B}^L$, $\mathcal{D}^L$, $\mathcal{F}^L$, and $\mathcal{E}^L$ are STF Cartesian tensors [93, 94] which should not be confused with the external multipoles entering the local metric tensor. These coefficients are defined on the world line $\mathcal{W}$ of the origin of the local coordinates and depend only on the coordinate time t of the global coordinates taken on $\mathcal{W}$. Explicit form of the coefficients $\mathcal{B}^L$, $\mathcal{D}^L$, $\mathcal{F}^L$ is obtained in the process of matching after substituting (5.12) and (5.13) and corresponding expressions for the scalar field and the metric tensor to equations (5.3)–(5.5). More details on the process of solving the matching equations is given in textbook [93] and in our review paper [158]. The matching solution is given in the next section.

5.2 Matching solution

Post-Newtonian coordinate transformation

We match the local coordinates, w^α, built inside and near the extended body B, with the global coordinates, x^α. Parameterized post-Newtonian transformation between the local and global coordinates is given by two equations [93, 101],

$$u = t - \epsilon^2(\mathcal{A} + v_B^k R_B^k) + \epsilon^4\left[\mathcal{B} + \left(\frac{1}{3}v_B^k a_B^k - \frac{1}{6}\dot{\bar{U}}(\boldsymbol{x}_B) + \frac{1}{6}\dot{\mathcal{Q}}\right)R_B^2 - \frac{1}{10}\dot{a}_B^k R_B^k R_B^2 + \sum_{l=1}^{\infty}\frac{1}{l!}\mathcal{B}^L R_B^L\right] + \mathcal{O}\left(\epsilon^5\right), \tag{5.16}$$

$$w^i = R_B^i + \epsilon^2\left[\left(\frac{1}{2}v_B^i v_B^k + \delta^{ik}\gamma\bar{U}(\boldsymbol{x}_B) - \delta^{ik}A + F^{ik}\right)R_B^k + a_B^k R_B^i R_B^k - \frac{1}{2}a_B^i R_B^2\right] + \mathcal{O}\left(\epsilon^4\right), \tag{5.17}$$

where

$$R_B^i = x^i - x_B^i, \tag{5.18}$$

is the (global) coordinate distance taken on the hypersurface of constant time t between the point of matching, x^i, and the origin of the local coordinates, $x_B^i = x_B^i(t)$.

Functions $\mathcal{A}$ and $\mathcal{B}$ depend on the global coordinate time t and obey the ordinary differential equations,

$$\frac{d\mathcal{A}}{dt} = \frac{1}{2}v_B^2 + \bar{U}(\boldsymbol{x}_B) - \mathcal{Q}, \tag{5.19}$$

$$\frac{d\mathcal{B}}{dt} = -\frac{1}{8}v_B^4 - \left(\gamma + \frac{1}{2}\right)v_B^2\bar{U}(\boldsymbol{x}_B) + \frac{1}{2}\bar{U}^2(\boldsymbol{x}_B) + 2(1+\gamma)v_B^k\bar{U}^k(\boldsymbol{x}_B) - \bar{\Psi}(\boldsymbol{x}_B) + \frac{1}{2}\bar{\chi}_{,tt}(\boldsymbol{x}_B) + \mathcal{Q}\left[-\frac{1}{2}v_B^2 + \frac{1}{2}\mathcal{Q} - \bar{U}(\boldsymbol{x}_B)\right], \tag{5.20}$$

that describe the post-Newtonian transformation between the coordinate time u of the local coordinates and the coordinate time t of the global coordinates. The other functions entering (5.16) and (5.17) are defined by algebraic relations

$$\mathcal{B}^i = 2(1+\gamma)\bar{U}^i(\boldsymbol{x}_B) - (1+2\gamma)v_B^i\bar{U}(\boldsymbol{x}_B) - \frac{1}{2}v_B^i v_B^2 - \mathcal{Q}v_B^i, \tag{5.21}$$

$$\mathcal{B}^{ij} = Z^{ij} + 2(1+\gamma)\bar{U}^{\langle i,j\rangle}(\boldsymbol{x}_B) - 2(1+\gamma)v_B^{\langle i}\bar{U}^{,j\rangle}(\boldsymbol{x}_B) + 2a_B^{\langle i}a_B^{j\rangle}, \tag{5.22}$$

$$\mathcal{B}^{iL} = Z^{\langle iL\rangle} + 2(1+\gamma)\bar{U}^{\langle i,L\rangle}(\boldsymbol{x}_B) - 2(1+\gamma)v_B^{\langle i}\bar{U}^{,L\rangle}(\boldsymbol{x}_B) \quad (l \geq 2), \tag{5.23}$$

where a comma denotes a partial derivative with respect to spatial global coordinates taken as many times as the number of indices standing after it, and some residual gauge freedom parameterized by STF Cartesian tensors Z^L has been left out and is explicitly shown. Functions Z^L can be chosen, for example, to make the coefficients $\mathcal{B}^L$ ($l \geq 2$) nil but we do not make this choice and leave Z^L arbitrary.

The external (with respect to body B) potentials $\bar{U}$, $\bar{U}^i$, $\bar{\Psi}$, $\bar{\chi}$ are obtained by subtracting the local gravitational potentials of the body B from the total integrals taken over all bodies of the N-body system [94],

$$\bar{U}(\boldsymbol{x}_B) = \sum_{A \neq B} U_A(\boldsymbol{x}_B) \tag{5.24a}$$

$$\bar{U}^i(\boldsymbol{x}_B) = \sum_{A \neq B} U^i_A(\boldsymbol{x}_B)\,, \tag{5.24b}$$

$$\bar{\Psi}(\boldsymbol{x}_B) = \sum_{A \neq B} \Psi_A(\boldsymbol{x}_B) \tag{5.24c}$$

$$\bar{\chi}(\boldsymbol{x}_B) = \sum_{A \neq B} \chi_A(\boldsymbol{x}_B)\,, \tag{5.24d}$$

where the potentials U_A, U^i_A, Ψ_A, χ_A are given by integrals (4.13)–(4.20), respectively. Each of the potentials, $\bar{U}_A$, $\bar{U}^i_A$, and $\bar{\chi}_A$ is a linear functional of mass density ρ^* of body A alone. On the other hand, the potential $\bar{\Psi}_A$ is split in five parts two of which, Ψ_{A2} and Ψ_{A3}, given by (4.17) and (4.18), respectively, depend on the gravitational potential of external (with respect to body A) bodies which includes the potential of body B. The gravitational potential of body B must be left in the integrands of Ψ_{A2} and Ψ_{A3} as it describes the back action of the body B on the external gravitational field of body A. This subtle point in the correct formulation of the external gravitational field for a massive body in the N-body system, was pointed out by Fichtengoltz [70] and emphasized in [5, 85, 107].

The antisymmetric rotational matrix $F^{ik} \equiv \varepsilon^{ipk}\mathcal{F}^p$ is a solution of ordinary differential equation [94, 158]

$$\frac{dF^{ik}}{dt} = -2(1+\gamma)\bar{U}^{[i,k]}(\boldsymbol{x}_B) + (1+2\gamma)v_B^{[i}\bar{U}^{,k]}(\boldsymbol{x}_B) + v_B^{[i}Q^{k]}\,, \tag{5.25}$$

The first term on the right-hand side of (5.25) describes the Lense–Thirring (gravitomagnetic) precession, the second term describes the de-Sitter (geodetic) precession in the scalar-tensor theory of gravity, and the third term describes the Thomas precession depending on the local (nongeodesic) acceleration $Q^k = \delta^{kp}Q_p$ of the origin of the local coordinates with respect to a geodesic world line of a freely falling particle (see Section 6.4). In the scalar-tensor theory both the Lense–Thirring and the de-Sitter precession depend on the PPN parameter γ while the Thomas precession does not. The reason is that the Thomas precession is generically a special relativistic effect [115] that cannot depend on any particular choice of a specific gravitational theory. The presence of matrix F^{ik} in the spatial part of the post-Newtonian transformation (5.17) means that the spatial axes of the local coordinates has a kinematic rotation with respect to spatial axes of the global coordinates.

Bootstrap effect and self-forces

It is important to emphasize that all internal self-forces inside the bodies cancelled out in the process of matching of the external and internal solutions of the metric tensor. It means that the internal potentials characterizing the strength of gravitational and/or scalar field inside the bodies do not affect the relative motion of the bodies with respect to each other. This fact is an important ingredient of the post-Newtonian theory of motion of extended bodies telling us that there is no body's *bootstrap effect* (internal self-forces cancelled out) that is the third Newton's law is not violated in the post-Newtonian approximation[2]. The bootstrap effect is absent in conservative approximations of general relativity and scalar-tensor theory of gravity. However, in radiative approximations taking into account emission of gravitational waves the bootstrap effect may be present as the gravitational radiation–reaction force is basically a self-force – the retarded interaction of the body with its own radiative field [81, 82, 127] – which may lead to self-accelerated solution of the post-Newtonian equations of motion depending on higher order time derivatives. One of us (S.K.) studied the origin of the bootstrap effect with the help of delay equations that model field retardation effects and predict runaway modes [30]. It was shown that when retardation effects are small, the physically significant solutions belong to the so-called *slow manifold* of the system which is identified with the attractor in the state space of the delay equation. It was also demonstrated via an example that when retardation effects are no longer small, the motion could exhibit bifurcation phenomena that are not contained in the local equations of motion. These kind of effects are absent in the post-Newtonian approximation, and we do not discuss them in the rest of the chapter.

World line of the origin of the local coordinates

Matching the metric tensors in the local and global coordinates yields the equations of translational motion of the origin of the local coordinates, $x^i_B = x^i_B(t)$, with respect to the global coordinates [93, 94]. Namely,

$$\begin{aligned} a^i_B = {} & \bar{U}_{,i}(\boldsymbol{x}_B) - Q^i + \\ & \epsilon^2 \Big\{ \bar{\Psi}_{,i}(\boldsymbol{x}_B) - \frac{1}{2}\bar{\chi}_{,itt}(\boldsymbol{x}_B) + 2(\gamma+1)\dot{\bar{U}}^i(\boldsymbol{x}_B) - 2(\gamma+1)v^k_B \bar{U}^{k,i}(\boldsymbol{x}_B) \\ & - (2\gamma+1)v^i_B \dot{\bar{U}}(\boldsymbol{x}_B) - 2(\beta+\gamma)\bar{U}(\boldsymbol{x}_B)\bar{U}_{,i}(\boldsymbol{x}_B) + \gamma v^2_B \bar{U}_{,i}(\boldsymbol{x}_B) - v^i_B v^k_B \bar{U}_{,k}(\boldsymbol{x}_B) \\ & + \left(F^{ik} + \frac{1}{2}v^i_B v^k_B\right) Q_k + [A - 2Q + v^2_B + (\gamma+2)\bar{U}(\boldsymbol{x}_B)]Q_i \Big\} + \mathcal{O}\left(\epsilon^4\right), \end{aligned} \tag{5.26}$$

2 The term *bootstrap effect* is attributed to R. E. Raspe's story "The surprising adventures of Baron Münchausen," where the main character pulled himself along with his horse out of a swamp by his hair pigtail, thus, surpassing the third Newton's law in this tall-tale story.

where a dot above function denoted the total derivative with respect to time t, $a_B^i = d^2x_B^i/dt^2$ is a coordinate acceleration of the origin of the local coordinates, Q^i is a dipole moment ($l = 1$) of the external solution for the metric tensor in the local coordinates that appears in h_{00} component of the metric tensor perturbation (4.43), and comma denotes a partial derivative with respect to a spatial global coordinates. It should be noticed that the post-Newtonian terms in (5.26) depending on $\bar{U}_{,i}$ and Q_i can be always reshuffled by making use of the Newtonian approximation of this equation, $Q_i = \bar{U}_{,i} - a_B^i$.

The acceleration, a_B^i, is explicitly expressed in terms of the external gravitational potentials, $\bar{U}, \bar{U}^i$, etc., and the dipole moment $Q^i = \delta^{ij}Q_j$. It is remarkable that function $Q^i = Q^i(u)$ is not limited by the gauge conditions and can be chosen arbitrary. Physically, it determines the magnitude and direction of the inertial force acting in the local coordinates on a test particle being in a free fall [115, 117]. Only after fixing the choice of Q^i formula (5.26) becomes an ordinary differential equation which can be solved to find out the world line $\mathcal{W}$ of the origin of the local coordinates on the spacetime manifold.

A trivial choice of $Q^i = 0$ looks attractive as it immediately converts (5.26) to a fully determined differential equations. It is this choice that has been made by Dixon [62] but it means that the origin of the local coordinates moves along a geodesic world line of a test particle falling freely in the background spacetime defined by the external part of the local metric tensor. Unfortunately, this choice does not allow us to keep the origin of the local coordinates at the center of mass of the body B possessing nonvanishing internal multipole moments $\mathcal{I}^L$. This is because the internal moments of the body B interact with the tidal gravitational field of other bodies from the N-body system that forces the body B move along an accelerated (nongeodesic) world line [101]. Thus, the world line $\mathcal{Z}$ of the center of mass of the body B is not a geodesic, and $Q^i \neq 0$ must be chosen to ensure that the center of mass of the body B is at the origin of the local coordinates associated with this body at any instant of time. This is equivalent to solving the internal problem of motion of the body's center of mass with respect to the local coordinates. Solution of this problem has been found in [93, 94, 101]) and will be discussed in Section 6.4.

The external multipoles

Matching determines the external multipoles in terms of partial derivatives from the gravitational potentials of the external bodies [93, 94]. The external multipoles of the scalar field are

$$P_L = \bar{\varphi}_{,L}(\boldsymbol{x}_B) + \mathcal{O}\left(\epsilon^2\right) , \qquad (l \geq 0) , \tag{5.27}$$

where the external scalar field $\bar{\varphi}$ coincides in this approximation with the external Newtonian potential $\bar{U}$ and is computed at the origin of the local coordinates, $x_B^i(t)$, at the instant of time t. We remind that the scalar field perturbation is coupled with the factor $\gamma - 1$, so that all physically observed scalar-field effects must be proportional to this factor. We also notice that the external monopole ($l = 0$) and dipole ($l = 1$) of the scalar field cannot be removed from the observable gravitational effects by making coordinate transformation to a freely falling frame because if a scalar field presents in one coordinate frame it must be present in any other coordinates. In other words, scalar fields do not obey the principle of equivalence. It was the primary reason why Einstein abandoned a pure scalar theory of gravity.

External mass-type multipoles Q^L for $l \geq 2$ are defined by the following equation[3]

$$
\begin{aligned}
Q^L &= \bar{U}^{,\langle L\rangle}(\boldsymbol{x}_B) \\
&+ \epsilon^2 \Big\{ \bar{\Psi}^{,\langle L\rangle}(\boldsymbol{x}_B) - \frac{1}{2}\bar{\chi}^{,tt\langle L\rangle}(\boldsymbol{x}_B) + 2(1+\gamma)\dot{\bar{U}}^{\langle i_l, L-1\rangle}(\boldsymbol{x}_B) \\
&- 2(1+\gamma)v_B^k \bar{U}^{k,\langle L\rangle}(\boldsymbol{x}_B) + (l - 2\gamma - 2)v_B^{\langle i_l}\dot{\bar{U}}^{,L-1\rangle}(\boldsymbol{x}_B) \\
&+ (1+\gamma)v_B^2 \bar{U}^{,\langle L\rangle}(\boldsymbol{x}_B) - \frac{l}{2}v_B^k v_B^{\langle i_l}\bar{U}^{,L-1>k}(\boldsymbol{x}_B) - l\gamma\bar{U}(\boldsymbol{x}_B)\bar{U}^{,\langle L\rangle}(\boldsymbol{x}_B) \\
&- (l^2 - l + 2\gamma + 2)a_B^{\langle i_l}\bar{U}^{,L-1\rangle}(\boldsymbol{x}_B) - lF^{k\langle i_l}\bar{U}^{,L-1>k}(\boldsymbol{x}_B) + K^L \\
&+ \dot{Z}^L + lA\bar{U}^{,\langle L\rangle}(\boldsymbol{x}_B)\Big\} + \mathcal{O}\left(\epsilon^2\right) ,
\end{aligned}
\tag{5.28}
$$

where we have used notations

$$K^{ij} \equiv 3a_B^{\langle i}a_B^{j\rangle} , \tag{5.29}$$

$$K^L \equiv 0, \qquad (l \geq 3) . \tag{5.30}$$

External current-type multipoles C^L for $l \geq 2$ are given by

$$
\begin{aligned}
\varepsilon^{ipj}C^{pL-1} &= \frac{4l(1+\gamma)}{l+1}\Big\{ v_B^{[i}\bar{U}^{,j]L-1}(\boldsymbol{x}_B) - \bar{U}^{[i,j]L-1}(\boldsymbol{x}_B) \\
&- \frac{l-1}{l}\delta^{i_{l-1}[i}\dot{\bar{U}}^{,j]L-2}(\boldsymbol{x}_B)\Big\} + \mathcal{O}\left(\epsilon^2\right),
\end{aligned}
\tag{5.31}
$$

where the dot denotes the time derivative with respect to time t. The external multipole moments Q_L and C_L are analogs of Dixon's multipole moments $A_{\alpha_1\ldots\alpha_l\mu\nu}$ and $B_{\alpha_1\ldots\alpha_l\mu\nu}$, respectively (see (1.13) and (1.14)). We shall use the above expressions for the external multipoles in derivation of equations of motion of extended bodies in the next section.

3 We remind that the spatial indices are raised and lowered with the Kronecker symbol δ^{ij}.

6 Post-Newtonian equations of motion of extended bodies in local coordinates

Matching equations yield equations of motion of the world line $\mathcal{W}$ of the origin of the local coordinates with respect to the global coordinates as a function of yet undetermined, external dipole moment Q_i. This moment is not arbitrary but must be derived from the condition that the world line $\mathcal{W}$ coincides with the world line $\mathcal{Z}$ of the body's center of mass. To solve this problem we will have to find out the translational equations of motion of the center of mass of the body B. The most natural approach to derive these equations is to employ the local coordinates. Derivation of the equations of motion of the body's center of mass in the local coordinates can be done in two different ways which include:

(1) method of integration of microscopic equations of motion of matter proposed by Fock [73, 126] and Papapetrou [123–125];
(2) method of surface integrals introduced by Einstein, Infeld and Hoffmann (EIH) [68] and significantly improved by Thorne and Hartle [147]. This EIH method is thoroughly explained in textbook [4].

The Fock–Papapetrou method is operating with a continuous distribution of matter and will be employed in this chapter. We define the center of mass and linear momentum of the body B, derive the post-Newtonian microscopic equations of motion of matter of the body in the local coordinates and, then, integrate them over the volume of the body B in order to get the law of evolution of the linear momentum of the body. As soon as the law of evolution is set up, the external dipole moment Q_i follows from the condition that the linear momentum is conserved. We explain this procedure below in more detail.

6.1 Microscopic post-Newtonian equations of motion

The microscopic post-Newtonian equations of motion of matter include: (1) equation of continuity, (2) thermodynamic equation relating the elastic energy, Π, to the stress tensor, $\pi_{\alpha\beta}$, and (3) the Navier–Stokes equation.

The equation of continuity in the local coordinates $w^\alpha = (cu, \boldsymbol{w})$ has the most simple form if we use the invariant density ρ^*, defined in (3.18). It reads

$$\frac{\partial \rho^*}{\partial u} + \frac{\partial\left(\rho^* \nu^i\right)}{\partial w^i} = 0\,, \tag{6.1}$$

where $\nu^i = dw^i/du$ is a coordinate velocity of matter in the local coordinates. This equation is exact with all post-Newtonian corrections taken into account.

The thermodynamic equation relating the internal elastic energy, Π, and the stress tensor, $\pi_{\alpha\beta}$, is required only in a linearized approximation where the stress tensor is completely characterized by its spatial components π_{ij}. Hence, one has from Equation (2.10) the following differential equation:

$$\frac{d\Pi}{du}+\frac{\pi_{ij}}{\rho^*}\frac{\partial v^i}{\partial w^j}=\mathcal{O}\left(\epsilon^2\right), \tag{6.2}$$

where the operator of convective derivative is $d/du \equiv \partial/\partial u + v^i\partial/\partial w^i$.

The Navier–Stokes equation follows from the spatial component of the law of conservation of the energy–momentum tensor, $T^{i\nu}{}_{;\nu}=0$. It yields

$$\begin{aligned}
&\rho^*\frac{d}{du}\left\{v^i+\epsilon^2\left[\left(\frac{1}{2}v^2+\Pi+\frac{1}{2}\hat{h}_{00}+\frac{\hat{h}}{3}\right)v^i+\hat{h}_{0i}\right]\right\}+\epsilon^2\frac{\partial(\pi_{ij}v^j)}{\partial u}\\
&=\frac{1}{2}\rho^*\frac{\partial\hat{h}_{00}}{\partial w^i}-\frac{\partial\pi_{ij}}{\partial w^j}+\epsilon^2\left\{\rho^*\left[\frac{1}{2}\frac{\partial\hat{l}_{00}}{\partial w^i}+\frac{1}{4}(v^2+2\Pi+\hat{h}_{00})\frac{\partial\hat{h}_{00}}{\partial w^i}\right.\right.\\
&\quad\left.+\frac{1}{6}v^2\frac{\partial\hat{h}}{\partial w^i}+v^k\frac{\partial\hat{h}_{0k}}{\partial w^i}\right]+\frac{1}{6}\pi_{kk}\frac{\partial\hat{h}}{\partial w^i}+\frac{1}{2}\pi_{ik}\left(\frac{\partial\hat{h}_{00}}{\partial w^k}-\frac{1}{3}\frac{\partial\hat{h}}{\partial w^k}\right)\\
&\quad\left.+\frac{1}{2}\left(\hat{h}_{00}-\frac{1}{3}\hat{h}\right)\frac{\partial\pi_{ik}}{\partial w^k}\right\}+\mathcal{O}\left(\epsilon^4\right),
\end{aligned} \tag{6.3}$$

where the metric tensor perturbations $\hat{h}_{00}$, $\hat{l}_{00}$, $\hat{h}_{0i}$, $\hat{h}_{ij}$ and $\hat{h}=\hat{h}_{ii}$ in the local coordinates have been defined in Section 4.2.

6.2 Post-Newtonian mass of an extended body

There are two algebraically independent definitions of the post-Newtonian mass in the scalar-tensor theory – the *active* mass and the *conformal* mass which are defined, respectively, by equations (4.51) and (4.54) for multipolar index $l=0$.

More specifically, the active mass of body B covered by the local coordinates, is [93, 94]

$$\begin{aligned}
\mathcal{M}=\mathrm{M}_{\mathrm{GR}}+\epsilon^2&\left\{\frac{1}{6}(\gamma-1)\ddot{\mathcal{J}}^{(2)}-\frac{1}{2}\eta\int\limits_{V_B}\rho^*\hat{U}_B d^3w-[A+(2\beta-\gamma-1)P]\mathrm{M}_{\mathrm{GR}}\right.\\
&\left.-\sum_{l=1}^{\infty}\frac{1}{l!}[(\gamma l+1)Q_L+2(\beta-1)P_L]\mathcal{I}^{\langle L\rangle}\right\}+\mathcal{O}\left(\epsilon^4\right),
\end{aligned} \tag{6.4}$$

where

$$\mathrm{M}_{\mathrm{GR}} = \int_{V_B} \rho^* \left[1 + \epsilon^2 \left(\frac{1}{2}\nu^2 + \Pi - \frac{1}{2}\hat{U}_B\right)\right] d^3w + \mathcal{O}\left(\epsilon^4\right) \tag{6.5}$$

is general-relativistic definition of the post-Newtonian mass of the body [153], and

$$\mathcal{I}^{(2)} = \int_{V_B} \rho^* w^2 d^3 w \,, \tag{6.6}$$

is the rotational moment of inertia of the body B with respect to the origin of the local coordinates, $\mathcal{I}^{\langle L\rangle}$ are *active* multipole moments of the body B defined in (4.51), two dots above $\mathcal{I}^{(2)}$ denote a second derivative with respect to time u.

Mass, M_{GR}, depends only on the internal distribution of mass, kinetic, thermal, and gravitational energy densities. In case of a single, isolated body residing in asymptotically flat spacetime, this mass coincides with the post-Newtonian expansion of the Tolman mass [97, 148]. If the body is isolated, the post-Newtonian mass M_{GR} is conserved. However, in the N-body system the gravitational interaction of the body B with external bodies causes tides which change the internal distribution of matter and shape of the body B, thus, leading to dependence of M_{GR} on time. It is governed by the ordinary differential equation [93]

$$\dot{\mathrm{M}}_{\mathrm{GR}} = \epsilon^2 \sum_{l=1}^{\infty} \frac{1}{l!} Q_L \dot{\mathcal{I}}^{\langle L\rangle} \,. \tag{6.7}$$

The conformal mass of the body B is [93, 94]

$$M = \mathrm{M}_{\mathrm{GR}} - \epsilon^2 \left\{ [A + (1-\gamma)P]\mathrm{M}_{\mathrm{GR}} + \sum_{l=1}^{\infty} \frac{l+1}{l!} Q_L \mathcal{I}^{\langle L\rangle} \right\} + \mathcal{O}\left(\epsilon^4\right) , \tag{6.8}$$

Relation between the active and conformal masses is obtained by comparing (6.4) and (6.8) and reads [93, 94]

$$\begin{aligned} M = \mathcal{M} + \epsilon^2 \Bigg\{ & \frac{1}{2}\eta \int_{V_B} \rho^* \hat{U}_B d^3 w - \frac{1}{6}(\gamma - 1)\ddot{\mathcal{I}}^{(2)} \\ & + 2(\beta - 1)\left(\mathcal{M}P + \sum_{l=1}^{\infty} \frac{1}{l!} P_L \mathcal{I}^{\langle L\rangle} \right) + (\gamma - 1) \sum_{l=1}^{\infty} \frac{1}{(l-1)!} Q_L \mathcal{I}^{\langle L\rangle} \Bigg\} + \mathcal{O}\left(\epsilon^4\right) , \end{aligned} \tag{6.9}$$

where $\eta = 4\beta - \gamma - 3$ is called the Nordtvedt parameter [153]. The numerical value of this parameter is known with the precision better than $|\eta| < 2 \times 10^{-4}$ from LLR experiment [120].

One can see that in the scalar-tensor theory of gravity the conformal mass of the body differs from its active mass. It was noticed by Dicke [52, 54] and Nordtvedt [118, 153] who has derived the integral term being proportional to the Nordvedt parameter η on the right-hand side of (6.9). The actual difference is more complicated and includes the term with the time derivative of the rotational moment of inertia of the body as well as the tidal contributions. If an extended body were completely isolated from external gravitational field, the difference between the active and conformal masses would be due to the Nordtvedt effect caused by $\eta \neq 0$, and the time-dependence of the body's rotational moment of inertia (for example, because of radial oscillations, etc.). In case when the presence of gravitational field of the external bodies of the N-body system cannot be ignored, we also have to account for the gravitational coupling between the external gravitational field multipoles, P_L and Q_L, with the internal multipole moments $\mathcal{I}^L$ of the body B which yields additional contribution to the difference between the masses.

It is interesting to notice that a rigorous post-Newtonian definition (6.8) of conformal mass M of an extended body in the N-body system contains a post-Newtonian contribution of the coupling terms $\sim Q_L \mathcal{I}^{\langle L \rangle}$ which can be interpreted as a reminiscence of Mach's principle stating that mass of each body originates from its gravitational interaction with the external universe. Our calculations do not confirm Mach's principle completely but indicate that Mach's idea has got a partial support and justification in general theory of relativity.

6.3 Post-Newtonian center of mass and linear momentum of an extended body

Functional form of equations of motion of extended bodies in the N-body system depends crucially on the choice of the reference point inside the body defining its center of mass. Position of the center of mass of the body is determined by the value of the internal dipole moment of the body. In scalar-tensor theory of gravity there are two possible definitions of the dipole moment: active dipole moment $\mathcal{I}^i$ and conformal dipole moment I^i. It is difficult to foresee which definition is the best. Only after completing direct calculation of equations of motion it becomes clear that it is the conformal dipole moment yields the most optimal choice of the post-Newtonian center of mass of each body [93, 94, 158]. This is because after double differentiation with respect to time, only the conformal dipole moment leads to conservation of the individual body's linear momentum, p^i, while the post-Newtonian scalar or active dipole moments do not bear such a property.

Thus, we define the post-Newtonian center of mass of the body by making use of (4.54) for the multipolar index $l = 1$. After some simplifications and rearrangements of terms in the integrand, the conformal dipole moment reads

$$I^i = \int_{V_B} \rho^* w^i \left[1 + \epsilon^2 \left(\frac{1}{2}\nu^2 + \Pi - \frac{1}{2}\hat{U}_B\right)\right] d^3w$$
$$- \epsilon^2 \left\{ [A + (1-\gamma)P] \int_{V_B} \rho^* w^i d^3w + \sum_{l=1}^{\infty} \frac{l+1}{l!} Q_L \mathcal{I}^{\langle iL \rangle} \right.$$
$$\left. + \frac{1}{2} \sum_{l=0}^{\infty} \frac{1}{(2l+3)l!} Q_{iL} \mathcal{N}^L \right\} + \mathcal{O}\left(\epsilon^4\right) , \qquad (6.10)$$

where

$$\mathcal{N}^L = \int_{V_B} \rho^* w^2 w^{\langle L \rangle} d^3w , \qquad (l \geq 0) \qquad (6.11)$$

are the moments generalizing the rotational moment of inertia. Indeed, for $l = 0$ the scalar function $\mathcal{N} = \mathcal{I}^{(2)}$. Dipole moment is a function of the local coordinate time u only and it defines the displacement of the center of mass of the body from the origin of the local coordinates. Hence, if the origin of the local coordinates coincides with the center of mass of the body, the dipole moment vanishes $I^i = 0$.

The post-Newtonian definitions of mass and the center of mass of the body depend not only on the distribution of density of matter, its velocity, and stresses inside the body but also on other terms describing mutual gravitational coupling of internal and external multipoles. Thorne and Hartle [147] believed that these terms introduce ambiguity to the definitions of mass and the center of mass as they can be either included or excluded from these definitions. We have shown [93, 94, 158] that this ambiguity is fictitious and does not exists. The coupling terms must be included to the definitions of mass, center of mass, and other internal multipoles to achieve the most economic form of the equations of motion depending only on the mass-type and spin-type internal multipole moments of the bodies. It is also true that only under such definitions we can get EIH equations of motion in the limiting case of spherically symmetric and nonrotating bodies.

The post-Newtonian linear momentum of the body, p^i, is defined as the first derivative of the dipole moment (6.10) with respect to the local time u,

$$p^i = \dot{I}^i(u) , \qquad (6.12)$$

where the dot denotes the time derivative with respect to u. After taking the derivative and using the local equations of motion of matter (6.3), we obtain

$$p^i = \int_{V_B} \rho^* \nu^i \left[1 + \epsilon^2 \left(\frac{1}{2}\nu^2 + \Pi - \frac{1}{2}\hat{U}_B\right)\right] d^3w \qquad (6.13)$$

$$+\epsilon^2 \int\limits_{V_B} \left[\pi_{ik}v^k - \frac{1}{2}\rho^* \hat{W}_B^i\right] d^3w$$

$$-\epsilon^2 \frac{d}{du}\left\{[A + (1-\gamma)P]I^i + \sum_{l=1}^{\infty} \frac{l+1}{l!} Q_L \mathcal{J}^{\langle iL\rangle} + \frac{1}{2}\sum_{l=0}^{\infty} \frac{1}{(2l+3)l!} Q_{iL}\mathcal{N}^L\right\}$$

$$+\epsilon^2 \sum_{l=1}^{\infty} \frac{1}{l!}\left[Q_L \dot{\mathcal{J}}^{\langle iL\rangle} + \frac{l}{2l+1} Q_{iL-1}\dot{\mathcal{N}}^{L-1} - Q_L \int\limits_{V_B} \rho^* \dot{v}^i w^{\langle L\rangle} d^3w\right] + \mathcal{O}\left(\epsilon^4\right),$$

where the new internal potential

$$\hat{W}_B^i = G \int\limits_{V_B} \frac{\rho^*(u, \boldsymbol{w}')\nu'^k (w^k - w'^k)(w^i - w'^i)}{|\boldsymbol{w} - \boldsymbol{w}'|^3} d^3w' . \tag{6.14}$$

Until now the point $\boldsymbol{x}_B(t)$ has represented a location of the origin of the local coordinate system in the global coordinates taken at the time t. In general, the origin of the local coordinates may not coincide with the center of mass of the body B which can move with respect to the local coordinates with some velocity and acceleration. In order to be able to keep the center of mass of the body at the origin of the local coordinates we have to prove that for any instant of time the dipole moment defined by (6.10) and its time derivative, that is, the linear momentum of the body given by (6.13), can be kept equal to zero. This requirement can be satisfied, if and only if, the second time derivative of the dipole moment with respect to the local coordinate time u vanishes, that is

$$\dot{p}^i(u) = 0 . \tag{6.15}$$

It is remarkable that this equation expressing conservation of the linear momentum, can be satisfied after making an appropriate choice of the external dipole moment Q_i that characterizes a locally measurable acceleration of the origin of the local coordinates with respect to a geodesic world line of a freely falling test particle. We prove it in the next section.

6.4 Translational equation of motion in the local coordinates

Translational equations of motion of the center of mass of body B with respect to the local coordinates w^α, which cover the body and its neighborhood, are derived by making use of definition (6.13) of its linear momentum, p^i. Differentiating (6.13) one time with respect to the local coordinate time u, making use of the microscopic equations of motion (6.1)–(6.3), and integrating by parts to rearrange a number of terms, one obtains [93, 158]

$$
\begin{aligned}
\dot{p}^i = {} & \mathcal{M}Q^i + \sum_{l=1}^{\infty}\frac{1}{l!}Q_{iL}\mathcal{I}^{\langle L\rangle} \\
& - \epsilon^2\Bigg\{\sum_{l=2}^{\infty}\frac{1}{(l+1)!}[(l^2+l+4)Q_L + 2(\gamma-1)P_L]\ddot{\mathcal{I}}^{\langle iL\rangle} \\
& + \sum_{l=2}^{\infty}\frac{2l+1}{(l+1)(l+1)!}[(l^2+2l+5)\dot{Q}_L + 2(\gamma-1)\dot{P}_L]\dot{\mathcal{I}}^{\langle iL\rangle} \\
& + \sum_{l=2}^{\infty}\frac{2l+1}{(2l+3)(l+1)!}[(l^2+3l+6)\ddot{Q}_L + 2(\gamma-1)\ddot{P}_L]\mathcal{I}^{\langle iL\rangle} \\
& + [3Q_k + (\gamma-1)P_k]\ddot{\mathcal{I}}^{\langle ik\rangle} + \frac{3}{2}[4\dot{Q}_k + (\gamma-1)\dot{P}_k]\dot{\mathcal{I}}^{\langle ik\rangle} \\
& + \frac{3}{5}[5\ddot{Q}_k + (\gamma-1)\ddot{P}_k]\mathcal{I}^{\langle ik\rangle} + \sum_{l=2}^{\infty}\frac{1}{l!}\dot{Z}_{iL}\mathcal{I}^{\langle L\rangle} \\
& + \sum_{l=1}^{\infty}\frac{1}{(l+1)!}\varepsilon_{ipq}\left[\dot{C}_{pL}\mathcal{I}^{\langle qL\rangle} + \frac{l+2}{l+1}C_{pL}\dot{\mathcal{I}}^{\langle qL\rangle}\right] \\
& - 2\sum_{l=1}^{\infty}\frac{l+1}{(l+2)!}\varepsilon_{ipq}\Bigg[(2Q_L + (\gamma-1)P_{pL})\dot{\mathcal{S}}^{\langle qL\rangle} \\
& + \frac{l+1}{l+2}(2\dot{Q}_{pL} + (\gamma-1)\dot{P}_{pL})\mathcal{S}^{\langle qL\rangle}\Bigg] - \sum_{l=1}^{\infty}\frac{l(l+2)}{(l+1)(l+1)!}C_{iL}\mathcal{S}^{\langle L\rangle} \\
& - \frac{1}{2}\varepsilon_{ikq}\left[(4Q_k + 2(\gamma-1)P_k)\dot{\mathcal{S}}^q + (2\dot{Q}_k + (\gamma-1)\dot{P}_k)\mathcal{S}^q\right] \\
& + (P^i - Q^i)\Bigg[\frac{1}{2}\eta\int_{V_B}\rho^*\hat{U}^{(B)}d^3w - \frac{1}{6}(\gamma-1)\ddot{\mathcal{I}}^{(2)} \\
& + 2(\beta-1)\left(\mathcal{M}P + \sum_{l=1}^{\infty}\frac{1}{l!}P_L\mathcal{I}^{\langle L\rangle}\right) + (\gamma-1)\sum_{l=1}^{\infty}\frac{1}{(l-1)!}Q_L\mathcal{I}^{\langle L\rangle}\Bigg]\Bigg\},
\end{aligned}
\tag{6.16}
$$

where the spatial indices are raised and lowered with the Kronecker symbol, the matching condition $P_L = Q_L + \mathcal{O}(\epsilon^2)$ $(l \geq 2)$ has been used, and all terms which are proportional to I^i and/or p^i vanish when the origin of the local coordinates coincides with the center of mass of body B at any instant of time, thus, making $I^i = 0$ and $p^i = 0$. We shall use these conditions in the derivation of translational equations of motion of the center of mass of the body B. The omitted terms can be found in [93, Equation 6.19].

It is important to notice that the right-hand side of (6.16) is the force exerted on the body B due to the gravitational coupling of its internal *active* multipole moments, $\mathcal{I}^L$, $\mathcal{S}^L$ with the external multipole moments Q_L, P_L, C_L. The right-hand side of (6.16) also contains the inertial force, $\mathcal{M}Q_i$, due to the nongeodesic motion of the origin of the local coordinates. Hence, in the most general case, $\dot{p}^i \neq 0$ which means that the center of mass of the body B accelerates with respect to the origin of local coordinates and the world line $\mathcal{Z}$ of the center of mass does not coincide with the world line $\mathcal{W}$ of the origin of the local coordinates. However, the external dipole moment Q_i is yet unconstrained function of time and we can always reach consensus with the condition (6.15) by choosing Q_i in such a way that it compensates the gravitational coupling force. To eliminate acceleration of the world line $\mathcal{Z}$ with respect to $\mathcal{W}$ and to fix the center of mass of the body B at the origin of the local coordinates we, first, impose condition (6.15) and solve (6.16) with respect to Q_i. It constrains the inertial force and yields

$$
\begin{aligned}
MQ^i &= (M-\mathcal{M})\,P^i - \sum_{l=1}^{\infty}\frac{1}{l!}Q_{iL}\mathcal{J}^{\langle L\rangle} \\
&+\epsilon^2\Bigg\{\sum_{l=2}^{\infty}\frac{1}{(l+1)!}[(l^2+l+4)Q_L+2(\gamma-1)P_L]\ddot{\mathcal{J}}^{\langle iL\rangle} \\
&+\sum_{l=2}^{\infty}\frac{2l+1}{(l+1)(l+1)!}\left[(l^2+2l+5)\dot{Q}_L+2(\gamma-1)\dot{P}_L\right]\dot{\mathcal{J}}^{\langle iL\rangle} \\
&+\sum_{l=2}^{\infty}\frac{2l+1}{(2l+3)(l+1)!}\left[(l^2+3l+6)\ddot{Q}_L+2(\gamma-1)\ddot{P}_L\right]\mathcal{J}^{\langle iL\rangle} \\
&+[3Q_k+(\gamma-1)P_k]\ddot{\mathcal{J}}^{\langle ik\rangle}+\frac{3}{2}\left[4\dot{Q}_k+(\gamma-1)\dot{P}_k\right]\dot{\mathcal{J}}^{\langle ik\rangle} \\
&+\frac{3}{5}\left[5\ddot{Q}_k+(\gamma-1)\ddot{P}_k\right]\mathcal{J}^{\langle ik\rangle}+\sum_{l=2}^{\infty}\frac{1}{l!}\dot{Z}_{iL}\mathcal{J}^{\langle L\rangle} \\
&+\sum_{l=1}^{\infty}\frac{1}{(l+1)!}\varepsilon_{ipq}\left[\dot{C}_{pL}\mathcal{J}^{\langle qL\rangle}+\frac{l+2}{l+1}C_{pL}\dot{\mathcal{J}}^{\langle qL\rangle}\right] \\
&-2\sum_{l=1}^{\infty}\frac{l+1}{(l+2)!}\varepsilon_{ipq}\Bigg[(2Q_L+(\gamma-1)P_{pL})\dot{\mathcal{S}}^{\langle qL\rangle} \\
&+\frac{l+1}{l+2}(2\dot{Q}_{pL}+(\gamma-1)\dot{P}_{pL})\mathcal{S}^{\langle qL\rangle}\Bigg]-\sum_{l=1}^{\infty}\frac{l(l+2)}{(l+1)(l+1)!}C_{iL}\mathcal{S}^{\langle L\rangle} \\
&-\frac{1}{2}\varepsilon_{ikq}\left[(4Q_k+2(\gamma-1)P_k)\dot{\mathcal{S}}^{q}+\left(2\dot{Q}_k+(\gamma-1)\dot{P}_k\right)\mathcal{S}^{q}\right]\Bigg\}\,,
\end{aligned}
\tag{6.17}
$$

where M and $\mathcal{M}$ are the *conformal* and *active* gravitational masses of the body B, respectively, and we again omitted terms which are proportional to the dipole moment I^i and linear momentum p^i of the body like we did it in (6.16). The two masses, M and $\mathcal{M}$, are not equal according to (6.9). The difference between them plays a role of a "charge" of the scalar field ϕ which couples with the dipole moment $P^i = \bar{U}_{,i}$ of the external scalar field leading to the Dicke–Nordtvedt effect [52, 54, 153]. In general relativity, $M = \mathcal{M}$, and the first term on the right-hand side of (6.17) vanishes.

Equation (6.17) is a condition for the fulfilment of the law of conservation of linear momentum (6.15) in local coordinates. It ensures that the world line $\mathcal{W}$ of the origin of local coordinates does not accelerate with respect to the world line $\mathcal{Z}$ of the center of mass of body B. Equation (6.17) does not warranty, however, that $\mathcal{W}$ and $\mathcal{Z}$ coincides. The origin of the local coordinates still can move uniformly with respect to the center of mass of the body. To eliminate this uniform motion we impose condition, $p^i = 0$. The freedom which remains is a constant relative displacement of the origin of the local coordinates with respect to the center of mass of the body. This constant displacement is removed by additional constrain imposed on the internal (conformal) dipole moment of the body, $I^i = 0$. This procedure ensures that the world lines $\mathcal{W}$ and $\mathcal{Z}$ coincide.

The right-hand side of (6.17), divided by M, must be substituted for Q^i in the equations of motion of the origin of the local coordinates (5.26) to convert it to the equations of motion of the center of mass of the body B in the global coordinates. These equations still contain the external gravitational potentials like $\bar{U}$, $\bar{\Psi}$, $\bar{U}^i$, etc., which are given in the form of integrals expressed in the global coordinates. These integrals should be explicitly expanded with respect to the internal multipoles of the bodies in order to complete the theory. We shall conduct this calculation in the next section and derive equations of motion of extended bodies in the N-body system in terms of their internal multipoles as well as coordinates and velocities of their centers of mass.

7 Post-Newtonian equations of motion of extended bodies in global coordinates

7.1 STF expansions of the external gravitational potentials in terms of the internal multipoles

External gravitational potentials $\bar{U}$, $\bar{\Psi}$, $\bar{U}^i$, $\bar{\chi}$ are given in the form of volume integrals (5.24) which are not convenient for practical calculations in relativistic celestial mechanics. The practice is to expand the gravitational potentials of external bodies in terms of their own internal multipole moments – mass, dipole, quadrupole, etc. The expansion itself is rather straightforward but the problem here is that the external potentials are given as integrals in the global coordinates, x^α, and the multipolar expansion will be obtained in terms of the internal multipole moments of the bodies expressed in the global coordinates. However, the internal multipoles of an extended body makes physical sense only when they are expressed in the local coordinates, w^α, associated with that body. Therefore, we have to convert the internal multipoles of each body from the global to local coordinates associated with this body.

We have already introduced the local coordinates, $w^\alpha = (cu, w^i)$, associated with body B. In a similar fashion, the local coordinates can be introduced near any other extended body $C \neq B$. We shall use the letter C as a subindex to label the bodies as well as the local coordinates and functions associated with them. The local coordinates of the body C are denoted $w^\alpha_C = (cu_C, w^i_C)$. Coordinate transformation between w^α_C and the global coordinates x^α is similar to (5.16) and (5.17) except that now we have to pin the label C to all quantities,

$$u_C = t - \epsilon^2(\mathcal{A}_C + v^k_C R^k_C) + \epsilon^4 \left[\mathcal{B}_C + \left(\frac{1}{3} v^k_C a^k_C - \frac{1}{6}\dot{\bar{U}}(\boldsymbol{x}_C) + \frac{1}{6}\dot{\mathcal{Q}}_C \right) R^2_C \right.$$

$$\left. - \frac{1}{10}\dot{a}^k_C R^k_C R^2_C + \sum_{l=1}^{\infty} \frac{1}{l!} \mathcal{B}^L_C R^L_C \right] + \mathcal{O}\left(\epsilon^5\right) , \tag{7.1}$$

$$w^i_C = R^i_C + \epsilon^2 \left[\left(\frac{1}{2} v^i_C v^k_C + \delta^{ik} \gamma \bar{U}(\boldsymbol{x}_C) - \delta^{ik} A_C + F^{ik}_C \right) R^k_C \right.$$

$$\left. + \; a^k_C R^i_C R^k_C - \frac{1}{2} a^i_C R^2_C \right] + \mathcal{O}\left(\epsilon^4\right) , \tag{7.2}$$

where

$$R^i_C = x^i - x^i_C , \tag{7.3}$$

and $x^i_C = x^i_C(t)$ marks the global coordinates of the origin of the local coordinates of the body C, $v^i_C = dx^i_C/dt$, $a^i_C = dv^i_C/dt$. Defining equations for all other functions entering (7.1) and (7.2) remain similar to corresponding equations shown in Section 5.2.

Let us introduce the following notations:

$$D^{ij}_C \equiv \delta^{ik} \gamma \bar{U}_C(\boldsymbol{x}_C) - \delta^{ik} A_C , \tag{7.4}$$

$$D^{ijk}_C \equiv \frac{1}{2}\left(a^j_C \delta^{ik} + a^k_C \delta^{ij} - a^i_C \delta^{jk} \right) , \tag{7.5}$$

that will allow us to shorten subsequent equations.

The external gravitational potentials are integrals of the global coordinates which integrands contain a kernel, $1/|\boldsymbol{x} - \boldsymbol{x}'|$, of Green's function of the Laplace equation. This kernel is expanded into multipolar series as follows:

$$\frac{1}{|\boldsymbol{x} - \boldsymbol{x}'|} = \frac{1}{|\boldsymbol{x} - \boldsymbol{x}_C - (\boldsymbol{x}' - \boldsymbol{x}_C)|} = \sum_{l=0}^{\infty} \frac{(-1)^l}{l!} \left(\frac{1}{R_C} \right)_{,L} R'^{\langle L \rangle}_C , \tag{7.6}$$

where

$$R'^i_C = x'^i - x^i_C , \tag{7.7}$$

and $(1/R_C)_{,L}$ denotes a partial derivative of lth order with respect to spatial coordinates $(1/R_C)_{,L} \equiv \partial_L(1/R_C) \equiv \partial_{i_1...i_l}(1/R_C)$ where each partial derivative $\partial_i = \partial/\partial x^i$. Post-Newtonian transformation of $R_C'^i$ to the local coordinates, w_C^i, is slightly different from (7.2) because the points, x^i and x'^i are lying on a hypersurface of constant time t, while the points w_C^i and $w_C'^i$ are lying on a hypersurface of constant local time u_C. It requires additional Lie transport of $w_C'^i$ from one hypersurface to another which shows that the corresponding post-Newtonian transformation of $R_C'^i$ reads [25, 93, 102]

$$R_C'^i = w_C'^i - \epsilon^2 \left[\left(\frac{1}{2} v_C^i v_C^k + D_C^{ik} + F_C^{ik} \right) w_C'^k + D_C^{ijk} w_C'^j w_C'^k \right] + \epsilon^2 \mathrm{v}_C'^i v_C^k \left(w_C^k - w_C'^k \right) + \mathcal{O}\left(\epsilon^4\right) , \tag{7.8}$$

where

$$\mathrm{v}_C'^i = v'^i - v_C^i + \mathcal{O}\left(\epsilon^2\right) , \tag{7.9}$$

is a relative velocity of matter of the body C with respect to the origin of its own local coordinates.

Equation (7.8) allows us to transform the Newtonian counterpart of the internal multipole moments in the global coordinates to the local coordinates as follows [93, 94, 158]:

$$\begin{aligned} \int\limits_{V_C} \rho^{*\prime}_C R_C'^{\langle L \rangle} d^3x' = \int\limits_{V_C} \rho_C^{*\prime} w_C'^{\langle L \rangle} d^3 w_C' + \epsilon^2 \Bigg(& -\frac{l}{2} v_C^k v_C^{\langle i_l} \mathcal{I}_C^{L-1\rangle k} \\ & + l F_C^{k<i_l} \mathcal{I}_C^{L-1>k} - l D_C^{k<i_l} \mathcal{I}_C^{L-1>k} - l \mathcal{I}_C^{jk<L-1} D_C^{i_l>jk} \\ & - v_C^k \dot{\mathcal{J}}_C^{k\langle L \rangle} + v_C^k R_C^k \dot{\mathcal{J}}_C^{\langle L \rangle} + v_C^k \int\limits_{V_C} \rho_C^{*\prime} \mathrm{v}_C'^k w_C'^{\langle L \rangle} d^3 w_C' \Bigg) + \mathcal{O}\left(\epsilon^4\right) , \end{aligned} \tag{7.10}$$

where we have used the fact that the product of the invariant density, ρ^* with three-dimensional coordinate volume is Lie invariant and does not change when transported from one hypersurface to another along world lines of matter and transformed from the global to local coordinates, that is $\rho^*(t, \boldsymbol{x}) d^3x = \rho^*(u_C, \boldsymbol{w}_C) d^3 w_C \equiv \rho_C^* d^3 w_C$ exactly [93].

The Newtonian potential of body C in the global coordinates is transformed with the help of (7.6)–(7.10) to the local coordinates as follows [158]:

$$
\begin{aligned}
U_C(t,\boldsymbol{x}) &= G\int\limits_{V_C}\frac{\rho^*(t,\boldsymbol{x}')}{|\boldsymbol{x}-\boldsymbol{x}'|}d^3x' \\
&= G\sum_{l=0}^{\infty}\frac{(-1)^l}{l!}\left(\frac{1}{R_C}\right)_{,L}\Bigg\{\mathcal{I}_C^{\langle L\rangle}+\epsilon^2\left[A_C+(2\beta-\gamma-1)P_C\right]\mathcal{I}_C^{\langle L\rangle} \\
&\quad -\epsilon^2\int\limits_{V_C}\rho_C^{*\prime}\left[\left(\gamma+\frac{1}{2}\right)v_C^{\prime 2}+\Pi_C^{\prime}+\gamma\frac{\pi_C^{\prime kk}}{\rho_C^{*\prime}}-(2\beta-1)\hat{U}_C^{\prime}\right]w_C^{\prime\langle L\rangle}d^3w_C^{\prime} \\
&\quad -\frac{\epsilon^2}{2(2l+3)}\left[\ddot{\mathcal{N}}_C^{\langle L\rangle}-4(1+\gamma)\frac{2l+1}{l+1}\dot{\mathcal{R}}_C^{\langle L\rangle}\right] \\
&\quad +\epsilon^2\sum_{k=1}^{\infty}\frac{1}{k!}\left[Q_C^K+2(\beta-1)P_C^K\right]\int\limits_C\rho_C^{*\prime}w_C^{\prime\langle K\rangle}w_C^{\prime\langle L\rangle}d^3w_C^{\prime} \\
&\quad +\epsilon^2\left(-\frac{l}{2}v_C^k v_C^{\langle i_l}\mathcal{I}_C^{L-1\rangle k}+lF_C^{k<i_l}\mathcal{I}_C^{L-1>k}-lD_C^{k<i_l}\mathcal{I}_C^{L-1>k}\right. \\
&\quad \left.-\,l\mathcal{I}_C^{jk<L-1}D_C^{i_l>jk}-v_C^k\dot{\mathcal{I}}_C^{k\langle L\rangle}+v_C^kR_C^k\dot{\mathcal{I}}_C^{\langle L\rangle}\right)\Bigg\} \\
&\quad +\epsilon^2G\sum_{l=1}^{\infty}\frac{(-1)^l}{(l+1)!}\left(\frac{1}{R_C}\right)_{,L}v_C^k\dot{\mathcal{I}}_C^{\langle kL\rangle} \\
&\quad -\epsilon^2G\sum_{l=1}^{\infty}\frac{(-1)^l l}{(l+1)!}\varepsilon_{kpq}v_C^k\left(\frac{1}{R_C}\right)_{,pL-1}\mathcal{S}_C^{\langle qL-1\rangle} \\
&\quad +\epsilon^2G\sum_{l=1}^{\infty}\frac{(-1)^l(2l-1)}{(2l+1)l!}v_C^k\left(\frac{1}{R_C}\right)_{,kL-1}\mathcal{R}_C^{\langle L-1\rangle}+\mathcal{O}\left(\epsilon^4\right),
\end{aligned}
\tag{7.11}
$$

where two additional multipole moments

$$
\mathcal{N}_C^L=\int\limits_C\rho_C^* w_C^2 w_C^{\langle L\rangle}d^3w_C\,, \tag{7.12}
$$

$$
\mathcal{R}_C^L=\int\limits_C\rho_C^* v_C^k w_C^{\langle kL\rangle}d^3w_C\,. \tag{7.13}
$$

They were called "bad" moments in Ref. [47] because they are not reduced to the "canonical" multipole moments, $\mathcal{I}^L$ and $\mathcal{S}^L$, which appear in the multipolar expansion of the metric tensor in the local coordinates. There are also explicit dependence of expansion (7.11) on the integrals involving interior physical quantities such as velocity of matter v^i_C, potential energy Π_C, stresses π^{ij}_C, and self-potential $\hat{U}_C$. We have proved that these "interior structure" integrals do not appear in the final equations of motion, because they cancel out with the same integrals from the multipolar expansions of other external potentials that are shown below. The "bad" multipoles, $\mathcal{N}^L_C$ and $\mathcal{R}^L_C$, will not show up in the final form of equations of motion for the same reason – the mutual cancellation with similar terms.

External vector potential is expressed in terms of the internal multipole moments as follows:

$$
\begin{aligned}
U^i_C(t,\boldsymbol{x}) &= G\int\limits_{V_C} \frac{\rho^*(t,\boldsymbol{x}')v'^i}{|\boldsymbol{x}-\boldsymbol{x}'|}d^3x' \\
&= G\sum_{l=0}^{\infty}\frac{(-1)^l}{l!}\left(\frac{1}{R_C}\right)_{,L}\mathcal{I}^{\langle L\rangle}_C v^i_C + G\sum_{l=1}^{\infty}\frac{(-1)^l}{(l+1)!}\left(\frac{1}{R_C}\right)_{,L}\dot{\mathcal{I}}^{\langle iL\rangle}_C \\
&\quad - G\sum_{l=1}^{\infty}\frac{(-1)^l l}{(l+1)!}\varepsilon_{ipq}\left(\frac{1}{R_C}\right)_{,pL-1}\mathcal{S}^{\langle qL-1\rangle}_C \\
&\quad + G\sum_{l=1}^{\infty}\frac{(-1)^l(2l-1)}{(2l+1)l!}\left(\frac{1}{R_C}\right)_{,iL-1}\mathcal{R}^{\langle L-1\rangle}_C + \mathcal{O}\left(\epsilon^2\right) .
\end{aligned}
\tag{7.14}
$$

Multipolar expansion of the other external potentials is given by

$$
\begin{aligned}
\Psi_{C1}(t,\boldsymbol{x}) &= G\int\limits_{V_C} \frac{\rho^*(t,\boldsymbol{x}')v'^2}{|\boldsymbol{x}-\boldsymbol{x}'|}d^3x' \\
&= G\sum_{l=0}^{\infty}\frac{(-1)^l}{l!}\left(\frac{1}{R_C}\right)_{,L}\left(\mathcal{I}^{\langle L\rangle}_C v^2_C + \int\limits_{V_C}\rho^{*\prime}_C v'^2_C w'^{\langle L\rangle}_C d^3w'_C\right) \\
&\quad + 2G\sum_{l=1}^{\infty}\frac{(-1)^l}{(l+1)!}\left(\frac{1}{R_C}\right)_{,L} v^m_C\dot{\mathcal{I}}^{\langle mL\rangle}_C \\
&\quad - 2G\sum_{l=1}^{\infty}\frac{(-1)^l l}{(l+1)!}\varepsilon_{mpq}v^m_C\left(\frac{1}{R_C}\right)_{,pL-1}\mathcal{S}^{\langle qL-1\rangle}_C \\
&\quad + 2G\sum_{l=1}^{\infty}\frac{(-1)^l(2l-1)}{(2l+1)l!}v^m_C\left(\frac{1}{R_C}\right)_{,mL-1}\mathcal{R}^{\langle L-1\rangle}_C + \mathcal{O}\left(\epsilon^2\right) ,
\end{aligned}
\tag{7.15}
$$

$$\Psi_{C3}(t,\boldsymbol{x}) = G\int\limits_{V_C}\frac{\rho^*(t,\boldsymbol{x}')\varphi(t,\boldsymbol{x}')}{|\boldsymbol{x}-\boldsymbol{x}'|}d^3x' \tag{7.16}$$

$$= G\sum_{l=0}^{\infty}\frac{(-1)^l}{l!}\left(\frac{1}{R_C}\right)_{,L}\int\limits_{V_C}\rho_C^{*\prime}U_C'w_C'^{\langle L\rangle}d^3w_C'$$

$$+ G^2\sum_{l=0}^{\infty}\sum_{k=0}^{\infty}\sum_{p=0}^{\infty}\frac{(-1)^{l+p}}{l!\,k!\,p!}\left(\frac{1}{R_C}\right)_{,L}\left(\frac{1}{R_{BC}}\right)_{,KP}\mathcal{I}_B^{\langle K\rangle}$$

$$\times\int\limits_{V_C}\rho_C^{*\prime}w_C'^{\langle L\rangle}w_C'^{\langle P\rangle}d^3w_C'$$

$$+ G^2\sum_{D\neq C}\sum_{l=0}^{\infty}\sum_{k=0}^{\infty}\sum_{p=0}^{\infty}\frac{(-1)^{l+k}}{l!\,k!\,p!}\left(\frac{1}{R_C}\right)_{,L}\left(\frac{1}{R_{CD}}\right)_{,KP}\mathcal{I}_D^{\langle K\rangle}$$

$$\times\int\limits_{V_C}\rho_C^{*\prime}w_C'^{\langle L\rangle}w_C'^{\langle P\rangle}d^3w_C' + \mathcal{O}\left(\epsilon^2\right),$$

$$\Psi_{C4}(t,\boldsymbol{x}) = G\int\limits_{V_C}\frac{\rho^*(t,\boldsymbol{x}')\Pi(t,\boldsymbol{x}')}{|\boldsymbol{x}-\boldsymbol{x}'|}d^3x' \tag{7.17}$$

$$= G\sum_{l=0}^{\infty}\frac{(-1)^l}{l!}\left(\frac{1}{R_C}\right)_{,L}\int\limits_{V_C}\rho_C^{*\prime}\Pi_C'w_C'^{\langle L\rangle}d^3w_C' + \mathcal{O}\left(\epsilon^2\right),$$

$$\Psi_{C5}(t,\boldsymbol{x}) = G\int\limits_{V_C}\frac{\pi^{kk}(t,\boldsymbol{x}')}{|\boldsymbol{x}-\boldsymbol{x}'|}d^3x' \tag{7.18}$$

$$= G\sum_{l=0}^{\infty}\frac{(-1)^l}{l!}\left(\frac{1}{R_C}\right)_{,L}\int\limits_{V_C}\pi_C'^{kk}w_C'^{\langle L\rangle}d^3w_C' + \mathcal{O}\left(\epsilon^2\right),$$

where we have also used index D to numerate the bodies in (7.16). One more potential

$$\Psi_{C2} = 6\gamma\Psi_{C3}\,. \tag{7.19}$$

The above expansions contain a number of integrals depending on the internal structure of the bodies explicitly. They will mutually cancel out by similar terms after substitution of these expansions to equations of motion (5.26).

7.2 Translational equations of motion

Translational equations of motion of the center of mass of body B in the global coordinates follows directly from (5.26) and (6.17) and the multipolar decompositions of external potentials provided in Section 7.1. The equations have the following form:

$$M_B a_B^i = F_N^i + \epsilon^2 F_{pN}^i + \mathcal{O}\left(\epsilon^4\right), \tag{7.20}$$

where M_B is the inertial (conformal) mass of the body, F^i_N is the Newtonian gravitational force, and F^i_{pN} is the post-Newtonian gravitational force.

The Newtonian gravitational force, F^i_N, is given by a linear superposition of gravitational forces exerted on the body B by external masses of the N-body system,

$$F^i_N = -\sum_{C\neq B}\sum_{j=0}^{\infty}\sum_{l=0}^{\infty}\frac{(-1)^j(2j+2l+1)!!}{j!\,l!}\frac{G\mathcal{I}^{\langle J\rangle}_B\mathcal{I}^{\langle L\rangle}_C}{R^{2j+2l+3}_{BC}}R^{\langle iJL\rangle}_{BC}\,, \tag{7.21}$$

where $\mathcal{I}^{\langle J\rangle}_B = \mathcal{I}^{\langle p_1\dots p_j\rangle}_B$ are active STF multipole moments of the body B, $\mathcal{I}^{\langle L\rangle}_C = \mathcal{I}^{\langle q_1\dots q_l\rangle}_C$ are active STF multipole moments of the external body C, $R_{BC} = |\boldsymbol{R}_{BC}| = \delta ijR^i_{BC}R^j_{BC}$,

$$R^i_{BC} = R^i_B - R^i_C\,, \tag{7.22}$$

is the coordinate distance between the centers of mass of the bodies defined by the condition of vanishing conformal dipole moment, $I^i = 0$, of the body in its own local coordinates (see Section 6.3), $R^{\langle iJL\rangle}_{BC} = R^{ip_1\dots p_jq_1\dots q_l}$, and repeated indices mean the Einstein summation rule.

We emphasize that the Newtonian gravitational force (7.21) in scalar-tensor theory of gravity depends on the active multipole moments and has a post-Newtonian contribution from the active dipole moments $\mathcal{I}^i$ of the bodies (terms with $l = 1$ and $j = 1$) which do not vanish even if the center of mass of the body is located at the origin of the local coordinates. This is because the center of mass was given in terms of the conformal dipole moment by condition $I^i = 0$. However, the active dipole moment $\mathcal{I}^i \neq I^i$ [93, Equation 6.26].

Additional notice is that the inertial mass, M_B, on the left-hand side of (7.20) is the conformal mass of the body B while the gravitational force (7.21) actually depends on the active masses of the bodies, $\mathcal{M}$ which are the terms with $l = 0$ and $j = 0$ in (7.21). The two masses do not coincide as follows from (6.9). It leads to violation of the strong principle of equivalence in scalar-tensor theory of gravity.

Post-Newtonian gravitational force

The post-Newtonian gravitational force, F^i_{pN}, in (7.20) can be split in three parts,

$$F^i_{pN} = F^i_{pN\mathcal{I}} + F^i_{pN\mathcal{S}} + \mathcal{F}^i_{pN}\,, \tag{7.23}$$

where $F^i_{pN\mathcal{I}}$ is caused by the mass-type multipole moments, $F^i_{pN\mathcal{S}}$ is associated with the spin-type multipole moments, and $\mathcal{F}^i_{pN}$ originates from the post-Newtonian transformation of the internal multipole moments, that appear in calculations of Section 7.1, from the global to local coordinates coordinates.

The mass-type multipole force $F^i_{pN\mathcal{I}}$ consists of various terms describing gravitational coupling between two, three, four, and five internal multipole moments of the extended bodies in the N-body system. The force depends on the first and second time-derivatives of the moments as well. It has the following structure:

$$
\begin{aligned}
F^i_{pN\mathcal{I}} = {} & F^i_{\mathcal{I}\mathcal{I}} + F^i_{\mathcal{I}\dot{\mathcal{I}}} + F^i_{\mathcal{I}\ddot{\mathcal{I}}} + F^i_{\dot{\mathcal{I}}\dot{\mathcal{I}}} + F^i_{\mathcal{I}\mathcal{I}\mathcal{I}} + F^i_{\mathcal{I}\mathcal{I}\mathcal{I}\mathcal{I}} + \\
& F^i_{q\mathcal{I}\mathcal{I}} + F^i_{q\mathcal{I}\dot{\mathcal{I}}} + F^i_{q\mathcal{I}\ddot{\mathcal{I}}} + F^i_{q\dot{\mathcal{I}}\dot{\mathcal{I}}} + F^i_{\dot{q}\mathcal{I}\mathcal{I}} + F^i_{\dot{q}\mathcal{I}\dot{\mathcal{I}}} + F^i_{\ddot{q}\mathcal{I}\mathcal{I}} + F^i_{q\mathcal{I}\mathcal{I}\mathcal{I}\mathcal{I}} \,,
\end{aligned}
\tag{7.24}
$$

where each particular term denotes the number of the moments corresponding to the coupling order. The terms in the first line of (7.24) describe the direct gravitational coupling of the multipole moments. The terms in the second line of (7.24) labeled by a letter "q" (quadrupole) originate from an indirect coupling of the internal quadrupole moments of a body, $\mathcal{I}^{\langle ij\rangle}$ with the corresponding external dipole moment Q_i, which describes acceleration (6.17) of the local coordinates with respect to a geodesic passing through the origin of the coordinates. Thus, all terms labeled with 'q' would be absent had we chosen the local coordinates moving along time-like geodesics.

Specific expressions for different terms in (7.24) are given below in terms of the coordinate distances (7.22) between the bodies and the corresponding multipolar tensor coefficients $D_{\mathcal{I}\mathcal{I}}, D_{\mathcal{I}\dot{\mathcal{I}}}, D_{\mathcal{I}\ddot{\mathcal{I}}}$, etc., which are given in Section 7.2 below. The force components read

$$
\begin{aligned}
F^i_{\mathcal{I}\mathcal{I}} = G \sum_{C\neq B}\sum_{j=0}^{\infty}\sum_{l=0}^{\infty} \Bigg(& D^{JL}_{\mathcal{I}\mathcal{I}} \frac{R^{\langle iJL\rangle}_{BC}}{R^{2j+2l+3}_{BC}} + D^{ikJL}_{\mathcal{I}\mathcal{I}} \frac{R^{\langle kJL\rangle}_{BC}}{R^{2j+2l+3}_{BC}} \\
& + D^{kJL}_{\mathcal{I}\mathcal{I}} \frac{R^{\langle ikJL\rangle}_{BC}}{R^{2j+2l+5}_{BC}} + D^{ikmJL}_{\mathcal{I}\mathcal{I}} \frac{R^{\langle kmJL\rangle}_{BC}}{R^{2j+2l+5}_{BC}} \\
& + {}^{(1)}D^{ikmJL}_{\mathcal{I}\mathcal{I}} \frac{R^{\langle ikmJL\rangle}_{BC}}{R^{2j+2l+5}_{BC}} + {}^{(2)}D^{ikmJL}_{\mathcal{I}\mathcal{I}} \frac{R^{\langle ikmJL\rangle}_{BC}}{R^{2j+2l+7}_{BC}} \Bigg) \,,
\end{aligned}
\tag{7.25}
$$

$$
\begin{aligned}
F^i_{\mathcal{I}\dot{\mathcal{I}}} = G \sum_{C\neq B}\sum_{j=0}^{\infty}\sum_{l=0}^{\infty} \Bigg(& D^{iJL}_{\mathcal{I}\dot{\mathcal{I}}} \frac{R^{\langle JL\rangle}_{BC}}{R^{2j+2l+1}_{BC}} + D^{JL}_{\mathcal{I}\dot{\mathcal{I}}} \frac{R^{\langle iJL\rangle}_{BC}}{R^{2j+2l+3}_{BC}} \\
& + D^{ikJL}_{\mathcal{I}\dot{\mathcal{I}}} \frac{R^{\langle kJL\rangle}_{BC}}{R^{2j+2l+3}_{BC}} + {}^{(1)}D^{kJL}_{\mathcal{I}\dot{\mathcal{I}}} \frac{R^{\langle ikJL\rangle}_{BC}}{R^{2j+2l+3}_{BC}} + {}^{(2)}D^{kJL}_{\mathcal{I}\dot{\mathcal{I}}} \frac{R^{\langle ikJL\rangle}_{BC}}{R^{2j+2l+5}_{BC}} \Bigg) \,,
\end{aligned}
\tag{7.26}
$$

$$
\begin{aligned}
F^i_{\mathcal{I}\ddot{\mathcal{I}}} = G \sum_{C\neq B}\sum_{j=0}^{\infty}\sum_{l=0}^{\infty} \Bigg(& D^{iJL}_{\mathcal{I}\ddot{\mathcal{I}}} \frac{R^{\langle JL\rangle}_{BC}}{R^{2j+2l+1}_{BC}} \\
& + {}^{(1)}D^{JL}_{\mathcal{I}\ddot{\mathcal{I}}} \frac{R^{\langle iJL\rangle}_{BC}}{R^{2j+2l+1}_{BC}} + {}^{(2)}D^{JL}_{\mathcal{I}\ddot{\mathcal{I}}} \frac{R^{\langle iJL\rangle}_{BC}}{R^{2j+2l+3}_{BC}} \Bigg) \,,
\end{aligned}
\tag{7.27}
$$

$$
F^i_{\dot{\mathcal{I}}\dot{\mathcal{I}}} = G \sum_{C\neq B}\sum_{j=0}^{\infty}\sum_{l=0}^{\infty} \left(D^{iJL}_{\dot{\mathcal{I}}\dot{\mathcal{I}}} \frac{R^{\langle JL\rangle}_{BC}}{R^{2j+2l+1}_{BC}} + D^{JL}_{\dot{\mathcal{I}}\dot{\mathcal{I}}} \frac{R^{\langle iJL\rangle}_{BC}}{R^{2j+2l+3}_{BC}} \right) \,,
\tag{7.28}
$$

$$
F^i_{q\mathcal{I}\mathcal{I}} = G \sum_{C\neq B}\sum_{j=0}^{\infty}\sum_{l=0}^{\infty} D^{ikmnJL}_{q\mathcal{I}\mathcal{I}} \frac{R^{\langle kmnJL\rangle}_{BC}}{R^{2j+2l+7}_{BC}} \,,
\tag{7.29}
$$

$$F^{i}_{q\mathcal{J}\dot{\mathcal{J}}} = G \sum_{C\neq B} \sum_{j=0}^{\infty} \sum_{l=0}^{\infty} D^{ikmJL}_{q\mathcal{J}\dot{\mathcal{J}}} \frac{R^{\langle kmJL\rangle}_{BC}}{R^{2j+2l+5}_{BC}}\,, \tag{7.30}$$

$$F^{i}_{q\mathcal{J}\ddot{\mathcal{J}}} = G \sum_{C\neq B} \sum_{j=0}^{\infty} \sum_{l=0}^{\infty} D^{ikJL}_{q\mathcal{J}\ddot{\mathcal{J}}} \frac{R^{\langle kJL\rangle}_{BC}}{R^{2j+2l+3}_{BC}}\,, \tag{7.31}$$

$$F^{i}_{q\dot{\mathcal{J}}\dot{\mathcal{J}}} = G \sum_{C\neq B} \sum_{j=0}^{\infty} \sum_{l=0}^{\infty} D^{ikJL}_{q\dot{\mathcal{J}}\dot{\mathcal{J}}} \frac{R^{\langle kJL\rangle}_{BC}}{R^{2j+2l+3}_{BC}}\,, \tag{7.32}$$

$$F^{i}_{\dot{q}\mathcal{J}\mathcal{J}} = G \sum_{C\neq B} \sum_{j=0}^{\infty} \sum_{l=0}^{\infty} D^{ikmJL}_{\dot{q}\mathcal{J}\mathcal{J}} \frac{R^{\langle kmJL\rangle}_{BC}}{R^{2j+2l+5}_{BC}}\,, \tag{7.33}$$

$$F^{i}_{\dot{q}\mathcal{J}\dot{\mathcal{J}}} = G \sum_{C\neq B} \sum_{j=0}^{\infty} \sum_{l=0}^{\infty} D^{ikJL}_{\dot{q}\mathcal{J}\dot{\mathcal{J}}} \frac{R^{\langle kJL\rangle}_{BC}}{R^{2j+2l+3}_{BC}}\,, \tag{7.34}$$

$$F^{i}_{\ddot{q}\mathcal{J}\mathcal{J}} = G \sum_{C\neq B} \sum_{j=0}^{\infty} \sum_{l=0}^{\infty} D^{ikJL}_{\ddot{q}\mathcal{J}\mathcal{J}} \frac{R^{\langle kJL\rangle}_{BC}}{R^{2j+2l+3}_{BC}}\,, \tag{7.35}$$

$$\begin{aligned} F^{i}_{\mathcal{J}\mathcal{J}\mathcal{J}} = {} & G^2 \sum_{C\neq B} \sum_{D\neq C} \sum_{j=0}^{\infty} \sum_{l=0}^{\infty} \sum_{k=0}^{\infty} {}^{(1)}D^{JLK}_{\mathcal{J}\mathcal{J}\mathcal{J}} \frac{R^{\langle iJL\rangle}_{BC} R^{\langle K\rangle}_{CD}}{R^{2j+2l+3}_{BC} R^{2k+1}_{CD}} \\ & + G^2 \sum_{C\neq B} \sum_{D\neq B} \sum_{j=0}^{\infty} \sum_{l=0}^{\infty} \sum_{k=0}^{\infty} {}^{(2)}D^{JLK}_{\mathcal{J}\mathcal{J}\mathcal{J}} \frac{R^{\langle iJL\rangle}_{BC} R^{\langle K\rangle}_{BD}}{R^{2j+2l+3}_{BC} R^{2k+1}_{BD}}\,, \end{aligned} \tag{7.36}$$

$$\begin{aligned} F^{i}_{\mathcal{J}\mathcal{J}\mathcal{J}\mathcal{J}} = {} & G^2 \sum_{C\neq B} \sum_{D\neq C} \sum_{j=0}^{\infty} \sum_{l=0}^{\infty} \sum_{k=0}^{\infty} \sum_{s=0}^{\infty} \left({}^{(1)}D^{JLSK}_{\mathcal{J}\mathcal{J}\mathcal{J}\mathcal{J}} \frac{R^{\langle JL\rangle}_{BC} R^{\langle iKS\rangle}_{CD}}{R^{2j+2l+1}_{BC} R^{2k+2s+3}_{CD}} \right. \\ & + {}^{(2)}D^{JLSK}_{\mathcal{J}\mathcal{J}\mathcal{J}\mathcal{J}} \frac{R^{\langle imJL\rangle}_{BC} R^{\langle mKS\rangle}_{CD}}{R^{2j+2l+3}_{BC} R^{2k+2s+3}_{CD}} + {}^{(3)}D^{JLSK}_{\mathcal{J}\mathcal{J}\mathcal{J}\mathcal{J}} \frac{R^{\langle imJL\rangle}_{BC} R^{\langle mKS\rangle}_{CD}}{R^{2j+2l+5}_{BC} R^{2k+2s+3}_{CD}} \\ & + D^{mJLSK}_{\mathcal{J}\mathcal{J}\mathcal{J}\mathcal{J}} \frac{R^{\langle iJL\rangle}_{BC} R^{\langle mKS\rangle}_{CD}}{R^{2j+2l+3}_{BC} R^{2k+2s+3}_{CD}} + D^{JiLSK}_{\mathcal{J}\mathcal{J}\mathcal{J}\mathcal{J}} \frac{R^{\langle mJL\rangle}_{BC} R^{\langle mKS\rangle}_{CD}}{R^{2j+2l+3}_{BC} R^{2k+2s+3}_{CD}} \\ & \left. + D^{JmLSK}_{\mathcal{J}\mathcal{J}\mathcal{J}\mathcal{J}} \frac{R^{\langle iJL\rangle}_{BC} R^{\langle mKS\rangle}_{CD}}{R^{2j+2l+3}_{BC} R^{2k+2s+3}_{CD}} + D^{iJLSK}_{\mathcal{J}\mathcal{J}\mathcal{J}\mathcal{J}} \frac{R^{\langle mJL\rangle}_{BC} R^{\langle mKS\rangle}_{CD}}{R^{2j+2l+3}_{BC} R^{2k+2s+3}_{CD}} \right) \\ & + G^2 \sum_{C\neq B} \sum_{D\neq B} \sum_{j=0}^{\infty} \sum_{l=0}^{\infty} \sum_{k=0}^{\infty} \sum_{s=0}^{\infty} \left(D^{JSLK}_{\mathcal{J}\mathcal{J}\mathcal{J}\mathcal{J}} \frac{R^{\langle JL\rangle}_{BC} R^{\langle iKS\rangle}_{BD}}{R^{2j+2l+1}_{BC} R^{2k+2s+3}_{BD}} \right. \\ & \left. + D^{mJSLK}_{\mathcal{J}\mathcal{J}\mathcal{J}\mathcal{J}} \frac{R^{\langle iJL\rangle}_{BC} R^{\langle mKS\rangle}_{BD}}{R^{2j+2l+3}_{BC} R^{2k+2s+3}_{BD}} + D^{iJSLK}_{\mathcal{J}\mathcal{J}\mathcal{J}\mathcal{J}} \frac{R^{\langle mJL\rangle}_{BC} R^{\langle mKS\rangle}_{BD}}{R^{2j+2l+3}_{BC} R^{2k+2s+3}_{BD}} \right)\,, \end{aligned} \tag{7.37}$$

$$F^i_{q\mathcal{I}\mathcal{I}\mathcal{I}\mathcal{I}} = G^2 \sum_{C\neq B} \sum_{D\neq C} \sum_{j=0}^{\infty} \sum_{l=0}^{\infty} \sum_{k=0}^{\infty} \sum_{s=0}^{\infty} D^{ipJLSK}_{q\mathcal{I}\mathcal{I}\mathcal{I}\mathcal{I}} \frac{R^{\langle pmJL\rangle}_{BC} R^{\langle mKS\rangle}_{CD}}{R^{2j+2l+5}_{BC} R^{2k+2s+3}_{CD}}$$
$$+ G^2 \sum_{C\neq B} \sum_{D\neq B} \sum_{j=0}^{\infty} \sum_{l=0}^{\infty} \sum_{k=0}^{\infty} \sum_{s=0}^{\infty} D^{ipJSLK}_{q\mathcal{I}\mathcal{I}\mathcal{I}\mathcal{I}} \frac{R^{\langle pqJL\rangle}_{BC} R^{\langle qKS\rangle}_{BD}}{R^{2j+2l+5}_{BC} R^{2k+2s+3}_{BD}} . \tag{7.38}$$

The spin-type post-Newtonian force has the following structure:

$$F^i_{\mathcal{S}} = F^i_{\mathcal{S}\mathcal{I}} + F^i_{\dot{\mathcal{S}}\mathcal{I}} + F^i_{\mathcal{S}\dot{\mathcal{I}}} + F^i_{\mathcal{S}\mathcal{S}} + F^i_{s\mathcal{I}\mathcal{I}} + F^i_{s\mathcal{I}\dot{\mathcal{I}}} + F^i_{\dot{s}\mathcal{I}\mathcal{I}} , \tag{7.39}$$

where each component of the force is expressed in terms of the corresponding multipolar tensor coefficients $D_{\mathcal{S}\mathcal{I}}$, $D_{\dot{\mathcal{S}}\mathcal{I}}$, etc. The coefficients are given in Section 7.2 below. The terms labeled with "s" stem from the cross product, $\varepsilon_{ipq}S^kQ^q$, of body's spin S^p and the external multipole moment Q^q after replacing Q^q with the right-hand side of (6.17) and making use of the rotational equations of motion for spin of body ($B = 1, 2, \ldots, N$) [93, see section 6.2.2]

$$\dot{\mathcal{S}}^i_B = \sum_{l=0}^{\infty} \frac{1}{l!} \varepsilon_{ijk} Q_{kL} \mathcal{I}^{jL}_B + \mathcal{O}\left(\epsilon^2\right) . \tag{7.40}$$

The spin-type force components are

$$F^i_{\mathcal{S}\mathcal{I}} = G \sum_{C\neq B} \sum_{j=0}^{\infty} \sum_{l=0}^{\infty} \left({}^{(1)}D^{ipJL}_{\mathcal{S}\mathcal{I}} \frac{R^{\langle pJL\rangle}_{BC}}{R^{2j+2l+3}_{BC}} + {}^{(2)}D^{pJL}_{\mathcal{S}\mathcal{I}} \frac{R^{\langle ipJL\rangle}_{BC}}{R^{2j+2l+5}_{BC}} \right.$$
$$\left. + {}^{(3)}D^{ikpJL}_{\mathcal{S}\mathcal{I}} \frac{R^{\langle kpJL\rangle}_{BC}}{R^{2j+2l+5}_{BC}} \right) , \tag{7.41}$$

$$F^i_{\dot{\mathcal{S}}\mathcal{I}} = G \sum_{C\neq B} \sum_{j=0}^{\infty} \sum_{l=0}^{\infty} D^{ipJL}_{\dot{\mathcal{S}}\mathcal{I}} \frac{R^{\langle pJL\rangle}_{BC}}{R^{2j+2l+3}_{BC}} , \tag{7.42}$$

$$F^i_{\mathcal{S}\dot{\mathcal{I}}} = G \sum_{C\neq B} \sum_{j=0}^{\infty} \sum_{l=0}^{\infty} \left(D^{ipJL}_{\mathcal{S}\dot{\mathcal{I}}} \frac{R^{\langle pJL\rangle}_{BC}}{R^{2j+2l+3}_{BC}} + D^{pJL}_{\mathcal{S}\dot{\mathcal{I}}} \frac{R^{\langle ipJL\rangle}_{BC}}{R^{2j+2l+5}_{BC}} \right) , \tag{7.43}$$

$$F^i_{\mathcal{S}\mathcal{S}} = G \sum_{C\neq B} \sum_{j=0}^{\infty} \sum_{l=0}^{\infty} D^{pJqL}_{\mathcal{S}\mathcal{S}} \frac{R^{\langle ipqJL\rangle}_{BC}}{R^{2j+2l+7}_{BC}} , \tag{7.44}$$

$$F^i_{s\mathcal{I}\mathcal{I}} = G \sum_{C\neq B} \sum_{j=0}^{\infty} \sum_{l=0}^{\infty} D^{ikpJL}_{s\mathcal{I}\mathcal{I}} \frac{R^{\langle kpJL\rangle}_{BC}}{R^{2j+2l+5}_{BC}} , \tag{7.45}$$

$$F^i_{s\mathcal{I}\dot{\mathcal{I}}} = G \sum_{C\neq B} \sum_{j=0}^{\infty} \sum_{l=0}^{\infty} D^{ipJL}_{s\mathcal{I}\dot{\mathcal{I}}} \frac{R^{\langle pJL\rangle}_{BC}}{R^{2j+2l+3}_{BC}} , \tag{7.46}$$

$$F^i_{\dot{s}\mathcal{I}\mathcal{I}} = G \sum_{C\neq B} \sum_{j=0}^{\infty} \sum_{l=0}^{\infty} D^{ipJL}_{\dot{s}\mathcal{I}\mathcal{I}} \frac{R^{\langle pJL\rangle}_{BC}}{R^{2j+2l+3}_{BC}} . \tag{7.47}$$

Finally, the force

$$\mathcal{F}^i_{pN} = G \sum_{C \neq B} \sum_{j=0}^{\infty} \sum_{l=0}^{\infty} \left({}^{(1)}D^{JL}_{\mathcal{F}} \frac{R^{\langle iJL \rangle}_{BC}}{R^{2j+2l+3}_{BC}} + {}^{(1)}D^{kJL}_{\mathcal{F}} \frac{R^{\langle ikJL \rangle}_{BC}}{R^{2j+2l+5}_{BC}} + D^{ikmJL}_{\mathcal{F}} \frac{R^{\langle kmJL \rangle}_{BC}}{R^{2j+2l+5}_{BC}} \right) + G \sum_{C \neq B} \sum_{j=1}^{\infty} \sum_{l=0}^{\infty} \left({}^{(2)}D^{JL}_{\mathcal{F}} \frac{R^{\langle iJL \rangle}_{BC}}{R^{2j+2l+3}_{BC}} + {}^{(2)}D^{kJL}_{\mathcal{F}} \frac{R^{\langle ikJL \rangle}_{BC}}{R^{2j+2l+5}_{BC}} \right), \tag{7.48}$$

is also expressed in terms of the corresponding multipolar tensor coefficients $D_{\mathcal{F}}$ given at the end of Section 7.2 below.

The post-Newtonian force in translational equations of motion has been calculated in this chapter for the system of N extended bodies with arbitrary internal structure, shape, and density distribution. It includes the coupling of all internal multipoles of the bodies in the post-Newtonian approximation. The force converges to Einstein–Infeld–Hoffman (EIH) equations of motion [68, 73, 126] for monopole point-like particles which can be considered as massive, spherically symmetric bodies which size is negligibly small compared with the distances between the bodies. The force (7.23) also yields the correct analytic expression for the Lense–Thirring (gravitomagnetic) force due to the gravitational coupling of the body's spin with its orbital angular momentum and spins of the other bodies in the system [19, 32].

We have compared our result for the post-Newtonian gravitational force (7.23) with the post-Newtonian force calculated by Xu et al., [159] for binary pulsars which are considered as monopole-spin-quadrupole particles. We found (almost) perfect agreement of the Xu–Wu–Schäfer force with our expression (7.23) in general-relativistic limit of the PPN parameters $\beta = \gamma = 1$. The only difference was found for the force component $F^i_{\mathcal{S}\mathcal{J}}$ in the second term of the tensor coefficient $D^{ipJL}_{\mathcal{S}\mathcal{J}}$ given by equation (7.91) in the next section. This term is missed in [159].

We also confirmed the post-Newtonian equations of motion derived by Racine and Flanagan [130] by means of the surface-integral technique [4, 76, 77, 147] in general relativity after taking into account the omissions pointed out by Vines et al. in the erratum paper [131]. In particular, the term omitted in [130, Equation 6.12c] and recovered in [131, Equation 1.1] is given by our tensor coefficient ${}^{(2)}D^{JLSK}_{\mathcal{J}\mathcal{J}\mathcal{J}\mathcal{J}}$ in equation (7.76) which enters our expression (7.37) for the post-Newtonian force $F^i_{\mathcal{J}\mathcal{J}\mathcal{J}\mathcal{J}}$. Notice that we give our tensor coefficients for the gravitational force while Racine, Vines, and Flanagan operates with the coefficients for acceleration. Therefore, our tensor coefficients must be divided by mass M_B of the body in order to get the Racine–Vines–Flanagan coefficients. Our results extend the general-relativistic calculation of papers [130, 131] to the post-Newtonian realm of scalar-tensor theory of gravity parameterized with two PPN parameters, β and γ.

We have to make a remark on the post-Newtonian force for spherically symmetric bodies calculated by one of us (S.K.) in [93, Section 6.3.4]. According to this chapter the post-Newtonian force (7.23) depends only on STF multipole moments which are supposed to vanish for spherically symmetric configurations. The reader may think that it should reduce (7.23) to EIH force for monopole particles while the textbook [93, Equations 6.84 and 6.85] shows that besides the EIH force, the post-Newtonian force for spherically symmetric bodies also depends on the rotational moments of inertia of the bodies of the order four, and higher. There is no contradiction here as the definition of the internal multipole moments (including mass) depends on the interior mass density distribution which includes the potential energy of the tidal gravitational field of external bodies (see (4.51)). This makes the internal density distribution depending on the post-Newtonian term being proportional to $Q_K w^{\langle K\rangle}$ which is isotropic but not spatially homogeneous. It contributes to the STF internal mass-type moment of multipolarity l the following term, $Q_K \int_{V_B} \sigma w^{\langle K\rangle} w^{\langle L\rangle} d^3 w$, which should be integrated in the local coordinates (see the last line in (4.51)). This term does not vanish even if the body has a spherical shape and the density distribution σ is spherically symmetric. This is why the finite-size effects will appear in the post-Newtonian equations of motion even for spherically symmetric bodies as discussed in more detail in textbook [93, Section 6.3]. This explains why the force (7.23) does depend on the rotational moments of inertia of spherically symmetric bodies having finite size.

Discussion of the finite size post-Newtonian effects in equations of motion of spherically symmetric bodies has been also given by a number of other researchers including Brumberg [19], Spyrou [139–142], Caporali [26, 27], Dallas [33], Vincent [150] and, more recently, by Arminjon [2]. The post-Newtonian finite-size effects obtained in these papers depend on the rotational moments of inertia of the second order which is inconsistent with our derivation of the post-Newtonian force and represent spurious, coordinate-dependent effect in general relativity which can be removed by the appropriate choice of the center of mass and a proper definition of body's quadrupole moments as discussed in [93, Section 6.3.4] and in [119]. Nonetheless, the gravitational force between spherically symmetric bodies can depend on the rotational moments of inertia of the forth order and higher.

Multipolar tensor coefficients

The multipolar tensor coefficients are expressed in terms of the internal (active) multipole moments and their derivatives. They also depend on velocities v^i_A, v^i_B, v^i_C of the bodies and on their relative velocities,

$$v^i_{BC} = v^i_B - v^i_C \,. \tag{7.49}$$

The multipolar tensor coefficients are given by the following expressions:

$$D^{JL}_{\mathcal{J}\mathcal{J}} = \frac{(-1)^j(2j+2l+1)!!}{j!\,l!}\mathcal{J}^{\langle J\rangle}_B\mathcal{J}^{\langle L\rangle}_C \times \left[-\gamma v_B^2 + 2(\gamma+1)v_B^k v_C^k - \frac{2(2\gamma+1)(j+l)+10\gamma+7}{2(2j+2l+5)}v_C^2\right], \tag{7.50}$$

$$D^{ikJL}_{\mathcal{J}\mathcal{J}} = \frac{(-1)^j(2j+2l+1)!!}{j!\,l!}\mathcal{J}^{\langle J\rangle}_B\mathcal{J}^{\langle L\rangle}_C \times \left[v_B^i v_C^k - \frac{2}{2j+2l+5}v_C^{ik} + 2(\gamma+1)v_{BC}^i v_{BC}^k\right], \tag{7.51}$$

$$D^{kJL}_{\mathcal{J}\mathcal{J}} = \frac{(-1)^j(2j+2l+3)!!}{j!\,l!}\left\{\frac{2j+2l+3}{2(2j+2l+7)}\mathcal{J}^{\langle J\rangle}_B\mathcal{J}^{\langle mL\rangle}_C v_C^k v_C^m + \mathcal{J}^{\langle mJ\rangle}_B\mathcal{J}^{\langle L\rangle}_C\left[\frac{1}{2}v_B^k v_B^m - v_B^m v_C^k + \frac{2}{2j+2l+7}v_C^k v_C^m\right]\right\}, \tag{7.52}$$

$$D^{ikmJL}_{\mathcal{J}\mathcal{J}} = \frac{(-1)^j(2j+2l+3)!!}{j!\,l!}\left\{-\frac{1}{2j+2l+7}\mathcal{J}^{\langle J\rangle}_B\mathcal{J}^{\langle iL\rangle}_C v_C^k v_C^m + \mathcal{J}^{\langle iJ\rangle}_B\mathcal{J}^{\langle L\rangle}_C\left[\frac{1}{2j+3}v_B^k v_B^m - \frac{2}{2j+3}v_B^k v_C^m + \frac{2(2j+l+5)}{(2j+3)(2j+2l+7)}v_C^k v_C^m - \frac{2j^2+5j+4}{2j+3}v_{BC}^k v_{BC}^m\right]\right\}, \tag{7.53}$$

$${}^{(1)}D^{ikmJL}_{\mathcal{J}\mathcal{J}} = \frac{(-1)^j(2j+2l+3)!!}{2j!\,l!}\mathcal{J}^{\langle J\rangle}_B\mathcal{J}^{\langle L\rangle}_C v_C^k v_C^m, \tag{7.54}$$

$${}^{(2)}D^{ikmJL}_{\mathcal{J}\mathcal{J}} = \frac{(-1)^j(2j+2l+5)!!}{(2j+2l+9)j!\,l!}\mathcal{J}^{\langle aJ\rangle}_B\mathcal{J}^{\langle aL\rangle}_C v_C^k v_C^m, \tag{7.55}$$

$$D^{iJL}_{\mathcal{J}\dot{\mathcal{J}}} = \frac{(-1)^j(2j+2l-1)!!}{j!\,l!}\mathcal{J}^{\langle J\rangle}_B\dot{\mathcal{J}}^{\langle L\rangle}_C \times \left\{-(1+2\gamma)v_B^i + \frac{2[2(\gamma+1)(j+l)+3\gamma+2]}{(2j+2l+3)}v_C^i\right\} - 2(\gamma+1)\frac{(-1)^j(2j+2l-1)!!}{j!\,l!}\dot{\mathcal{J}}^{\langle J\rangle}_B\mathcal{J}^{\langle L\rangle}_C v_{BC}^i, \tag{7.56}$$

$$\begin{aligned} D^{JL}_{\mathcal{I}\dot{\mathcal{I}}} &= \frac{(-1)^j(2j+2l+1)!!}{j!\,l!}\Bigg\{ -\mathcal{I}_B^{\langle J\rangle}\dot{\mathcal{I}}_C^{\langle L\rangle} v_C^k v_{BC}^k \\ &\quad + \mathcal{I}_B^{\langle J\rangle}\dot{\mathcal{I}}_C^{\langle kL\rangle}\left[\frac{2j+2l+3}{2j+2l+5}v_C^k + \frac{2(\gamma+1)}{l+1}v_{BC}^k\right] \\ &\quad - \mathcal{I}_B^{\langle kJ\rangle}\dot{\mathcal{I}}_C^{\langle L\rangle}\left[\frac{2j+2l+3}{2j+2l+5}v_C^k + v_{BC}^k\right]\Bigg\} \\ &\quad - 2(\gamma+1)\frac{(-1)^j(2j+2l+1)!!}{(j+1)!\,l!}\dot{\mathcal{I}}_B^{\langle kJ\rangle}\mathcal{I}_C^{\langle L\rangle} v_{BC}^k\,, \end{aligned} \tag{7.57}$$

$$\begin{aligned} D^{ikJL}_{\mathcal{I}\dot{\mathcal{I}}} &= \frac{(-1)^j(2j+2l+1)!!}{j!\,l!} \\ &\quad \times\Bigg\{ -\mathcal{I}_B^{\langle J\rangle}\dot{\mathcal{I}}_C^{\langle iL\rangle}\left[\frac{2}{2j+2l+5}v_C^k + \frac{2(\gamma+1)}{l+1}v_{BC}^k\right] \\ &\quad + \mathcal{I}_B^{\langle iJ\rangle}\dot{\mathcal{I}}_C^{\langle L\rangle}\left[\frac{2}{2j+2l+5}v_C^k + 2(j+1)v_{BC}^k\right]\Bigg\} \\ &\quad + \frac{(-1)^j(2j^2+3j+2\gamma+3)(2j+2l+1)!!}{(j+1)!\,l!}\dot{\mathcal{I}}_B^{\langle iJ\rangle}\mathcal{I}_C^{\langle L\rangle} v_{BC}^k\,, \end{aligned} \tag{7.58}$$

$${}^{(1)}D^{kJL}_{\mathcal{I}\dot{\mathcal{I}}} = \frac{(-1)^j(2j+2l+1)!!}{j!\,l!}\mathcal{I}_B^{\langle J\rangle}\dot{\mathcal{I}}_C^{\langle L\rangle} v_C^k\,, \tag{7.59}$$

$${}^{(2)}D^{kJL}_{\mathcal{I}\dot{\mathcal{I}}} = \frac{2(-1)^j(2j+2l+3)!!}{(2j+2l+7)j!\,l!}\mathcal{I}_B^{\langle mJ\rangle}\dot{\mathcal{I}}_C^{\langle mL\rangle} v_C^k\,, \tag{7.60}$$

$${}^{(1)}D^{JL}_{\mathcal{I}\ddot{\mathcal{I}}} = \frac{(-1)^j(2j+2l-1)!!}{2j!\,l!}\mathcal{I}_B^{\langle J\rangle}\ddot{\mathcal{I}}_C^{\langle L\rangle}\,, \tag{7.61}$$

$$\begin{aligned} D^{iJL}_{\mathcal{I}\ddot{\mathcal{I}}} &= \frac{(-1)^j(2j+2l-1)!!}{j!\,l!} \\ &\quad \times\Bigg\{\frac{4(\gamma+1)j+(4\gamma+3)l+6\gamma+5}{(2j+2l+3)(l+1)}\mathcal{I}_B^{\langle J\rangle}\ddot{\mathcal{I}}_C^{\langle iL\rangle} \\ &\quad - \frac{1}{(2j+3)}\left[(j+2)(2j+1)+\frac{2l}{2j+2l+3}\right]\mathcal{I}_B^{\langle iJ\rangle}\ddot{\mathcal{I}}_C^{\langle L\rangle}\Bigg\} \\ &\quad - \frac{(-1)^j(j^2+j+2\gamma+2)(2j+2l-1)!!}{(j+1)!\,l!}\ddot{\mathcal{I}}_B^{\langle iJ\rangle}\mathcal{I}_C^{\langle L\rangle}\,, \end{aligned} \tag{7.62}$$

$${}^{(2)}D^{JL}_{\mathcal{I}\ddot{\mathcal{I}}} = \frac{(-1)^j(2j+2l+1)!!}{(2j+2l+5)j!\,l!}\mathcal{I}_B^{\langle kJ\rangle}\ddot{\mathcal{I}}_C^{\langle kL\rangle}\,, \tag{7.63}$$

$$\begin{aligned} D^{iJL}_{\dot{\mathcal{I}}\dot{\mathcal{I}}} &= \frac{(-1)^j(2j+2l-1)!!}{j!\,l!} \\ &\quad \times\left[\frac{2(\gamma+1)}{l+1}\dot{\mathcal{I}}_B^{\langle J\rangle}\dot{\mathcal{I}}_C^{\langle iL\rangle} - \frac{2j^2+3j+2\gamma+3}{j+1}\dot{\mathcal{I}}_B^{\langle iJ\rangle}\dot{\mathcal{I}}_C^{\langle L\rangle}\right], \end{aligned} \tag{7.64}$$

$$D^{JL}_{\dot{\mathcal{I}}\dot{\mathcal{I}}} = 2(\gamma+1)\frac{(-1)^j(2j+2l+1)!!}{(j+1)!(l+1)!}\dot{\mathcal{I}}^{\langle kJ\rangle}_B\dot{\mathcal{I}}^{\langle kL\rangle}_C\,, \tag{7.65}$$

$$D^{ikmnJL}_{q\mathcal{I}\mathcal{I}} = -3\frac{(-1)^j(2j+2l+5)!!}{j!\,l!}\frac{\mathcal{I}^{\langle ik\rangle}_B}{\mathcal{M}_B}\mathcal{I}^{\langle J\rangle}_B\mathcal{I}^{\langle L\rangle}_C v^{mn}_{BC}\,, \tag{7.66}$$

$$D^{ikmJL}_{q\mathcal{I}\dot{\mathcal{I}}} = 6\frac{(-1)^j(2j+2l+3)!!}{j!\,l!}\frac{\mathcal{I}^{\langle ik\rangle}_B}{\mathcal{M}_B}(\mathcal{I}^{\langle J\rangle}_B\dot{\mathcal{I}}^{\langle L\rangle}_C+\dot{\mathcal{I}}^{\langle J\rangle}_B\mathcal{I}^{\langle L\rangle}_C)v^{m}_{BC}\,, \tag{7.67}$$

$$D^{ikJL}_{q\mathcal{I}\ddot{\mathcal{I}}} = -3\frac{(-1)^j(2j+2l+1)!!}{j!\,l!}\frac{\mathcal{I}^{\langle ik\rangle}_B}{\mathcal{M}_B}\left(\mathcal{I}^{\langle J\rangle}_B\ddot{\mathcal{I}}^{\langle L\rangle}_C+\ddot{\mathcal{I}}^{\langle J\rangle}_B\mathcal{I}^{\langle L\rangle}_C\right)\,, \tag{7.68}$$

$$D^{ikJL}_{q\dot{\mathcal{I}}\dot{\mathcal{I}}} = -6\frac{(-1)^j(2j+2l+1)!!}{j!\,l!}\frac{\mathcal{I}^{\langle ik\rangle}_B}{\mathcal{M}_B}\dot{\mathcal{I}}^{\langle J\rangle}_B\dot{\mathcal{I}}^{\langle L\rangle}_C\,, \tag{7.69}$$

$$D^{ikmJL}_{\dot{q}\mathcal{I}\mathcal{I}} = 6\frac{(-1)^j(2j+2l+3)!!}{j!\,l!}\frac{\dot{\mathcal{I}}^{\langle ik\rangle}_B}{\mathcal{M}_B}\mathcal{I}^{\langle J\rangle}_B\mathcal{I}^{\langle L\rangle}_C v^{m}_{BC}\,, \tag{7.70}$$

$$D^{ikJL}_{\dot{q}\mathcal{I}\dot{\mathcal{I}}}\dot{q}\mathcal{I}\dot{\mathcal{I}} = -6\frac{(-1)^j(2j+2l+1)!!}{j!\,l!}\frac{\dot{\mathcal{I}}^{\langle ik\rangle}_B}{\mathcal{M}_B}(\mathcal{I}^{\langle J\rangle}_B\dot{\mathcal{I}}^{\langle L\rangle}_C+\dot{\mathcal{I}}^{\langle J\rangle}_B\mathcal{I}^{\langle L\rangle}_C)\,, \tag{7.71}$$

$$D^{ikJL}_{\ddot{q}\mathcal{I}\mathcal{I}} = -3\frac{(-1)^j(2j+2l+1)!!}{j!\,l!}\frac{\ddot{\mathcal{I}}^{\langle ik\rangle}_B}{\mathcal{M}_B}\mathcal{I}^{\langle J\rangle}_B\mathcal{I}^{\langle L\rangle}_C\,, \tag{7.72}$$

$${}^{(1)}D^{JLK}_{\mathcal{I}\mathcal{I}\mathcal{I}} = \gamma\frac{(-1)^j(l+1)(2j+2l+1)!!(2k-1)!!}{j!\,l!k!}\mathcal{I}^{\langle J\rangle}_B\mathcal{I}^{\langle L\rangle}_C\mathcal{I}^{\langle K\rangle}_D\,, \tag{7.73}$$

$${}^{(2)}D^{JLK}_{\mathcal{I}\mathcal{I}\mathcal{I}} = \frac{(-1)^j(\gamma j+\gamma-1)(2j+2l+1)!!(2k-1)!!}{j!\,l!k!}\mathcal{I}^{\langle J\rangle}_B\mathcal{I}^{\langle L\rangle}_C\mathcal{I}^{\langle K\rangle}_D\,, \tag{7.74}$$

$$\begin{aligned}{}^{(1)}D^{JLSK}_{\mathcal{I}\mathcal{I}\mathcal{I}\mathcal{I}} &= -\frac{(-1)^{j+s}[4(\gamma+1)(j+l)+6\gamma+5](2j+2l-1)!!(2k+2s+1)!!}{(2j+2l+3)j!\,l!k!s!\mathcal{M}_C}\\ &\quad\times\mathcal{I}^{\langle J\rangle}_B\mathcal{I}^{\langle L\rangle}_C\mathcal{I}^{\langle S\rangle}_C\mathcal{I}^{\langle K\rangle}_D\,,\end{aligned} \tag{7.75}$$

$${}^{(2)}D^{JLSK}_{\mathcal{I}\mathcal{I}\mathcal{I}\mathcal{I}} = -\frac{(-1)^{j+s}(2j+2l+1)!!(2k+2s+1)!!}{2j!\,l!k!s!\mathcal{M}_C}\mathcal{I}^{\langle J\rangle}_B\mathcal{I}^{\langle L\rangle}_C\mathcal{I}^{\langle S\rangle}_C\mathcal{I}^{\langle K\rangle}_D\,, \tag{7.76}$$

$$\begin{aligned}D^{iJLSK}_{\mathcal{I}\mathcal{I}\mathcal{I}\mathcal{I}} &= \frac{(-1)^{j+s}(2j^2+2jl+7j+2l+4)(2j+2l+1)!!(2k+2s+1)!!}{(2j+2l+5)j!\,l!k!s!\mathcal{M}_C}\\ &\quad\times\mathcal{I}^{\langle iJ\rangle}_B\mathcal{I}^{\langle L\rangle}_C\mathcal{I}^{\langle S\rangle}_C\mathcal{I}^{\langle K\rangle}_D\,,\end{aligned} \tag{7.77}$$

$$D^{mJLSK}_{\mathcal{I}\mathcal{I}\mathcal{I}\mathcal{I}} = -\frac{(-1)^{j+s}(2j+2l+1)!!(2k+2s+1)!!}{(2j+2l+5)j!\,l!k!s!\mathcal{M}_C}\mathcal{I}^{\langle mJ\rangle}_B\mathcal{I}^{\langle L\rangle}_C\mathcal{I}^{\langle S\rangle}_C\mathcal{I}^{\langle K\rangle}_D\,, \tag{7.78}$$

$$D^{JiLSK}_{\mathcal{I}\mathcal{I}\mathcal{I}\mathcal{I}} = \frac{(-1)^{j+s}(2j+2l+1)!!(2k+2s+1)!!}{(2j+2l+5)j!\,l!k!s!\mathcal{M}_C}\mathcal{I}^{\langle J\rangle}_B\mathcal{I}^{\langle iL\rangle}_C\mathcal{I}^{\langle S\rangle}_C\mathcal{I}^{\langle K\rangle}_D\,, \tag{7.79}$$

$$\begin{aligned}D^{JmLSK}_{\mathcal{I}\mathcal{I}\mathcal{I}\mathcal{I}} &= -\frac{(-1)^{j+s}(2l^2+2jl+2j+7l+4)(2j+2l+1)!!(2k+2s+1)!!}{(2j+2l+5)j!\,l!k!s!\mathcal{M}_C}\\ &\quad\times\mathcal{I}^{\langle J\rangle}_B\mathcal{I}^{\langle mL\rangle}_C\mathcal{I}^{\langle S\rangle}_C\mathcal{I}^{\langle K\rangle}_D\,,\end{aligned} \tag{7.80}$$

$$ {}^{(3)}D^{JLSK}_{\mathcal{IIII}} = -\frac{(-1)^{j+s}(2j+2l+3)!!(2k+2s+1)!!}{(2j+2l+7)j!\,l!k!s!\mathcal{M}_C}\mathcal{I}^{\langle aJ\rangle}_B\mathcal{I}^{\langle aL\rangle}_C\mathcal{I}^{\langle S\rangle}_C\mathcal{I}^{\langle K\rangle}_D\,, \tag{7.81} $$

$$ D^{JSLK}_{\mathcal{IIII}} = \frac{(-1)^{j+s}(j+2\gamma+2)(2j+2l-1)!!(2k+2s+1)!!}{j!\,l!k!s!\mathcal{M}_B} \times \mathcal{I}^{\langle J\rangle}_B\mathcal{I}^{\langle S\rangle}_B\mathcal{I}^{\langle L\rangle}_C\mathcal{I}^{\langle K\rangle}_D\,, \tag{7.82} $$

$$ D^{mJSLK}_{\mathcal{IIII}} = -\frac{(-1)^{j+s}(j+1)(2j+2l+1)!!(2k+2s+1)!!}{j!\,l!k!s!\mathcal{M}_B} \times \mathcal{I}^{\langle mJ\rangle}_B\mathcal{I}^{\langle S\rangle}_B\mathcal{I}^{\langle L\rangle}_C\mathcal{I}^{\langle K\rangle}_D\,, \tag{7.83} $$

$$ {}^{(10)}D^{iJSLK}_{\mathcal{IIII}} = -\frac{(-1)^{j+s}j(2j+2l+1)!!(2k+2s+1)!!}{j!\,l!k!s!\mathcal{M}_B} \times \mathcal{I}^{\langle iJ\rangle}_B\mathcal{I}^{\langle S\rangle}_B\mathcal{I}^{\langle L\rangle}_C\mathcal{I}^{\langle K\rangle}_D\,, \tag{7.84} $$

$$ D^{ipJLSK}_{q\mathcal{IIII}} = 3\frac{(-1)^{j+s}(2j+2l+3)!!(2k+2s+1)!!}{j!\,l!k!s!}\frac{\mathcal{I}^{\langle ip\rangle}_B}{\mathcal{M}_B\mathcal{M}_C} \times \mathcal{I}^{\langle J\rangle}_B\mathcal{I}^{\langle L\rangle}_C\mathcal{I}^{\langle S\rangle}_C\mathcal{I}^{\langle K\rangle}_D\,, \tag{7.85} $$

$$ D^{ipJSLK}_{q\mathcal{IIII}} = -3\frac{(-1)^{j+s}(2j+2l+3)!!(2k+2s+1)!!}{j!\,l!k!s!}\frac{\mathcal{I}^{\langle ip\rangle}_B}{\mathcal{M}_B^2} \times \mathcal{I}^{\langle J\rangle}_B\mathcal{I}^{\langle S\rangle}_B\mathcal{I}^{\langle L\rangle}_C\mathcal{I}^{\langle K\rangle}_D\,, \tag{7.86} $$

$$ {}^{(1)}D^{ipJL}_{\mathcal{SI}} = -2(\gamma+1)\frac{(-1)^j j(2j+2l+1)!!}{(j+1)^2 j!\,l!}\mathcal{S}^{\langle J\rangle}_B\mathcal{I}^{\langle L\rangle}_C\,\varepsilon_{ipq}v^q_{BC}\,, \tag{7.87} $$

$$ {}^{(2)}D^{pJL}_{\mathcal{SI}} = -2(\gamma+1)\frac{(-1)^j(j+1)(2j+2l+3)!!}{(j+2)^2 j!\,l!}\mathcal{S}^{\langle qJ\rangle}_B\mathcal{I}^{\langle L\rangle}_C\,\varepsilon_{kpq}v^k_{BC} - 2(\gamma+1)\frac{(-1)^j(2j+2l+3)!!}{(l+2)j!\,l!}\mathcal{I}^{\langle J\rangle}_B\mathcal{S}^{\langle qL\rangle}_C\,\varepsilon_{kpq}v^k_{BC}\,, \tag{7.88} $$

$$ {}^{(3)}D^{ikpJL}_{\mathcal{SI}} = 2(\gamma+1)\frac{(-1)^j(j+1)(2j+2l+3)!!}{(j+2)^2 j!\,l!}\mathcal{S}^{\langle qJ\rangle}_B\mathcal{I}^{\langle L\rangle}_C\,\varepsilon_{ipq}v^k_{BC} + 2(\gamma+1)\frac{(-1)^j(2j+2l+3)!!}{(l+2)j!\,l!}\mathcal{I}^{\langle J\rangle}_B\mathcal{S}^{\langle qL\rangle}_C\,\varepsilon_{ipq}v^k_{BC}\,, \tag{7.89} $$

$$ D^{ipJL}_{\dot{\mathcal{S}}\mathcal{I}} = -2(\gamma+1)\frac{(-1)^j(2j+2l+1)!!}{(j+2)j!\,l!}\varepsilon_{ipq}\dot{\mathcal{S}}^{\langle qJ\rangle}_B\mathcal{I}^{\langle L\rangle}_C - 2(\gamma+1)\frac{(-1)^j(2j+2l+1)!!}{(l+2)j!\,l!}\varepsilon_{ipq}\mathcal{I}^{\langle J\rangle}_B\dot{\mathcal{S}}^{\langle qL\rangle}_C\,, \tag{7.90} $$

$$D^{ipJL}_{\mathcal{S}\dot{\mathcal{J}}} = 2(\gamma+1)\frac{(-1)^j(2j+2l+1)!!}{j!\,l!}\varepsilon_{ipq} \times\left[\frac{j}{(j+1)^2(l+1)}\mathcal{S}_B^{\langle J\rangle}\dot{\mathcal{J}}_C^{\langle qL\rangle} - \frac{j+1}{(j+2)^2}\mathcal{S}_B^{\langle qJ\rangle}\dot{\mathcal{J}}_C^{\langle L\rangle}\right] - 2(\gamma+1)\frac{(-1)^j(2j+2l+1)!!}{(l+2)j!\,l!}\varepsilon_{ipq}\dot{\mathcal{J}}_B^{\langle J\rangle}\mathcal{S}_C^{\langle qL\rangle}\,, \tag{7.91}$$

$$D^{pJL}_{\mathcal{S}\dot{\mathcal{J}}} = 2(\gamma+1)\frac{(-1)^j(j+1)(2j+2l+3)!!}{(j+2)^2 j!(l+1)!}\varepsilon_{kpq}\mathcal{S}_B^{\langle qJ\rangle}\dot{\mathcal{J}}_C^{\langle kL\rangle} - 2(\gamma+1)\frac{(-1)^j(2j+2l+3)!!}{(l+2)(j+1)!\,l!}\varepsilon_{kpq}\dot{\mathcal{J}}_B^{\langle kJ\rangle}\mathcal{S}_C^{\langle qL\rangle}\,, \tag{7.92}$$

$$D^{pJqL}_{\mathcal{S}\mathcal{S}} = 2(\gamma+1)\frac{(-1)^j(2j+2l+5)!!}{(j+2)(l+2)j!\,l!}\mathcal{S}_B^{\langle pJ\rangle}\mathcal{S}_C^{\langle qL\rangle}\,, \tag{7.93}$$

$$D^{ikpJL}_{s\mathcal{J}\mathcal{J}} = -\frac{(-1)^j(2j+2l+3)!!}{j!\,l!}\frac{\mathcal{S}^q_B}{\mathcal{M}_B}\mathcal{J}_B^{\langle J\rangle}\mathcal{J}_C^{\langle L\rangle}\varepsilon_{ipq}v^k_{BC}\,, \tag{7.94}$$

$$D^{ipJL}_{s\mathcal{J}\dot{\mathcal{J}}} = \frac{(-1)^j(2j+2l+1)!!}{j!\,l!}\frac{\mathcal{S}^q_B}{\mathcal{M}_B}\varepsilon_{ipq}\left(\mathcal{J}_B^{\langle J\rangle}\dot{\mathcal{J}}_C^{\langle L\rangle} + \dot{\mathcal{J}}_B^{\langle J\rangle}\mathcal{J}_C^{\langle L\rangle}\right)\,, \tag{7.95}$$

$$D^{ipJL}_{\dot{s}\mathcal{J}\mathcal{J}} = 2\frac{(-1)^j(2j+2l+1)!!}{j!\,l!}\frac{\dot{\mathcal{S}}^q_B}{\mathcal{M}_B}\varepsilon_{ipq}\mathcal{J}_B^{\langle J\rangle}\mathcal{J}_C^{\langle L\rangle}\,, \tag{7.96}$$

$$^{(1)}D^{JL}_{\mathcal{F}} = -\frac{(-1)^j(2j+2l+1)!!}{j!\,l!}\mathcal{J}_B^{\langle J\rangle}\mathcal{J}_C^{\langle L\rangle}\left[(j+1)A_B + (l+1)A_C\right]\,, \tag{7.97}$$

$$^{(1)}D^{kJL}_{\mathcal{F}} = -\frac{(-1)^j(2j+2l+3)!!}{j!\,l!}\left(\mathcal{J}_B^{\langle mJ\rangle}\mathcal{J}_C^{\langle L\rangle}F^{km}_B + \mathcal{J}_B^{\langle J\rangle}\mathcal{J}_C^{\langle mL\rangle}F^{mk}_C\right)\,, \tag{7.98}$$

$$D^{ikmJL}_{\mathcal{F}} = \frac{2(-1)^j(2j+2l+3)!!}{(2j+3)j!\,l!}\mathcal{J}_B^{\langle iJ\rangle}\mathcal{J}_C^{\langle L\rangle}F^{km}_B\,, \tag{7.99}$$

$$^{(2)}D^{JL}_{\mathcal{F}} = -\frac{(-1)^j(2j+2l+1)!!}{j!\,l!}\mathcal{A}_B\left(\dot{\mathcal{J}}_B^{\langle J\rangle}\mathcal{J}_C^{\langle L\rangle} + \mathcal{J}_B^{\langle J\rangle}\dot{\mathcal{J}}_C^{\langle L\rangle}\right)\,, \tag{7.100}$$

$$^{(2)}D^{kJL}_{\mathcal{F}} = \frac{(-1)^j(2j+2l+3)!!}{j!\,l!}\mathcal{A}_B\mathcal{J}_B^{\langle J\rangle}\mathcal{J}_C^{\langle L\rangle}v^k_{BC}\,. \tag{7.101}$$

This completes derivation of the translational equations of motion of extended bodies in the global coordinates.

8 Covariant equations of translational motion of extended bodies

8.1 Effective background manifold

Equations of translational motion of extended bodies derived in the previous section, can be reformulated in a covariant form that is independent of the choice of coordinates and/or gauge condition. The covariant form of the equations is formulated on the effective background spacetime manifold, $\bar{\mathcal{M}}$, which emerge naturally in geometric description of world line $\mathcal{Z}$ of an extended body B that is considered as massive particle endowed with mass M, mass-type $\mathcal{I}^L$, and spin-type $\mathcal{S}^L$, multipoles. In this section, we assume that the body is always at the origin of the local coordinates which world line $\mathcal{W}$ coincides with $\mathcal{Z}$ so that they are indistinguishable. We shall also use a convention, $c = 1$, to avoid appearance of awkward combinations of symbols.

Manifold $\bar{\mathcal{M}}$ is endowed with the effective background metric tensor, $\bar{g}_{\alpha\beta}$, which is originally defined in the global coordinates x^α in terms of the external potentials $\bar{U}, \bar{U}^i, \bar{\Psi}, \bar{\chi}$ given in (5.24). The background metric is determined uniquely by the matching of the external and internal solutions for the metric tensor and scalar field which effectively cancel out all internal forces and potentials from the equations of motion of the origin of the local coordinates (5.26). These equations of motion can be represented in the form of equation of geodesic of the effective background metric that is disturbed by some external force exerted on the "particle" located at the origin of the local coordinates (see Section 8.2). The effective metric is given by the following equation (cf. [5, 107, 147]):

$$\bar{g}_{00}(t, \boldsymbol{x}) = -1 + 2\bar{U} + 2\left(\bar{\Psi} - \beta\bar{U}^2 - \frac{1}{2}\bar{\chi}_{,tt}\right) + \mathcal{O}(5) , \tag{8.1}$$

$$\bar{g}_{0i}(t, \bar{x}) = -2(1 + \gamma)\bar{U}^i + \mathcal{O}(5) , \tag{8.2}$$

$$\bar{g}_{ij}(t, \boldsymbol{x}) = \delta_{ij}\left(1 + 2\gamma\bar{U}\right) + \mathcal{O}(4) , \tag{8.3}$$

where all potentials on the right-hand side of (8.1)–(8.3) are functions of $x^\alpha = (ct, \boldsymbol{x})$, and the symbol $\mathcal{O}(n) = \mathcal{O}(\epsilon^n)$. The background metric in arbitrary coordinates can be obtained from (8.1)–(8.3) by performing a coordinate transformation.

The background metric, $\bar{g}_{\alpha\beta}$, is a starting point of the subsequent covariant development. It has the Christoffel symbols

$$\bar{\Gamma}^\alpha_{\mu\nu} = \frac{1}{2}\bar{g}^{\alpha\beta}\left(\bar{g}_{\beta\mu,\nu} + \bar{g}_{\beta\nu,\mu} - \bar{g}_{\mu\nu,\beta}\right) , \tag{8.4}$$

which can be directly calculated in the global coordinates, x^α, by taking partial derivatives from the metric components (8.1)–(8.3). In what follows, we shall make use of a covariant derivative defined on the background manifold $\bar{\mathcal{M}}$ with the help of the Christoffel symbols $\bar{\Gamma}^\alpha_{\mu\nu}$. The covariant derivative on the background manifold, $\bar{\mathcal{M}}$, is denoted with a vertical bar in order to distinguish it from the covariant derivative defined on the original spacetime manifold, $\mathcal{M}$, denoted with a semicolon or with ∇. For example, the parallel transport of any differentiable vector field V^α is defined on the background manifold by the following equation:

$$V^\alpha{}_{|\beta} = V^\alpha{}_{,\beta} + \bar{\Gamma}^\alpha_{\mu\beta} V^\mu \,, \tag{8.5}$$

which is naturally extended to tensor fields in a standard way [93]. It is straightforward to define other geometric objects on the background manifold like the Riemann tensor

$$\bar{R}^\alpha{}_{\mu\beta\nu} = \bar{\Gamma}^\alpha_{\mu\nu,\beta} - \bar{\Gamma}^\alpha_{\mu\beta,\nu} + \bar{\Gamma}^\alpha_{\sigma\beta}\bar{\Gamma}^\sigma_{\mu\nu} - \bar{\Gamma}^\alpha_{\sigma\nu}\bar{\Gamma}^\sigma_{\mu\beta} \,, \tag{8.6}$$

and its contractions – the Ricci tensor $\bar{R}_{\mu\nu} = \bar{R}^\alpha{}_{\mu\alpha\nu}$, and the Ricci scalar $\bar{R} = \bar{g}^{\mu\nu}\bar{R}_{\mu\nu}$. Tensor indices on the background manifold are raised and lowered with the help of the metric $\bar{g}_{\alpha\beta}$.

The local coordinates, $w^\alpha = (u, w^i)$, of a body B on the effective background manifold are reduced to the local inertial coordinates of a point-like particle with mass M placed at the origin of the coordinates and moving along world line $\mathcal{Z}$. The time coordinate u measured at the origin of the local coordinates on the effective background manifold is now identical to a proper time counted along the world line of the point-like particle. We shall denote this proper time $s \equiv u(\boldsymbol{x} = \boldsymbol{x}_B)$. The background metric in the local coordinates is reduced to

$$\hat{\bar{g}}_{00} = -1 + 2Q + \sum_{l=1}^{\infty} \frac{1}{l!} Q_L w^L + \mathcal{O}(4) \,, \tag{8.7}$$

$$\hat{\bar{g}}_{0i} = \mathcal{O}(3) \,, \tag{8.8}$$

$$\hat{\bar{g}}_{ij} = \delta_{ij}\left(1 + 2A + \sum_{l=1}^{\infty} Q_L w^L\right) + \mathcal{O}(4) \,, \tag{8.9}$$

where the external multipoles $Q_L = \bar{U}_{,i_1\ldots i_l}(\boldsymbol{x}_B)$, and the omitted post-Newtonian terms are identical with $\hat{h}^{\rm ext}_{\alpha\beta}$ given in (4.46)–(4.49). Post-Newtonian transformation from the global, x^α, to local, w^α, coordinates has been provided in Section 5.2. It smoothly matches the two forms of the background metrics in these coordinates. In what follows, we will need a matrix of transformation taken on the world line of the origin of the local coordinates

$$\Lambda^\alpha{}_\beta \equiv \left[\frac{\partial w^\alpha}{\partial x^\beta}\right]_{x^i = x^i_B} \,. \tag{8.10}$$

The components of this matrix are [93, Section 5.1.3]

$$\Lambda^0{}_0 = 1 + \frac{1}{2}v_B^2 - \bar{U}(\boldsymbol{x}_B) - Q + \mathcal{O}(4) \,, \tag{8.11}$$

$$\Lambda^0{}_i = -v^i_B + \mathcal{O}(3) \,, \tag{8.12}$$

$$\Lambda^i{}_0 = -v_B^i + \mathcal{O}(3)\,, \tag{8.13}$$

$$\Lambda^i{}_j = \delta^{ij}\left[1 - A + \gamma\bar{U}(\boldsymbol{x}_B)\right] + \frac{1}{2}v_B^i v_B^j + F^{ij} + \mathcal{O}(4)\,, \tag{8.14}$$

where F^{ij} is the matrix of relativistic precession and Q and A are defined by the adopted system of units of time and space measurements in astronomy [89]. The residual terms shown in (8.11)–(8.14) are not important in the derivation given below but, if necessary, can be found in [93, Section 5.1.3].

We will also need the inverse matrix of transformation between the local and global coordinates taken on the world line $\mathcal{Z}$ of the origin of the local coordinates. We shall denote this matrix as

$$\Omega^\alpha{}_\beta \equiv \left[\frac{\partial x^\alpha}{\partial w^\beta}\right]_{x^i = x_B^i}\,. \tag{8.15}$$

It is obvious that

$$\Lambda^\alpha{}_\beta \Omega^\beta{}_\gamma = \delta^\alpha_\gamma\,, \tag{8.16}$$

which complies with the definition of inverse matrix.

Matrices $\Lambda^\alpha{}_\beta$ and $\Omega^\alpha{}_\beta$ allow us to transform the internal multipole moments of the bodies and external gravitational field from the local to global coordinates and opposite. We should be careful about distinguishing the components of the multipole moments in the two coordinate charts. The convention is that all multipole moments, both internal and external, are defined in such a way that they have only spatial components in the local coordinate chart which means that they are orthogonal to 4-velocity of the world line $\mathcal{Z}$. For this reason, only the spatial components (indexed with Roman letters) of the moments participate in the transformation from local to global coordinates. Thus, the multipole moments with Roman indices belongs to the local chart and the same object with Greek indices belongs to the global coordinates.

In order to arrive to covariant formulation of the translational equations of motion we, first, formulate equations of motion in the local coordinates and, then, transform them to the global coordinates with the help of the transformation matrices. After having applied (8.16) it turned out that all transformation matrices cancel out at the final form of the equations which acquire a covariant four-dimensional form without any reference to an original coordinate charts that were used in the intermediate calculations. We carry out these calculations below.

8.2 Geodesic motion and 4-force

Let us introduce a 4-velocity u^α of the center of mass of body B. In the global coordinates, x^α, the world line $\mathcal{Z}$ of the body's center of mass is described parametrically by $x_B^0 = ct$, and $x_B^i(t)$. The 4-velocity is defined by

$$u^\alpha = \frac{dx_B^\alpha}{ds}\,, \tag{8.17}$$

where s is the proper time along the world line $\mathcal{Z}$. The increment ds of the proper time is related to the increments dx^α of the global coordinates by equation,

$$ds^2 = -\bar{g}_{\alpha\beta} dx^\alpha dx^\beta \,, \tag{8.18}$$

which tells us that the 4-velocity (8.17) is normalized to unity, $u_\alpha u^\alpha = \bar{g}_{\alpha\beta} u^\alpha u^\beta = -1$. In the local coordinates the world line $\mathcal{Z}$ is given by $(u = s,\ w^i = 0)$, and the 4-velocity has components $u^\alpha = (1, 0, 0, 0)$. In the global coordinates the components of the 4-velocity are, $u^\alpha = (dt/ds,\ dx^i_B/ds)$, which yields three-dimensional velocity of the body's center of mass, $v^i = u^i/u^0 = dx^i_B/dt$.

The extended body is treated as a "particle" endowed with mass M, mass-type multipole moments $\mathcal{I}^L$ and spin-type multipole moments $\mathcal{S}^L$ attached to the origin of the local coordinates at any time. This set of multipoles fully characterize the internal structure of the body. The multipoles can depend on time including the mass which is not constant in the most general case (see (6.7)). We postulate that the covariant equations of motion of body B have the following form:

$$M \frac{du^\alpha}{ds} = F^\alpha \,, \tag{8.19}$$

where $du^\alpha/ds \equiv u^\beta u^\alpha{}_{|\beta}$ is four acceleration of the center of mass of the body B, and F^α is four-force that causes the world line $\mathcal{Z}$ of the center of mass of the body to deviate from geodesic of the background manifold $\bar{\mathcal{M}}$. We have to introduce this force to equation (8.19) because the origin of the local coordinates experiences acceleration Q_i given by (6.17). In what follows it is more convenient to operate with a four-force per unit mass defined by $f^\alpha \equiv F^\alpha/M$. The force f^α is orthogonal to 4-velocity, $u_\alpha f^\alpha = 0$. Hence, in arbitrary coordinates the time (co-vector) component of the force is related to its spatial components: $f_0 = -v^i_B f_i$. Contravariant time component of the force relates to its spatial components as follows:

$$f^0 = -\frac{\bar{g}_{ij}}{\bar{g}_{00}} v^j_B f^i \,. \tag{8.20}$$

Our task is to prove that (8.19) matches equations of motion (5.26) of the center of mass of body B derived in the global coordinates by making use of matching of the external and internal solutions of the field equations.

To this end we parameterize (8.19) by coordinate time t instead of the proper time s, which yields

$$\begin{aligned} a^i_B = & -\bar{\Gamma}^i_{00} - 2\bar{\Gamma}^i_{0p} v^p_B - \bar{\Gamma}^i_{pq} v^p_B v^q_B \\ & + \left(\bar{\Gamma}^0_{00} + 2\bar{\Gamma}^0_{0p} v^p_B + \bar{\Gamma}^0_{pq} v^p_B v^q_B\right) v^i_B + \left(f^i - f^0 v^i_B\right) \left(\frac{ds}{dt}\right)^2 \,, \end{aligned} \tag{8.21}$$

where $v^i_B = dx^i_B/dt$ and $a^i_B = dv^i_B/dt$ are coordinate velocity and acceleration of the body's center of mass with respect to the global coordinates. We calculate the Christoffel symbols, $\bar{\Gamma}^\alpha_{\mu\nu}$, the derivative ds/dt, substitute them to (8.21) along with (8.20), and retain only the Newtonian and post-Newtonian terms. It yields

$$
\begin{aligned}
a_B^i = {} & \bar{U}_{,i}(\boldsymbol{x}_B) + \bar{\Psi}_{,i}(\boldsymbol{x}_B) - \frac{1}{2}\bar{\chi}_{,itt}(\boldsymbol{x}_B) + 2(\gamma+1)\dot{\bar{U}}^i(\boldsymbol{x}_B) \\
& - 2(\gamma+1)v_B^k\bar{U}^{k,i}(\boldsymbol{x}_B) - (2\gamma+1)v_B^i\dot{\bar{U}}(\boldsymbol{x}_B) \\
& - 2(\beta+\gamma)\bar{U}(\boldsymbol{x}_B)\bar{U}_{,i}(\boldsymbol{x}_B) + \gamma v_B^2\bar{U}_{,i}(\boldsymbol{x}_B) - v_B^i v_B^k \bar{U}_{,k}(\boldsymbol{x}_B) \\
& + f^i - v_B^i v_B^k f^k - \left[2\bar{U}(\boldsymbol{x}_B) + v_B^2 - 2Q\right] f^i \,.
\end{aligned} \tag{8.22}
$$

This equation matches the equation of motion (5.26) if the spatial components of the force are

$$
f^i = -Q^i - \frac{1}{2}v_B^i v_B^k Q^k + F^{ik}Q^k + \left[A + \gamma\bar{U}(\boldsymbol{x}_B)\right]Q^i \,, \tag{8.23}
$$

By simple inspection we can prove that the post-Newtonian force (8.23) can be written down in a covariant form

$$
f^\alpha = -\bar{g}^{\alpha\beta}\Lambda^i{}_\beta Q_i \,, \tag{8.24}
$$

where $\Lambda^i{}_\beta$ is given above in (8.11)–(8.14), and Q_i is a vector of four acceleration in the local coordinates. The quantity $\Lambda^i{}_\beta Q_i = Q_\beta$ defines components of the four acceleration in the global coordinates with the four acceleration Q_α being orthogonal to 4-velocity, $u^\alpha Q_\alpha = 0$ (see discussion at the end of Section 8.1).

Explicit form of Q_i in the local coordinates is given in (6.17) and should be used in (8.24) along with the covariant form of the external, Q_L, C_L, P_L, and internal, $\mathcal{I}^L$, $\mathcal{S}^L$, moments. This is a matter of discussion in the next section.

8.3 Four-dimensional form of multipole moments

Internal multipole moments

The mathematical procedure that was used in construction of the local coordinates w^α around each extended body in the N-body system indicates that all multipole moments defined in the local coordinates are the STF Cartesian tensors which have only spatial components with their time components being identically nil. It means that the moments are orthogonal to the 4-velocity u^α of the world line $\mathcal{Z}$ of the center of mass of the body. Four-dimensional form of the internal multipole moments is defined by the law of transformation from local to global coordinates,

$$
\mathcal{I}^{\langle\alpha_1\alpha_2\ldots\alpha_l\rangle} \equiv \Omega^{\alpha_1}{}_{i_1}\Omega^{\alpha_2}{}_{i_2}\ldots\Omega^{\alpha_l}{}_{i_l}\mathcal{I}^{\langle i_1 i_2\ldots i_l\rangle} \,, \tag{8.25}
$$

$$
\mathcal{S}^{\langle\alpha_1\alpha_2\ldots\alpha_l\rangle} \equiv \Omega^{\alpha_1}{}_{i_1}\Omega^{\alpha_2}{}_{i_2}\ldots\Omega^{\alpha_l}{}_{i_l}\mathcal{S}^{\langle i_1 i_2\ldots i_l\rangle} \,, \tag{8.26}
$$

and the condition of orthogonality to the 4-velocity,

$$
u_{\alpha_1}\mathcal{I}^{\langle\alpha_1\alpha_2\ldots\alpha_l\rangle} = 0 \quad , \quad u_{\alpha_1}\mathcal{S}^{\langle\alpha_1\alpha_2\ldots\alpha_l\rangle} = 0 \,, \tag{8.27}
$$

where (8.27) is applied to each index. Notice that the matrix of transformation (8.15) has been used in (8.25) and (8.26). It reflects the fact that the internal multipoles were obtained as integrals from the products of coordinates which increments behave under coordinate transformations as vectors, $dx^\alpha = (\partial x^\alpha/\partial w^\beta)dw^\beta$.

External multipole moments

The external multipole moments has been defined by solutions of the field equations in such a way that they have only spatial components in the local coordinates. It means the external multipoles are orthogonal to the 4-velocity of the world line of the center of mass of the body,

$$u^{\alpha_1}Q_{\langle\alpha_1\alpha_2\ldots\alpha_l\rangle} = 0\,, \qquad u^{\alpha_1}P_{\langle\alpha_1\alpha_2\ldots\alpha_l\rangle} = 0\,, \qquad u^{\alpha_1}C_{\langle\alpha_1\alpha_2\ldots\alpha_l\rangle} = 0\,. \tag{8.28}$$

These orthogonality conditions allow us to find out the time components of the multipole moments from their spatial components in any coordinates other than the local chart. This property makes it possible to extend the local three-dimensional definition of the external multipole moments to the four-dimensional tensors by making use of the matrix of transformation (8.10). It yields

$$Q_{\langle\alpha_1\alpha_2\ldots\alpha_l\rangle} \equiv \Lambda^{i_1}{}_{\alpha_1}\Lambda^{i_2}{}_{\alpha_2}\ldots\Lambda^{i_l}{}_{\alpha_l}Q_{\langle i_1i_2\ldots i_l\rangle}\,, \tag{8.29}$$

$$C_{\langle\alpha_1\alpha_2\ldots\alpha_l\rangle} \equiv \Lambda^{i_1}{}_{\alpha_1}\Lambda^{i_2}{}_{\alpha_2}\ldots\Lambda^{i_l}{}_{\alpha_l}C_{\langle i_1i_2\ldots i_l\rangle}\,, \tag{8.30}$$

$$P_{\langle\alpha_1\alpha_2\ldots\alpha_l\rangle} \equiv \Lambda^{i_1}{}_{\alpha_1}\Lambda^{i_2}{}_{\alpha_2}\ldots\Lambda^{i_l}{}_{\alpha_l}P_{\langle i_1i_2\ldots i_l\rangle}\,. \tag{8.31}$$

We have used over here the matrix of transformation (8.10) because the external multipole moments are defined in terms of partial derivatives from the external potentials $\bar{U}$, $\bar{\Psi}$, etc., which behave under coordinate transformations like co-vectors. Four-dimensional form of the gauge-dependent multipoles Z_L is defined similarly

$$Z_{\langle\alpha_1\alpha_2\ldots\alpha_l\rangle} \equiv \Lambda^{i_1}{}_{\alpha_1}\Lambda^{i_2}{}_{\alpha_2}\ldots\Lambda^{i_l}{}_{\alpha_l}Z_{\langle i_1i_2\ldots i_l\rangle}\,. \tag{8.32}$$

It is known that in general relativity the external multipole moments, Q_L and C_L are defined in local coordinates by the Riemann tensor, $\bar{R}^\alpha{}_{\mu\beta\nu}$, of the background metric (8.7) and its spatial derivatives taken at the origin of the local coordinates [127, 143, 147, 161]. This definition remains valid in the scalar-tensor theory of gravity. In the local coordinates the mass-type multipole moments $Q_L = Q_{\langle i_1\ldots i_l\rangle}$, are expressed in terms of the Riemann tensor $R_{0i_10i_2}$ for $l = 2$ and its spatial derivatives $\bar{R}_{0i_10i_2,i_3\ldots i_l}$ for $l \geq 2$. The spin-type multipoles $C_L = C_{\langle i_1i_2\ldots i_l\rangle}$ are given in the local coordinates by the contraction of the Levi–Civita symbol and the Riemann tensor, $\bar{R}_{pq0<i_1}\varepsilon_{i_2>pq}$ for $l = 2$, and by partial derivatives $\bar{R}_{pq0<i_1,i_2\ldots i_{l-1}}\varepsilon_{i_l>pq}$ for $l \geq 2$. The scalar field external multipoles, P_L, are not related in any way to the Riemann tensor because they are expressed in terms of the partial derivatives of the scalar field $\bar{\varphi}$.

Four-dimensional formulation of the external multipole moments is achieved by contracting the Riemann tensor with a vector of 4-velocity, u^α, and taking covariant derivatives projected on the hyperplane being orthogonal to the 4-velocity. The projection is fulfilled with the help of the tensor operator of projection

$$\mathcal{P}^\alpha_\beta \equiv \delta^\alpha_\beta + u^\alpha u_\beta \,, \tag{8.33}$$

which satisfies to $\mathcal{P}^\alpha_\gamma \mathcal{P}^\gamma_\beta = \mathcal{P}^\alpha_\beta$. The operator of projection has only three independent components and reduces to a three-dimensional Kronecker symbol on the world line $\mathcal{Z}$ of the body's center of mass (which coincides with the origin of the local coordinates).

Four-dimensional scalar field external multipoles are defined by (5.27) and (8.31) and reads

$$P_{\langle a_1 \dots a_l \rangle} = \mathcal{P}^{\beta_1}_{<\alpha_1} \cdots \mathcal{P}^{\beta_l}_{\alpha_l>} \bar{\varphi}_{|\beta_1 \cdots \beta_l} + \mathcal{O}(2) \,, \tag{8.34}$$

where $\bar{\varphi}$ is the scalar field perturbation caused by external with respect to the body B masses. In the global coordinates $\bar{\varphi} = \bar{U}$ defined in (5.24a).

Tedious but straightforward calculations prove that the four-dimensional form of the mass-type external multipoles Q_L and C_L are

$$\begin{aligned} Q_{\langle a_1 \dots a_l \rangle} &= E_{\langle a_1 \dots a_l \rangle} + \dot{Z}_{\langle a_1 \dots a_l \rangle} \\ &+ 3 \sum_{s=0}^{l-2} \frac{(l-2)!}{s!(l-2-s)!} E_{<\alpha_1 \dots \alpha_{s+1}} E_{\alpha_{s+2} \dots \alpha_l>} \\ &+ 2 \sum_{s=0}^{l-3} \frac{(l-2)!}{s!(l-2-s)!} E_{<\alpha_1 \dots \alpha_{s+2}} E_{\alpha_{s+3} \dots \alpha_l>} \\ &+ 2 \sum_{k=0}^{l-3} \sum_{s=0}^{k} \frac{(l-2-k)k!}{s!(k-s)!} E_{<\alpha_1 \dots \alpha_{s+1}} E_{\alpha_{s+2} \dots \alpha_l>} \\ &+ 2 \sum_{s=0}^{l-2} \frac{(l-2)!}{s!(l-2-s)!} E_{<\alpha_1 \dots \alpha_{s+1}} \Phi_{\alpha_{s+2} \dots \alpha_l>} \\ &+ 2 \sum_{k=0}^{l-3} \sum_{s=0}^{k} \frac{(l-k-1)k!}{s!(k-s)!} \Phi_{<\alpha_1 \dots \alpha_{s+1}} E_{\alpha_{s+2} \dots \alpha_l>} \\ &+ 2 \sum_{s=0}^{l-2} \frac{(l-2)!}{s!(l-2-s)!} \Theta_{<\alpha_1 \dots \alpha_{s+2}} \Phi_{\alpha_{s+3} \dots \alpha_l>} \\ &+ 2 \sum_{s=0}^{l-2} \frac{(l-2)!}{s!(l-2-s)!} \Theta_{<\alpha_1 \dots \alpha_{s+1}} \Phi_{\alpha_{s+2} \dots \alpha_l>} + \mathcal{O}(4) \,, \end{aligned} \tag{8.35}$$

$$C_{\langle a_1 \dots a_l \rangle} = \frac{l}{l+1} B_{\langle a_1 \dots a_l \rangle} + \mathcal{O}(2) \,, \tag{8.36}$$

where a dot denotes a covariant derivative defined later in (8.41) and we have introduced the following abbreviations:

$$\Phi_{\langle a_1\dots\alpha_l\rangle} \equiv (\gamma-1)P_{\langle a_1\dots\alpha_l\rangle}\,, \tag{8.37}$$

$$\Theta_{\langle a_1\dots\alpha_l\rangle} \equiv \frac{\beta-1}{\gamma-1}P_{\langle a_1\dots\alpha_l\rangle}\,, \tag{8.38}$$

$$E_{\langle a_1\dots\alpha_l\rangle} \equiv -\mathcal{P}^{\beta_1}_{<\alpha_1}\cdots\mathcal{P}^{\beta_l}_{\alpha_l>}u^\mu u^\nu \bar{R}_{\mu\beta_1\nu\beta_2|\beta_3\cdots\beta_l}\,, \tag{8.39}$$

$$B_{\langle a_1\dots\alpha_l\rangle} \equiv -\varepsilon_{\mu\beta_1\rho\sigma}\mathcal{P}^{\rho\delta}\mathcal{P}^{\sigma\gamma}\mathcal{P}^{\beta_1}_{<\alpha_1}\cdots\mathcal{P}^{\beta_l}_{\alpha_l>}u^\mu u^\nu \bar{R}_{\delta\gamma\beta_2\nu|\beta_3\cdots\beta_l}\,, \tag{8.40}$$

with $\varepsilon_{\alpha\beta\gamma\delta}$ being a four-dimensional Levi–Civita tensor. It can be checked that in the local coordinates the right-hand sides of (8.35) and (8.36) are reduced to Q_L and C_L, respectively.

Four-diemnsional definitions of the multipole moments given in this section allows us to transform the products of the moments given in the local coordinates to their covariant counterparts, for example, $Q_L\mathcal{J}^L = Q_{\langle i_1\dots i_l\rangle}\mathcal{J}^{\langle i_1\dots i_l\rangle} = Q_{\langle\alpha_1\dots\alpha_l\rangle}\mathcal{J}^{\langle\alpha_1\dots\alpha_l\rangle}$, etc. In all such products the matrices of transformation cancel out giving rise to covariant expressions being independent of the coordinate choice.

8.4 Covariant translational equations of motion

The covariant equations of translational motion of the center of mass of an extended body B are given by (8.19) with the force f^α defined in (8.24). We use (6.17) to replace Q_i in the expression (8.24) for the force, and a covariant version of equation (7.40) to replace the time derivative of body's spins. We also replace all products of the multipole moments in the local coordinates to their four-dimensional forms in accordance with the definitions given above in Section 8.3. The time derivatives of the multipole moments in the local coordinates are transformed to covariant derivatives taken on the background manifold along the direction of the 4-velocity vector u^α. For instance, for a tensor multipole $\mathcal{J}$ (the indices are suppressed) the first time derivative mapping is

$$\dot{\mathcal{J}} = \frac{d\mathcal{J}}{ds} = u^\alpha\mathcal{J}_{|\alpha}\,. \tag{8.41}$$

and the second time derivative becomes

$$\ddot{\mathcal{J}} = \frac{d}{ds}\frac{d\mathcal{J}}{ds} = u^\alpha u^\alpha\mathcal{J}_{|\alpha\beta} + u^\alpha u^\alpha{}_{|\beta}\mathcal{J}_{|\alpha} = u^\alpha u^\alpha\mathcal{J}_{|\alpha\beta} + f^\alpha\mathcal{J}_{|\alpha}\,, \tag{8.42}$$

where we have used the equation of motion (8.19) in order to replace the covariant derivative from the 4-velocity.

Linear momentum

We have noticed that a number of terms in the covariant expression for the four-force, f^α, can be grouped together to form a total time derivative. Following Fock [73], it is more natural to put all total time derivative terms to the left-hand side of the translational equations of motion and combine it with the four-acceleration $u^\alpha u^\mu{}_{|\beta}$. This procedure introduces a four-momentum of the extended body B in the following form (envisaged by Dixon [62, Equation 83]):

$$\mathfrak{p}^\mu = M n^\mu \,, \tag{8.43}$$

where the *dynamical* velocity, n^α of the body is an algebraic sum of the kinematic velocity, u^α and post-Newtonian corrections:

$$\begin{aligned}
n^\mu &= u^\mu \\
&+ \frac{1}{M}\sum_{l=2}^{\infty}\frac{(l^2+l+4)}{(l+1)!}E_{\langle\alpha_1\dots\alpha_l\rangle}u^\nu \mathcal{J}^{\langle\mu\alpha_1\dots\alpha_l\rangle}{}_{|\nu} \\
&+ \frac{1}{M}\sum_{l=2}^{\infty}\frac{(2l+1)(l^2+3l+6)}{(2l+3)(l+1)!}E_{\langle\alpha_1\dots\alpha_l\rangle|\nu}u^\nu \mathcal{J}^{\langle\mu\alpha_1\dots\alpha_l\rangle} \\
&+ \frac{1}{M}\sum_{l=1}^{\infty}\frac{1}{(l+2)l!}\varepsilon^{\mu\rho}{}_{\nu\sigma}B_{\langle\rho\alpha_1\dots\alpha_l\rangle}u^\nu \mathcal{J}^{\langle\sigma\alpha_1\dots\alpha_l\rangle} \\
&- \frac{4}{M}\sum_{l=1}^{\infty}\frac{(l+1)^2}{(l+2)(l+2)!}\varepsilon^{\mu\rho}{}_{\nu\sigma}E_{\langle\rho\alpha_1\dots\alpha_l\rangle}u^\nu \mathcal{S}^{\langle\sigma\alpha_1\dots\alpha_l\rangle} \\
&+ \frac{2}{M}\sum_{l=1}^{\infty}\frac{1}{(l+1)!}\Phi_{\langle\alpha_1\dots\alpha_l\rangle}u^\nu \mathcal{J}^{\langle\mu\alpha_1\dots\alpha_l\rangle}{}_{|\nu} \\
&+ \frac{2}{M}\sum_{l=1}^{\infty}\frac{(2l+1)}{(2l+3)(l+1)!}\Phi_{\langle\alpha_1\dots\alpha_l\rangle|\nu}u^\nu \mathcal{J}^{\langle\mu\alpha_1\dots\alpha_l\rangle} \\
&- \frac{2}{M}\sum_{l=0}^{\infty}\frac{(l+1)^2}{(l+2)(l+2)!}\varepsilon^{\mu\rho}{}_{\nu\sigma}\Phi_{\langle\rho\alpha_1\dots\alpha_l\rangle}u^\nu \mathcal{S}^{\langle\sigma\alpha_1\dots\alpha_l\rangle}\,, \\
&- \frac{3}{M^2}\sum_{l=1}^{\infty}\frac{1}{l!}E_{\langle\sigma\alpha_1\dots\alpha_l\rangle|\nu}u^\nu \mathcal{J}^{\langle\alpha_1\dots\alpha_l\rangle}\mathcal{J}^{\langle\mu\sigma\rangle} \\
&- \frac{3}{M^2}\sum_{l=1}^{\infty}\frac{1}{l!}E_{\langle\sigma\alpha_1\dots\alpha_l\rangle}u^\nu \mathcal{J}^{\langle\alpha_1\dots\alpha_l\rangle}{}_{|\nu}\mathcal{J}^{\langle\mu\sigma\rangle} \\
&- \frac{3}{M^2}\sum_{l=1}^{\infty}\frac{1}{l!}E_{\langle\sigma\alpha_1\dots\alpha_l\rangle}u^\nu \mathcal{J}^{\langle\alpha_1\dots\alpha_l\rangle}\mathcal{J}^{\langle\mu\sigma\rangle}{}_{|\nu} \\
&+ \frac{1}{M^2}\sum_{l=1}^{\infty}\frac{1}{l!}\varepsilon^{\mu\rho}{}_{\nu\sigma}E_{\langle\rho\alpha_1\dots\alpha_l\rangle}u^\nu \mathcal{J}^{\langle\alpha_1\dots\alpha_l\rangle}\mathcal{S}^\sigma\,,
\end{aligned} \tag{8.44}$$

which provides the post-Newtonian momentum–velocity relation that has not been established in Dixon's paper [62] to full extent. Partial progress in finding this relation has been made by Ehlers and Rudolph [67] but their momentum–velocity relation is incomplete and, furthermore, goes beyond the post-Newtonian approximation without justification.

It is straightforward to check that the dynamical velocity (8.44) is orthogonal to the *kinematic* velocity u^μ,

$$n^\alpha u_\alpha = -1\,, \tag{8.45}$$

which is a consequence of the orthogonality conditions (8.27) and (8.28). The dynamical velocity is normalized $n^\alpha n_\alpha = -1$ which yields for the linear momentum

$$\mathfrak{p}^2 = \bar{g}_{\mu\nu}\mathfrak{p}^\mu\mathfrak{p}^\nu = -M^2\,. \tag{8.46}$$

We notice that the body's four-momentum is not conserved, $\mathfrak{p}^2 \neq$ const., because of nonconservation of mass (6.7).

Gravitational force

Reshuffling all total time derivative terms to the definition of linear momentum $\mathfrak{p}^\mu$, we find out that all second time derivative terms are eliminated from the four-force. It brings equation (8.19) of translational motion of body B to the following covariant form:

$$\frac{d\mathfrak{p}^\mu}{ds} = F^\mu_M + F^\mu_D + F^\mu_E + F^\mu_B + F^\mu_\Phi + F^\mu_{EE} + F^\mu_{EB} + F^\mu_{E\Phi} + F^\mu_{\Phi\Phi} \tag{8.47}$$

where the first term on the right-hand side of (8.47) is caused by the change in the mass of the body

$$F^\mu_M = \dot{M}u^\mu\,, \tag{8.48}$$

where $\dot{M}$ is derived by taking a time derivative from the definition of mass (6.8) and making use of (6.7). It yields

$$\dot{M} = -\sum_{l=1}^{\infty}\frac{1}{(l-1)!}\mathrm{Q}_{\langle a_1\ldots a_l\rangle}\dot{\mathfrak{J}}^{\langle a_1\ldots a_l\rangle} - \sum_{l=1}^{\infty}\frac{l+1}{l!}\dot{\mathrm{Q}}_{\langle a_1\ldots a_l\rangle}\mathfrak{J}^{\langle a_1\ldots a_l\rangle} + \mathcal{O}(4)\,, \tag{8.49}$$

where the factor $1/c^2$ in front of the two sums was omitted since we work in this section under assumption that $c = 1$. We emphasize that formula (8.49) is valid in general relativity as well as in scalar-tensor theory of gravity.

The second term on the right-hand side of (8.47) describes the post-Newtonian Dicke's force caused by violation of the strong principle of equivalence (SEP)

$$F^\mu_D = (\mathcal{M} - M)\,P^\mu\,, \tag{8.50}$$

where

$$P^\mu = \bar{g}^{\beta\gamma}\mathcal{P}^\mu_{\ \beta}\bar{\varphi}_{|\gamma}\,, \tag{8.51}$$

is an external scalar field dipole moment. The difference between the active, $\mathcal{M}$, and conformal, M, masses is given in (6.9). It plays a role of scalar "charge" of body B which interacts with external scalar field like an electric charge interacts with an external electric field in electrodynamics.

The other forces on the right-hand side of (8.47) correspond to the gravitational interaction between internal multipole moments of the body B and their derivatives with the external multipole moments describing tidal gravitational field,

$$F^{\mu}_{E} = \sum_{l=1}^{\infty} \frac{1}{l!} \bar{g}^{\mu\nu} E_{\langle \nu\alpha_1\dots\alpha_l\rangle} \mathcal{I}^{\langle \alpha_1\dots\alpha_l\rangle} \tag{8.52}$$

$$+ \sum_{l=2}^{\infty} \frac{1}{l!} u^{\nu} u^{\mu} E_{\langle \alpha_1\dots\alpha_l\rangle} \mathcal{I}^{\langle \alpha_1\dots\alpha_l\rangle}{}_{|\nu} \tag{8.53}$$

$$- \sum_{l=2}^{\infty} \frac{(l^2+2l-3)}{(2l+3)(l+1)(l+1)!} u^{\nu} u^{\beta} E_{\langle \alpha_1\dots\alpha_l\rangle|\nu} \mathcal{I}^{\langle \mu\alpha_1\dots\alpha_l\rangle}{}_{|\beta}$$

$$+ 4 \sum_{l=1}^{\infty} \frac{(l+1)}{(l+2)(l+2)!} u^{\nu} u^{\beta} \varepsilon^{\mu\rho}{}_{\beta\sigma} E_{\langle \rho\alpha_1\dots\alpha_l\rangle} \mathcal{S}^{\langle \sigma\alpha_1\dots\alpha_l\rangle}{}_{|\nu}$$

$$- \frac{1}{\mathcal{M}} \sum_{l=1}^{\infty} \frac{1}{l!} u^{\nu} u^{\beta} \varepsilon^{\mu\rho}{}_{\beta\sigma} E_{\langle \rho\alpha_1\dots\alpha_l\rangle} \mathcal{I}^{\langle \alpha_1\dots\alpha_l\rangle} \mathcal{S}^{\sigma}{}_{|\nu}$$

$$F^{\mu}_{B} = \sum_{l=1}^{\infty} \frac{l}{(l+1)!} \bar{g}^{\mu\nu} B_{\langle \nu\alpha_1\dots\alpha_l\rangle} \mathcal{S}^{\langle \alpha_1\dots\alpha_l\rangle} \tag{8.54}$$

$$+ \sum_{l=1}^{\infty} \frac{1}{(l+2)!} u^{\nu} u^{\beta} \varepsilon^{\mu}{}_{\beta\rho\sigma} B_{\langle \rho\alpha_1\dots\alpha_l\rangle} \mathcal{I}^{\langle \sigma\alpha_1\dots\alpha_l\rangle}{}_{|\nu}$$

$$F^{\mu}_{\Phi} = 2 \sum_{l=1}^{\infty} \frac{1}{(2l+3)(l+1)(l+1)!} u^{\nu} u^{\beta} \Phi_{\langle \alpha_1\dots\alpha_l\rangle|\nu} \mathcal{I}^{\langle \mu\alpha_1\dots\alpha_l\rangle}{}_{|\beta} \tag{8.55}$$

$$+ 2 \sum_{l=0}^{\infty} \frac{(l+1)}{(l+2)(l+2)!} u^{\nu} u^{\beta} \varepsilon^{\mu\rho}{}_{\beta\sigma} \Phi_{\langle \rho\alpha_1\dots\alpha_l\rangle} \mathcal{S}^{\langle \sigma\alpha_1\dots\alpha_l\rangle}{}_{|\nu},$$

$$F^{\mu}_{EE} = 3\sum_{l=1}^{\infty}\sum_{s=0}^{l-1}\frac{1}{s!(l-1-s)!l}\bar{g}^{\mu\nu}E_{<\alpha_1\dots a_{s+1}}E_{\nu\alpha_{s+2}\dots\alpha_l>}\mathcal{I}^{\langle\alpha_1\dots\alpha_l\rangle} \tag{8.56}$$
$$+2\sum_{l=1}^{\infty}\sum_{s=0}^{l-2}\frac{1}{s!(l-1-s)!l}\bar{g}^{\mu\nu}E_{<\alpha_1\dots a_{s+2}}E_{\nu\alpha_{s+3}\dots\alpha_l>}\mathcal{I}^{\langle\alpha_1\dots\alpha_l\rangle}$$
$$+2\sum_{l=1}^{\infty}\sum_{k=0}^{l-2}\sum_{s=0}^{k}\frac{(l-1-k)k!}{l!s!(k-s)!}\bar{g}^{\mu\nu}E_{<\alpha_1\dots a_{s+1}}E_{\nu\alpha_{s+2}\dots\alpha_l>}\mathcal{I}^{\langle\alpha_1\dots\alpha_l\rangle}$$
$$+\frac{4}{\mathcal{M}}\sum_{l=1}^{\infty}\sum_{k=1}^{\infty}\frac{(l+1)^2}{(l+2)(l+2)!k!}\varepsilon^{\mu\nu}{}_{\rho\sigma}E_{\langle\rho\alpha_1\dots\alpha_l\rangle}E_{\langle\nu\beta_1\dots\beta_k\rangle}\mathcal{I}^{\langle\beta_1\dots\beta_k\rangle}\mathcal{S}^{\langle\sigma\alpha_1\dots\alpha_l\rangle}$$
$$+\frac{1}{\mathcal{M}^2}\sum_{l=1}^{\infty}\sum_{k=1}^{\infty}\frac{1}{l!k!}\varepsilon^{\mu\rho\nu}{}_{\sigma}E_{\langle\nu\beta_1\dots\beta_k\rangle}E_{\langle\rho\alpha_1\dots\alpha_l\rangle}\mathcal{I}^{\langle\beta_1\dots\beta_k\rangle}\mathcal{I}^{\langle\alpha_1\dots\alpha_l\rangle}\mathcal{S}^{\sigma}$$
$$F^{\mu}_{EB} = \frac{1}{\mathcal{M}}\sum_{l=1}^{\infty}\sum_{k=1}^{\infty}\frac{1}{(l+2)l!k!}\varepsilon^{\mu\rho\nu}{}_{\sigma}E_{\langle\nu\beta_1\dots\beta_k\rangle}B_{\langle\rho\alpha_1\dots\alpha_l\rangle}\mathcal{I}^{\langle\beta_1\dots\beta_k\rangle}\mathcal{I}^{\langle\sigma\alpha_1\dots\alpha_l\rangle}\,, \tag{8.57}$$
$$F^{\mu}_{E\Phi} = 2\sum_{l=1}^{\infty}\frac{\gamma-1}{l!}\bar{g}^{\mu\nu}\Theta E_{\langle\nu\alpha_1\dots\alpha_l\rangle}\mathcal{I}^{\langle\alpha_1\dots\alpha_l\rangle} \tag{8.58}$$
$$+2\sum_{l=1}^{\infty}\sum_{s=0}^{l-1}\frac{1}{s!(l-1-s)!l}\bar{g}^{\mu\nu}E_{<\alpha_1\dots\alpha_{s+1}}\Phi_{\nu\alpha_{s+2}\dots\alpha_l>}\mathcal{I}^{\langle\alpha_1\dots\alpha_l\rangle}$$
$$+2\sum_{l=1}^{\infty}\sum_{k=0}^{l-2}\sum_{s=0}^{k}\frac{(l-k)k!}{l!s!(k-s)!}\bar{g}^{\mu\nu}\Phi_{<\alpha_1\dots\alpha_{s+1}}E_{\nu\alpha_{s+2}\dots\alpha_l>}\mathcal{I}^{\langle\alpha_1\dots\alpha_l\rangle}$$
$$+\frac{2}{\mathcal{M}}\sum_{l=0}^{\infty}\sum_{k=1}^{\infty}\frac{(l+1)^2}{(l+2)(l+2)!k!}\varepsilon^{\mu\nu\rho}{}_{\sigma}E_{\langle\nu\beta_1\dots\beta_k\rangle}\mathcal{I}^{\langle\beta_1\dots\beta_k\rangle}$$
$$\times\Phi_{\langle\rho\alpha_1\dots\alpha_l\rangle}\mathcal{S}^{\langle\sigma\alpha_1\dots\alpha_l\rangle}\,,$$
$$F^{\mu}_{\Phi\Phi} = -2\sum_{l=1}^{\infty}\frac{1}{l!}\bar{g}^{\mu\nu}\Theta\Phi_{\langle\nu\alpha_1\dots\alpha_l\rangle}\mathcal{I}^{\langle\alpha_1\dots\alpha_l\rangle} \tag{8.59}$$
$$+2\sum_{l=1}^{\infty}\sum_{s=0}^{l-1}\frac{1}{s!(l-1-s)!l}\bar{g}^{\mu\nu}\Theta_{<\alpha_1\dots\alpha_{s+2}}\Phi_{\nu\alpha_{s+3}\dots\alpha_l>}\mathcal{I}^{\langle\alpha_1\dots\alpha_l\rangle}$$
$$+2\sum_{l=1}^{\infty}\sum_{s=0}^{l-1}\frac{1}{s!(l-1-s)!l}\bar{g}^{\mu\nu}\Theta_{<\alpha_1\dots\alpha_{s+1}}\Phi_{\nu\alpha_{s+2}\dots\alpha_l>}\mathcal{I}^{\langle\alpha_1\dots\alpha_l\rangle}$$

Herein, functions Φ and Θ without indices correspond to definitions (8.37) and (8.38) with indices omitted,

$$\Phi = (\gamma-1)\bar{\varphi}\,, \qquad \Theta = \frac{\beta-1}{\gamma-1}\bar{\varphi}\,. \tag{8.60}$$

8.5 Comparison with Dixon's translational equations of motion

The force F^{μ}_{M} is absent at Dixon's paper [62] because he assumed that the mass M of the body is conserved but this is a rather strong constrain on the time dependence of the internal and external multipoles of the body which cannot be satisfied in the most general case. Dixon's paper [62] does not contain forces F^{μ}_{D}, F^{μ}_{Φ}, $F^{\mu}_{E\Phi}$, and $F^{\mu}_{\Phi\Phi}$ because he worked in the framework of general relativity where these forces are nil by definition because there is no a long-range scalar field in general relativity. Dixon has also neglected general-relativistic forces F^{μ}_{EE} and an essential number of terms in F^{μ}_{E} and F^{μ}_{EB}. Dixon's equations of translational motion (1.13) correspond to the following, severely truncated version of our equations (8.47)

$$\frac{d\mathfrak{p}^{\mu}}{ds} = \frac{1}{2}\bar{g}^{\mu\nu}B_{\langle\nu\alpha\rangle}\mathcal{S}^{\alpha} + \sum_{l=2}^{\infty}\frac{1}{l!}\bar{g}^{\mu\nu}E_{\langle\nu\alpha_1\ldots\alpha_l\rangle}\mathcal{I}^{\langle\alpha_1\ldots\alpha_l\rangle}\,, \tag{8.61}$$

where we have kept only the spin-dipole term in F^{μ}_{B} and the first term on the right-hand side of (8.52) in F^{μ}_{E}.

Comparison with Dixon's equation (1.13) allows us to make the following identifications between our and Dixon's notations for multipole moments ($l \geq 2$):

$$S^{\mu\nu} \equiv \varepsilon^{\mu\nu\alpha\beta}u_{\alpha}\mathcal{S}_{\beta}\,, \tag{8.62a}$$

$$J^{\alpha_1\ldots\alpha_l\mu\nu} \equiv \frac{l+1}{l-1}u^{[\mu}\mathcal{I}^{<\alpha_1]\alpha_2\ldots\alpha_{l-1}[\alpha_l>}u^{\nu]}\,, \tag{8.62b}$$

$$A_{\alpha_1\ldots\alpha_l\mu\nu} \equiv 2\frac{l-1}{l+1}\bar{R}_{\mu\langle\alpha_1/\nu/\alpha_2|\alpha_3\ldots\alpha_l\rangle}\,, \tag{8.62c}$$

which agrees with Dixon's definitions and a forward backslashes around index ν means that it is excluded from STF symmetrization indicated by the angular brackets.

References

[1] T. Appourchaux, R. Burston, Y. Chen, M. Cruise, H. Dittus, B. Foulon, P. Gill, L. Gizon, H. Klein, S. Klioner, S. Kopeikin, H. Krüger, C. Lämmerzahl, A. Lobo, X. Luo, H. Margolis, W.-T. Ni, A. P. Patón, Q. Peng, A. Peters, E. Rasel, A. Rüdiger, É. Samain, H. Selig, D. Shaul, T. Sumner, S. Theil, P. Touboul, S. Turyshev, H. Wang, L. Wang, L. Wen, A. Wicht, J. Wu, X. Zhang and C. Zhao, Astrodynamical Space Test of Relativity Using Optical Devices I (ASTROD I) – A class-M fundamental physics mission proposal for Cosmic Vision 2015–2025, *Experimental Astronomy* **23** (2009), 491–527.

[2] M. Arminjon, Equations of motion according to the asymptotic post-Newtonian scheme for general relativity in the harmonic gauge, *Physical Review D* **72** (2005), 084002.

[3] V. I. Arnold, *Mathematical Methods of Classical Mechanics*, Berlin: Springer, 1995.

[4] H. Asada, T. Futamase and P. Hogan, *Equations of Motion in General Relativity*, Oxford: Oxford University Press, 2011.

[5] N. Ashby and B. Bertotti, Relativistic effects in local inertial frames, *Physical Review D* **34** (1986), 2246–2259.

[6] I. Bailey and W. Israel, Relativistic dynamics of extended bodies and polarized media: An eccentric approach, *Annals of Physics* **130** (1980), 188–214.
[7] W. Beiglböck, The center-of-mass in Einstein's theory of gravitation, *Communications in Mathematical Physics* **5** (1967), 106–130.
[8] J. D. Bekenstein, Relativistic gravitation theory for the modified Newtonian dynamics paradigm, *Phys. Rev. D* **70** (2004), 083509.
[9] D. Bini, C. Cherubini, A. Geralico and A. Ortolan, Dixon's extended bodies and weak gravitational waves, *General Relativity and Gravitation* **41** (2009), 105–116.
[10] L. Blanchet, Gravitational Radiation from Post-Newtonian Sources and Inspiralling Compact Binaries, *Living Reviews in Relativity* **9** (2006), 4, URL (cited in Apr 27, 2014) www.livingreviews.org/lrr-2006-4.
[11] L. Blanchet and T. Damour, Radiative gravitational fields in general relativity. I – General structure of the field outside the source, *Royal Society of London Philosophical Transactions Series A* **320** (1986), 379–430.
[12] L. Blanchet and T. Damour, Post-Newtonian generation of gravitational waves, *Ann. Inst. H. Poincaré* **50** (1989), 337–408.
[13] L. Blanchet, T. Damour and G. Esposito-Farèse, Dimensional regularization of the third post-Newtonian dynamics of point particles in harmonic coordinates, *Physical Review D* **69** (2004), 124007.
[14] L. Blanchet, T. Damour and B. R. Iyer, Surface-integral expressions for the multipole moments of post-Newtonian sources and the boosted Schwarzschild solution, *Classical and Quantum Gravity* **22** (2005), 155–181.
[15] L. Blanchet and G. Faye, Lorentzian regularization and the problem of point-like particles in general relativity, *Journal of Mathematical Physics* **42** (2001), 4391–4418.
[16] L. Blanchet and B. R. Iyer, Hadamard regularization of the third post-Newtonian gravitational wave generation of two point masses, *Physical Review D* **71** (2005), 024004.
[17] C. Brans and R. H. Dicke, Mach's principle and a relativistic theory of gravitation, *Physical Review* **124** (1961), 925–935.
[18] R. A. Breuer and E. Rudolph, The force law for the dynamic two-body problem in the second post-Newtonian approximation of general relativity, *General Relativity and Gravitation* **14** (1982), 181–211.
[19] V. A. Brumberg, *Relativistic Celestial Mechanics*, Moscow: Nauka (in Russian), 1972.
[20] V. A. Brumberg, Relativistic reduction of astronomical measurements and reference frames, in: *ASSL Vol. 86: IAU Colloq. 56: Reference Coordinate Systems for Earth Dynamics* (E. M. Gaposchkin and B. Kolaczek, eds.), pp. 283–294, 1981.
[21] V. A. Brumberg, *Essential Relativistic Celestial Mechanics*, New York: Adam Hilger, 1991.
[22] V. A. Brumberg and A. M. Finkelshtein, The problem of comparing the observation results with the theoretical data to check on relativistic effects in the solar system, *Soviet Journal of Experimental and Theoretical Physics (JETP)* **49** (1979), 749–755.
[23] V. A. Brumberg and S. M. Kopeikin, Relativistic time scales in the solar system, *Celestial Mechanics and Dynamical Astronomy* **48** (1990), 23–44.
[24] V. A. Brumberg and S. M. Kopejkin, Relativistic theory of celestial reference frames, in: *ASSL Vol. 154: Reference Frames* (J. Kovalevsky, I. I. Mueller and B. Kolaczek, eds.), pp. 115–141, 1989a.
[25] V. A. Brumberg and S. M. Kopejkin, Relativistic Reference Systems and Motion of Test Bodies in the Vicinity of the Earth, *Nuovo Cimento B Serie* **103** (1989b), 63–98.
[26] A. Caporali, A reformulation of the post-Newtonian approximation to general relativity. I – The metric and the local equations of motion, *Nuovo Cimento B Serie* **61** (1981), 181–212.

[27] A. Caporali, A reformulation of the post-Newtonian approximation to general relativity. II. Post-Newtonian equations of motion for extended bodies, *Nuovo Cimento B Serie* **61** (1981), 205–212.
[28] S. Chandrasekhar, *Ellipsoidal figures of equilibrium*, New York: Dover, 1987.
[29] Z. Cheng, Relation between the Love numbers and the Earth models., *Journal Nanjing Univ.* **27** (1991), 234–242.
[30] C. Chicone, S. M. Kopeikin, B. Mashhoon and D. G. Retzloff, Delay equations and radiation damping, *Physics Letters A* **285** (2001), 17–26.
[31] I. Ciufolini, Frame-dragging, gravitomagnetism and Lunar Laser Ranging, *New Astronomy* **15** (2010), 332–337.
[32] I. Ciufolini and J. A. Wheeler, *Gravitation and Inertia*, Princeton University Press: Princeton, NJ, 1995.
[33] S. S. Dallas, Equations of motion for rotating finite bodies in the extended PPN formalism, *Celestial Mechanics* **15** (1977), 111–123.
[34] T. Damour, Gravitational radiation and the motion of compact bodies, in: *Gravitational Radiation* (N. Deruelle and T. Piran, eds.), pp. 59–144, 1983.
[35] T. Damour, The problem of motion in Newtonian and Einsteinian gravity, in: *Three Hundred Years of Gravitation* (S. W. Hawking, and W. Israel, eds.), pp. 128–198, Cambridge: Cambridge University Press, 1989.
[36] T. Damour, Introductory Lectures on the Effective One Body Formalism, *International Journal of Modern Physics A* **23** (2008), 1130–1148.
[37] T. Damour, Binary Systems as Test-Beds of Gravity Theories, in: *Astrophysics and Space Science Library* (M. Colpi, P. Casella, V. Gorini, U. Moschella and A. Possenti, eds.), Astrophysics and Space Science Library **359**, pp. 1–4020, 2009.
[38] T. Damour and G. Esposito-Farese, Tensor-multi-scalar theories of gravitation , *Classical and Quantum Gravity* **9** (1992), 2093–2176.
[39] T. Damour and G. Esposito-Farese, Nonperturbative strong-field effects in tensor-scalar theories of gravitation, *Physical Review Letters* **70** (1993), 2220–2223.
[40] T. Damour and G. Esposito-Farèse, Gravitational-wave versus binary-pulsar tests of strong-field gravity, *Physical Review D* **58** (1998), 042001.
[41] T. Damour, L. Grishchuk, S. Kopejkin and G. Schäfer, Higher-order relativistic dynamics of binary systems, in: *Proceedings of the Fifth Marcel Grossmann Meeting, part A* (D. Blair and M. J. Buckingham, eds.), pp. 451–459, World Scientific, Singapore, 1989.
[42] T. Damour and B. R. Iyer, Multipole analysis for electromagnetism and linearized gravity with irreducible Cartesian tensors, *Physical Review D* **43** (1991), 3259–3272.
[43] T. Damour and K. Nordtvedt, General relativity as a cosmological attractor of tensor-scalar theories, *Physical Review Letters* **70** (1993), 2217–2219.
[44] T. Damour and G. Schaefer, New tests of the strong equivalence principle using binary-pulsar data, *Physical Review Letters* **66** (1991), 2549–2552.
[45] T. Damour and G. Schäfer, Lagrangians for n point masses at the second post-Newtonian approximation of general relativity., *General Relativity and Gravitation* **17** (1985), 879–905.
[46] T. Damour, M. Soffel and C. Xu, General-relativistic celestial mechanics. I. Method and definition of reference systems, *Physical Review D* **43** (1991), 3273–3307.
[47] T. Damour, M. Soffel and C. Xu, General-relativistic celestial mechanics. II. Translational equations of motion, *Physical Review D* **45** (1992), 1017–1044.
[48] P. D. D'Eath, Dynamics of a small black hole in a background universe, *Physical Review D* **11** (1975a), 1387–1403.
[49] P. D. D'Eath, Interaction of two black holes in the slow-motion limit, *Physical Review D* **12** (1975b), 2183–2199.

[50] H. Dehnen, H. Frommert and F. Ghaboussi, Higgs field and a new scalar-tensor theory of gravity, *International Journal of Theoretical Physics* **31** (1992), 109–114.
[51] S. Dell'Agnello, M. Maiello, D. G. Currie, A. Boni, S. Berardi, C. Cantone, G. O. Delle Monache, N. Intaglietta, C. Lops, M. Garattini, M. Martini, G. Patrizi, L. Porcelli, M. Tibuzzi, R. Vittori, G. Bianco, A. Coradini, C. Dionisio, R. March, G. Bellettini, R. Tauraso and J. Chandler, Probing general relativity and new physics with lunar laser ranging, *Nuclear Instruments and Methods in Physics Research A* **692** (2012), 275–279.
[52] R. H. Dicke, Long-Range Scalar Interaction, *Physical Review* **126** (1962), 1875–1877.
[53] R. H. Dicke, Mach's principle and invariance under transformation of units, *Physical Review* **125** (1962), 2163–2167.
[54] R. H. Dicke, The weak and strong principles of equivalence, *Annals of Physics* **31** (1965), 235–239.
[55] S. Dittmaier and M. Schumacher, The Higgs boson in the Standard Model – From LEP to LHC: Expectations, Searches, and Discovery of a Candidate, *Progress in Particle and Nuclear Physics* **70** (2013), 1–54.
[56] H. Dittus, C. Lämmerzahl and S. G. Turyshev (eds.), *Lasers, Clocks and Drag-Free Control: Exploration of Relativistic Gravity in Space*, Astrophysics and Space Science Library **349**, 2008.
[57] W. G. Dixon, Dynamics of extended bodies in general relativity. I. Momentum and angular momentum, *Royal Society of London Proceedings Series A* **314** (1970), 499–527.
[58] W. G. Dixon, Dynamics of Extended Bodies in General Relativity. II. Moments of the Charge-Current Vector, *Royal Society of London Proceedings Series A* **319** (1970), 509–547.
[59] W. G. Dixon, The definition of multipole moments for extended bodies, *General Relativity and Gravitation* **4** (1973), 199–209.
[60] W. G. Dixon, Dynamics of Extended Bodies in General Relativity. III. Equations of Motion, *Royal Society of London Philosophical Transactions Series A* **277** (1974), 59–119.
[61] W. G. Dixon, Slow motion approximations in general relativity II. Gauge transformations, *General Relativity and Gravitation* **5** (1974), 119–137.
[62] W. G. Dixon, Extended bodies in general relativity: their description and motion., in: *Isolated Gravitating Systems in General Relativity* (J. Ehlers, ed.), pp. 156–219, 1979.
[63] B. A. Dubrovin, A. T. Fomenko and S. P. Novikov, *Modern Geometry. Part II.*, New York: Springer-Verlag, February 1984.
[64] M. Efroimsky and P. Goldreich, Gauge symmetry of the N-body problem in the Hamilton-Jacobi approach, *Journal of Mathematical Physics* **44** (2003), 5958–5977.
[65] M. Efroimsky and P. Goldreich, Gauge freedom in the N-body problem of celestial mechanics, *Astronomy & Astrophysics* **415** (2004), 1187–1199.
[66] J. Ehlers, Isolated systems in general relativity, in: *Ninth Texas Symposium on Relativistic Astrophysics* (J. Ehlers, J. J. Perry and M. Walker, eds.), Annals of the New York Academy of Sciences **336**, pp. 279–294, February 1980.
[67] J. Ehlers and E. Rudolph, Dynamics of extended bodies in general relativity center-of-mass description and quasirigidity, *General Relativity and Gravitation* **8** (1977), 197–217.
[68] A. Einstein, L. Infeld and B. Hoffmann, The gravitational equations and the problem of motion, *The Annals of Mathematics* **39** (1938), 65–100.
[69] L. P. Eisenhart, *Differential Geometry*, Princeton: Princeton University Press, 1947.
[70] I. G. Fichtengoltz, Lagrangian form of equations of motion in the second approximation of Einstein's theory of gravity, *Journal of Experimental and Theoretical Physics (JETP)* **20** (1950), 233–242.
[71] M. Fierz, On the physical interpretation of P. Jordan's extended theory of gravitation, *Helv. Phys. Acta* **29** (1956), 128–134.
[72] É. É. Flanagan and E. Rosenthal, Can Gravity Probe B usefully constrain torsion gravity theories?, *Physical Review D* **75** (2007), 124016.

[73] V. A. Fock, *The Theory of Space, Time and Gravitation*, New York: Pergamon Press, February 1959.
[74] J. Frauendiener, Conformal infinity, *Living Reviews in Relativity* **7** (2004), 1.
[75] T. Futamase, P. A. Hogan and Y. Itoh, Equations of motion in general relativity of a small charged black hole, *Physical Review D* **78** (2008), 104014.
[76] T. Futamase and Y. Itoh, The Post-Newtonian Approximation for Relativistic Compact Binaries, *Living Reviews in Relativity* **10** (2007), 2.
[77] T. Futamase and B. F. Schutz, Gravitational radiation and the validity of the far-zone quadrupole formula in the Newtonian limit of general relativity, *Physical Review D* **32** (1985), 2557–2565.
[78] J. Getino, Perturbed nutations, Love numbers and elastic energy of deformation for Earth models 1066A and 1066B, *Zeitschrift Angewandte Mathematik und Physik* **44** (1993), 998–1021.
[79] D. Gorbonos and B. Kol, A dialogue of multipoles: Matched asymptotic expansion for caged black holes, *Journal of High Energy Physics* **6** (2004), 53.
[80] D. Gorbonos and B. Kol, Matched asymptotic expansion for caged black holes: regularization of the post-Newtonian order, *Classical and Quantum Gravity* **22** (2005), 3935–3959.
[81] L. P. Grishchuk and S. M. Kopeikin, The Motion of a Pair of Gravitating Bodies Including the Radiation Reaction Force, *Soviet Astronomy Letters* **9** (1983), 230–232.
[82] L. P. Grishchuk and S. M. Kopeikin, Equations of motion for isolated bodies with relativistic corrections including the radiation reaction force, in: *Relativity in Celestial Mechanics and Astrometry. High Precision Dynamical Theories and Observational Verifications* (J. Kovalevsky and V. A. Brumberg, eds.), IAU Symposium **114**, pp. 19–33, 1986.
[83] F. W. Hehl, Y. N. Obukhov and D. Puetzfeld, On Poincaré gauge theory of gravity, its equations of motion, and Gravity Probe B, *Physics Letters A* **377** (2013), 1775–1781.
[84] S. Hergt and G. Schäfer, Higher-order-in-spin interaction Hamiltonians for binary black holes from source terms of Kerr geometry in approximate ADM coordinates, *Physical Review D* **77** (2008), 104001.
[85] L. Infeld and J. Plebański, *Motion and Relativity*, Oxford: Oxford University Press, 1960.
[86] P. Jordan, Formation of the stars and development of the Universe, *Nature* **164** (1949), 637–640.
[87] P. Jordan, Zum gegenwärtigen Stand der Diracschen kosmologischen Hypothesen, *Zeitschrift fur Physik* **157** (1959), 112–121.
[88] R. E. Kates and L. S. Kegeles, Nonanalytic terms in the slow-motion expansion of a radiating scalar field on a Schwarzschild background., *Physical Review D* **25** (1982), 2030–2037.
[89] S. A. Klioner, Relativistic scaling of astronomical quantities and the system of astronomical units, *Astronomy & Astrophysics* **478** (2008), 951–958.
[90] S. A. Klioner and M. H. Soffel, Relativistic celestial mechanics with PPN parameters, *Physical Review D* **62** (2000), 024019.
[91] I. Kolář, P. W. Michor and J. Slovák, *Natural Operations In Differential Geometry*, Springer: Berlin, 1993.
[92] S. Kopeikin, *The gravitomagnetic influence on Earth-orbiting spacecrafts and on the lunar orbit*, Astrophysics and Space Science Library **367**, pp. 337–343, Springer: Dordrecht, 2010.
[93] S. Kopeikin, M. Efroimsky and G. Kaplan, *Relativistic Celestial Mechanics of the Solar System*, Weinheim: Wiley-VCH, September 2011.
[94] S. Kopeikin and I. Vlasov, Parametrized post-Newtonian theory of reference frames, multipolar expansions and equations of motion in the N-body problem, *Phys. Rep.* **400** (2004), 209–318.
[95] S. Kopeikin and I. Vlasov, The Effacing Principle in the Post-Newtonian Celestial Mechanics, in: *The Eleventh Marcel Grossmann Meeting On Recent Developments in Theoretical and*

Experimental General Relativity, Gravitation and Relativistic Field Theories (H. Kleinert, R. T. Jantzen and R. Ruffini, eds.), pp. 2475–2477, 2008.
[96] S. Kopeikin and Y. Xie, Celestial reference frames and the gauge freedom in the post-Newtonian mechanics of the Earth–Moon system, *Celestial Mechanics and Dynamical Astronomy* **108** (2010), 245–263.
[97] S. M. Kopeikin, General relativistic equations of binary motion for extended bodies with conservative corrections and radiation damping, *Soviet Astronomy* **29** (1985), 516–524.
[98] S. M. Kopeikin, Asymptotic matching of gravitational fields in the solar system, *Soviet Astronomy* **33** (1989), 665–672.
[99] S. M. Kopeikin, Comment on "Gravitomagnetic Influence on Gyroscopes and on the Lunar Orbit", *Physical Review Letters* **98** (2007), 229001.
[100] S. M. Kopeikin, E. Pavlis, D. Pavlis, V. A. Brumberg, A. Escapa, J. Getino, A. Gusev, J. Müller, W.-T. Ni and N. Petrova, Prospects in the orbital and rotational dynamics of the Moon with the advent of sub-centimeter lunar laser ranging, *Advances in Space Research* **42** (2008), 1378–1390.
[101] S. M. Kopejkin, Celestial Coordinate Reference Systems in Curved Spacetime, *Celestial Mechanics* **44** (1988), 87–115.
[102] S. M. Kopejkin, Relativistic reference systems in the solar system., *Itogi Nauki i Tekhniki Seriia Astronomiia* **41** (1991), 87–146.
[103] A. Kostelecky, Theory and Status of Lorentz Violation, *APS Meeting Abstracts* (2008), I3.00001.
[104] J. Kovalevsky and I. I. Mueller, Reference frames in astronomy and geophysics: Introduction, in: *Reference Frames in Astronomy and Geophysics*, pp. 1–12, 1989.
[105] J. Kovalevsky and P. K. Seidelmann, *Fundamentals of Astrometry*, Cambridge: Cambridge University Press, 2004.
[106] M. Kramer and N. Wex, The double pulsar system: a unique laboratory for gravity, *Classical and Quantum Gravity* **26** (2009), 073001.
[107] L. D. Landau and E. M. Lifshitz, *The classical theory of fields*, Oxford: Pergamon Press, 1975.
[108] D. R. Lorimer and P. C. C. Freire, New Limits on the Strong Equivalence Principle from Two Long-period Circular-orbit Binary Pulsars, in: *Binary Radio Pulsars* (F. A. Rasio and I. H. Stairs, eds.), Astronomical Society of the Pacific Conference Series **328**, p. 19, July 2005.
[109] Y. Mao, M. Tegmark, A. H. Guth and S. Cabi, Constraining torsion with Gravity Probe B, *Physical Review D* **76** (2007), 104029.
[110] R. March, G. Bellettini, R. Tauraso and S. Dell'Agnello, Constraining spacetime torsion with LAGEOS, *General Relativity and Gravitation* **43** (2011), 3099–3126.
[111] M. Mathisson, Republication of: New mechanics of material systems, *General Relativity and Gravitation* **42** (2010), 1011–1048.
[112] M. Mathisson, Republication of: The mechanics of matter particles in general relativity, *General Relativity and Gravitation* **42** (2010), 989–1010.
[113] D. Mattingly, Modern Tests of Lorentz Invariance, *Living Reviews in Relativity* **8** (2005), 5.
[114] R.-M. Memmesheimer and G. Schäfer, Third post-Newtonian constrained canonical dynamics for binary point masses in harmonic coordinates, *Physical Review D* **71** (2005), 044021.
[115] C. W. Misner, K. S. Thorne and J. A. Wheeler, *Gravitation*, San Francisco: W. H. Freeman and Co., 1973.
[116] T. Mitchell and C. M. Will, Post-Newtonian gravitational radiation and equations of motion via direct integration of the relaxed Einstein equations. V. Evidence for the strong equivalence principle to second post-Newtonian order, *Physical Review D* **75** (2007), 124025.
[117] W.-T. Ni and M. Zimmermann, Inertial and gravitational effects in the proper reference frame of an accelerated, rotating observer, *Physical Review D* **17** (1978), 1473–1476.

[118] K. Nordtvedt, Post-Newtonian Gravitational Effects in Lunar Laser Ranging, *Physical Review D* **7** (1973), 2347–2356.

[119] K. Nordtvedt, Gravitational equation of motion of spherical extended bodies, *Physical Review D* **49** (1994), 5165–5172.

[120] K. Nordtvedt, 30 years of lunar laser ranging and the gravitational interaction, *Classical and Quantum Gravity* **16** (1999), A101–A112.

[121] Y. Nutku, The Post-Newtonian Equations of Hydrodynamics in the Brans-Dicke Theory, *Astrophys. J.* **155** (1969a), 999–1007.

[122] Y. Nutku, The energy–momentum complex in the Brans–Dicke theory, *Astrophys. J.* **158** (1969b), 991–996.

[123] A. Papapetrou, Equations of motion in general relativity, *Proceedings of the Physical Society A* **64** (1951), 57–75.

[124] A. Papapetrou, Equations of Motion in General Relativity: II The Coordinate Condition, *Proceedings of the Physical Society A* **64** (1951), 302–310.

[125] A. Papapetrou, Spinning test-particles in general relativity. I, *Proceedings of the Royal Society of London. Series A. Mathematical and Physical Sciences* **209** (1951), 248–258.

[126] N. M. Petrova, On equations of motion and tensor of matter for a system of finite masses in general theory of relativity, *Zh. Exp. Theor. Phys.* **19** (1949), 989–999.

[127] E. Poisson, A. Pound and I. Vega, The motion of point particles in curved spacetime, *Living Reviews in Relativity* **14** (2011).

[128] D. Puetzfeld and Y. N. Obukhov, Equations of motion in gravity theories with nonminimal coupling: A loophole to detect torsion macroscopically?, *Physical Review D* **88** (2013), 064025.

[129] D. Puetzfeld and Y. N. Obukhov, Unraveling gravity beyond Einstein with extended test bodies, *Physics Letters A* **377** (2013), 2447–2449.

[130] É. Racine and É. É. Flanagan, Post-1-Newtonian equations of motion for systems of arbitrarily structured bodies, *Physical Review D* **71** (2005), 044010.

[131] É. Racine, J. E. Vines and É. É. Flanagan, Erratum: Post-1-Newtonian equations of motion for systems of arbitrarily structured bodies [*Phys. Rev. D* **71**, 044010 (2005)], *Physical Review D* **88** (2013), 089903.

[132] A. D. Rendall, On the definition of post-Newtonian approximations, *Royal Society of London Proceedings Series A* **438** (1992), 341–360.

[133] T. Sauer and A. Trautman, Myron Mathisson: what little we know of his life, *Acta Physica Polonica B Proceedings Supplement* **1** (2008), 7–26.

[134] G. Schäfer, *Post-Newtonian Methods: Analytic Results on the Binary Problem*, pp. 167–210, 2011.

[135] R. Schattner, The center of mass in general relativity, *General Relativity and Gravitation* **10** (1979), 377–393.

[136] G. E. Shilov, *Generalized Functions and Partial Differential Equations: Mathematics and its Applications*, translated from Russian by B. Seckler, Gordon & Breach Science Publishers, 1968.

[137] M. Soffel, S. A. Klioner, G. Petit, P. Wolf, S. M. Kopeikin, P. Bretagnon, V. A. Brumberg, N. Capitaine, T. Damour, T. Fukushima, B. Guinot, T.-Y. Huang, L. Lindegren, C. Ma, K. Nordtvedt, J. C. Ries, P. K. Seidelmann, D. Vokrouhlický, C. M. Will and C. Xu, The IAU 2000 Resolutions for Astrometry, Celestial Mechanics, and Metrology in the Relativistic Framework: Explanatory Supplement, *Astron. J.* **126** (2003), 2687–2706.

[138] M. H. Soffel, *Relativity in Astrometry, Celestial Mechanics and Geodesy*, Berlin: Springer, 1989.

[139] N. Spyrou, The N-body problem in general relativity, *The Astrophysical Journal* **197** (1975), 725–743.

[140] N. Spyrou, Relativistic equations of motion of extended bodies, *General Relativity and Gravitation* **9** (1978), 519–530.
[141] N. Spyrou, Relativistic effects in many-body systems of finite size, internal structure, and internal motions. I – 'Self-acceleration' of astrophysical systems, *General Relativity and Gravitation* **10** (1979), 581–598.
[142] N. Spyrou, Relativistic effects in many-body systems of finite size, internal structure, and internal motions. II – The determination of the inertial and rest masses of binary stars, *General Relativity and Gravitation* **13** (1981), 487–493.
[143] W.-M. Suen, Multipole moments for stationary, non-asymptotically-flat systems in general relativity, *Physical Review D* **34** (1986), 3617–3632.
[144] J. L. Synge, *Relativity: The general theory*, Series in Physics, Amsterdam: North-Holland Publication Co., 1964.
[145] M. Tessmer, J. Hartung and G. Schäfer, Aligned spins: orbital elements, decaying orbits, and last stable circular orbit to high post-Newtonian orders, *Classical and Quantum Gravity* **30** (2013), 015007.
[146] K. S. Thorne, Multipole expansions of gravitational radiation, *Rev. Mod. Phys.* **52** (1980), 299–340.
[147] K. S. Thorne and J. B. Hartle, Laws of motion and precession for black holes and other bodies, *Physical Review D* **31** (1985), 1815–1837.
[148] R. C. Tolman, On the use of the energy–momentum principle in general relativity, *Phys. Rev.* **35** (1930), 875–895.
[149] W. Tulczyjew, Motion of multipole particles in general relativity theory, *Acta Physica Polonica* **18** (1959), 393—409.
[150] M. A. Vincent, The relativistic equations of motion for a satellite in orbit about a finite-size, rotating earth, *Celestial Mechanics* **39** (1986), 15–21.
[151] J. M. Weisberg, D. J. Nice and J. H. Taylor, Timing Measurements of the Relativistic Binary Pulsar PSR B1913+16, *The Astrophysical Journal* **722** (2010), 1030–1034.
[152] N. Wex, New limits on the violation of the strong equivalence principle in strong field regimes., *Astronomy & Astrophysics* **317** (1997), 976–980.
[153] C. M. Will, *Theory and Experiment in Gravitational Physics*, Cambridge: Cambridge University Press, March 1993.
[154] C. M. Will, The confrontation between general relativity and experiment, *Living Reviews in Relativity* **9** (2006), 3.
[155] C. M. Will, The Confrontation between General Relativity and Experiment, *Living Reviews in Relativity* **9** (2006), 3 (cited on February 10, 2009).
[156] J. G. Williams, S. G. Turyshev and D. H. Boggs, Progress in lunar laser ranging tests of relativistic gravity, *Physical Review Letters* **93** (2004), 261101.
[157] J. G. Williams, S. G. Turyshev and D. H. Boggs, Lunar laser ranging tests of the equivalence principle, *Classical and Quantum Gravity* **29** (2012), 184004.
[158] Y. Xie and S. Kopeikin, Post-Newtonian Reference Frames for Advanced Theory of the Lunar Motion and a New Generation of Lunar Laser Ranging, *Acta Physica Slovaca* **60** (2010), 393–495.
[159] C. Xu, X. Wu and G. Schäfer, Binary systems with monopole, spin, and quadrupole moments, *Physical Review D* **55** (1997), 528–539.
[160] P. B. Yasskin and W. R. Stoeger, Propagation equations for test bodies with spin and rotation in theories of gravity with torsion, *Physical Review D* **21** (1980), 2081–2094.
[161] X.-H. Zhang, Multipole expansions of the general-relativistic gravitational field of the external universe, *Physical Review D* **34** (1986), 991–1004.
[162] V. N. Zharkov and V. P. Trubitsyn, *Physics of planetary interiors*, Astronomy and Astrophysics Series, Tucson: Pachart, 1978.

Michael Soffel

On the DSX-framework

1 Introduction

In the beautiful city of Tübingen in the southwest part of Germany, I worked with my good old friend Chongming Xu on the problem of relativistic celestial mechanics for several years. We carefully studied the literature on that subject and thought about possible improvements. When we first saw the paper by Brumberg and Kopeikin [1] we got really excited and realized that it presented a major breakthrough. During the 5th Marcel Grossmann meeting 1988 in Perth, Australia, I met Thibault (Damour) and learned that he had good ideas to push our subject forward. We agreed that he would visit us in Tübingen. Finally he came with a huge pile of notes on a new formulation of relativistic celestial mechanics. It was not too different from the Brumberg–Kopeikin approach but contained important improvements in various directions. Moreover, Thibault's approach was constructive in the sense that very little was assumed from the very beginning and one proceeds with a series of Lemmas and Theorems. The whole structure had a great intrinsic beauty and mathematical rigor which, I think, was really new in that field. So we started working out details of this new approach by proving more Lemmas and Theorems and formulating a new and improved theory of relativistic reference systems. Finally we published a series of papers (Damour et al. [2–5]), on what later was called the DSX-framework.

This contribution focuses on the first DSX paper [2], where the foundations of this new approach was worked out and, as a first application, a new and improved derivation of the well-known post-Newtonian Einstein–Infeld–Hoffmann equations for a system of mass-monopoles was presented. The following major part of the Introduction was taken literally from [2]. I would like to apologize for that but improving this text, that was essentially formulated by Thibault, is not an easy job.

The problem of describing the dynamics of N gravitationally interacting extended bodies is the cardinal problem of any theory of relativity. Within the framework of Newton's theory this problem, called "celestial mechanics," has been thoroughly investigated (see, e.g. [6]). Very shortly after the discovery of Einstein's theory of gravity, Einstein [7], Droste [8], de Sitter [9]), and Lorentz and Droste [10] devised an approximation method (called "post-Newtonian") which allowed them to compare general relativity with Newton's theory of gravity, and to predict several "relativistic effects" in celestial mechanics, such as the relativistic advance of the perihelion of planets, and

Michael Soffel: Lohrmann Observatory TU Dresden, Planetary Geodesy, Mommsenstr 13, 01062 Dresden, Germany

the relativistic precession of the Moon's orbit. This post-Newtonian approach to general-relativistic celestial mechanics was subsequently developed by many authors, notably by Fock [11] (for a review of the development of the problem of motion in general relativity, see, e.g. [12]).

However, to match the high precision of modern observational techniques such as satellite laser ranging, lunar laser ranging, very long baseline interferometry, radar ranging to spacecraft and planets etc., one needs a correspondingly accurate relativistic theory of celestial mechanics able to describe both the global gravitational dynamics of a system of N extended bodies, the local gravitational structure of each, and the way each of these N local structures meshes into the global one. The traditional post-Newtonian approach to relativistic celestial mechanics fails, for both conceptual and technical reasons, to bring a satisfactory answer to this problem. This traditional post-Newtonian approach uses only one global coordinate system $x^\mu = (ct, x, y, z)$, to describe an N-body system, and models itself on the Newtonian approach to celestial mechanics consisting of decomposing the problem into two subproblems [6, 11]: (i) *the external problem*: to determine the motion of the centers of mass of the N bodies; (ii) *the internal problem*: to determine the motion of each body around its center of mass. However, the treatments of both subproblems in the traditional post-Newtonian approach are unsatisfactory.

The *external problem* is attacked by introducing some collective variable, say $z^i(t), i = 1, 2, 3$, generalizing the Newtonian center of mass, i.e. describing the overall motion of each body as seen in the global coordinate system x^μ. Then, one attempts to derive some (translational) "equations of motion" for $z^i(t)$ by integrating over each considered body the local law of balance of energy and momentum, i.e. the covariant conservation of the stress–energy tensor,

$$\nabla_\nu T^{\mu\nu} = 0 \, .$$

However, the various definitions of the position in the global coordinate system of the center of mass z^i used in post-Newtonian investigations have never been quite satisfactory, especially when considering rotating bodies. Moreover, the final equations of motion for $z^i(t)$ contain various other collective variables ("spin" and higher "multipole moments") describing the gravitational structure of each body as seen in the global system x^μ, which are not related in a simply way to any physical "local" multipole moments, defined, say, in an operational way by the motion of artificial satellites around each body.

Concerning the treatment of the *internal problem* in the usual post-Newtonian approach, it is even more unsatisfactory for the following reasons. In Newtonian celestial mechanics the introduction of nonrotating accelerated "mass-centered frames" associated with each body, i.e. of local coordinates

$$X^i = x^i - z^i(t) \, , \tag{1.1}$$

where $i = 1, 2, 3$, and z^i denote the global coordinates of the center of mass, serves both a kinematical and a dynamical purpose. The kinematical usefulness of the local coordinates X^i stems from the fact that they are "comoving" with the considered body, while their dynamical usefulness comes from the fact that they succeed in decoupling, to a large degree, the "internal" from the "external" problem. Indeed, with respect to these local frames X^i the external gravitational field is greatly "effaced" [12] in the sense that the effective external gravitational potential acting locally on the body and its environment,

$$U^{\mathrm{eff}}(X^i) = U^{\mathrm{ext}}(z^i + X^i) - U^{\mathrm{ext}}(z^i) - \frac{d^2 z^i}{dt^2} X^i ,$$

is essentially reduced to tidal forces.

For a long time, the relativistic internal problem has been given only little attention, and many authors working in the global post-Newtonian framework have, more or less implicitly, assumed that the usual Newtonian formula (1.1) was sufficient for defining a useful "mass-centered frame" in Einsteinian gravity. In principle, this view is admissible because the coordinate systems are arbitrary in general relativity, and the definition (1.1) is as kinematically useful as in Newtonian gravity. However, the formula (1.1) does not define a dynamically useful mass-centered frame in general relativity, in the sense that it does not efface the external gravitational field down to tidal effects, but, instead introduces in the description of the internal dynamics of the body many external "relativistic" effects proportional to the square of the orbital velocity or the external gravitational potential. As discussed in [12] the latter effects come from the fact that the external description (x^μ-coordinate representation) of each body contains many "apparent deformations" which are not intrinsic to the body (notably the "Lorentz contraction," linked with the orbital velocity, and the "Einstein contraction," linked with the external gravitational potential).

As emphasized by Damour [12], those technical defects of the usual global post-Newtonian approach are partly rooted in, and certainly further enhanced by, the conceptual defect of surreptitiously introducing a kind of "neo-Newtonian" [13] interpretation of general relativity, by which the global coordinates $t \equiv x^0/c, (x, y, z) \equiv (x^i), i = 1, 2, 3$ are implicitly identified with the absolute time and space of Newtonian theory. This implicit conceptual reduction of Einstein's theory to the Procustean bed of Newton's framework is liable to cause technical mistakes when one forgets the existence of the "apparent deformations" alluded to above. In recent years, several authors have tried to remedy some of the defects of the traditional post-Newtonian approach to the N-body problem. For instance, Martin et al. [14] and Hellings [15] have tried, in an essentially heuristic manner, to explicitly include the main apparent deformations due to the use of an external coordinate representation. A more ambitious approach consists of defining a local comoving frame by using, not a kinematical criterion (like in Equation (1.1)), but a dynamical one: i.e. to find a useful relativistic definition of an accelerated frame of reference with respect to which the external gravitational effects are strongly effaced. In the simple case of negligibly self-gravitating test bodies moving in a background gravitational field (e.g. an artificial satellite around the Earth) such "external-gravitational-field-effacing" frames are the well-known "locally inertial frames" which can be explicitly constructed by means of Fermi coordinates based on the center-of-mass world line (see, e.g. [16, 17]).

In the more subtle case of (possibly strong) self-gravitating test bodies (i.e. of mass much smaller than the masses of the other bodies) it has been argued as early as 1921 by Weyl [18] that such frames should exist, and be the locally inertial frames (or Fermi frames) of the "external spacetime" generated by the masses of the other bodies only. Weyl [18] used this argument to conclude that test bodies (even self-gravitating ones) follow geodesics of the "external spacetime." This heuristic reasoning has been later taken up [19–21] although it never became clear what could be rigorously proven with its help (because of the lack of mathematical control on the limiting process which defines what one means by "test-body" and "external universe").

Concerning nontest bodies (of mass comparable to the masses of the other bodies), some authors (in particular Bertotti [22]) remarked that, at the first post-Newtonian approximation, the orbital motion, according to the Lorentz–Droste–Einstein–Infeld–Hoffmann equations of motion, of one self-gravitating body member of an N-body system, could be interpreted as the motion of a test body in some effective external gravitational field. This remark, together with the previous results for test bodies, suggests that it should be always possible to define good "external-gravitational-field-effacing frames" around any body A, abstracted from an N-body system, by constructing some "locally inertial coordinate system" in some "effective external gravitational field." At the heuristic level, such a construction has been more or less explicitly assumed by many authors [17, 23–27]. More explicit results on such local external-field-effacing frames have been obtained in the study of the motion of strongly self-gravitating bodies (neutron stars or black holes), because this was a problem where the standard only global-frame approach was definitively inadequate to derive results needed for astrophysical applications. In particular, D'Eath [20] and Damour [28], in their studies of binary systems of gravitationally condensed bodies, have made an explicit use of local external-field-effacing coordinate systems $X^\alpha = (cT, X^a)$ (one for each body), linked with the global coordinate system x^μ, covering the binary system, by transformation formulas of the type ($a = 1, 2, 3$)

$$x^\mu(T, X^a) = z^\mu(T) + e^\mu_a(T)X^a + \mathcal{O}\left((X^a)^2\right) + \cdots ,$$

and have derived the constraints on the functions $z^\mu(T), e^\mu_a(T)$, imposed by the requirement of a suitable effacement in the X^α system. Other explicit results about such good local frames were also obtained in the study of weakly self-gravitating bodies, treated at the first post-Newtonian approximation, notably through the introduction of "generalized Fermi coordinates" by Ashby and Bertotti [29, 30] (see also [31], and the contributions of Bertotti, Boucher, Fukushima, Fujimoto, Kinoshita, Aoki, and others in [32]).

More recently a notable progress in the theory of such local relativistic frames (at the post-Newtonian approximation, relevant to systems of N weakly self-gravitating bodies) has been achieved by Brumberg and Kopeikin in a series of publications [1, 33, 34]; see also Voinov [35]). Their approach combines the usual post-Newtonian-type expansions with the multipole expansion formalisms for internally generated [36–38], and externally generated [27, 39, 40] gravitational fields, and with asymptotic matching techniques [20, 28].

A further step forward was made by Damour, Soffel, and Xu in a series of papers [2–5]. Improvements of the Brumberg–Kopeikin formalism are:

- the form of the energy–momentum tensor of matter composing the N bodies is left completely open;
- whereas the spatial coordinates in the global and the N local coordinate systems are taken to be harmonic the gauge of the various time coordinates is left open within the post-Newtonian approximation;
- a compact notation for the metric tensor in the $N+1$ coordinate systems; in each of these systems the metric tensor is completely specified by two metric potentials only, a scalar potential describing gravitoelectric type effects, and a vector potential related to gravitomagnetic type effects;
- the use of post-Newtonian physically meaningful mass- and spin-multipole moments (Blanchet–Damour moments) in each of the local systems.

This contribution to the Festschrift discusses the most important features of this DSX (Damour–Soffel–Xu) formalism.

2 The post-Newtonian formalism

2.1 The general form of the metric

The idea of the post-Newtonian formalism is to employ the fact that in the solar system velocities of astronomical bodies are small and gravitational fields are weak. The PN-formalism is a *slow motion, weak field approximation* to Einstein's theory of gravity. The small parameter of this expansion is

$$\epsilon = \frac{v}{c}$$

and everywhere in the solar system we have

$$\frac{U}{c^2} \sim \frac{p}{\rho c^2} \sim \Pi \sim \left(\frac{1}{c}\right)^2 \sim \epsilon^2 < 10^{-5} .$$

Here, p denotes the pressure, ρ is the density of matter and Π is the specific internal energy density (internal energy density divided by the rest energy density). An upper limit for U/c^2 will be given by $GM/(c^2R)$, where M is the mass of some gravitating body and R its radius. For example, (E: Earth, S: Sun)

$$\frac{GM_{\rm E}}{c^2R_{\rm E}} \simeq 6.9 \times 10^{-10}\,, \qquad \frac{GM_{\rm S}}{c^2R_{\rm S}} \simeq 2.1 \times 10^{-6}\,.$$

For the orbital velocity of the Earth about the Sun we have

$$\left(\frac{v_{\rm E}}{c}\right)^2 \simeq \left(\frac{30}{300\,000}\right)^2 \simeq 10^{-8}\,.$$

We will now assume the existence of certain coordinates $x^\mu = (ct, x^i)$ such that the following PN-assumptions hold:

$$g_{00} = -1 + \mathcal{O}(c^{-2}), \quad g_{0i} = \mathcal{O}(c^{-3}), \quad g_{ij} = \delta_{ij} + \mathcal{O}(c^{-2})\,. \tag{2.1}$$

In the following we will frequently encounter such order symbols that we will often abbreviate by

$$\mathcal{O}_n \equiv \mathcal{O}(c^{-n})\,.$$

Moreover, we will assume, in accordance with the results from special relativity, that

$$T^{00} = \mathcal{O}(c^2), \quad T^{0i} = \mathcal{O}(c^{+1}), \quad T^{ij} = \mathcal{O}(c^0) \tag{2.2}$$

and

$$\partial_0 \equiv \frac{\partial}{\partial ct} = \mathcal{O}_1 \cdot \partial_i\,.$$

With assumptions (2.1) we can write the metric tensor in the form

$$\begin{aligned} g_{00} &= -\exp\left(-\frac{2}{c^2}w\right) \\ g_{0i} &= -\frac{4}{c^3}w_i \\ g_{ij} &= \gamma_{ij}\exp\left(+\frac{2}{c^2}w\right) \end{aligned} \tag{2.3}$$

with

$$\gamma_{ij} = \delta_{ij} + \mathcal{O}_2\,.$$

Below we will show that the spatial field equations

$$G_{ij} = \kappa T_{ij} \tag{2.4}$$

will be satisfied by

$$\gamma_{ij} = \delta_{ij} + \mathcal{O}_4\,. \tag{2.5}$$

Our canonical form of the metric therefore reads

$$\begin{aligned} g_{00} &= -\exp\left(-\frac{2}{c^2}w\right) = -1 + \frac{2w}{c^2} - \frac{2w^2}{c^4} + \mathcal{O}_6 \\ g_{0i} &= -\frac{4}{c^3}w_i \\ g_{ij} &= \delta_{ij}\left(1 + \frac{2}{c^2}w\right) + \mathcal{O}_4 . \end{aligned} \tag{2.6}$$

Note, that the metric tensor is completely specified by two potentials: a (gravitoelectric) scalar potential $w(t, \mathbf{x})$ and a (gravitomagnetic) vector potential $w_i(t, \mathbf{x})$ just as in case of Maxwell's theory of electromagnetism. From the "Newtonian" limit one finds that the scalar potential w generalizes the Newtonian gravitational potential U. As will become clear from the following a split of w into a Newtonian and some post-Newtonian part is not meaningful; it only gives rise to confusion. The potential w determines the time–time and the space–space part of the metric tensor. The time–space component, g_{0i}, that is determined by the vector potential w_i describes gravitomagnetic type effects, i.e. effects that arise from matter currents (moving or rotating masses).

We will now compute the differential geometrical quantities to post-Newtonian accuracy for celestial mechanical problems:

$$\begin{aligned} g &\equiv -\det g_{\mu\nu} = e^{4w/c^2} + \mathcal{O}_4 \\ \sqrt{g} &= e^{2w/c^2} + \mathcal{O}_4 . \end{aligned} \tag{2.7}$$

The *inverse* metric $g^{\mu\nu}$ is given by

$$\begin{aligned} g^{00} &= -\exp\left(+\frac{2}{c^2}w\right) = -1 - \frac{2w}{c^2} - \frac{2w^2}{c^4} + \mathcal{O}_6 \\ g^{0i} &= -\frac{4}{c^3}w_i + \mathcal{O}_5 \\ g^{ij} &= \delta_{ij}\exp\left(-\frac{2}{c^2}w\right) + \mathcal{O}_4 = \delta_{ij}\left(1 - \frac{2}{c^2}w\right) + \mathcal{O}_4 . \end{aligned} \tag{2.8}$$

For the Christoffel symbols

$$\Gamma^\alpha{}_{\beta\gamma} = \frac{1}{2}g^{\alpha\lambda}(g_{\lambda\beta,\gamma} + g_{\lambda\gamma,\beta} - g_{\beta\gamma,\lambda})$$

one finds

$$\begin{aligned} \Gamma^0{}_{00} &= -\frac{w_{,0}}{c^2} + \mathcal{O}_5 \\ \Gamma^0{}_{0i} &= -\frac{w_{,i}}{c^2} + \mathcal{O}_6 \\ \Gamma^0{}_{ij} &= \delta_{ij}\frac{w_{,0}}{c^2} + \frac{4}{c^3}w_{(i,j)} + \mathcal{O}_5 \end{aligned}$$

$$\begin{aligned}
\Gamma^i_{\ 00} &= -\frac{w_{,i}}{c^2} + 4\frac{w\,w_{,i}}{c^4} - \frac{4}{c^3}w_{i,0} + \mathcal{O}_6 \\
\Gamma^i_{\ 0j} &= -\frac{4}{c^3}w_{[i,j]} + \frac{w_{,0}}{c^2}\delta_{ij} + \mathcal{O}_5 \\
\Gamma^i_{\ jk} &= \delta_{ij}\frac{w_{,k}}{c^2} + \delta_{ik}\frac{w_{,j}}{c^2} - \delta_{jk}\frac{w_{,i}}{c^2} + \mathcal{O}_4\,,
\end{aligned} \tag{2.9}$$

where

$$\begin{aligned}
w_{(i,j)} &\equiv \frac{1}{2}(w_{i,j} + w_{j,i}) \\
w_{[i,j]} &\equiv \frac{1}{2}(w_{i,j} - w_{j,i})\,.
\end{aligned}$$

For the components of the Riemann curvature tensor one obtains

$$\begin{aligned}
R_{0i0j} &= -\frac{1}{c^2}w_{,ij} - \frac{1}{c^4}\left[-2ww_{,ij} + \delta_{ij}w_{,tt} + 2\partial_t(w_{i,j} + w_{j,i})\right. \\
&\quad \left. - 3w_{,i}w_{,j} + \delta_{ij}w_{,k}w_{,k}\right] + \mathcal{O}_6 \\
R_{0ijk} &= \frac{1}{c^3}\left[\delta_{ij}w_{,tk} - \delta_{ik}w_{,tj} + 2\partial_i(w_{j,k} - w_{k,j})\right] + \mathcal{O}_5 \\
R_{ijkl} &= \frac{1}{c^2}\left[\delta_{il}w_{,jk} - \delta_{jl}w_{,ik} + \delta_{kj}w_{,il} - \delta_{ki}w_{,jl}\right] + \mathcal{O}_4
\end{aligned} \tag{2.10}$$

and

$$\begin{aligned}
R^0_{\ ijk} &= -R_{0ijk} + \mathcal{O}_5 = R^i_{\ 0jk} + \mathcal{O}_5 \\
R^i_{\ jkl} &= R_{ijkl} + \mathcal{O}_4 \\
R^0_{\ i0j} &= \frac{w_{,ij}}{c^2} + \frac{1}{c^4}\left[\delta_{ij}w_{,tt} + 2\partial_t(w_{i,j} + w_{j,i})\right. \\
&\quad \left. - 3w_{,i}w_{,j} + \delta_{ij}w_{,k}w_{,k}\right] + \mathcal{O}_6 \\
R^i_{\ 0j0} &= -\frac{1}{c^2}w_{,ij} - \frac{1}{c^4}\left[-4ww_{,ij} + \delta_{ij}w_{,tt} + 2\partial_t(w_{i,j} + w_{j,i})\right. \\
&\quad \left. - 3w_{,i}w_{,j} + \delta_{ij}w_{,k}w_{,k}\right] + \mathcal{O}_6\,.
\end{aligned} \tag{2.11}$$

From this one derive the components of the Ricci tensor

$$R_{\mu\nu} = R^\sigma_{\ \mu\sigma\nu}$$

$$\begin{aligned}
R_{00} &= -\frac{1}{c^2}\Delta w - \frac{1}{c^4}\left[-4w\,\Delta w + 3w_{,tt} + 4\partial_{ti}w_i\right] + \mathcal{O}_6 \\
R_{0i} &= \frac{1}{c^3}\left[-2w_{,ti} + 2(\Delta w - \partial_{ij}w_j)\right] + \mathcal{O}_5 \\
R_{ij} &= -\frac{1}{c^2}\delta_{ij}\,\Delta w + \mathcal{O}_4
\end{aligned} \tag{2.12}$$

and

$$\begin{aligned} R^{00} &= -\frac{1}{c^2}\Delta w - \frac{3}{c^4}w_{,tt} - \frac{4}{c^4}\partial_{ti}w_i + \mathcal{O}_6 \\ R^{0i} &= -\frac{2}{c^3}\left[\Delta w_i - \partial_{ij}w_j - \partial_{ti}w\right] + \mathcal{O}_5 \\ R^{ij} &= -\frac{1}{c^2}\delta_{ij}\,\Delta w + \mathcal{O}_4\,. \end{aligned} \tag{2.13}$$

From this we find the curvature scalar

$$R = g^{\mu\nu}R_{\mu\nu} = g^{00}R_{00} + 2g^{0i}R_{0i} + g^{ij}R_{ij}\,,$$

i.e.

$$R = -\frac{2}{c^2}\Delta w + \mathcal{O}_4\,. \tag{2.14}$$

3 Field equations and the gauge problem

We will now show that the spatial field equations are satisfied to $\mathcal{O}_4$ by $\gamma_{ij} = \delta_{ij} + \mathcal{O}_4$. According to (2.2) $T_{ij} = \mathcal{O}_0$ and we have to prove that

$$R_{ij} - \frac{1}{2}g_{ij}R = \frac{8\pi G}{c^4}T_{ij} = \mathcal{O}_4\,.$$

The left-hand side of this equation reads

$$-\frac{1}{c^2}\delta_{ij}\,\Delta w + \mathcal{O}_4 + \frac{1}{2}\delta_{ij}\,\frac{2}{c^2}\Delta w + \mathcal{O}_4 = \mathcal{O}_4$$

as was to be shown. This implies that only the field equations

$$G_{00} = \kappa\,T_{00}, \qquad G_{0i} = \kappa\,T_{0i}$$

or

$$\begin{aligned} R^{00} &= \frac{8\pi\,G}{c^4}\left(T^{00} - \frac{1}{2}g^{00}g_{\alpha\beta}T^{\alpha\beta}\right) \\ R^{0i} &= \frac{8\pi\,G}{c^4}\left(T^{0i} - \frac{1}{2}g^{0i}g_{\alpha\beta}T^{\alpha\beta}\right) \end{aligned}$$

remain to be solved. Taking into account the order of $T^{\alpha\beta}$ one finds

$$\begin{aligned} R^{00} &= \frac{4\pi G}{c^4}(T^{00} + T^{ss}) + \mathcal{O}_6 \\ R^{0i} &= \frac{8\pi G}{c^4}T^{0i} + \mathcal{O}_5\,. \end{aligned}$$

Inserting expressions (2.13) we get the field equations in the form:

$$\begin{aligned}\Delta w + \frac{3}{c^2} w_{,tt} + \frac{4}{c^2} \partial_{ti} w_i &= -4\pi G\sigma + \mathcal{O}_4 \\ \Delta w_i - \partial_{ij} w_j - \partial_{ti} w &= -4\pi G\sigma^i + \mathcal{O}_2\end{aligned} \tag{3.1}$$

with

$$\sigma \equiv \frac{T^{00} + T^{ss}}{c^2}\,; \qquad \sigma^i \equiv \frac{T^{0i}}{c}\,.$$

Here, σ acts as active gravitational mass–energy density generalizing the density ρ in Newton's theory of gravity. σ^i is the active gravitational mass–current density that does not act as a field generating source in Newton's theory. The field equations (3.1) have a very remarkable property: they are *linear* in the metric potentials

$$w^\mu \equiv (w, w_i)\,.$$

This results from the special form of the metric tensor and the various approximations involved. Note that Einstein's field equations are nonlinear in general and the achieved linearity simplifies the formalism tremendously. The choice of spatially isotropic coordinates with $\gamma_{ij} = \delta_{ij} + \mathcal{O}_4$ corresponds to three gauge conditions for the three spatial coordinates x^i.

A function $f(x^\mu)$ is called *harmonic* if

$$\Box_g f \equiv \frac{1}{\sqrt{g}} \left(\sqrt{g} g^{\mu\nu} f_{,\nu}\right)_{,\mu} = 0\,.$$

Note, that in three-dimensional Euclidean space, $\boldsymbol{R}^3$, the operator $\Box_g$ reduces to the usual Laplacian Δ thus $\Box_g$ might be called the covariant Laplacian.

By direct calculation one can show that the spatial coordinates x^i of the metric (2.6) are harmonic up to terms of $\mathcal{O}_4$, i.e.

$$\Box_g x^i = \mathcal{O}_4\,. \tag{3.2}$$

Because of this only the gauge freedom of the time coordinate is left. The field equations simplify further if also the time coordinate is chosen to be harmonic, i.e. if

$$\Box_g x^0 = \mathcal{O}_5\,.$$

We have

$$\begin{aligned}\sqrt{g}\Box_g x^0 &= \left(\sqrt{g} g^{\mu 0}\right)_{,\mu} = \left(\sqrt{g} g^{00}\right)_{,0} + \left(\sqrt{g} g^{i0}\right)_{,i} \\ &= -\frac{4 w_{,0}}{c^2} + \frac{4 w_{i,i}}{c^3} + \mathcal{O}_5\,.\end{aligned}$$

Hence, we can write the condition for the time coordinate t to be harmonic in the form

$$0 = -\frac{4}{c^3}\left(\partial_t w + \partial_i w_i\right) + \mathcal{O}_5\,. \tag{3.3}$$

With this harmonic gauge the field equations take the form

$$\Box w = -4\pi G\sigma + \mathcal{O}_4 \tag{3.4}$$

$$\Delta w_i = -4\pi G\sigma^i + \mathcal{O}_2\,, \tag{3.5}$$

where $\Box$ is the flat space d'Alembertian

$$\Box \equiv -\frac{1}{c^2}\frac{\partial^2}{\partial t^2} + \Delta\,.$$

We can even combine the source and the field variables

$$\sigma_\mu \equiv (\sigma, \sigma_i); \qquad w_\mu \equiv (w, w_i)$$

and write in obvious notation

$$\Box w_\mu = -4\pi G\sigma_\mu + \mathcal{O}(4,2)\,.$$

If we consider one isolated system (e.g. an idealized solar system) with no gravitational sources outside this system we can require

$$g_{\mu\nu} \longrightarrow \eta_{\mu\nu}\,, \qquad |\mathbf{x}| \to \infty$$

i.e. we consider our spacetime manifold to be asymptotically flat. For our potentials this implies

$$(w, w_i) \longrightarrow 0 \qquad |\mathbf{x}| \to \infty\,.$$

With such a condition for asymptotic flatness the retarded solution of our field equations reads

$$w^\mu_{\mathrm{ret}}(t,\mathbf{x}) = G\int d^3x'\,\frac{\sigma^\mu(t_{\mathrm{ret}};\mathbf{x}')}{|\mathbf{x}-\mathbf{x}'|}\,, \tag{3.6}$$

where t_{ret} is retarded time

$$t_{\mathrm{ret}} \equiv t - \frac{|\mathbf{x}-\mathbf{x}'|}{c}\,.$$

Another possible solution is the advanced one with t_{ret} being replaced by t_{adv} with

$$t_{\mathrm{adv}} \equiv t + \frac{|\mathbf{x}-\mathbf{x}'|}{c}\,.$$

Still another solution is

$$w^\mu_{\mathrm{mixed}}(t,\mathbf{x}) = \frac{1}{2}[w^\mu_{\mathrm{ret}}(t,\mathbf{x}) + w^\mu_{\mathrm{adv}}(t,\mathbf{x})]\,.$$

This mixed solution that might also be called time-symmetric solution is in fact usually used in post-Newtonian theories. The reason for this is the following: if we expand σ^μ around the coordinate time t we encounter a sequence of time derivatives and the first time derivative is related with irreversible processes such as the emission of gravity waves that do not occur in the first post-Newtonian approximation to Einstein's theory of gravity. With $r \equiv |\mathbf{x} - \mathbf{x}'|$ we get

$$\begin{aligned} w^{\mu}_{\text{ret/adv}}(t, \mathbf{x}) &= G \int d^3x' \, \frac{\sigma^{\mu}(t \mp r/c; \mathbf{x}')}{r} \\ &= G \int d^3x' \, \frac{\sigma^{\mu}(t; \mathbf{x}')}{r} \\ &\mp \frac{G}{c} \frac{\partial}{\partial t} \int d^3x' \, \sigma^{\mu}(t; \mathbf{x}') + \frac{G}{2c^2} \frac{\partial^2}{\partial t^2} \int d^3x' \, \sigma^{\mu}(t; \mathbf{x}') r \, . \end{aligned}$$

If we take the time symmetric solution then in the expansion the first time derivative terms cancel automatically, i.e.

$$w^{\mu}_{\text{mixed}}(t, \mathbf{x}) = G \int d^3x' \, \frac{\sigma^{\mu}(t; \mathbf{x}')}{|\mathbf{x} - \mathbf{x}'|} + \frac{G}{2c^2} \frac{\partial^2}{\partial t^2} \int d^3x' \, \sigma^{\mu}(t; \mathbf{x}') |\mathbf{x} - \mathbf{x}'| \, . \tag{3.7}$$

Note that the retarded, advanced, or mixed solutions of the harmonic field equations are not the only ones. If w, w_i solve these equations then also

$$w' = w - \frac{1}{c^2} \partial_t \lambda; \quad w_i' = w_i + \frac{1}{4} \partial_i \lambda$$

corresponding to a change of the time variable of the form

$$t \longrightarrow t' : \qquad t' = t - c^{-4} \lambda \, .$$

If $\Delta \lambda = \mathcal{O}_2$ then w'_{μ} will be another harmonic solution of the PN field equations.

4 The gravitational field of a body

As first application of the post-Newtonian formalism we will discuss the gravitational field of some isolated matter distribution.

4.1 Post-Newtonian multipole moments

Let us consider first the case of a single static body. From the PN field equations (3.4) and (3.5), we get

$$w(t, \mathbf{x}) = G \int d^3x' \, \frac{\sigma(t, \mathbf{x}')}{|\mathbf{x} - \mathbf{x}'|} + \mathcal{O}_5 \, ; \quad w^i(t, \mathbf{x}) = 0 \, .$$

Usually the term $|\mathbf{x} - \mathbf{x}'|^{-1}$ is expanded in terms of spherical harmonics. In relativity, partly due to the important role of Lorentz transformations, usually another expansion, related with Cartesian tensors, is employed.

A Cartesian l-tensor is a set of real or complex numbers $T_{i_1 i_2 \ldots i_l}$ with l different indices i_1 to i_l, each taking the values $1, 2, 3$, or equivalently (x, y, z). For the sake of compactness often a set of l Cartesian indices is abbreviated by a multi-index, e.g. $L \equiv i_1 i_2 \cdots i_l$ etc. Usually Einstein's summation convention is assumed, i.e. if some index appears twice a summation over that index is implied automatically, e.g.

$$A_L B_L \equiv A_{i_1 i_2 \cdots i_l} B_{i_1 i_2 \cdots i_l} \equiv \sum_{i_1 \ldots i_l = 1}^{3} A_{i_1 i_2 \cdots i_l} B_{i_1 i_2 \cdots i_l} .$$

Given a Cartesian tensor T_L, we denote its symmetric part by parentheses

$$T_{(L)} \equiv T_{(i_1 \cdots i_l)} \equiv \frac{1}{l!} \sum_\sigma T_{i_{\sigma(1)} \cdots i_{\sigma(l)}} ,$$

where σ runs over all $l!$ permutations of $(12 \cdots l)$. If T_L is a Cartesian l-tensor; each quantity where we put two arbitrary indices equal with a subsequent summation is called a trace of T_L. If every trace of T_L vanishes it is called trace-free. Of great importance are symmetric and trace-free (STF) Cartesian tensors. The STF-part of T_L is denoted indifferently by $\hat{T}_L \equiv T_{<L>} \equiv T_{\langle i_1 \cdots i_l \rangle}$. The explicit expression of the STF part reads (e.g. [36])

$$\hat{T}_L = \sum_{k=0}^{[\frac{1}{2}l]} a_k^l \delta_{(i_1 i_2} \cdots \delta_{i_{2k-1} i_{2k}} S_{i_{2k+1} \cdots i_l) a_1 a_1 \cdots a_k a_k}$$

where

$$S_L = T_{(L)} ,$$

$$a_k^l = \frac{l!}{(2l-1)!!} \frac{(-1)^k (2l-2k-1)!!}{(l-2k)!(2k)!!} ,$$

$[\frac{1}{2}l]$ denoting the integer part of $\frac{1}{2}l$. For instance, $\hat{T}_{ij} = T_{(ij)} - \frac{1}{3}\delta_{ij} T_{aa}$; $\hat{T}_{ijk} = T_{(ijk)} - \frac{1}{5}[\delta_{ij} T_{(kaa)} + \delta_{jk} T_{(iaa)} + \delta_{ki} T_{(jaa)}]$.

A Taylor expansion of $|\mathbf{x} - \mathbf{x}'|^{-1}$ in the gravitoelectric potential $w(t, \mathbf{x})$ then yields

$$w(t, \mathbf{x}) = G \sum_{l \geq 0} \frac{(-1)^l}{l!} \int d^3x' \, \sigma' x'^{i_1} \cdots x'^{i_l} \partial_{i_1 \cdots i_l} \left(\frac{1}{r} \right) ,$$

where $r \equiv |\mathbf{x}|$. Let

$$\varphi_{i_1 \cdots i_l} \equiv \partial_{i_1 \cdots i_l} \left(\frac{1}{r} \right) \equiv \frac{\partial^l}{\partial x^{i_1} \cdots x^{i_l}} \left(\frac{1}{r} \right) .$$

Then because of

$$\Delta \left(\frac{1}{r} \right) = 0$$

the symmetric Cartesian (every index takes the values x, y, z) tensor $\varphi_{i_1 \cdots i_l}$ is trace-free, i.e. a Cartesian STF-tensor. By induction one can prove that

$$\varphi_L = (-1)^l (2l-1)!! \frac{\hat{n}_L}{r^{l+1}} ,$$

where

$$n_{i_1\cdots i_l} \equiv \frac{x^{i_1}}{r}\cdots\frac{x^{i_l}}{r},$$

so that

$$w(t,\mathbf{x}) = G\sum_{l\geq 0}\frac{(2l-1)!!}{l!}M_L\frac{\hat{n}_L}{r^{l+1}} \tag{4.1}$$

with

$$M_L = \hat{M}_L = \int d^3x'\,\sigma'\hat{x}'^L\,.$$

The following theorem is not restricted to the static case.

Theorem 1 (Blanchet and Damour [41]). *Outside of some isolated matter distribution the functions*

$$w_\mu(t,\mathbf{x}) = G\int d^3x'\,\frac{\sigma_\mu(t_\pm;\mathbf{x}')}{|\mathbf{x}-\mathbf{x}'|}$$

with

$$t_\pm \equiv t_{\rm mixed} = \frac{1}{2}(t_{\rm ret}+t_{\rm adv})\,.$$

admit an expansion of the form ($r = |\mathbf{x}|$)

$$w(t,\mathbf{x}) = G\sum_{l\geq 0}\frac{(-1)^l}{l!}\partial_L\left[r^{-1}M_L(t_\pm)\right] + \frac{1}{c^2}\partial_t\Lambda + \mathcal{O}_4 \tag{4.2}$$

$$\begin{aligned} w_i(t,\mathbf{x}) = -G\sum_{l\geq 1}\frac{(-1)^l}{l!}&\left[\partial_{L-1}\left(r^{-1}\frac{d}{dt}M_{iL-1}\right)\right.\\ &\left.+\frac{l}{l+1}\epsilon_{ijk}\partial_{jL-1}\left(r^{-1}S_{kL-1}\right)\right] - \frac{1}{4}\partial_i\Lambda + \mathcal{O}_2\,. \end{aligned} \tag{4.3}$$

Here,

$$\Lambda \equiv 4G\sum_{l\geq 0}\frac{(-1)^l}{(l+1)!}\frac{2l+1}{2l+3}\partial_L(r^{-1}\mu_L(t_\pm))$$

$$\mu_L \equiv \int d^3x\,\hat{x}^{iL}\sigma^i(t,\mathbf{x})$$

and

$$\begin{aligned} M_L(t) \equiv& \int d^3x\,\hat{x}^L\sigma + \frac{1}{2(2l+3)c^2}\frac{d^2}{dt^2}\int d^3x\,\hat{x}^L\mathbf{x}^2\sigma\\ &-\frac{4(2l+1)}{(l+1)(2l+3)c^2}\frac{d}{dt}\int d^3x\,\hat{x}^{iL}\sigma^i \qquad (l\geq 0) \end{aligned} \tag{4.4}$$

$$S_L(t) \equiv \int d^3x\,\epsilon^{ij<k_l}\hat{x}^{L-1>i}\sigma^j\,. \qquad (l\geq 1) \tag{4.5}$$

The Cartesian STF-tensors M_L and S_L are the post-Newtonian mass- and spin-multipole moments that characterize the gravitational action of the matter distribution in the outside vacuum region. They have been first introduced in a paper by Blanchet and Damour [41] and are called BD-moments. M is the post-Newtonian mass of the matter distribution, M_i is its mass dipole, M_{ij} its mass quadrupole, etc. S^i with

$$S^i = \int d^3x\, \epsilon^{jki} x^j \sigma^k = \int d^3x\, (\mathbf{x} \times \boldsymbol{\sigma})^i$$

is the spin vector (total angular momentum) of the matter distribution (note that for an ideal fluid $\sigma^i = \rho v^i + \mathcal{O}_2$).

The proof of the BD-theorem can be taken from [41].

It is not difficult to see that Λ is a harmonic function, i.e. $\Delta\Lambda = 0$. For that reason we can remove the Λ-terms by means of a gauge transformation; in such a *skeletonized harmonic gauge* the potentials w and w_i also satisfy our harmonic field equations.

For many applications we can neglect the time derivatives of M_L. That is the case for bodies with approximate axial symmetry rotating approximately around their symmetry axis. We will also neglect the higher spin moments S_L for $l > 1$. With these approximations in the skeletonized gauge the metric potentials read [42]

$$\begin{aligned} w(t, \mathbf{x}) &\simeq G \sum_{l \geq 0} \frac{(-1)^l}{l!} M_L \partial_L (r^{-1}) + \mathcal{O}_4 \\ w_i(t, \mathbf{x}) &\simeq -\frac{G}{2} \frac{(\mathbf{x} \times \mathbf{S})^i}{r^3} + \mathcal{O}_2 . \end{aligned} \tag{4.6}$$

This result for w is remarkable indeed because *formally* it agrees with the Newtonian one. Note, however, that here we are dealing with solutions of the post-Newtonian field equations. In other words, the post-Newtonian BD-mass moments M_L have been introduced in such a clever way that the solution w of a relativistic field equation practically looks Newtonian. If one thinks about a parameter formalism whose parameters are determined from observational data then the BD moments are the quantities that parameterizes the gravitational field outside some matter distribution in our post-Newtonian framework; they are directly measurable (e.g. by an analysis of satellite orbits). Especially it makes no sense to split the M_L's into some "Newtonian-part" plus c^{-2}-corrections.

5 Geodesic motion in the PN-Schwarzschild field

We now come to the problem of motion of some satellite or planet or Moon in the gravitational field of some spherically symmetric central mass that is described by

$w_i = 0$ and

$$w = \frac{GM}{r} \equiv \frac{\mu}{r} .$$

There are many possibilities to formulate such a post-Newtonian orbit [43]; here we concentrate on the Brumberg [34] form.

As we will see later the motion of a test body in some external gravitational field is along a geodesic determined by the external metric, that satisfies an equation of the form

$$\frac{d^2x^\mu}{d\lambda^2} + \Gamma^\mu_{\nu\sigma} \frac{dx^\nu}{d\lambda} \frac{dx^\sigma}{d\lambda} = 0 , \tag{5.1}$$

where λ parameterizes the geodetic curve $x^\mu(\lambda)$. Replacing λ by the coordinate time t by using the geodetic equation (5.1) with $\mu = 0$, the geodesic equation takes the form

$$\frac{d^2x^i}{dt^2} = -c^2 \left\{ \Gamma^i_{\ 00} + 2\Gamma^i_{\ 0j} \frac{v^j}{c} + \Gamma^i_{\ jk} \frac{v^j}{c} \frac{v^k}{c} - \left[\Gamma^0_{\ 00} + 2\Gamma^0_{\ 0j} \frac{v^j}{c} + \Gamma^0_{\ jk} \frac{v^j}{c} \frac{v^k}{c} \right] \frac{v^i}{c} \right\} .$$

The relevant nonvanishing Christoffel-symbols read to PN-order

$$\Gamma^i_{\ 00} = \frac{\mu}{c^2 r^2} \frac{x^i}{r} \left(1 - 4\frac{\mu}{c^2 r} \right) ,$$

$$\Gamma^i_{\ jk} = -\frac{\mu}{c^2 r^3} \left(x^k \delta_{ij} + x^j \delta_{ik} - x^i \delta_{jk} \right) ,$$

$$\Gamma^0_{\ 0j} = \frac{\mu}{c^2 r^2} \frac{x^j}{r} ,$$

so that the equation of motion takes the form

$$\frac{d^2x^i}{dt^2} = -\mu \frac{x^i}{r^3} + \frac{\mu}{c^2 r} \left[4\mu \frac{x^i}{r^3} - \frac{x^i}{r^2} \mathbf{v}^2 + 4 \frac{v^i}{r^2} (\mathbf{x} \cdot \mathbf{v}) \right] . \tag{5.2}$$

It has several advantages to derive this equation of motion from a Lagrangian $\mathcal{L}$. The geodesic equation satisfies an extremal principle of the form

$$0 = \delta \int ds = \delta \int \left(\frac{ds}{dt} \right) dt$$

and instead of ds we can equally well use the proper time interval $d\tau$. For that reason one usually defines a Lagrangian $\mathcal{L}$ via

$$\frac{d\tau}{dt} = 1 - c^{-2} \mathcal{L} .$$

For the PN Schwarzschild field the Lagrangian takes the form

$$\mathcal{L} = \frac{\mu}{r} + \frac{1}{2} \mathbf{v}^2 - \frac{1}{2c^2} \left(\frac{\mu}{r} \right)^2 + \frac{3}{2} \frac{\mu}{r} \left(\frac{v}{c} \right)^2 + \frac{1}{8} \frac{v^4}{c^2} \tag{5.3}$$

up to terms of $\mathcal{O}_4$.

The main advantage of the Lagrangian is the possibility to construct explicitly first integrals of motion connected with the conservation of (specific) energy $\mathcal{E}$ and angular momentum $\mathbf{J}$ (e.g. [34, 43]). These two conserved quantities are given by

$$\begin{aligned}\mathcal{E} &= \mathbf{v}\frac{\partial \mathcal{L}}{\partial \mathbf{v}} - \mathcal{L} \\ &= \frac{1}{2}\mathbf{v}^2 - \frac{\mu}{r} + \frac{3}{8}\mathbf{v}^2\left(\frac{\mathbf{v}}{c}\right)^2 + \frac{\mu}{2c^2 r}\left(\frac{\mu}{r} + 3\mathbf{v}^2\right)\end{aligned} \tag{5.4}$$

and

$$\begin{aligned}\mathbf{J} &= \mathbf{x} \times \frac{\partial \mathcal{L}}{\partial \mathbf{v}} \\ &= (\mathbf{x} \times \mathbf{v})\left(1 + \frac{1}{2}\frac{\mathbf{v}^2}{c^2} + \frac{3\mu}{c^2 r}\right) .\end{aligned} \tag{5.5}$$

Note that the conservation of angular momentum implies that the orbit is confined to a coordinate plane. Using polar coordinates in the orbital plane with

$$\mathbf{x} = r\,\mathbf{e}_r\,, \qquad \dot{\mathbf{x}} = \dot{r}\,\mathbf{e}_r + r\dot{\phi}\,\mathbf{e}_\phi$$

and

$$|\mathbf{x} \times \dot{\mathbf{x}}| = r^2\dot{\phi}\,, \qquad \mathbf{v}^2 = \dot{r}^2 + r^2\dot{\phi}^2$$

the specific energy $\mathcal{E}$ and the absolute value of the orbital angular momentum $\mathcal{J} = |\mathbf{J}|$ can be written as

$$\begin{aligned}\mathcal{E} &= \frac{1}{2}\left(\dot{r}^2 + r^2\dot{\phi}^2\right) - \frac{\mu}{r} + \frac{3}{8c^2}\left(\dot{r}^2 + r^2\dot{\phi}^2\right)^2 \\ &\quad + \frac{\mu}{2c^2 r}\left(\frac{\mu}{r} + 3\left(\dot{r}^2 + r^2\dot{\phi}^2\right)\right) \\ \mathcal{J} &= r^2\dot{\phi}\left(1 + \frac{1}{2c^2}\left(\dot{r}^2 + r^2\dot{\phi}^2\right) + \frac{3\mu}{c^2 r}\right) .\end{aligned}$$

This leads us to first-order differential equations of motion

$$\begin{aligned}r^2\dot{\phi} &= \mathcal{J}\left(1 - \frac{1}{2c^2}\left(\dot{r}^2 + r^2\dot{\phi}^2\right) - \frac{3\mu}{c^2 r}\right) \\ &= \mathcal{J}\left(1 - \frac{\mathcal{E}}{c^2} - \frac{4\mu}{c^2 r}\right)\end{aligned} \tag{5.6}$$

and

$$\begin{aligned}\dot{r}^2 &= -r^2\dot{\phi}^2 + \frac{2\mu}{r} + 2\mathcal{E} - \frac{3}{4c^2}\left(\dot{r}^2 + r^2\dot{\phi}^2\right)^2 - \frac{3\mu}{c^2 r}\left(\dot{r}^2 + r^2\dot{\phi}^2\right) - \frac{\mu^2}{c^2 r^2} \\ &= -r^2\dot{\phi}^2 + \frac{2\mu}{r} + 2\,\mathcal{E} - 3\,\frac{\mathcal{E}^2}{c^2} - 12\,\frac{\mu}{r}\frac{\mathcal{E}}{c^2} - 10\,\frac{\mu}{r}\frac{\mu}{c^2 r} .\end{aligned} \tag{5.7}$$

Eliminating the $\dot{\phi}^2$ term from the last equation we get

$$\dot{r}^2 = A + \frac{2B}{r} + \frac{C}{r^2} + \frac{D}{r^3} \tag{5.8}$$

with

$$\begin{aligned}
A &= 2\mathcal{E} - 3\frac{\mathcal{E}^2}{c^2} \\
B &= \mu - 6\,\mu\frac{\mathcal{E}}{c^2} \\
C &= -\mathcal{J}^2\left(1 - \frac{2\mathcal{E}}{c^2}\right) - 10\,\frac{\mu^2}{c^2} \\
D &= 8\,\mathcal{J}^2\frac{\mu}{c^2}\,.
\end{aligned}$$

Using

$$\begin{aligned}
\dot{r}^2 &= \left(\frac{dr(\phi(t))}{dt}\right)^2 = \left(\frac{dr}{d\phi}\dot{\phi}\right)^2 \\
&= \mathcal{J}^2\left(\frac{d(1/r)}{d\phi}\right)^2\left(1 - 8\,\frac{\mu}{c^2 r} + \frac{\mathcal{E}}{c^2}\right)
\end{aligned}$$

the radial equation can be written in the form

$$\left(\frac{d(1/r)}{d\phi}\right)^2 = A' + \frac{2B'}{r} + \frac{C'}{r^2} \tag{5.9}$$

with

$$\begin{aligned}
A' &= \frac{2\mathcal{E}}{\mathcal{J}^2}\left(1 + \frac{1}{2}\frac{\mathcal{E}}{c^2}\right) \\
B' &= \frac{\mu}{\mathcal{J}^2}\left(1 + 4\,\frac{\mathcal{E}}{c^2}\right) \\
C' &= -1 + 6\,\frac{\mu^2}{c^2\mathcal{J}^2}\,.
\end{aligned}$$

Let us now write the radial equation as

$$\left(\frac{d(1/r)}{d\phi}\right)^2 = \mathcal{C}\left(\frac{1}{r} - \frac{1}{a(1+e)}\right)\left(\frac{1}{a(1-e)} - \frac{1}{r}\right) \tag{5.10}$$

with

$$\mathcal{C} = 1 - 6\frac{\mu^2}{c^2\mathcal{J}^2}\,. \tag{5.11}$$

From this representation we see that $r_{\pm} = a(1 \pm e)$ represent the minimal and maximal values for r, i.e. a and e have the usual meaning as semimajor axis and numerical eccentricity of the post-Newtonian orbit and might be considered as integration constants alternative to $\mathcal{E}$ and $\mathcal{J}$. We have

$$A' = -\frac{1}{a^2(1-e^2)}\mathcal{C}$$

and

$$B' = \frac{\mathcal{C}}{a(1-e^2)}$$

from which we derive

$$\mathcal{J}^2 = \mu a(1-e^2)\left[1 - 2\frac{\mu}{c^2 a} + \frac{6}{1-e^2}\frac{\mu}{c^2 a}\right]$$
$$\mathcal{E} = -\frac{\mu}{2a}\left[1 - \frac{7\mu}{4c^2 a}\right].$$

The solution of (5.10) is then simply given by

$$r = \frac{a(1-e^2)}{1+e\cos f} \tag{5.12}$$

where the "true anomaly" f obeys the relation

$$\left(\frac{df}{d\phi}\right)^2 = \mathcal{C} = 1 - 6\frac{\mu}{a(1-e^2)c^2},$$

i.e.

$$f = \left[1 - 3\frac{\mu}{a(1-e^2)c^2}\right](\phi - \phi_0). \tag{5.13}$$

This implies that the post-Newtonian orbit is that of a precessing ellipse, the secular drift of the argument of periastron per revolution being given by

$$\Delta\phi = 2\pi\frac{3\mu}{a(1-e^2)c^2}. \tag{5.14}$$

Eliminating $\mathcal{E}$ and $\mathcal{J}$ from (5.6) we obtain

$$r^2\dot{\phi} = \sqrt{\mu a(1-e^2)}\left[1 + \left(-\frac{1}{2} + \frac{3}{(1-e^2)} - 4\frac{a}{r}\right)\frac{\mu}{c^2 a}\right] \tag{5.15}$$

or together with (5)

$$\sqrt{\mu a(1-e^2)}dt = r^2 df\left[1 + 4\frac{\mu}{c^2 r} + \frac{1}{2}\frac{\mu}{c^2 a}\right]. \tag{5.16}$$

For a circular orbit with $r = a$, $e = 0$ we have

$$\dot{\phi}^2 \equiv n^2 = \frac{\mu}{a^3}\left(1 - 3\frac{\mu}{c^2 a}\right), \tag{5.17}$$

defining the mean motion n of the post-Newtonian orbit and the mean anomaly M via the relation

$$M = n(t - t_0) = nt + M_0 \,.$$

If one introduces as in the usual Newtonian Kepler theory an eccentric anomaly E via

$$\sin f = \frac{(1 - e^2)^{1/2} \sin E}{1 - e \cos E} \,; \qquad \cos f = \frac{\cos E - e}{1 - e \cos E}$$

so that

$$\frac{df}{dE} = \frac{\sqrt{1 - e^2}}{1 - e \cos E}$$

and

$$r = a(1 - e \cos E)$$

then the integration of (5.16) leads to a corresponding Kepler equation in the form

$$M = \left(1 + 3 \frac{\mu}{c^2 a}\right) E - \left[1 - \frac{\mu}{c^2 a}\right] e \sin E \,. \tag{5.18}$$

The time dependence of the orbital point is then obtained by means of the Kepler equation via $t \to M \to E \to f$.

6 Astronomical reference frames

6.1 Transformation between global and local systems: first results

For the description of the gravitational N-body problem, we will consider a total of $N + 1$ different coordinate systems: one global coordinate system $x^\mu = (ct, x^i)$ in which all N bodies are contained and in which the global dynamics of the system can be described and N local charts $X_A = (cT_A, X_A^a)$, $A = 1, \ldots, N$, where the system X_A^α is assumed to move with body A of the system.

In the following, we will often speak about two coordinates systems, a global one, $\Sigma_{\rm glob}$, with coordinates $x^\mu = (ct, x^i)$ and a local one, $\Sigma_{\rm loc}$, with coordinates $X^\alpha = (cT, X^a)$. Mostly the global one will be the Barycentric Celestial Reference System (BCRS) and the local one will be the Geocentric Celestial Reference System (GCRS). We now assume the corresponding metric tensors to be of the following canonical form:

$$\begin{aligned} g_{00} &= -\exp(-2w/c^2) \\ g_{0i} &= -\frac{4}{c^3} w_i \\ g_{ij} &= \delta_{ij} \exp(+2w/c^2) + \mathcal{O}_4 \end{aligned}$$

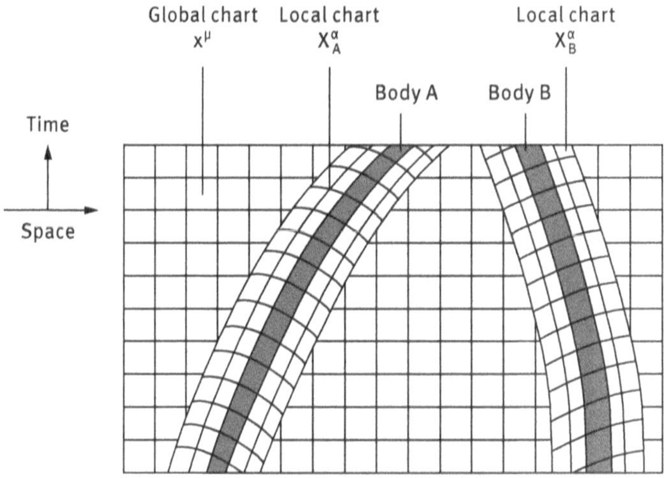

Fig. 1. One global and N local coordinate systems are used for the description of the gravitational N-body problem (from [44]).

in Σ_{glob} and

$$
\begin{aligned}
G_{00} &= -\exp(-2W/c^2) \\
G_{0a} &= -\frac{4}{c^3}W_a \\
G_{ab} &= \delta_{ab}\exp(+2W/c^2) + \mathcal{O}_4
\end{aligned}
$$

in Σ_{loc}, i.e. we assume the metric tensors to be of the same form but with metric potentials $w^\mu \equiv (w, w_i)$ in the global system and different ones $W^\alpha \equiv (W, W^a)$ in the local system. Moreover, we assume the usual conditions (2.2) for the energy–momentum tensor in all $N+1$ coordinate systems. We write the transformation $X^\alpha \to x^\mu$ in the general form

$$x^\mu(X^\alpha) = z^\mu(T) + e^\mu_a(T)X^a + \xi^\mu(T, X^a)\,, \tag{6.1}$$

where ξ^μ is at least quadratic in X^a. Here, $z^\mu(T)$ describes the world line of some suitably selected point associated with the body under consideration. This world line will be called the central world line of the corresponding body; later it will be chosen as the body's post-Newtonian center of mass.

Lemma 1. *Let $A^\mu_\alpha \equiv \partial x^\mu/\partial X^\alpha$ and assume*

$$A^0_0 = 1 + \mathcal{O}_2\,, \quad A^i_0 = \mathcal{O}_1\,, \quad A^0_a = \mathcal{O}_1\,, \quad A^i_a = \mathcal{O}_0\,.$$

Then from the form of the metric tensors in the two coordinate systems we have

$$e_0^0(T) \equiv \frac{1}{c}\frac{dz^0}{dT} = 1 + \mathcal{O}_2\,,$$

$$e_a^0(T) = \frac{1}{c}\frac{dz^i}{dT}e_a^i + \mathcal{O}_3\,,$$

$$e_0^0(T)e_a^i(T) = \left(1 + \frac{1}{2c^2}\mathbf{v}^2\right)\left(\delta_{ij} + \frac{1}{2c^2}v^i v^j\right)R_a^j(T) + \mathcal{O}_4\,,$$

$$\xi^0(T, X^a) = \mathcal{O}_3\,,$$

$$\xi^i(T, X^a) = \frac{1}{c^2}e_a^i(T)\left[\frac{1}{2}A_a\mathbf{X}^2 - X^a(\mathbf{A}\cdot\mathbf{X})\right] + \mathcal{O}_4\,,$$

where

$$v^i = \frac{dz^i}{dt} = \frac{dz^i}{dT} + \mathcal{O}_2\,,$$

$$A_a = e_a^i\frac{d^2z^i}{dT^2} + \mathcal{O}_2 = e_a^i\frac{d^2z^i}{dt^2} + \mathcal{O}_2$$

and $R_a^i(T)$ *is a slowly time-dependent rotation matrix with*

$$R_a^i R_a^j = \delta_{ij}\,, \qquad R_a^i R_b^i = \delta_{ab}$$

and

$$\frac{dR_a^i}{dT} = \mathcal{O}_2\,.$$

For a proof of Lemma 1 see [2].

Let us now consider the transformation law for the metric potentials.

Theorem 2. *From the transformation of the metric tensors one finds*

$$\begin{aligned} w &= \left(1 + \frac{2\mathbf{V}^2}{c^2}\right)W + \frac{4}{c^2}V^a W_a + \frac{c^2}{2}\ln\left[A_0^0A_0^0 - A_a^0A_a^0\right] + \mathcal{O}_4 \\ w^i &= R_a^i W^a + v^i W + \frac{c^3}{4}(A_0^0A_0^i - A_a^0A_a^i) + \mathcal{O}_2\,, \end{aligned} \tag{6.2}$$

where

$$v^i \equiv R_a^i V^a \qquad \text{or} \quad V^a \equiv R_i^a v^i\,.$$

The proof follows directly from $g^{\mu\nu}(x) = A_\alpha^\mu A_\beta^\nu G^{\alpha\beta}(X)$.

Note that the post-Newtonian transformation of metric potentials $(W, W^a) \to (w, w^i)$ *is linear*, i.e. of the form

$$w^\mu(x) = \mathcal{A}^\mu_\alpha(T) W^\alpha(X) + \mathcal{B}^\mu(X) .$$

Here,

$$\mathcal{A}^0_0 = 1 + \frac{2\mathbf{V}^2}{c^2} ; \quad \mathcal{A}^0_a = \frac{4}{c^2} V^a ; \quad \mathcal{A}^i_0 = v^i ; \quad \mathcal{A}^i_a = R^i_a$$

and

$$\mathcal{B}^0 = \frac{c^2}{2} \ln \left[A^0_0 A^0_0 - A^0_a A^0_a \right] ; \quad \mathcal{B}^i = \frac{c^3}{4} (A^0_0 A^i_0 - A^0_a A^i_a) .$$

For several purposes one also needs the inverse of (6.2) namely

$$W_\alpha = \mathcal{A}^{-1}_{\alpha\mu} (w_\mu - \mathcal{B}_\mu) ,$$

which reads explicitly ($\mathcal{B}_\mu = \mathcal{B}^\mu$)

$$\begin{aligned} W &= \left[1 + \frac{2}{c^2} \mathbf{V}^2 \right] (w - \mathcal{B}) - \frac{4}{c^2} v^i (w_i - \mathcal{B}_i) + \mathcal{O}_4 \\ W_a &= -V^a (w - \mathcal{B}) + R^i_a (w_i - \mathcal{B}_i) + \mathcal{O}_2 . \end{aligned} \tag{6.3}$$

6.2 Split of local potentials, multipole moments

Let us consider the metric tensor $G_{\alpha\beta}$ in the local system defined by the metric potentials $W_\alpha \equiv (W, W_a)$. In the gravitational N-body system these local potentials can be split into two parts

$$\begin{aligned} W(T, \mathbf{X}) &= W_{\text{self}}(T, \mathbf{X}) + W_{\text{ext}}(T, \mathbf{X}) \\ W^a(T, \mathbf{X}) &= W^a_{\text{self}}(T, \mathbf{X}) + W^a_{\text{ext}}(T, \mathbf{X}) . \end{aligned} \tag{6.4}$$

In the following, we will often use the notation

$$\overline{W} \equiv W_{\text{ext}} ; \quad \overline{W}^a \equiv W^a_{\text{ext}} .$$

If the local system is associated with body E (e.g. the Earth) then the self parts of the metric potentials, W^α_{self}, result from the gravitational action of body E itself whereas the external parts of the potentials, W^α_{ext}, result from the action of all other bodies of the system and inertial terms that appear in the local E-system. In mathematical terms the self-parts of the local metric potentials are defined by

$$\begin{aligned} W_{\text{self}}(T, \mathbf{X}) &= G \int_E d^3 X' \frac{\Sigma(T, \mathbf{X}')}{|\mathbf{X} - \mathbf{X}'|} + \frac{G}{2c^2} \frac{\partial^2}{\partial T^2} \int_E d^3 X' \Sigma(T, \mathbf{X}') |\mathbf{X} - \mathbf{X}'| , \\ W^a_{\text{self}}(T, \mathbf{X}) &= G \int_E d^3 X' \frac{\Sigma^a(T, \mathbf{X}')}{|\mathbf{X} - \mathbf{X}'|} , \end{aligned} \tag{6.5}$$

where the integrals extend over the support of body E only.

Theorem 3 (Extended Blanchet–Damour Theorem). *In the local system of body* E, *outside of* E, *the self-parts of the metric potential,* $W^\alpha_{\rm self}$, *admit a convergent expansion of the form* ($R = |\mathbf{X}|$)

$$W_{\rm self}(T,\mathbf{X}) = G\sum_{l\ge 0}\frac{(-1)^l}{l!}\partial_L[R^{-1}M^{\rm E}_L(T_\pm)] + \frac{1}{c^2}\partial_T\Lambda_{\rm E} + \mathcal{O}_4$$

$$W^a_{\rm self}(T,\mathbf{X}) = -G\sum_{l\ge 1}\frac{(-1)^l}{l!}\left[\partial_{L-1}\left(R^{-1}\frac{d}{dT}M^{\rm E}_{aL-1}\right)\right.$$
$$\left. + \frac{l}{l+1}\epsilon_{abc}\partial_{bL-1}(R^{-1}S^{\rm E}_{cL-1})\right] - \frac{1}{4}\partial_a\Lambda^{\rm E} + \mathcal{O}_2\,. \qquad (6.6)$$

Here,

$$\Lambda^{\rm E} \equiv 4G\sum_{l\ge 0}\frac{(-1)^l}{(l+1)!}\frac{2l+1}{2l+3}\partial_L(R^{-1}\mu^{\rm E}_L(T_\pm))$$

$$\mu^{\rm E}_L \equiv \int_{\rm E} d^3X\,\hat{X}^{bL}\Sigma^b(T,\mathbf{X})$$

and

$$M^{\rm E}_L(T) = \int_{\rm E} d^3X\,\hat{X}^L\Sigma + \frac{1}{2(2l+3)c^2}\frac{d^2}{dT^2}\int_{\rm E} d^3X\,\hat{X}^L\mathbf{X}^2\Sigma$$
$$-\frac{4(2l+1)}{(l+1)(2l+3)c^2}\frac{d}{dT}\int_{\rm E} d^3X\,\hat{X}^{aL}\Sigma^a \qquad (l\ge 0)$$

$$S^{\rm E}_L(T) = \int_{\rm E} d^3X\,\epsilon^{ab<c_l}\hat{X}^{L-1>a}\Sigma^b\,, \qquad (l\ge 1)\,, \qquad (6.7)$$

where $f(T_\pm) = [f(T+R/c) + f(T-R/c)]/2$.

The proof is analogous to the one for the Blanchet–Damour theorem.

$M^{\rm E}_L$ and $S^{\rm E}_L$ are the BD mass- and spin-multipole moments of body E that reduce to the corresponding moments in case that there is only one single body. Note that a post-Newtonian center of mass of body E can be introduced by the vanishing of the BD mass-dipole, i.e. by

$$M^{\rm E}_a = 0\,.$$

Theorem 4. *Let*

$$w_{\rm E}(t,\mathbf{x}) = G\int_{\rm E} d^3x'\frac{\sigma(t,\mathbf{x}')}{|\mathbf{x}-\mathbf{x}'|} + \frac{G}{2c^2}\frac{\partial^2}{\partial t^2}\int_{\rm E} d^3x'\sigma(t,\mathbf{x}')|\mathbf{x}-\mathbf{x}'|\,,$$

$$w^i_{\rm E}(t,\mathbf{x}) = G\int_{\rm E} d^3x'\frac{\sigma^i(t,\mathbf{x}')}{|\mathbf{x}-\mathbf{x}'|}$$

be the metric potentials in the global system induced by body E, then

$$w^\mu_{\rm E} = \mathcal{A}_{\mu\alpha}(T) W^\alpha_{\rm self,E} + \mathcal{O}(4,2)\,,$$

or explicitly

$$\begin{aligned} w_{\rm E} &= \left(1 + \frac{2\mathbf{v}^2}{c^2}\right) W_{\rm self,E} + \frac{4}{c^2} V^a W^a_{\rm self,E} + \mathcal{O}_4 \\ w^i_{\rm E} &= R^i_a W^a_{\rm self,E} + v^i W_{\rm self,E} + \mathcal{O}_2\,, \end{aligned} \tag{6.8}$$

where $W^\alpha_{\rm self,E}$ are the self-parts of the metric potentials in the local E-system and the velocity $\mathbf{v}$ refers to the central point of that system (e.g. the barycentric velocity of the geocenter).

Damour, Soffel, and Xu have found several independent proofs for this central Theorem. One of them might be called the physicist's proof, where the transformation rules for the energy–momentum tensor and the coordinate transformations are used to transform the integrals defining $W_{\rm self}$ and $W^a_{\rm self}$ into the global system (see also [1]). Another proof, that might be called the proof of a field theoretician, can be found in Damour et al. [2].

6.3 Tetrad induced local coordinates

So far the quantities e^μ_α have been constrained but not fully specified. This will now be done by requiring these quantities to represent an orthonormal tetrad along the world line $\mathcal{L}_{\rm E}$ of the origin of the local E-system given by $X^a = 0$ with respect to the external metric, i.e.

$$g^{\rm ext}_{\mu\nu} e^\mu_\alpha e^\nu_\beta |_{\mathcal{L}_{\rm E}} = \eta_{\alpha\beta}\,. \tag{6.9}$$

Here, the external metric tensor, $g^{\rm ext}_{\mu\nu}$, is defined by

$$\begin{aligned} g^{\rm ext}_{00} &= -e^{-2\overline{w}/c^2}\,, \\ g^{\rm ext}_{0i} &= -\frac{4}{c^3}\overline{w}^i\,, \\ g^{\rm ext}_{ij} &= \delta_{ij} e^{2\overline{w}/c^2} + \mathcal{O}_4 \end{aligned}$$

with

$$\overline{w}^\mu = \sum_{A \neq {\rm E}} w^\mu_A\,.$$

Thus, considering the external part of the metric only, the local coordinates X^α will be chosen as tetrad induced coordinates. One consequence of this is

$$W^\alpha_{\rm ext}(T, \mathbf{0}) = 0 \tag{6.10}$$

and the external part of the metric is Minkowskian at the spatial coordinate origin, i.e.

$$G^{\rm ext}_{\alpha\beta}|_{X^a=0} = \eta_{\alpha\beta}\,.$$

The gravitational influence of external bodies is *effaced* and relation (6.10) is sometimes called *weak effacement condition.*

Theorem 5. *From the tetrad condition* (6.9) *one infers that*

$$\begin{aligned}
e^0_0(T) &= 1 + \frac{1}{c^2}\left(\frac{1}{2}\mathbf{v}_{\rm E}^2 + \overline{w}(\mathbf{z}_{\rm E})\right)\\
&\quad + \frac{1}{c^4}\left(\frac{3}{8}\mathbf{v}_{\rm E}^4 + \frac{1}{2}(\overline{w}(\mathbf{z}_{\rm E}))^2 + \frac{5}{2}\overline{w}(\mathbf{z}_{\rm E})\mathbf{v}_{\rm E}^2 - 4\overline{w}^i(\mathbf{z}_{\rm E})v^i_{\rm E}\right) + \mathcal{O}_6\\
e^0_a(T) &= R^i_a\left[\frac{v^i_{\rm E}}{c}\left(1 + \frac{1}{c^2}\left\{\frac{1}{2}\mathbf{v}_{\rm E}^2 + 3\overline{w}(\mathbf{z}_{\rm E})\right\}\right) - \frac{4}{c^3}\overline{w}^i(\mathbf{z}_{\rm E})\right] + \mathcal{O}_5\\
e^i_a(T) &= \left(1 - \frac{1}{c^2}\overline{w}(\mathbf{z}_{\rm E})\right)\left(\delta^{ij} + \frac{1}{2c^2}v^i_{\rm E}v^j_{\rm E}\right)R^j_a + \mathcal{O}_4\,.
\end{aligned} \tag{6.11}$$

6.4 The standard transformation between global and local coordinates

According to (6.1) the transformation between global coordinates $x^\mu = (ct, x^i)$ and local ones $X^\alpha = (cT, X^a)$ for one and the same event is given by

$$\begin{aligned}
ct &= z^0(T) + e^0_a(T)X^a + \xi^0\,,\\
x^i &= z^i(T) + e^i_a(T)X^a + \xi^i\,.
\end{aligned} \tag{6.12}$$

The tetrad components, $e^\mu_a(T)$, are given by Theorem 5 and ξ^i is given by Lemma 1.

The quantity ξ^0, that fixes the $1/c^4$-part in the $T \leftrightarrow t$ transformation is left open in Damour et al. [2]. If the harmonic gauge is chosen both in the global as well as in the local system the condition for ξ^0 reads

$$\Delta\xi^0 = \frac{1}{c}\frac{de^0_0}{dT} + \frac{1}{c^2}\frac{d^2e^0_a}{dT^2}X^a + \mathcal{O}_5\,.$$

Taking any solution of this equation we can then always add a function ξ^0_L which satisfies the Laplace equation $\Delta_X\xi^i_L = 0$. From this one finds that ξ^0 for harmonic coordinates might be chosen in the form

$$\begin{aligned}
c^3\xi^0(T,\mathbf{X}) &= -\,2\overline{w}_{,ab}X^aX^b - \frac{1}{2}\overline{w}_{,t}\mathbf{X}^2 + \mathbf{v}\cdot\mathbf{X}(\overline{w}_{,a}X^a) + \mathbf{v}\cdot\mathbf{X}(\overline{w}_{,a} - a_a)X^a\\
&\quad + \frac{1}{2}(\mathbf{v}\cdot\mathbf{a} - v^a\overline{w}_{,a})\mathbf{X}^2 + \frac{1}{10}(\dot{\mathbf{a}}\cdot\mathbf{X})\,\mathbf{X}^2\,.
\end{aligned} \tag{6.13}$$

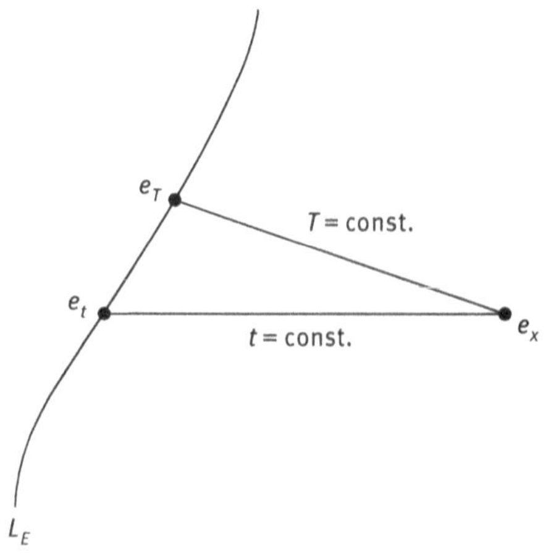

Fig. 2. Three events of importance for the inversion of the coordinate transformation, e_X, e_t, and e_T.

Since solar-system ephemerides are given in global coordinates for practical applications it is useful to invert these relations in the form $X^\alpha = X^\alpha(x^\mu)$. To this end let us consider the three events in Figure 2 denoted by e_X, e_T and e_t. In the local E-system e_X has coordinates (cT, X^a) related to (ct, x^i) by the general transformation rule (6.1), $e_T(e_t)$ denotes the intersection of the T = const. (t = const.) hyper-surface through e_X with the world line of the origin of the local E-system (e.g. the geocenter), given by $X^a = 0$. These two events have coordinates

$$\begin{aligned} e_T:\ & (t_{\rm sim}, z^i_{\rm E}(t_{\rm sim}) \qquad (T, \mathbf{0})\,, \\ e_t:\ & (t, z^i_{\rm E}(t)) \qquad (T_{\rm sim}, \mathbf{0})\,. \end{aligned}$$

Using the general transformation rule (6.1) one finds that

$$\begin{aligned} T_{\rm sim} &= T + \frac{1}{c}\frac{e^0_a}{e^0_0}X^a + \mathcal{O}_4 \\ t_{\rm sim} &= t - \frac{1}{c}e^0_a X^a + \mathcal{O}_4 \end{aligned}$$

and with $\mathbf{z}_{\rm sim}(t) \equiv \mathbf{z}(T_{\rm sim})$

$$x^i - z^i_{\rm sim}(t) = \left(e^i_a - \frac{e^i_0 e^0_a}{e^0_0}\right)_T X^a + \xi^i(T, X^a) + \mathcal{O}_4\,. \tag{6.14}$$

Now, $e^i_0(T) = dz^i(T)/(cdT)$ and, therefore, $e^i_0/e^0_0 = dz^i/(c\,dt)$, we get by inserting the expressions for e^i_a, e^i_0 and ξ^i and solving for X^a

$$X^a = R_{ai}\left\{r^i + \frac{1}{c^2}\left[\frac{1}{2}v^i(\mathbf{v}\cdot\mathbf{r}) + \overline{w}(\mathbf{z}_{\rm E})r^i + r^i(\mathbf{a}\cdot\mathbf{r}) - \frac{1}{2}a^i r^2\right]\right\} + \mathcal{O}_4\,, \tag{6.15}$$

where $\mathbf{r}(t) \equiv \mathbf{x} - \mathbf{z}_{\rm E}(t)$ and $\mathbf{a}(t)$ is the acceleration of $\mathbf{z}_{\rm E}$ in global coordinates.

The derivation of the $T = T(t)$ relation is more complicated. First, we will derive this relation for an event on the central world line, i.e. for $X^a = 0$ where we have $t = z^0(T)/c$ and therefore

$$\frac{dt}{dT} = \frac{1}{c}\frac{dz^0(T)}{dT} = e^0_0(T)$$

or

$$\frac{dT}{dt} = (e^0_0(T))^{-1}\,.$$

Let f be some function defined at the central world line, $\mathcal{L}_\mathrm{E}$, then for some event on $\mathcal{L}_\mathrm{E}$ we have $f(t) = f(T)$ though the values for t and T will differ to post-Newtonian order. From the last relation we therefore get for events at $X^a = 0$

$$\left.\frac{dT}{dt}\right|_{\mathcal{L}_\mathrm{E}} = 1 - \frac{1}{c^2}\frac{dA(t)}{dt} + \frac{1}{c^4}\frac{dB(t)}{dt} + \mathcal{O}_5$$

with

$$\frac{d}{dt}A(t) = \frac{1}{2}\mathbf{v}_\mathrm{E}^2 + \overline{w}(\mathbf{z}_\mathrm{E})\,,$$
$$\frac{d}{dt}B(t) = -\frac{1}{8}\mathbf{v}_\mathrm{E}^4 - \frac{3}{2}\overline{w}(\mathbf{z}_\mathrm{E}) + 4v^i_\mathrm{E}\overline{w}^i(\mathbf{z}_\mathrm{E}) + \frac{1}{2}\overline{w}^2(\mathbf{z}_\mathrm{E})\,.$$

Next we consider some event outside the central world line where we have

$$ct = z^0(T) + e^0_a(T)X^a + \xi^0(T, X^a)\,.$$

If f is again some function on the central world line we have

$$f(T) \simeq f(t - \mathbf{v}\cdot\mathbf{r}/c^2) = f(t) - f_{,t}\cdot(\mathbf{v}\cdot\mathbf{r}/c^2) + \mathcal{O}_3$$

and $\partial_T \simeq \partial_t + v^i\partial_i$. We then find that

$$e^0_a(T)X^a = \frac{1}{c}\mathbf{v}\cdot\mathbf{r} + \frac{1}{c^3}\left[(\mathbf{v}\cdot\mathbf{r})\mathbf{v}^2 + 4\overline{w}(\mathbf{v}\cdot\mathbf{r}) - \frac{1}{2}(\mathbf{v}\cdot\mathbf{a})r^2 - 4\overline{w}^i r^i\right] + \mathcal{O}_5\,,$$

where again the indices E have been dropped and the metric potentials have to be taken at $\mathcal{L}_\mathrm{E}$. Inserting expression (6.13) for ξ^0 we finally get [42]

$$T = t - \frac{1}{c^2}[A(t) + \mathbf{v}\cdot\mathbf{r}]$$
$$+ \frac{1}{c^4}[B(t) + B^i(t)r^i + B^{ij}(t)r^i r^j + C(t,\mathbf{x})] + \mathcal{O}_5\,, \qquad (6.16)$$

where

$$B^i(t) = -\frac{1}{2}\mathbf{v}^2 v^i + 4\overline{w}^i - 3v^i\overline{w}\,,$$
$$B^{ij}(t) = -v^i R_{aj}Q^a + 2\frac{\partial}{\partial x^j}\overline{w}^i - v^i\frac{\partial}{\partial x^j}\overline{w} + \frac{1}{2}\dot{\overline{w}}\delta^{ij}\,.$$

The dot on $\overline{w}$ indicated the total time derivative, i.e.

$$\dot{\overline{w}} \equiv \overline{w}_{,t} + v^i \overline{w}_{,i}$$

and

$$C(t, \mathbf{x}) = -\frac{1}{10} r^2 (\dot{\mathbf{a}} \cdot \mathbf{r}) ,$$

$$Q^a(t) = R_{ai} \left(\frac{\partial}{\partial x^i} \overline{w} - a^i \right) .$$

6.5 The description of tidal forces

Post-Newtonian tidal moments

We will now introduce a useful post-Newtonian generalization of the Newtonian tidal expansion of the effective potential describing the gravitational action of external bodies in a local system co-moving with some body E together with the inertial forces appearing in that system. As it will become obvious later in the discussion of the equations of motion of astronomical bodies the following external gravitoelectric and gravitomagnetic fields defined by

$$\overline{E}_a(T, \mathbf{x}) \equiv \partial_a \overline{W} + \frac{4}{c^2} \partial_T \overline{W}_a , \qquad \overline{B}_a(T, \mathbf{X}) = -4 \epsilon_{abc} \, \partial_b \overline{W}_c ,$$

will play a central role. For that reason they will be considered as post-Newtonian analogues of ∇U^{tidal}. In a more compact notation we write

$$\overline{\mathbf{E}} = \nabla \overline{W} + \frac{4}{c^2} \partial_T \overline{\mathbf{W}} , \qquad \overline{\mathbf{B}} = -4 \nabla \times \overline{\mathbf{W}} .$$

It is easy to see that the gravitoelectric and gravitomagnetic external fields, $\overline{\mathbf{E}}$ and $\overline{\mathbf{B}}$, are invariant under gauge transformations of the external metric potentials of the form

$$\overline{W}' = \overline{W} - \frac{1}{c^2} \partial_T \Lambda ; \qquad \overline{W}'_a = \overline{W}_a + \frac{1}{4} \partial \Lambda .$$

Under such gauge transformations the $\mathbf{E}$ and $\mathbf{B}$ fields transform according to

$$\overline{E}'_a = \partial_a \overline{W}' + \frac{4}{c^2} \partial_T \overline{W}'_a = \partial_a \overline{W} - \frac{1}{c^2} \partial_{Ta} \Lambda + \frac{4}{c^2} \partial_T \overline{W}_a + \frac{1}{c^2} \partial_{Ta} \Lambda = \overline{E}_a$$

and

$$\overline{B}'_a = \epsilon_{abc} \partial_b (-4 \overline{W}'_c) = -4 \epsilon_{abc} \partial_b \left(\overline{W}_c + \frac{1}{4} \partial_c \Lambda \right) = -\epsilon_{abc} \partial_b \overline{W}_c = \overline{B}_a ,$$

i.e. they are gauge invariant in the sense defined above.

Lemma 2. *In virtue of these field equations and the definitions of the external "gauge invariant"* $\overline{\mathbf{E}}$ *and* $\overline{\mathbf{B}}$ *fields, they satisfy the following homogeneous equations:*

$$\begin{aligned}
\nabla \times \overline{\mathbf{E}} &= -\frac{1}{c^2}\partial_T \overline{\mathbf{B}} \,, \\
\nabla \times \overline{\mathbf{B}} &= 4\partial_T \overline{\mathbf{E}} + \mathcal{O}_2 \,, \\
\nabla \cdot \overline{\mathbf{E}} &= -\frac{3}{c^2}\partial_T^2 \overline{W} + \mathcal{O}_4 \,, \\
\nabla \cdot \overline{\mathbf{B}} &= 0 \,, \\
\nabla^2 \overline{\mathbf{E}} &= \frac{1}{c^2}\partial_T^2 \overline{\mathbf{E}} + \mathcal{O}_4 \,, \\
\nabla^2 \overline{\mathbf{B}} &= \mathcal{O}_2 \,.
\end{aligned}$$

The proof follows by direct calculation.

Next we characterize these fields by two corresponding (gravitoelectric and gravitomagnetic) sets of post-Newtonian tidal moments

$$\begin{aligned}
G_L(T) &\equiv [\partial_{<L-1}\overline{E}_{a_l>}(T,\mathbf{X})]_{X^a=0} \qquad (l \geq 1) \,, \\
H_L(T) &\equiv [\partial_{<L-1}\overline{B}_{a_l>}(T,\mathbf{X})]_{X^a=0} \qquad (l \geq 1) \,.
\end{aligned}$$

Theorem 6. *The external gravitoelectric and gravitomagnetic fields,* $\overline{\mathbf{E}}$ *and* $\overline{\mathbf{B}}$*, are completely determined by the corresponding sets of tidal moments,* G_L *and* H_L *to post-Newtonian order. Explicitly they are given by*

$$\begin{aligned}
\overline{E}_a(T,\mathbf{X}) = \sum_{l\geq 0} \frac{1}{l!} \Bigg[& \hat{X}^L G_{aL}(T) + \frac{1}{2(2l+3)c^2}\mathbf{X}^2 \hat{X}^L \frac{d^2}{dT^2} G_{aL}(T) \\
& - \frac{7l-4}{(2l+1)c^2} \hat{X}^{aL-1} \frac{d^2}{dT^2} G_{L-1}(T) \\
& + \frac{l}{(l+1)c^2} \epsilon_{abc} \hat{X}^{bL-1} \frac{d}{dT} H_{cL-1}(T) \Bigg] + \mathcal{O}_4 \,, \\
\overline{B}_a(T,\mathbf{X}) = \sum_{l\geq 0} \Bigg[& \hat{X}^L H_{aL}(T) - \frac{4l}{l+1} \epsilon_{abc} \hat{X}^{bL-1} \frac{d}{dT} G_{cL-1}(T) \Bigg] + \mathcal{O}_2 \,,
\end{aligned} \tag{6.17}$$

where, by convention, we are assuming that any term which contains an undefined tidal moment has to be replaced by zero. For example, the term containing the factor $(7l-4)/(2l+1)$ *vanishes for* $l = 0$.

The proof is indicated in [2]. It involves the relation

$$X^L = \hat{X}^L + \frac{l(l-1)}{2(2l-1)} \mathbf{X}^2 \hat{X}^{(L-2}\delta^{a_{l-1}a_l)} + \mathcal{O}(\delta\delta) \,,$$

where $\mathcal{O}(\delta\delta)$ *denotes any term with (at least) two Kronecker-deltas and the STF decomposition of a tensor* $T_{a\langle L\rangle}$ *that is STF only with respect to the multi-index L:*

$$T_{a\langle L\rangle} = \hat{T}^{(+)}_{aL} + \epsilon_{ca<a_l}\hat{T}^{(0)}_{L-1>c} + \delta_{a<a_l}\hat{T}^{(-)}_{L-1>} \,,$$

in which each tensor $\hat{T}^{\pm,0}$ is STF:

$$\hat{T}^{(+)}_{L+1} = \mathrm{STF}_{L+1}(T_{La_{l+1}}) ,$$
$$\hat{T}^{(0)}_{L} = \mathrm{STF}_{L}\left[\frac{l}{l+1}\epsilon_{a_l bc}T_{bcL-1}\right] ,$$
$$\hat{T}^{(-)}_{L-1} = \frac{2l-1}{2l+1}T_{ccL-1} .$$

With similar techniques one can show that apart from simple gauge terms, given by a single function Λ, also the external potentials, $\overline{W}$ and $\overline{W}_a$, can be expressed in terms of the tidal-moments G_L and H_L (remember that $\overline{W}(X^a = 0) = \overline{W}_a(X^a = 0) = 0$).

Lemma 3. *The external metric potentials $\overline{W}$ and $\overline{W}_a$ can be written in terms of post-Newtonian tidal series in the form (DSX II, equations (4.15))*

$$\overline{W}(T,\mathbf{X}) = \sum_{l\geq 1}\frac{1}{l!}\left[\hat{X}^L G_L + \frac{1}{2(2l+3)c^2}\mathbf{X}^2\hat{X}^L\ddot{G}_L\right] + \frac{1}{c^2}\Lambda_{,T} + \mathcal{O}_4 ,$$
$$\overline{W}_a(T,\mathbf{X}) = \sum_{l\geq 1}\frac{1}{l!}\left[-\frac{2l+1}{(l+1)(2l+3)}\hat{X}^{aL}\dot{G}_L + \frac{l}{4(l+1)}\epsilon_{abc}\hat{X}^{bL-1}H_{cL-1}\right] - \frac{1}{4}\Lambda_{,a} + \mathcal{O}_2 . \quad (6.18)$$

The tidal moments, G^{E}_L and H^{E}_L, can be split into two parts: a part resulting from the gravitational action of some external body A≠ E, and some inertial part related with the inhomogeneous part in the transformation of metric potentials:

$$G^{\mathrm{E}}_L = \sum_{\mathrm{A}\neq\mathrm{E}} G^{\mathrm{A/E}}_L + G''_L$$
$$H^{\mathrm{E}}_L = \sum_{\mathrm{A}\neq\mathrm{E}} H^{\mathrm{A/E}}_L + H''_L .$$

The parts resulting from body A, as seen in the local E-system, result from the self-parts of body A in its own local system, $W^{+\mathrm{A}}$ and $W^{+\mathrm{A}}_a$, that are directly given in terms of the mass- and spin moments of A.

Lemma 4. *In the local* E*-system all nonvanishing inertial terms G''_L and H''_L are given by (DSX I, equations (6.30)):*

$$G''_a = -A_a + \mathcal{O}_4$$
$$G''_{ab} = \frac{3}{c^2}A_{<a}A_{b>} + \mathcal{O}_4$$
$$H''_a = \epsilon_{abc}\left[V_b A_c + c^2\frac{dR^i_b}{dT}R^i_c\right] + \mathcal{O}_2 , \quad (6.19)$$

where

$$A_a = \eta_{\mu\nu} e^\mu_a \frac{d^2 z^\nu_{\rm E}}{d\tau_f^2}$$

with

$$d\tau_f^2 = -c^{-2} \eta_{\mu\nu} dz^\mu_{\rm E} dz^\nu_{\rm E} .$$

Explicitly, G''_a *is given by*

$$G''_a = -R^i_a \frac{d^2 z^i_{\rm E}}{dt^2} + \frac{\overline{w}(\mathbf{z}_{\rm E})}{c^2} a^a_{\rm E} - \frac{1}{2c^2} (\mathbf{v}_{\rm E} \cdot \mathbf{a}_{\rm E}) v^a_{\rm E} - \frac{\mathbf{v}^2_{\rm E}}{c^2} a^a_A . \tag{6.20}$$

Thus the inertial gravitoelectric tidal dipole moment, G''_a contains the global coordinate acceleration of the center of mass of body E.

7 The gravitational N-body problem

7.1 Local evolution equations

The global equations of motion in the gravitational N-body problem result from the local evolution equation

$$T^\nu_{\mu;\nu} = T^\nu_{\mu,\nu} + \Gamma^\nu_{\nu\sigma} T^\sigma_\mu - \Gamma^\sigma_{\mu\nu} T^\nu_\sigma = 0 . \tag{7.1}$$

Lemma 5. *In any local system the local evolution equations* (7.1) *take the following form (DSX I, equations* (5.6)*):*

$$\frac{\partial}{\partial T}\Sigma + \frac{\partial}{\partial X^a}\Sigma^a = \frac{1}{c^2}\frac{\partial}{\partial T}T^{bb} - \frac{1}{c^2}\Sigma\frac{\partial}{\partial T}W + \mathcal{O}_4 . \tag{7.2}$$

This equation for $\mu = 0$ *is the energy equation. The* $\mu = a$ *equation reads*

$$\frac{\partial}{\partial T}\left[\left(1 + \frac{4}{c^2}W\right)\Sigma^a\right] + \frac{\partial}{\partial X^b}\left[\left(1 + \frac{4}{c^2}W\right)T^{ab}\right] = F^a + \mathcal{O}_4 . \tag{7.3}$$

This is the post-Newtonian Euler equation. Here,

$$F^a = \Sigma E_a + \frac{1}{c^2} B_{ab} \Sigma^b$$

is the gravitational Lorentz force, where

$$E_a = \partial_a W + \frac{4}{c^2}\partial_T W_a ,$$
$$B_{ab} = \epsilon_{abc} B_c = \partial_a(-4W_b) - \partial_b(-4W_a) .$$

Proof. We will show (7.3). The post-Newtonian energy equation (7.2) follows similarly. Since ($G \equiv -\det(G_{\alpha\beta})$)

$$\Gamma^{\nu}_{\nu\sigma} = \frac{1}{\sqrt{G}}\frac{\partial}{\partial X^{\sigma}}\sqrt{G}$$

the $\mu = a$ equation reads

$$\begin{aligned} T^{\nu}_{a;\nu} &= \frac{\partial}{\partial X^{\nu}}T^{\nu}_{a} + \left(\frac{1}{\sqrt{G}}\partial X^{\sigma}\sqrt{G}\right)T^{\sigma}_{a} - \Gamma^{\sigma}_{a\nu}T^{\nu}_{\sigma} \\ &= \frac{1}{\sqrt{G}}\frac{\partial}{\partial X^{\nu}}(\sqrt{G}T^{\nu}_{a}) - \Gamma^{\sigma}_{a\nu}T^{\nu}_{\sigma} = 0 \end{aligned}$$

or

$$\frac{\partial}{\partial X^{\nu}}(\sqrt{G}T^{\nu}_{a}) = \sqrt{G}G_{\sigma\nu}\Gamma^{\sigma}_{a\mu}T^{\mu\nu}\,.$$

Since

$$G_{\sigma\nu}\Gamma^{\sigma}_{a\mu}T^{\mu\nu} = \frac{1}{2}T^{\mu\nu}\partial_a G_{\mu\nu}$$

we obtain:

$$\frac{\partial}{\partial X^{\nu}}(\sqrt{G}T^{\nu}_{a}) = \frac{1}{2}\sqrt{G}T^{\mu\nu}\partial_a G_{\mu\nu}\,.$$

Since $\sqrt{G} = 1 + 2W/c^2 + \mathcal{O}_4$, the right-hand side of the last equation reads

$$\begin{aligned} &\frac{1}{2}\left(1+\frac{2W}{c^2}\right)\left[T^{00}\partial_a G_{00} + 2T^{0b}\partial_a G_{0b} + T^{bc}\partial_a G_{bc}\right] \\ &= \frac{1}{2}\left(1+\frac{2W}{c^2}\right)\left[2(T^{00}+T^{ss})\partial_a\frac{W}{c^2} - 8\frac{T^{0b}}{c^3}W^{b}_{,a}\right] + \mathcal{O}_4 \\ &= \Sigma W_{,a} - \frac{4}{c^2}\Sigma^{b}\partial_a W^{b} + \mathcal{O}_4\,. \end{aligned}$$

From this we get the Blanchet–Damour–Schäfer (BDS) form [45] of the post-Newtonian Euler equation:

$$\frac{\partial}{\partial T}\Pi^{a} + \frac{\partial}{\partial X^{b}}(\sqrt{G}T^{b}_{a}) = \Sigma W_{,a} - \frac{4}{c^2}\Sigma^{b}\partial_a W^{b} + \mathcal{O}_4\,, \tag{7.4}$$

where

$$\Pi^{a} = c^{-1}\sqrt{G}T^{0}_{a} = -\frac{4}{c^2}W^{a}\Sigma + \left(1+\frac{4W}{c^2}\right)\Sigma^{a} + \mathcal{O}_4\,.$$

Relation (7.3) then follows from the BDS-form with

$$\sqrt{G}T^{b}_{a} = \left(1+\frac{4W}{c^2}\right)T^{ab} - \frac{4}{c^2}W^{a}\Sigma^{b}$$

and

$$\frac{\partial}{\partial T}(W^{a}\Sigma) = \Sigma W^{a}_{,T} + W^{a}\Sigma_{,T} = \Sigma W^{a}_{,T} - W^{a}\Sigma^{b}_{,b} + \mathcal{O}_2\,.$$ □

7.2 The translational motion

We will now fix the central world line of a body A, member of gravitational N-body system, by choosing the origin of the local system, $X^a = 0$, to coincide with the post-Newtonian center of mass by

$$M_a^{\mathrm{A}}(T) = 0\,. \tag{7.5}$$

A d'Alembert criterion [46] will lead to the global translational equations of motion for the N-body problem.

Theorem 7. *From the local evolution equations* (7.2) *and* (7.3) *one obtains (DSX II,* (4.20) *and* (4.21)*):*

$$\frac{dM^{\mathrm{A}}}{dT} = F_0 + \mathcal{O}_4$$
$$\frac{d^2M_a^{\mathrm{A}}}{dT^2} = F_a + \mathcal{O}_4$$
$$\frac{dS_a^{\mathrm{A}}}{dT} = D_a + \mathcal{O}_2$$

with

$$F_0 = -\frac{1}{c^2}\sum_{l\geq 0}\frac{1}{l!}\left\{(l+1)M_L\frac{dG_L}{dT} + l\frac{dM_L}{dT}G_L\right\} + \mathcal{O}_4$$

$$\begin{aligned}F_a = \sum_{l\geq 0}\frac{1}{l!}\Bigg\{ & M_L G_{aL} + \frac{l}{c^2(l+1)}S_L H_{aL} + \frac{1}{c^2(l+2)}\epsilon_{abc}M_{bL}\frac{dH_{cL}}{dT} \\ & + \frac{1}{c^2(l+1)}\epsilon_{abc}\frac{dM_{bL}}{dT}H_{cL} - \frac{4(l+1)}{c^2(l+2)^2}\epsilon_{abc}S_{bL}\frac{dG_{cL}}{dT} - \frac{4}{c^2(l+2)}\epsilon_{abc}\frac{dS_{bL}}{dT}G_{cL} \\ & - \frac{2l^3+7l^2+15l+6}{c^2(l+1)(2l+3)}M_{aL}\frac{d^2G_L}{dT^2} - \frac{2l^3+5l^2+12l+5}{c^2(l+1)^2}\frac{dM_{aL}}{dT}\frac{dG_L}{dT} \\ & - \frac{l^2+l+4}{c^2(l+1)}\frac{d^2M_{aL}}{dT^2}G_L\Bigg\} + \mathcal{O}_4\,,\end{aligned}$$

$$D_a = \sum_{l\geq 0}\frac{1}{l!}\epsilon_{abc}M_{bL}G_{cL} + \mathcal{O}_2\,.$$

In the following, we will study the consequences of Theorem 7 for a system of N mass-monopoles, i.e. we will assume

$$M_L^{\mathrm{A}} = S_L^{\mathrm{A}} = 0\,, \quad \lambda \geq 1\,,$$

for all bodies A from the system. Since $G(T) = 0$, one finds that for the mass-monopole model

$$\frac{dM^{\mathrm{A}}}{dT} = \mathcal{O}_4\,,$$

i.e. the masses of the N bodies are conserved to post-Newtonian order. From (7.5) and Theorem 7 we then get from the d'Alembert (1743) equation

$$0 = \frac{d^2 M_a^{\rm A}}{dT^2} = F_a = M_{\rm A} G_a \,,$$

or

$$G_a = 0\,. \tag{7.6}$$

Implicitly the translational equations of motion is given by the vanishing of the external gravitoelectric tidal dipole moment G_a. It is not difficult to show that equation (7.6) is equivalent to the geodesic equation

$$\frac{d\overline{u}^\lambda}{d\overline{\tau}} + \overline{\Gamma}^\lambda_{\mu\nu}\overline{u}^\mu\overline{u}^\nu = \mathcal{O}_4 \tag{7.7}$$

in the external metric. Here,

$$\overline{u}^\mu = \overline{u}^\mu_{\rm E} = \frac{dz^\mu_{\rm E}}{d\overline{\tau}} = \frac{dz^\mu_{\rm E}}{dT}\,,$$

where we have assumed, according to the weak effacement condition, that $T = \overline{\tau}$ and $-c^2 d\overline{\tau}^2 = \overline{g}_{\mu\nu} dx^\mu dx^\nu|_{X=0}$.

Inserting the Christoffel symbols of the global external metric as seen by body A, given by $\overline{w}_{\rm A}$ and $\overline{w}_i^{\rm A}$, the translational equation of motion for the center of mass of body A in the mass-monopole model reads

$$\begin{aligned}\frac{d^2 z^i_{\rm A}}{dt^2} &= \left[1 - \frac{4}{c^2}\overline{w}_{\rm A} + \frac{1}{c^2}\mathbf{v}_{\rm A}^2\right]\partial_i\overline{w}_{\rm A} + \frac{4}{c^2}\partial_t\overline{w}_i^{\rm A} - \frac{4}{c^2}\left(\partial_i\overline{w}_j^{\rm A} - \partial_j\overline{w}_i^{\rm A}\right)v^j_{\rm A}\\ &\quad - \frac{1}{c^2}\left(3\partial_t\overline{w}_{\rm A} + 4v^j_{\rm A}\partial_j\overline{w}_{\rm A}\right)v^i_{\rm A} + \mathcal{O}_4\,,\end{aligned} \tag{7.8}$$

where $v^i_{\rm A}$ is the global coordinate velocity of body A. Finally we need the external metric potentials related with body A explicitly

$$\overline{w}_\mu^{\rm A} = \sum_{\rm B\neq A} w_\mu^{\rm B}\,.$$

In the local B-system we simply have

$$W^{\rm B}_{\rm self} = \frac{GM_{\rm B}}{R_{\rm B}}\,; \qquad W_a^{\rm B,self} = 0\,.$$

Transformation according to Theorem 4 yields

$$w^{\rm B} = \left(1 + \frac{2}{c^2}\mathbf{v}_{\rm B}^2\right)\frac{GM_{\rm B}}{R_{\rm B}}\,; \qquad w_i^{\rm B} = \frac{GM_{\rm B}}{R_{\rm B}}v^i_{\rm B}\,. \tag{7.9}$$

At this point we need the transformation rule for the inverse distance from the center of body B, that follows from (6.15) ($\mathbf{r}_B(t) \equiv \mathbf{x} - \mathbf{z}_B(t)$, $\mathbf{n}_B(t) \equiv \mathbf{r}_B(t)/|\mathbf{r}_B(t)|$, $\mathbf{a}_B \equiv d^2\mathbf{z}_B/dt^2$):

$$\frac{1}{R_B} = \frac{1}{r_B}\left[1 - \frac{\overline{w}(\mathbf{z}_B)}{c^2} - \frac{1}{2c^2}(\mathbf{v}_B \cdot \mathbf{n}_B)^2 - \frac{1}{2c^2}\mathbf{a}_B \cdot \mathbf{r}_B\right]$$

we get

$$w^B = \frac{GM_B}{r_B}\left[1 + 2\frac{\mathbf{v}_B^2}{c^2} - \frac{\overline{w}(\mathbf{z}_B)}{c^2} - \frac{1}{2c^2}(\mathbf{v}_B \cdot \mathbf{n}_B)^2 - \frac{1}{2c^2}\mathbf{a}_B \cdot \mathbf{r}_B\right]$$
$$w_i^B = \frac{GM_B}{r_B} v_B^i . \tag{7.10}$$

Inserting these potentials into (7.8) we finally end up with the Einstein–Infeld–Hoffmann (EIH) equations of motion for body A:

$$\frac{d^2 z_A^i}{dt^2} = a_A^{i(LD)}(\mathbf{z}_A, \mathbf{v}_A) + \mathcal{O}_4 , \tag{7.11}$$

where the Lorentz–Droste acceleration, $\mathbf{a}_A^{(LD)}$, is given by

$$\begin{aligned}\mathbf{a}_A^{(LD)} = &- \sum_{B\neq A} \frac{GM_B}{r_{AB}^2}\mathbf{n}_{AB}\left[1 + \frac{1}{c^2}\left[\mathbf{v}_A^2 + 2\mathbf{v}_B^2 - 4\mathbf{v}_A \cdot \mathbf{v}_B - \frac{3}{2}(\mathbf{n}_{AB} \cdot \mathbf{v}_B)^2\right]\right.\\
&\left. - 4\sum_{C\neq A}\frac{GM_C}{c^2 r_{AC}} - \sum_{C\neq B}\frac{GM_C}{c^2 r_{BC}}\left[1 + \frac{1}{2}\frac{r_{AB}}{r_{CB}}\mathbf{n}_{AB} \cdot \mathbf{n}_{CB}\right]\right]\\
&- \frac{7}{2}\sum_{B\neq A}\sum_{C\neq B}\mathbf{n}_{BC}\frac{G^2 M_B M_C}{c^2 r_{AB} r_{BC}^2}\\
&+ \sum_{B\neq A}(\mathbf{v}_A - \mathbf{v}_B)\frac{GM_B}{c^2 r_{AB}^2}(4\mathbf{n}_{AB} \cdot \mathbf{v}_A - 3\mathbf{n}_{AB} \cdot \mathbf{v}_B) ,\end{aligned} \tag{7.12}$$

where

$$r_{AB} \equiv |\mathbf{z}_A(t) - \mathbf{z}_B(t)| , \quad \mathbf{n}_{AB} \equiv [\mathbf{z}_A(t) - \mathbf{z}_B(t)]/r_{AB} .$$

8 Further developments

So far the central elements of Damour et al. [2] were exhibited and complemented with some so far unpublished material devoted to certain aspects of the whole issue. The DSX-formalism was further developed by three papers [3–5].

In the second paper [3] the problem of translational *laws of motion* for a system of N arbitrarily composed and shaped, weakly self-gravitating, rotating deformable bodies are obtained at the first post-Newtonian approximation to Einstein's theory of gravity. The full set of mass- and spin-multipole moments, M_L and S_L, was taken into account for each of the bodies. First complete and explicit results for the laws of motion of each body of the system were presented as an infinite series exhibiting the coupling of all the BD-moments to the post-Newtonian tidal moments, G_L and H_L, felt by this body. Finally explicit expressions of these tidal moments in terms of BD-moments of the other bodies are derived. For a derivation of corresponding *equations of motion* assumptions about the time dependence of BD-moments for each of the bodies have to be made. A rigidly rotating multipole model, that leads to a closed set of dynamical equations of motion, was presented in Klioner et al. [47].

The third paper of this series [4] deals with rotational equations of motion. It is shown how to associate with each body, in its own local frame, a post-Newtonian spin vector, whose local time evolution is entirely determined by the coupling between the BD-moments of that body and the tidal moments it experiences. The leading relativistic effects in the spin motion (geodetic-, Lense–Thirring- and Thomas precession) are discussed in detail as are the dominant relativistic contributions to the torque acting on a body in its local frame.

Finally the fourth paper of the series [5] formulates the basic translational equations of motion for artificial Earth satellites. These equations are given both in a local, geocentric system (the Geocentric Celestial Reference System, GCRS) and in the global, barycentric one (the Barycentric Celestial Reference System, BCRS).

Of course many other scientists have contributed important publications to the field of relativistic celestial mechanics (especially to the dynamics of compact binaries because of its relevance to the problem of gravity wave emission). For these developments the reader is referred to the corresponding literature (see e.g. [48] and the contributions by Thibault Damour by Gerhard Schäfer in this Festschrift and references cited therein). A generalization of the DSX-formalism by including the PPN-parameters β and γ was presented in [49].

An extended version of the Brumberg–Kopeikin formalism with PPN parameters included has been worked out in the review paper by Kopeikin and Vlasov [50] (see also the discussion in this volume, Chapter 3).

References

[1] Brumberg, V. A., Kopeikin, S. M., Nuovo Cimento, **103B**, 63, 1989.

[2] Damour, T., Soffel, M., Xu, C., Phys. Rev. D **43**, 3273, 1991 (DSX I).

[3] Damour, T., Soffel, M., Xu, C., Phys. Rev. D **45**, 1017, 1992 (DSX II).

[4] Damour, T., Soffel, M., Xu, C., Phys. Rev. D **47**, 3124, 1993 (DSX III).

[5] Damour, T., Soffel, M., Xu, C., Phys. Rev. D **49**, 618, 1994 (DSX IV).

[6] Tisserand, F., *Traité de Mécanique Céleste*, Vol. I (reprint of the first edition of 1889, Gauthier-Villars, Paris, 1960).

[7] Einstein, A., Preuss. Akad. Wiss. Berlin, Sitzber., 831, 1915.

[8] Droste, J., Versl. K. Akad. Wet. Amsterdam **19**, 447, 1916.
[9] de Sitter, W., Mon. Not. R. Astron. Soc. **76**, 699, 1916; **77**, 155, 1916.
[10] Lorentz, H. A., Droste, J., Versl. Akad. Wet. Amsterdam **26**, 392, 1917 (part I); **26**, 649, 1917 (part(II), English translation in Lorentz, H. A., *Collected Papers*, Vol. V, edited by Zeeman, P. and Fokker, A. D., Nijhoff, The Hague, 1937, pp. 330–355.
[11] Fock, V. A., 1959, *The Theory of Space, Time and Gravitation*, Pergamon, Oxford, 1959.
[12] Damour, T., in *300 Years of Gravitation*, edited by Hawking S. W. and Isreal W., Cambridge University Press, Cambridge, England, 1987.
[13] Eisenstaedt, J., Arch. Hist. Exact Sci. **35**, 115, 1986.
[14] Martin, C. F., Torrence, M. H., Misner, C. W., J. Geophys. Res. **90**, 9403, 1985.
[15] Hellings, R., Astron. J. **91**, 650, 1986.
[16] Synge, J. L., *Relativity: The General Theory*, North-Holland, Amsterdam, 1960.
[17] Misner, C. W., Thorne, K. S., Wheeler, J. A., *Gravitation*, Freeman, San Francisco, 1973.
[18] Weyl, H., *Raum, Zeit, Materie*, 4th edn., Springer, Berlin, 1921.
[19] Manasse, F. K., J. Math. Phys. **4**, 746, 1963.
[20] D'Eath, Phys. Rev. D **11**, 1387, 1975.
[21] Kates, R. E., Phys. Rev. D **22**, 1853, 1980.
[22] Bertotti, B., Nuovo Cimento **12**, 226, 1954.
[23] Will, C., in *Experimental Gravitation*, 56th Enrico Fermi School, Varenna, Italy, 1972; edited by Bertottir, B., Academic Press, New York, 1974
[24] Will, C., Eardley, D. M., Astrophys. J. **212**, L91, 1977.
[25] Mashhoon, B., Gen. Relativ. Gravit. **16**, 311, 1984.
[26] Mashhoon, B., Found. Phys. **15**, 497, 1985.
[27] Thorne, K. S., Hartle, J. B., Phys. Rev. D **31**, 1815, 1985.
[28] Damour, T., in *Gravitational Radiation*, Proceedings of the Les Houches Summer School, Les Houches, France, 1982, edited by Deruelle, N. and Piran, T., North-Holland, Amsterdam, 1983.
[29] Ashby, N., Bertotti, B., Phys. Rev. Lett. **52**, 485, 1984.
[30] Ashby, N., Bertotti, B., Phys. Rev. D **34**, 2246, 1986.
[31] Soffel, M. H., Ruder, H., Schneider, M., Astron. Astrophys. **157**, 357, 1986.
[32] Kovalevsky, J., Brumberg, V. A., *Relativity in Celestial Mechanics and Astronomy*, Proceedings of the Symposium, Leningrad, USSR, 1985, IAU Symposium No. **114**, Reidel, Dordrecht, 1986.
[33] Kopeikin, S. M., Celest. Mech. **44**, 87, 1988.
[34] Brumberg, V. A., *Essential Relativistic Celestial Mechanics*, Hilger, Bristol, 1991.
[35] Voinov, A. V., Celest. Mech. **42**, 293, 1988.
[36] Thorne, K., Rev. Mod. Phys. **52**, 299, 1980.
[37] Blanchet, L., Damour, T., Phil. Trans. Roy. Soc. London **A320**, 379, 1986.
[38] Damour, T., in *Gravitation in Astrophysics (Cargèse 1986)*, Preceedings of the NATO Advanced Study Institute, Cargese, France, 1986, edited by Carter, B. and Hartle, J. B., NATO ASI Series B: Physics, Vol. **156**, Plenum, New York, 1987, pp. 3–62.
[39] Zhang, X.-H., Phys. Rev. D **34**, 991, 1986.
[40] Suen, W. M., Phys. Rev. D **34**, 3617, 1986.
[41] Blanchet, L., Damour, T., Ann. Inst. H. Poincaré, **50**, 377, 1989.
[42] Soffel, M., Klioner, S., Petit, G., Wolf, P., Kopeikin, S., Bretagnon, P., Brumberg, V. A., Capitaine, N., Damour, T., Fukushima, T., Guinot, B., Huang, T., Lindegren, L., Ma, C., Nordtvedt, K., Ries, J., Seidelmann, P. K., Vokroughlický, D., Will, C., Cu, C., Astron. J. **126**, 2687, 2003.
[43] Soffel, M. H., *Relativity in Astrometry, Celestial Mechanics and Geodesy*, Springer, Berlin, 1989.
[44] Damour, T., Soffel, M., Xu, C., in *Relativistic Gravity Research, With Emphasis on Experiments and Observations*, Lecture Notes in Physics **410**, edited by Ehlers, J. and Schäfer, G., Springer, Berlin, 1992.

[45] Blanchet, L., Damour, T., Schäfer, G. Mon. Not. R. Astr. Soc. **242**, 289, 1990.
[46] D'Alembert, J., Traité de Dynamique, David, Paris, 1743.
[47] Klioner, S., Soffel. M., Xu, C., Wu, X., in: Proc. of Les Journées 2001, Systèmes de Référence Spatio-Temporels, edited by Capitaine, N., Brussels, Observatoire de Paris, 2001.
[48] Blanchet, L., Spallicci, A., Whiting, B., Mass and Motion in General Relativity, Fundamental Theories of Physics **162**, Springer, Berlin, 2011.
[49] Klioner, S., Soffel, M., Phys. Rev. D **62**, 024019, 2000.
[50] Kopeikin, S., Vlasov, I., Physics Reports, **400**, 209, 2004.

Pavel Korobkov and Sergei Kopeikin

General relativistic theory of light propagation in multipolar gravitational fields

1 Introduction

1.1 Statement of the problem

Direct experimental detection of gravitational waves is a fascinating but yet unsolved problem of modern fundamental physics. Enormous efforts have been undertaken to make progress toward its solution both by theorists and experimentalists [7–11]. The main theoretical efforts are presently focused on the calculation of templates of the gravitational waves emitted by coalescing binary systems comprised of neutron stars and/or black holes [12–17] as well as creation of improved filtering technique for gravitational wave detectors [18, 19] which will enable the extraction of the gravitational wave signal from all kinds of interferences present in the noisy data collected by the gravitational wave observatories. Direct experimental efforts have led to the construction of several ground-based optical interferometers with the length of arms reaching a few miles [20–24]. Some work is under way to build super-sensitive cryogenic-bar gravitational-wave detectors of Weber's type [25–28]. Spaceborne laser interferometric detectors like LISA [29], NGO [30], or ASTROD [31, 32] may significantly increase the sensitivity of the gravitational-wave detectors and revolutionize the field of gravitational physics.

There is no doubt that the detection of gravitational waves by the specialized gravitational wave antennas would provide the most direct evidence of the existence of these elusive ripples in the fabric of spacetime. On the other hand, there exist a number of astronomical phenomena which might be used for an indirect detection of gravitational waves and understanding gravitational physics of astrophysical systems emitting gravitational waves. It is worth emphasizing that the present-day ground-based gravitational wave detectors are sensitive to a rather narrow band of the gravitational wave spectrum ranging between 1000 and 1 Hz [33, 34]. Spaceborne gravitational wave detectors may bring the sensitivity band down to 1 mHz [30, 32]. In order to explore the ultralong gravitational wave phenomena much below 1 mHz we have to rely upon other astronomical techniques for example, pulsar timing [35, 36] or radiometry of po-

Pavel Korobkov: Solovetsky Monastery, Arkhangelsk Region 164070, Russia
Sergei Kopeikin: Department of Physics & Astronomy, University of Missouri, 322 Physics Bldg., Columbia, Missouri 65211, USA

larization of cosmic microwave background radiation [37, 38]. The short gravitational waves can be generated by coalescing binary systems of compact astrophysical objects like neutron stars and/or black holes [39–42]. The ultralong gravitational waves can be generated by localized gravitational sources and/or topological defects in the early universe [43–47].

Astronomical observations are conducted with electromagnetic waves (photons) of different frequencies over the spectrum spreading from very long radio waves to gamma rays. Therefore, in order to detect the effects of the gravitational waves by astronomical technique one has to solve the problem of propagation of electromagnetic waves through the field generated by a localized gravitationally bounded distribution of masses which motion is determined by general relativity. Notice that gravitational wave detectors are optical interferometers making use of light propagation. Therefore, correct understanding of the process of physical interaction of the laser beam of the interferometer with incoming gravitational waves is important for their unambiguous recognition and detection. Equations of propagation of electromagnetic signals must be derived in the framework of the same theory of gravity in order to keep description of gravitational and electromagnetic phenomena on the same theoretical ground. We draw attention of the reader that the parametrized post-Newtonian (PPN) formalism [48] does not comply with this requirement. The PPN formalism is constructed on the ground of plausible physical hypothesis and assumptions about alternatives to general relativity but it does not demand their overall conformity. Therefore, straightforward application of PPN formalism to discuss gravitational physics may eventually lead to incorrect results. As an example, we point out that PPN formalism does not produce the consistent post-Newtonian equations of motion of extended bodies even in a simplified case of two PPN parameters, β and γ, when more subtle effects of body's gravitational multipoles are taken into account [49]. PPN formalism applied to interpret gravitational light-ray deflection experiments by major planets of the solar system [50–52] created a notable "speed-of-light versus speed-of-gravity" controversy [53] which originates in the inability of PPN formalism to distinguish between physical effects of gravity and electromagnetism due to the noncovariant nature of PPN parametrization limited merely by the metric manifolds [54, 55].

In this chapter, we rely upon the Einstein theory of general relativity and assume that light propagates in vacuumm that is the interstellar medium has no impact on the speed of light propagation. General relativity is a geometrized theory of gravity and it assumes that both gravity and electromagnetic field propagate locally in vacuum with the same speed which is equal to the fundamental speed c in special theory of relativity [1]. Gravitational field is found as a solution of the Einstein field equations. The electromagnetic field is obtained by solving the Maxwell equations on the curved spacetime manifold derived at previous step from Einstein's equations. In the approximation of geometric optics the electromagnetic signals propagate along null geodesics of the metric tensor [1–3].

The problem of finding solutions of the null ray equations in curved spacetime attracted many researchers since the time of discovery of general relativity. Exact solutions of this problem were found in some, particularly simple cases of symmetric spacetimes like Schwarzschild's or Kerr's black hole, homogeneous and isotropic Friedman–Lemeître–Robertson–Walker (FLRW) cosmology, plane gravitational wave, etc. [1–3, 56]. For quite a long time these exact description of null geodesics was sufficient for the purposes of experimental gravitational physics. However, real physical spacetime has no symmetries and the solution of such current problems as direct detection of gravitational waves, interpretation of an anisotropy, and polarization of cosmic microwave background radiation (CMBR), exploration of inflationary models of the early universe, finding new experimental evidences in support of general relativity and quantum gravity, development of higher precision relativistic algorithms for space missions, and many others, cannot fully rely upon the mathematical techniques developed mainly in the symmetric spacetimes. It is especially important to find out a method of integration of equations of light geodesics in spacetimes having time-dependent gravitational perturbations of arbitrary multipole order.

In this chapter, we consider a case of an isolated astronomical system embedded to an asymptotically flat spacetime. This excludes Friedmann–Lemeître–Robertson–Walker spacetime which is not asymptotically flat (see Chapter 7). We assume that gravitational field of this system is characterized by an infinite set of gravitational multipoles emitting gravitational waves. Precise and coordinate-free mathematical definition of the asymptotic flatness is based on the concept of conformal infinity that was worked out in a series of papers [57–62] and we recommend the textbook of R. Wald [3] for a concise but comprehensive mathematical introduction to this subject. The asymptotic flatness implies the existence of a comformal compactification of the manifold but it may not exist for a particular case of an astrophysical system. There is also coordinate-free definition of gravitational multipoles of an isolated system given by Hansen [63]. However, they make sense only in stationary spacetimes but are not applicable for time-dependent gravitational fields which make them useless for practical analysis of real astronomical observations and for interpretation of relativistic effects in propagation of electromagnetic signals. We use neither Hansen's multipoles in the this chapter nor the comformal compactification technique. Moreover, because observable effects of gravitational waves are weak we shall consider only linear effects of gravitational multipoles so that relativistic effects produced by nonlinearity of the gravitational field will be ignored. The linearized approximation of general theory of relativity represents a straightforward and practically useful approach to description of the multipole structure of the gravitational field of a localized astronomical system. The multipolar gravitational formalism was developed by several scientists, most notably by Thorne [5] and Blanchet and Damour [64–68], and we use their results in this chapter.

Exact formulation of the problem under discussion in this chapter is as follows (see Figure 1). We assume (in a less restrictive sense than the comformal compactifi-

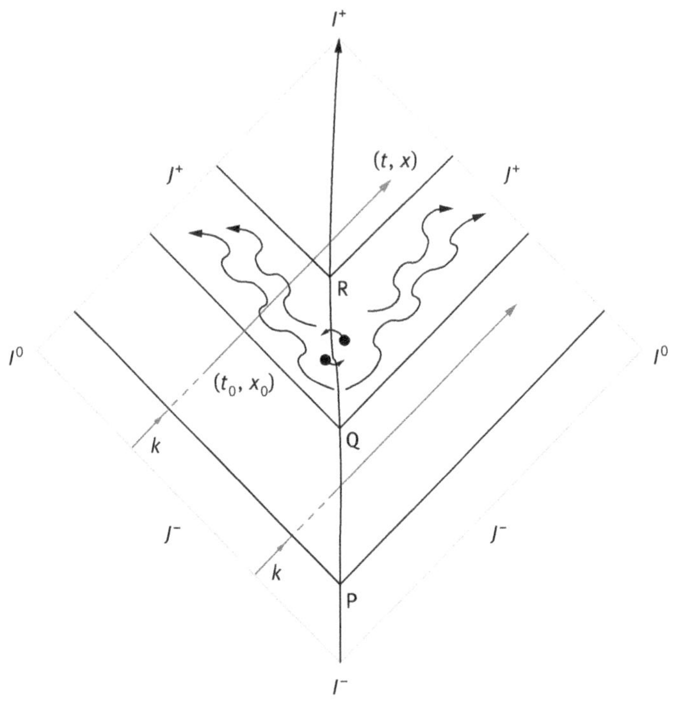

Fig. 1. The Penrose diagram shows a world line of the isolated system (binary star) originating at past timelike infinity I^- and ending at the future timelike infinity I^+. Both light and gravity propagate along null geodesics going from the past null infinity J^- to the future null infinity J^+. A particular light geodesic is emanating from the event $(t_0, \boldsymbol{x}_0)$ and ending at the event $(t, \boldsymbol{x})$. Extrapolation of this light geodesic to the past null infinity defines the null wave vector $k^\alpha = (k^0, \boldsymbol{k})$ of the electromagnetic wave under consideration.

cation technique does) that spacetime under consideration is asymptotically flat and covered globally by a single coordinate chart, $x^\alpha = (ct, \boldsymbol{x})$, where t is coordinate time and $\boldsymbol{x}$ are the spatial coordinates. An electromagnetic wave (photon) is emitted by a source of light at time t_0 and at point $\boldsymbol{x}_0$ toward an observer which receives the photon at the instant of time t and at the point $\boldsymbol{x}$. Photon propagates through the time-dependent gravitational field of the isolated astronomical system emitting gravitational waves. The structure of the gravitational field is described by a set of Blanchet–Damour multipole moments of the system which are functions of the retarded time $t - r/c$, where r is the distance from the isolated system to the field point and c is the fundamental speed. The retardation is physically due to the finite speed of gravity which coincides in general relativity with the speed of light in vacuum. Some confusion may arise in the interpretation of the observational effects because both gravity and light propagates in general relativity with the same speed. The discrimination between various effects is still possible because gravity and light propagate to observer from different sources and along different spatial directions (characteristics of the null cone in spacetime) [54].

Our task is to find out the relations connecting various physical parameters (direction of propagation, frequency, polarization, intensity, etc.) of the electromagnetic signal at the point of emission with those measured by the observer. Gravitational field affects propagation of the electromagnetic signal and changes its parameters along the light ray. Observing these changes allow us to study various properties of the gravitational field of the astronomical system and to detect gravitational waves with astronomical technique. The rest of the chapter is devoted to the mathematical solution of this problem.

1.2 Historical background

Propagation of electromagnetic signals through stationary gravitational field is a rather well-known subject having been originally discussed in classic textbooks by [69, 70]. Presently, almost any standard textbook on relativity describes solution of the null geodesic equations in the field of a spherically symmetric and rotating body or black hole. This solution is practical since it is used, for example, for interpretation of gravitational measurements and light-propagation experiments in the solar system [54]. Another application is gravitational lensing in our galaxy and in cosmology [71–73]. Continually growing accuracy of astronomical observations demands much better treatment of secondary effects in the propagation of light produced by the perturbations of the gravitational field associated with the higher order gravitational multipoles of planets and Sun [74]. Time-dependent multipoles emit gravitational waves which perturb propagation of light with the characteristic period of the gravitational wave. The influence of these periodic perturbations on light propagation parameters is important for developing correct strategy for understanding the principles of operation of gravitational wave detectors as well as for searching gravitational waves by astronomical techniques.

Among the most interesting sources of periodic gravitational waves with a well-predicted behavior are binary systems comprised of two stars orbiting each other around a common barycenter (center of mass). Indirect evidence in support of existence of gravitational waves emitted by binary pulsars was given by Joe Taylor with collaborators [75, 76]. However, direct observation of gravitational waves remains a challenging problem for experimental gravitational physics. The expected spectrum of gravitational waves extends from $\sim 10^4$ to 10^{-18} Hz [33, 34]. Within that range, the spectrum of periodic waves from known binary systems extends from about 10^{-3} Hz – the frequency of gravitational radiation from a contact white-dwarf binary [40], through the 10^{-4} to 10^{-6} Hz – the range of radiation from the main-sequence binaries [41], to the 10^{-7} to 10^{-9} Hz – the frequencies emitted by binary supermassive black holes presumably lurking in active galactic nuclei (AGN) [42]. The dimensionless strain of these waves at the Earth, h, may be as great as 10^{-21} at the highest frequencies, and as great as 3×10^{-15} at the lowest frequencies in the spectrum of gravitational waves [33, 34].

Sazhin [77] was the first who suggested the method of detection of gravitational waves emitted from a binary system by using timing observations of a background pulsar, more distant than the binary lying on the line of sight which passes sufficiently close in the sky to the binary. He had shown that the integrated time delay for the propagation of an electromagnetic pulse near the binary is proportional to $1/d^2$ where d is the impact parameterof the unperturbed trajectory of the signal. Similar idea was independently proposed by Detweiler [78] who has focused on discussing an application of pulsar timing for detection of a stochastic cosmological background of gravitational waves. More recently, Sazhin and Saphonova [79] have made estimates of the probability of observation of such an effect for pulsars in globular clusters and showed that the probability can be high, reaching 97%. Sazhin–Detweiler idea is currently used in pulsar timing arrays to detect gravitational waves in nano-Hertz frequency band [80–82].

Wahlquist [83] proposed another approach to the detection of periodic gravitational waves based on Doppler trackingof spacecraft traveling in deep space. His approach is restricted by the plane gravitational wave approximation developed earlier by Estabrook and Wahlquist [84]. Tinto ([85], and references therein) made the most recent theoretical contribution in this research area. The Doppler tracking technique is used in deep space missions for detection of gravitational waves by seeking for the characteristic triple signature in the continuously recorded phase of radio waves in the radio link between the ground station and spacecraft. The presence of this specific signature would indicate to the influence of the Doppler signal by a gravitational wave crossing the line of sight from the spacecraft to observer [86, 87].

Braginsky et al. [88, 89] raised a question about a possibility of using Very-Long Baseline Interferometry (VLBI) as a detector of stochastic gravitational waves produced in the early universe. This idea had also been investigated by Kaiser and Jaffe [90] and, the most prominently, by Pyne et al. [91] and Gwinn et al. [92] who showed that the overall effect in the time delay of VLBI signal is proportional to the strain of the metric perturbation, h, caused by the plane gravitational wave. They also calculated the pattern of proper motions of quasars over all the sky as an indicator of the presence of quadrupole and high-order harmonics of ultralong gravitational wave and set an observational limit on the energy density of such gravitational waves present in the early universe. Montanari [93] studied the perturbations of polarization of electromagnetic radiation propagating in the field of a plane gravitational wave and found that the effects are exceedingly small, unlikely to be observable.

Fakir ([94, 95] and references therein) has suggested to use astrometryto detect the periodic variations in apparent angular separations of appropriate light sources, caused by gravitational waves emitted by isolated sources of gravitational radiation. He was not able to develop a self-consistent approach to tackle the problem with a necessary completeness and rigor. For this reason, his estimate of the effect of the deflection of light caused by gravitational wave perturbation, is too optimistic. Another attempt to work out a more consistent approach to the calculation of the light deflection angle by the radiation field of an arbitrary source of gravitational waves has been undertaken by Durrer [96]. However, the calculations have been done only for the plane wave approximation. Nonetheless, the result obtained was extrapolated to the case of the localized source of gravitational waves without convincing justification. For this reason, the magnitude of the periodic changes of the light deflection angle was largely overestimated. The same misinterpretation of the effect of gravitational waves from localized sources can be found in the paper by Labeyrie [97] who studied a photometric modulation of background sources of light (stars) by gravitational waves emitted by fast-orbiting binary stars. Because of the erroneous predictions, the expected detection of gravitational waves from VLBI observations of a radio source GPS QSO 2022+171 undertaken by Pogrebenko et al. [98] was not based on firm theoretical ground and did not lead to success.

Damour and Esposito-Farèse [99] have studied the deflection of light and integrated time delay caused by the time-dependent gravitational field generated by a localized astrophysical source lying in the sky close to the line of sight to a background source of light. They worked in a quadrupole approximation and explicitly calculated the effects of the retarded gravitational field of the astrophysical source in its near, intermediate, and wave zones by making use of the Fourier-decomposition technique. Contrary to the claims of Fakir [94, 95] and Durrer [96] and in agreement with Sazhin's [77] calculations, they found that the contribution of the wave-zone and intermediate-zone fields to the deflection angle vanish exactly due to some remarkable mutual cancellations of different components of the gravitational field. The leading, total time-dependent deflection of light is created only by the quasi-static, near-zone quadrupolar part of the gravitational field.

Damour and Esposito-Farese [99] analyzed propagation of light under a simplifying condition that the impact parameter of the light ray is small with respect to the distances from observer and the source of light to the isolated system. We have found [6, 100] another way around to solve the problem of propagation of electromagnetic waves in the quadrupolar field of the gravitational waves emitted by the system without making any assumptions on mutual disposition of observer, source of light, and the system, thus, significantly improving and extending the result of [99]. At the same time the paper [100] did not answer the question about the impact of the other, higher order gravitational multipoles of the isolated system on the process of propagation of electromagnetic signals. This might be important if the effective gravitational wave emission of an octupole and/or higher order multipoles is equal or even exceeds that of the quadrupole as it may be in case of, for example, highly asymmetric stellar collapse [101], nearly head-on collision of two stars, or break-up of a binary system caused by a recoil of two black holes [102].

In this chapter we work out a systematic approach to the problem of propagation of light rays in the field of arbitrary gravitational multipole. While the most papers on light propagation consider both a light source and an observer as being located at infinity we do not need these assumptions. For this reason, our approach is generic and applicable for any mutual configuration of the source of light and observer with respect to the source of gravitational radiation. The integration technique which we use for finding solution of the equations of propagation of light rays was worked out in series of our papers [6, 100, 103, 104].

The metric tensor and coordinate systems involved in our calculations are described in Section 2 along with gauge conditions imposed on the metric tensor. The equations of propagation of electromagnetic waves in the geometric optics approximation are discussed in Section 3 and the method of their integration is given in Section 4. Exact solution of the equations of light propagation and the exact form of relativistic perturbations of the light trajectory and the coordinate speed of light are obtained in Section 5. Section 6 is devoted to the derivation of the primary observable relativistic effects – the integrated time delay, the deflection angle, the frequency shift, and the rotation of the plane of polarization of an electromagnetic wave. We discuss in Sections 7 and 8 two limiting cases of the most interesting relative configurations of the source of light, the observer, and the source of gravitational waves – the gravitational-lens configuration (Section 7) and the case of a plane gravitational wave (Section 8).

1.3 Notations and conventions

We consider a spacetime manifold which is asymptotically flat at infinity [115]. Metric tensor of the spacetime manifold is denoted by $g_{\alpha\beta}$ and its perturbation

$$h_{\alpha\beta} = g_{\alpha\beta} - \eta_{\alpha\beta} \,. \tag{1.1}$$

The determinant of the metric tensor is negative, and is denoted as $g = \det[g_{\alpha\beta}]$. A four-dimensional, fully antisymmetric Levi–Civita symbol $\epsilon_{\alpha\beta\gamma\delta}$ is defined in accordance with the convention $\epsilon_{0123} = +1$.

In this chapter we use a geometrodynamic system of units [2] such that the fundamental speed, c, and the universal gravitational constant, G, are equal to unity, that is $c = G = 1$. spacetime is assumed to be globally covered by a Cartesian-like coordinate system $(x^\alpha) \equiv (x^0, x^1, x^2, x^3) \equiv (t, x, y, z)$, where t and (x, y, z) are time and space coordinates, respectively. This coordinate system is reduced at infinity to the inertial Lorentz coordinates defined up to a global Lorentz–Poincare transformation [4]. Sometimes we shall use spherical coordinates (r, θ, ϕ) related to (x, y, z) by a standard transformation

$$x = r\sin\theta\cos\phi\,, \quad y = r\sin\theta\sin\phi\,, \quad z = r\cos\theta\,. \tag{1.2}$$

Spatial coordinates $(x^i) \equiv (x, y, z)$ in some equations will be denoted with a boldface font, $\boldsymbol{x} \equiv (x^i)$.

We shall operate with various geometric objects which have tensor indices. We agree that Greek (spacetime) indices $\alpha, \beta, \gamma, \ldots$ range from 0 to 3, and Latin (space) indices $a, b, c, \ldots$ run from 1 to 3. If not specifically stated the opposite, the Greek indices are raised and lowered by means of the Minkowski metric $\eta_{\alpha\beta} \equiv \mathrm{diag}(-1, 1, 1, 1)$, for example, $A^\alpha = \epsilon^{\alpha\beta} A_\beta$, $B_{\alpha\beta} = \epsilon_{\alpha\mu}\epsilon_{\beta\nu}B^{\mu\nu}$, and so on. The spatial indices are raised and lowered with the help of the Kronecker symbol (a unit matrix), $\delta_{ij} \equiv \mathrm{diag}(1, 1, 1)$. Regarding this rule the following conventions for the Cartesian coordinates hold: $x^i = x_i$ and $x^0 = -x_0$.

Repeated indices are summed over in accordance with Einstein's rule [2], for example,

$$A^{\alpha\beta\gamma}B_{\alpha\mu} \equiv A^{0\beta\gamma}B_{0\mu} + A^{1\beta\gamma}B_{1\mu} + A^{2\beta\gamma}B_{2\mu} + A^{3\beta\gamma}B_{3\mu} \,. \tag{1.3}$$

In the linearized (with respect to G) approximation of general relativity used in this chapter, there is no difference between spatial vectors and co-vectors nor between upper and lower space indices. Therefore, we do not distinguish between the superscript and subscript spatial indeses. For example, for a dot (scalar) product of two space vectors we have

$$A^iB_i = A_iB_i \equiv A_1B_1 + A_2B_2 + A_3B_3 \,. \tag{1.4}$$

In what follows, we shall commonly use the spatial multi-index notations for three-dimensional, Cartesian tensors [5] like this

$$\mathcal{I}_{A_l} \equiv \mathcal{I}_{a_1\ldots a_l} \,. \tag{1.5}$$

A tensor product of l identical spatial vectors k^i will be denoted as a three-dimensional tensor having l indices

$$k_{a_1}k_{a_2}\ldots k_{a_l} \equiv k_{a_1\ldots a_l} \,. \tag{1.6}$$

Full symmetrization with respect to a group of spatial indices of a Cartesian tensor will be denoted with round brackets embracing the indices

$$Q_{(a_1\ldots a_l)} \equiv \frac{1}{l!}\sum_\sigma Q_{\sigma(1)\ldots\sigma(l)} \,, \tag{1.7}$$

where σ is the set of all permutations of $(1, 2, \ldots, l)$ which makes $Q_{a_1\ldots a_l}$ fully symmetric in $a_1 \ldots a_l$.

It is convenient to introduce a special notation for symmetric trace-free (STF) Cartesian tensors by making use of angular brackets around STF indices. The explicit expression of the STF part of a tensor $Q_{a_1\ldots a_l}$ is [5, 64]

$$Q_{\langle a_1\ldots a_l\rangle} \equiv \sum_{k=0}^{[l/2]} a^l_k \delta_{(a_1a_2} \cdots \delta_{a_{2k-1}a_{2k}} S_{a_{2k+1}\ldots a_l)b_1b_1\ldots b_kb_k} \,, \tag{1.8}$$

where $[l/2]$ is the integer part of the number $l/2$,

$$S_{a_1...a_l} \equiv Q_{(a_1...a_l)} , \tag{1.9}$$

and the numerical coefficients

$$a_k^l = \frac{(-1)^k}{(2k)!!} \frac{l!}{(2l-1)!!} \frac{(2l-2k-1)!!}{(l-2k)!} . \tag{1.10}$$

We also assume that for any integer $l \geq 0$

$$l! \equiv l(l-1)\ldots 2\cdot 1 , \qquad 0! \equiv 1 , \tag{1.11}$$

and

$$l!! \equiv l(l-2)(l-4)\ldots(2 \text{ or } 1) , \qquad 0!! \equiv 1 . \tag{1.12}$$

One has, for example,

$$T_{\langle ab\rangle} = T_{(ab)} - \frac{1}{3}\delta_{ab}T_{cc} , \tag{1.13}$$

$$T_{\langle abc\rangle} = T_{(abc)} - \frac{1}{5}\delta_{ab}T_{(cjj)} - \frac{1}{5}\delta_{bc}T_{(ajj)} - \frac{1}{5}\delta_{ac}T_{(bjj)} , \tag{1.14}$$

and so on.

Cartesian tensors of the mass-type (mass) multipoles $\mathcal{I}_{\langle A_l\rangle}$ and spin-type (spin) multipoles $\mathcal{S}_{\langle A_l\rangle}$ entirely describing gravitational field outside of an isolated astronomical system are always STF objects that can be checked by inspection of the definition following from the multipolar decomposition of the metric tensor perturbation $h_{\alpha\beta}$ [5, 64]. For this reason, to avoid the appearance of overcomplicated index notations we shall omit the angular brackets around the spatial indices of these (and only these) Cartesian tensors, that is we adopt: $\mathcal{I}_{A_l} \equiv \mathcal{I}_{\langle A_l\rangle}$ and $\mathcal{S}_{A_l} \equiv \mathcal{S}_{\langle A_l\rangle}$.

We shall also use transverse (T) and transverse-traceless (TT) Cartesian tensors in our calculations [2, 5, 64]. These objects are defined by making use of the operator of projection

$$P_{jk} \equiv \delta_{jk} - k_{jk} , \tag{1.15}$$

onto the plane orthogonal to a unit vector k_j. This operator plays a role of a Kroneker symbol in the two-dimensional space in the sense that $P_{ij}P_{jk} = P_{ik}$, and $P_{ii} = 2$. Definitions of the transverse and transverse-traceless tensors is [5, 6]

$$Q^{\mathrm{T}}_{a_1...a_l} \equiv P_{a_1b_1}P_{a_2b_2}\cdots P_{a_lb_l}Q_{b_1...b_l} , \tag{1.16}$$

$$Q^{\mathrm{TT}}_{a_1...a_l} \equiv \sum_{k=0}^{[l/2]} b_k^l P_{(a_1a_2}\cdots P_{a_{2k-1}a_{2k}} W_{a_{2k+1}...a_l)b_1b_1...b_kb_k} , \tag{1.17}$$

where again $[l/2]$ is the integer part of $l/2$, $W_{a_1...a_l} \equiv Q^{\mathrm{T}}_{(a_1...a_l)}$, and the numerical coefficients

$$b_k^l = \frac{(-1)^k}{4^k} \frac{l(l-k-1)!!}{k!(l-2k)!} . \tag{1.18}$$

For instance,

$$Q^{\rm TT}_{ab} \equiv P_{i(a}P_{b)j}Q_{ij} - \frac{1}{2}P_{ab}P_{jk}Q_{jk}\,. \tag{1.19}$$

We shall also use the polynomial coefficients $C_l(p_1,\ldots,p_n)$ in some of our equations. They are defined by

$$C_l(p_1,\ldots,p_n) \equiv \frac{l!}{p_1!\ldots p_n!}\,, \tag{1.20}$$

where l and p_i are the positive integers such that $\sum_{i=1}^{n} p_i = l$. We introduce a Heaviside step function, $H(p-q)$, such that on the set of whole numbers

$$H(p-q) = \begin{cases} 0, & \text{if } p \le q\,,\\ 1, & \text{if } p > q\,. \end{cases} \tag{1.21}$$

Partial derivatives of any differentiable function, $f = f(t,\boldsymbol{x})$, are denoted as follows: $f_{,0} = \partial f/\partial t$ and $f_{,i} = \partial f/\partial x^i$. In general, comma standing after a function denotes a partial derivative with respect to a corresponding coordinate: $f_{,\alpha} \equiv \partial f(x)/\partial x^\alpha$. A dot above the function denotes a total derivative of the function with respect to time

$$\dot{f} \equiv df/dt = \partial f/\partial t + \dot{x}^i \partial f/\partial x^i\,, \tag{1.22}$$

where $\dot{x}^i$ denotes velocity along the integral curve $x^i = x^i(t)$ parametrized with coordinate time t. In this chapter the integral curves are light rays, and the derivatives $\dot{x}^i$ are taken along the light ray trajectory $(x^i) = \boldsymbol{x}(t)$. Sometimes the partial derivatives with respect to space coordinate x^i will be also denoted as $\partial_i \equiv \partial/\partial x^i$, and the partial time derivative will be denoted as $\partial_t \equiv \partial/\partial t$. A covariant derivative with respect to the coordinate x^α will be denoted as ∇_α.

We shall introduce and distinguish notations for integrals taken with respect to time at a fixed spatial point, from those taken along a light-ray trajectory. Specifically, the time integrals from a function $F(t,\boldsymbol{x})$, where $\boldsymbol{x}$ is a fixed point in space, are denoted as

$$F^{(-1)}(t,\boldsymbol{x}) \equiv \int_{-\infty}^{t} F(\tau,\boldsymbol{x})d\tau\,, \quad F^{(-2)}(t,\boldsymbol{x}) \equiv \int_{-\infty}^{t} F^{(-1)}(\tau,\boldsymbol{x})d\tau\,. \tag{1.23}$$

The time integrals from a function $F(t,\boldsymbol{x})$ taken on a light ray suggest that the spatial coordinate $\boldsymbol{x}$ is a function of time $\boldsymbol{x} \equiv \boldsymbol{x}(t)$, taken along the light ray. These integrals are denoted as

$$F^{[-1]}(t,\boldsymbol{x}) \equiv \int_{-\infty}^{t} F(\tau,\boldsymbol{x}(\tau))d\tau, \quad F^{[-2]}(t,\boldsymbol{x}) \equiv \int_{-\infty}^{t} F^{[-1]}(\tau,\boldsymbol{x}(\tau))d\tau\,, \tag{1.24}$$

where $\boldsymbol{x} \equiv \boldsymbol{x}(t)$ in the right side of these definitions. The integrals in (1.23) represent functions of time, t, and spatial, $\boldsymbol{x}$, coordinates. The integrals in (1.24) are functions of time, t, only.

Partial time derivative of the order p from a function $F(t, \boldsymbol{x})$ is denoted by

$$F^{(p)}(t, \boldsymbol{x}) = \frac{\partial^p F(t, \boldsymbol{x})}{\partial t^p}, \tag{1.25}$$

so that its action on the time integrals eliminates integration in the sense that

$$F^{(p)}(t, \boldsymbol{x}) = \frac{\partial^{p+1} F^{(-1)}(t, \boldsymbol{x})}{\partial t^{p+1}} = \frac{\partial^{p+2} F^{(-2)}(t, \boldsymbol{x})}{\partial t^{p+2}}. \tag{1.26}$$

Total time derivative of the order p from a function $F(t, \boldsymbol{x})$ is denoted by

$$F^{[p]}(t, \boldsymbol{x}) = \frac{d^p F(t, \boldsymbol{x})}{dt^p}. \tag{1.27}$$

The reader can easily confirm that

$$F^{[p]}(t, \boldsymbol{x}) = \frac{d^{p+1} F^{[-1]}(t, \boldsymbol{x})}{dt^{p+1}} = \frac{d^{p+2} F^{[-2]}(t, \boldsymbol{x})}{dt^{p+2}}. \tag{1.28}$$

In what follows, we shall denote spatial vectors by the bold italic letters, for instance, $A^i \equiv \boldsymbol{A}$, $k^i \equiv \boldsymbol{k}$, etc. The Euclidean dot product between two spatial vectors, for example $\boldsymbol{a}$ and $\boldsymbol{b}$, is denoted with a dot between them: $a^i b_i = \boldsymbol{a} \cdot \boldsymbol{b}$. The Euclidean wedge (cross) product between two spatial vectors is denoted with a symbol $\times$, that is $\epsilon_{ijk} a^j b^k = (\boldsymbol{a} \times \boldsymbol{b})^i$. Other particular notations will be introduced as soon as they appear in text.

2 The metric tensor, gauges and coordinates

2.1 The canonical form of the metric tensor perturbation

We consider an isolated astronomical system emitting gravitational waves and assume that gravitational field is weak everywhere so that the metric tensor can be expanded in a Taylor series with respect to the powers of gravitational constant G which labels the order of products of the metric tensor perturbations that are kept in the solution of the Einstein equations. We shall consider only a linearized post-Minkowskian approximation of general relativity and discard all terms of the order of G^2 and higher. The metric tensor is a linear combination of the Minkowski metric, $\eta_{\alpha\beta}$, and a small perturbation $h_{\alpha\beta}$

$$g_{\alpha\beta} = \eta_{\alpha\beta} + G h_{\alpha\beta} + O(G^2), \tag{2.1}$$

$$g^{\alpha\beta} = \eta_{\alpha\beta} - G h^{\alpha\beta} + O(G^2), \tag{2.2}$$

where $h_{\alpha\beta} \ll 1$ and we use $\eta_{\alpha\beta}$ to rise and lower indices so that, for example, $h^{\alpha\beta} = \eta^{\alpha\mu}\eta^{\beta\nu}h_{\mu\nu}$. In many cases, the origin of the coordinates is placed to the center of mass of the astrophysical system. It eliminates the dipole component of the gravitational field which is associated with a coordinate degree of freedom. In some cases, however, it is necessary to keep the dipole component unrestricted in order to determine position of the center of mass of the system under consideration with respect to another coordinate chart which is introduced independently for solving some other astronomical problems. This is important for unambiguous interpretation of gravitational experiments done with astrometric instruments [54, 104]. Fact of the matter is that the displacement of the center of mass of an astrophysical system from the origin of the coordinates induces translational deformations of the higher order multipole moments of the gravitational field which introduce a bias to the physical values of the multipoles. Therefore, physical interpretation of the observed values of the multipoles requires identification of the dipole moment and subtraction of the coordinate deformations caused by it. We shall discard the dipole component of the gravitational field in our solution.

The most general expression for the linearized perturbation of the metric tensor outside of the astronomical system emitting gravitational radiation was derived by Blanchet and Damour [64] by solving Einstein's equations. The perturbation is given in terms of the symmetric and trace-free (STF) mass and spin multipole moments(similar formulas were derived by Thorne [5]) and is described by the following expression:

$$h_{\alpha\beta} = h^{\text{can.}}_{\alpha\beta} + w_{\alpha,\beta} + w_{\beta,\alpha} , \tag{2.3}$$

where w_α are, the so-called, gauge functions describing the freedom in the choice of coordinates covering the manifold. The *canonical* perturbation, $h^{\text{can.}}_{\alpha\beta}$, obeys the homogeneous wave equation in vacuum

$$\Box h^{\text{can.}}_{\alpha\beta} = 0 , \tag{2.4}$$

which solution is chosen as

$$h^{\text{can.}}_{00}(t,\boldsymbol{x}) = \frac{2\mathcal{M}}{r} + 2\sum_{l=2}^{\infty}\frac{(-1)^l}{l!}\left[\frac{\mathcal{I}_{A_l}(t-r)}{r}\right]_{,A_l} , \tag{2.5}$$

$$h^{\text{can.}}_{0i}(t,\boldsymbol{x}) = -\frac{2\epsilon_{ipq}\mathcal{S}_p(t-r)N_q}{r^2} - 4\sum_{l=2}^{\infty}\frac{(-1)^l l}{(l+1)!}\left[\frac{\epsilon_{ipq}\mathcal{S}_{pA_{l-1}}(t-r)}{r}\right]_{,qA_{l-1}} + 4\sum_{l=2}^{\infty}\frac{(-1)^l}{l!}\left[\frac{\dot{\mathcal{I}}_{iA_{l-1}}(t-r)}{r}\right]_{,A_{l-1}} , \tag{2.6}$$

$$h^{\text{can.}}_{ij}(t,\boldsymbol{x}) = \delta_{ij}h^{\text{can.}}_{00}(t,\boldsymbol{x}) + q^{\text{can.}}_{ij}(t,\boldsymbol{x}) , \tag{2.7}$$

$$q^{\text{can.}}_{ij}(t,\boldsymbol{x}) = 4\sum_{l=2}^{\infty}\frac{(-1)^l}{l!}\left[\frac{\ddot{\mathcal{I}}_{ijA_{l-2}}(t-r)}{r}\right]_{,A_{l-2}} - 8\sum_{l=2}^{\infty}\frac{(-1)^l l}{(l+1)!}\left[\frac{\epsilon_{pq(i}\dot{\mathcal{S}}_{j)pA_{l-2}}(t-r)}{r}\right]_{,qA_{l-2}} . \tag{2.8}$$

Here $\mathcal{M}$ and $\mathcal{S}_i$ are the total mass and spin (angular momentum) of the system, and $\mathcal{I}_{A_l}$ and $\mathcal{S}_{A_l}$ are two independent sets of mass-type and spin-type multipole moments, $N^i = x^i/r$ is a unit vector directed from the origin of the coordinate system to the field point. Because the origin of the coordinate system has been chosen at the center of mass, expansions (2.5)–(2.8) do not depend on the mass-type dipole moment, $\mathcal{I}_i$, which is equal to zero by definition. We emphasize that in the linearized approximation the total mass $\mathcal{M}$ and spin $\mathcal{S}_i$ of the astronomical system are constant while all other multipoles are functions of time which temporal behavior obeys the equations of motion derived from the law of conservation of the stress–energy tensor of the system [1, 2]. Gravitational waves emitted by the system reduce its energy, linear and angular momenta. This effect does not appear in the linearized general relativity but in higher order approximations we would obtain where the mass, spin, and linear momentum of the system must be considered as functions of time like any other multipole. Higher order gravitational perturbations in the metric going beyond (2.5)–(2.8) are shown in a review paper by Blanchet [105]. They are not of concern in this chapter.

The *canonical* metric tensor (2.5)–(2.8) depends on the multipole moments $\mathcal{I}_{A_l}(t-r)$ and $\mathcal{S}_{A_l}(t-r)$ taken at the retarded instant of time. The retardation is explained by the finite speed of propagation of gravity (light propagation will be considered below). In the near zone of the isolated system the retardation due to the propagation of gravity is small and all functions of time in the metric tensor can be expanded in Taylor series around the present time t [5, 64]. This near-zone expansion of the metric tensor is called the post-Newtonian expansion leading to the post-Newtonian successive approximations [54]. The post-Newtonian expansion can be smoothly matched to the solution of the linearized Einstein equations in the domain of space being occupied by matter of the isolated system. The matching allows us to express the multipole moments in terms of matter variables [65]

$$\sigma \equiv T^{00} + T^{kk}\,, \quad \sigma^i \equiv T^{0i}\,, \tag{2.9}$$

where $T^{\alpha\beta}$ is the stress–energy tensor of matter bounded in space. In the first post-Newtonian approximation the multipole moments have a matter-compact support [65]

$$\mathcal{I}^{\rm 1PN}_{A_l} = \int\limits_V d^3\boldsymbol{x}\left\{x_{\langle A_l\rangle}\sigma + \frac{|\boldsymbol{x}|^2 x_{\langle A_l\rangle}}{2(2l+3)}\partial_t^2\sigma - \frac{4(2l+1)x_{\langle iA_l\rangle}}{(l+1)(2l+3)}\partial_t\sigma_i\right\} + O(c^{-4})\,, \tag{2.10}$$

$$\mathcal{S}^{\rm 1PN}_{A_l} = \int\limits_V d^3\boldsymbol{x}\epsilon_{pq<a_l}x_{A_{l-1}p>}\sigma_q + O(c^{-2})\,, \tag{2.11}$$

where notations have been explained in Section 1.3. In the higher post-Newtonian approximations the multipole moments have contributions coming directly from the stress–energy tensor of gravitational field (Landau–Lifshitz pseudotensor) which have noncompact support. Therefore, the multipole moments are expressed by more complicated functionals [105]. Radiative approximation of the *canonical* metric tensor reveals that contribution of the tails of gravitational waves must be added to the definitions of the multipole moments (2.10) and (2.11) so that the multipole moments in the radiative zone of the isolated system read [66, 67, 106]

$$\mathcal{I}_{A_l} = \mathcal{I}^{1\text{PN}}_{A_l} + 2\mathcal{M}\int_0^{+\infty} d\zeta \dddot{\mathcal{I}}^{1\text{PN}}_{A_l}(t-r-\zeta)\left[\ln\left(\frac{\zeta}{2b}\right) + \frac{2l^2+5l+4}{l(l+1)(l+2)} + \sum_{k=1}^{l-2}\frac{1}{k}\right], \tag{2.12}$$

$$\mathcal{S}_{A_l} = \mathcal{S}^{1\text{PN}}_{A_l} + 2\mathcal{M}\int_0^{+\infty} d\zeta \ddot{\mathcal{S}}^{1\text{PN}}_{A_l}(t-r-\zeta)\left[\ln\left(\frac{\zeta}{2b}\right) + \frac{l-1}{l(l+1)} + \sum_{k=1}^{l-1}\frac{1}{k}\right], \tag{2.13}$$

where b is the normalization constant which value is supposed to be absorbed to the definition of the origin of time scale in the radiative zone but this statement has not been checked so far.

2.2 The harmonic coordinates

Equation (2.3) holds in an arbitrary gauge imposed on the metric tensor. The harmonic gauge is defined by the condition [54]

$$2h^{\alpha\beta}{}_{,\beta} - h^{,\alpha} = 0\,, \tag{2.14}$$

where $h \equiv h^{\mu}{}_{\mu}$. The gauge condition (2.14) reduces the Einstein vacuum field equations to the wave equation (2.4) for the gravitational potentials $h^{\alpha\beta}$. Harmonic coordinates x^α are defined as solutions of the homogeneous wave equation $\Box x^\alpha = 0$ up to the gauge functions w^α. In particular, the harmonic *canonical* coordinates are defined by the condition that all gauge functions $w_\alpha = 0$. The *canonical* metric tensor (2.5)–(2.8) depends on two sets of multipole moments [5, 64] which reflects the existence of only two degrees of freedom of a free (detached from matter) gravitational field in general relativity [1–3]. At the same time one can obtain a generic expression for the harmonic metric tensor by making use of infinitesimal coordinate transformation

$$x'^\alpha = x^\alpha - w^\alpha \tag{2.15}$$

from the *canonical harmonic* coordinates x^α to arbitrary harmonic coordinates x'^α with the harmonic gauge functions w^α which satisfy to a homogeneous wave equation

$$\Box w^\alpha = 0\,. \tag{2.16}$$

The most general solution of this vector equation contains four sets of STF multipoles [5, 64]

$$w^0 = \sum_{l=0}^{\infty} \left[\frac{\mathcal{W}_{A_l}(t-r)}{r} \right]_{,A_l} , \tag{2.17}$$

$$w^i = \sum_{l=0}^{\infty} \left[\frac{\mathcal{X}_{A_l}(t-r)}{r} \right]_{,iA_l} + \sum_{l=1}^{\infty} \left[\frac{\mathcal{Y}_{iA_{l-1}}(t-r)}{r} \right]_{,A_{l-1}} \tag{2.18}$$
$$+ \sum_{l=1}^{\infty} \left[\epsilon_{ipq} \frac{\mathcal{Z}_{qA_{l-1}}(t-r)}{r} \right]_{,pA_{l-1}} ,$$

where $\mathcal{W}_{A_l}$, $\mathcal{X}_{A_l}$, $\mathcal{Y}_{iA_{l-1}}$, and $\mathcal{Z}_{qA_{l-1}}$ are Cartesian tensors depending on the retarded time. Their specific form is a matter of computational convenience (or the boundary conditions) for derivation and interpretation of observable effects but it does not affect the invariant quantities like the phase of electromagnetic wave propagating through the field of the multipoles.

The most convenient choice simplifying the structure of the metric tensor perturbations, is given by the following gauge functions:

$$w^0 = \sum_{l=2}^{\infty} \frac{(-1)^l}{l!} \left[\frac{\mathcal{I}^{(-1)}_{A_l}(t-r)}{r} \right]_{,A_l} , \tag{2.19}$$

$$w^i = \sum_{l=2}^{\infty} \frac{(-1)^l}{l!} \left[\frac{\mathcal{I}^{(-2)}_{A_l}(t-r)}{r} \right]_{,iA_l} \tag{2.20}$$
$$- 4 \sum_{l=2}^{\infty} \frac{(-1)^l}{l!} \left[\frac{\mathcal{I}_{iA_{l-1}}(t-r)}{r} \right]_{,A_{l-1}}$$
$$+ 4 \sum_{l=2}^{\infty} \frac{(-1)^l l}{(l+1)!} \left[\frac{\epsilon_{iba} \mathcal{S}^{(-1)}_{bA_{l-1}}(t-r)}{r} \right]_{,aA_{l-1}} .$$

These functions, after they are substituted to equation (2.3), transform the *canonical* metric tensor perturbation to a remarkably simple form

$$h_{00} = \frac{2\mathcal{M}}{r} , \tag{2.21}$$

$$h_{0i} = -\frac{2\epsilon_{ipq}\mathcal{S}_p N_q}{r^2} , \tag{2.22}$$

$$h_{ij} = \delta_{ij} h_{00} + h^{TT}_{ij} , \tag{2.23}$$

$$h^{TT}_{ij} = P_{ijkl} q^{\text{can.}}_{kl} , \tag{2.24}$$

where the TT-projection differential operator P_{ijkl}, applied to the symmetric tensors depending on both time and spatial coordinates, is given by

$$P_{ijkl} = (\delta_{ik} - \Delta^{-1}\partial_i\partial_k)(\delta_{jl} - \Delta^{-1}\partial_j\partial_l) - \frac{1}{2}(\delta_{ij} - \Delta^{-1}\partial_i\partial_j)(\delta_{kl} - \Delta^{-1}\partial_k\partial_l)\,, \tag{2.25}$$

and Δ and Δ^{-1} denote the Laplacian and the inverse Laplacian, respectively.

When comparing the *canonical* metric tensor with that given by equations (2.21)–(2.24) it is instructive to keep in mind that ${}^{(-2)}\ddot{\mathcal{I}}_{A_l}(t-r) = \mathcal{I}_{A_l}(t-r)$ and $\Delta(\mathcal{I}_{A_l}(t-r)/r) = \ddot{\mathcal{I}}_{A_l}(t-r)/r$ for $r \neq 0$. This is a consequence of the fact that function ${}^{(-2)}\ddot{\mathcal{I}}_{A_l}(t-r)$ is a solution of the homogeneous d'Lambert's equation, that is, $\Box[{}^{(-2)}\mathcal{I}_{A_l}(t-r)/r] = 0$ for $r \neq 0$. We also notice that $\mathcal{I}_{A_l}(t-r)/r = \Delta^{-1}[\ddot{\mathcal{I}}_{A_l}(t-r)/r]$ and ${}^{(-2)}\mathcal{I}_{A_l}(t-r)/r = \Delta^{-1}[\mathcal{I}_{A_l}(t-r)/r]$. The metric tensor harmonic perturbation (2.21)–(2.24) is similar to the Coulomb gauge in electrodynamics [107, 108].

2.3 The ADM coordinates

The Arnowitt–Deser–Misner (ADM) gauge condition in the linear approximation is given by two equations [61]

$$2h_{0i,i} - h_{ii,0} = 0\,, \qquad 3h_{ij,j} - h_{jj,i} = 0\,, \tag{2.26}$$

where the second equation holds exactly, and for any function $f = f(t, \boldsymbol{x})$ we use notations: $f_{,0} = \partial f/\partial t$ and $f_{,i} = \partial f/\partial x^i$. For comparison, the harmonic gauge condition (2.14) in the linear approximation reads

$$2h_{0i,i} - h_{ii,0} = h_{00,0}\,, \qquad 2h_{ij,j} - h_{jj,i} = -h_{00,i}\,. \tag{2.27}$$

The ADM gauge condition (2.26) brings the space-space component of the metric to the following form:

$$g_{ij} = \delta_{ij}\left(1 + \frac{1}{3}h_{kk}\right) + h_{ij}^{TT}\,, \tag{2.28}$$

where h_{ij}^{TT} denotes the transverse-traceless part of h_{ij} and $h_{kk} = 3h_{00}$. The ADM and harmonic gauge conditions are not compatible inside the regions occupied by matter. However, outside of matter they can co-exist simultaneously. Indeed, it is straightforward to check out that the metric tensor (2.21)–(2.24) satisfies both the harmonic and the ADM gauge conditions in the linear approximation along with the assumption that $\dot{\mathcal{M}} = 0$. This was first noticed in [100]. We call the coordinates in which the metric tensor is given by equations (2.21)–(2.24) as the ADM-harmonic coordinates.

The experimental problem of detection of gravitational waves is reduced to the observation of motion of test particles in the field of the incident or incoming gravitational wave. These test particles are photons in the electromagnetic wave used in observations and mirrors in ground-based gravitational-wave detectors or pulsars and

Earth in case of using a pulsar timing array. The gravitational wave affects propagation of photons and perturbs motion of the mirrors or pulsars and Earth. These perturbations must be explicitly calculated and clearly separated from noise to avoid possible misinterpretation of observable effects due to the gravitational wave. It turns out that the *canonical* form of the metric tensor (2.5)–(2.8) is well-adapted for performing an analytic integration of equations of light rays. At the same time, freely falling mirrors (or pulsars and Earth) experience influence of gravitational waves emitted by the isolated astronomical system and move with respect to the coordinate grid of the *canonical harmonic* coordinates in a complicated way. For this reason, the perturbations produced by the gravitational waves on the light propagation get mixed up with the motion of massive test particles in these coordinates.

Arnowitt et al. [61] showed that there exist *canonical ADM* coordinates which have a special property such that freely falling massive particles are not moving with respect to this coordinates despite that they are perturbed by the gravitational waves. This means that the ADM coordinates themselves are not inertial and, although have an advantage in treating motion of massive test particles, should be used with care in the interpretation of gravitational wave experiments. Making use of the *canonical ADM* coordinates simplifies analysis of the gravitational wave effects observed at gravitational wave observatories (LIGO, LISA, NGO, etc.) or by astronomical technique because the motion of observer (proof mass) is excluded from the equations. However, the mathematical structure of the metric tensor in the *canonical ADM* coordinates does not allow us to directly integrate equations for light rays analytically because it contains terms that are instantaneous functions of time. Integrals from these instantaneous functions of time cannot be performed explicitly [100].

The ADM-harmonic coordinates have the advantages of both harmonic and ADM coordinates. Thus, the ADM-harmonic coordinates allow us to get a full analytic solution of the light-ray equations and to eliminate the effects produced by the motion of observers with respect to the coordinate grid caused by the influence of gravitational waves. In other words, *all* physical effects produced by gravitational waves are contained merely in the solution of the equations of light propagation. This conclusion is, of course, valid in the linear approximation of general relativity and is not extended to the second approximation where gravitational-wave effects on light and motion of observers cannot be disentangled and have to be analyzed together.

Similar ideology based on the introduction of TT coordinates, has been earlier applied for analysis of the output signal of the gravitational-wave detectors with freely suspended masses [1, 2, 25] placed to the field of a plane gravitational wave, that is at the distance far away from the localized astronomical system emitting gravitational waves where the curvature of the gravitational-wave front is negligible. Our ADM-harmonic coordinates are an essential generalization of the standard TT coordinates because they can be constructed at an arbitrary distance from the astronomical system, thus, covering the near, intermediate, and radiative zones.

3 Equations of propagation of electromagnetic signals

3.1 Maxwell equations in curved spacetime

In this section, we treat gravitational field exactly without approximation. Therefore, all indices are raised and lowered by means of the metric tensor $g_{\alpha\beta}$ with $g^{\alpha\beta}$ defined in accordance with the standard rule $g^{\alpha\beta}g_{\beta\gamma} = \delta^{\alpha}_{\beta}$. The general formalism describing the behavior of electromagnetic radiation in an arbitrary gravitational field is well known and can be found, for example, in textbooks [1, 2, 109] or in reviews [110, 111]. Electromagnetic field is defined in terms of the (complex) electromagnetic tensor $F_{\alpha\beta}$ as a solution of the Maxwell equations. In the high-frequency limit one can approximate the electromagnetic tensor $F_{\alpha\beta}$ as [1, 2]

$$F_{\alpha\beta} = \{A_{\alpha\beta}\exp(i\varphi)\}\,, \tag{3.1}$$

where $A_{\alpha\beta}$ is a slowly varying (complex) amplitude and φ is a rapidly varying phase of the electromagnetic wave which is called *eikonal* [1, 112], and i is the imaginary unit, $i^2 = -1$. In the most general case of propagation of light in a transparent medium the eikonal is a complex function which real and imaginary parts are connected by the Kramers–Kronig dispersion relations [107]. We shall consider propagation of light in vacuum and neglect the imaginary part of the eikonal that is associated with absorption. Of course, the amplitude, $A_{\alpha\beta}$, and phase, φ, are functions of both time and spatial coordinates.

The source-free (vacuum) Maxwell equations are given by [1, 2]

$$\nabla_\alpha F_{\beta\gamma} + \nabla_\beta F_{\gamma\alpha} + \nabla_\gamma F_{\alpha\beta} = 0\,, \tag{3.2}$$

$$\nabla_\beta F^{\alpha\beta} = 0\,, \tag{3.3}$$

where ∇_α denotes covariant differentiation. Taking a covariant divergence from equation (3.2), using equation (3.3) and applying the rule of commutation of covariant derivatives of a tensor field of a second rank, we obtain the covariant wave equation for the electromagnetic field tensor

$$\Box_g F_{\alpha\beta} + R_{\alpha\beta\mu\nu}F^{\mu\nu} + R_{\mu\alpha}F_{\beta}{}^{\mu} - R_{\mu\beta}F_{\alpha}{}^{\mu} = 0\,, \tag{3.4}$$

where $\Box_g \equiv g^{\alpha\beta}\nabla_\alpha\nabla_\beta$, $R_{\alpha\beta\mu\nu}$ is the Riemann curvature tensor, and $R_{\alpha\beta} = R^{\gamma}{}_{\alpha\gamma\beta}$ is the Ricci tensor (definitions of the Riemann and Ricci tensors in this chapter are the same as in the textbook [3]). We consider the case of propagation of light in vacuum where the stress–energy tensor of matter, $T^{\alpha\beta}$, is absent. Due to the Einstein equations it yields $R_{\alpha\beta} = 0$. Hence, in our case (3.4) is reduced to a more simple form

$$\Box_g F_{\alpha\beta} + R_{\alpha\beta\mu\nu}F^{\mu\nu} = 0\,. \tag{3.5}$$

Differential operator $\square_g$ in (3.4) taken along with the Riemann and Ricci tensors is called de Rham's operator for the electromagnetic field [2?].

3.2 Maxwell equations in the geometric optics approximation

Let us now assume that the electromagnetic tensor $F_{\alpha\beta}$ shown in (3.1) can be expanded with respect to a small dimensionless parameter $\varepsilon = \lambda_{\rm em}/L$ where $\lambda_{\rm em}$ is a characteristic wavelength of the electromagnetic wave and L is a characteristic radius of spacetime curvature. The parameter ε is a bookkeeping parameter of the high-frequency approximation in expansion of the electromagnetic field beyond the limit of the geometric optics. More specifically, we assume that the expansion of the electromagnetic field given by equation (3.1) has the following form [2]:

$$F_{\alpha\beta} = \left(a_{\alpha\beta} + \varepsilon b_{\alpha\beta} + \varepsilon^2 c_{\alpha\beta} + \dots\right) \exp\left(\frac{i\varphi}{\varepsilon}\right) , \tag{3.6}$$

where $a_{\alpha\beta}$, $b_{\alpha\beta}$, $c_{\alpha\beta}$, etc. are functions of time and spatial coordinates.

Substituting expansion (3.6) into equation (3.2), taking into account a definition of the electromagnetic wave vector, $l_\alpha \equiv \partial\varphi/\partial x^\alpha$, and arranging the terms with similar powers of ε, lead to the chain of equations

$$l_\alpha\, a_{\beta\gamma} + l_\beta\, a_{\gamma\alpha} + l_\gamma\, a_{\alpha\beta} = 0 , \tag{3.7}$$

$$\nabla_\alpha\, a_{\beta\gamma} + \nabla_\beta\, a_{\gamma\alpha} + \nabla_\gamma\, a_{\alpha\beta} = -i\left(l_\alpha\, b_{\beta\gamma} + l_\beta\, b_{\gamma\alpha} + l_\gamma\, b_{\alpha\beta}\right) , \tag{3.8}$$

where we have neglected the effects of spacetime curvature which are of the order of $O(\varepsilon^2)$ that are too small to measure.

Similarly, equation (3.3) gives a chain of equations

$$l_\beta a^{\alpha\beta} = 0 , \tag{3.9}$$

$$\nabla_\beta a^{\alpha\beta} + i l_\beta\, b^{\alpha\beta} = 0 , \tag{3.10}$$

where we again neglected the effects of spacetime curvature.

Equation (3.9) implies that the amplitude, $a_{\alpha\beta}$, of the electromagnetic field tensor is orthogonal in the four-dimensional sense to a wave vector l_α, at least, in the first approximation. Contracting equation (3.7) with l_α and accounting for (3.9), we find that the wave vector l_α is null, that is

$$l_\alpha l^\alpha = 0 . \tag{3.11}$$

Taking a covariant derivative from this equation and using the fact that

$$\nabla_{[\beta}\, l_{\alpha]} = 0 , \tag{3.12}$$

because $l_\alpha = \nabla_\alpha \varphi$, one can show that the vector l_α obeys the null geodesic equation

$$l^\beta \nabla_\beta l^\alpha = 0 . \tag{3.13}$$

It means that the null vector l^α is parallel transported along itself in the curved spacetime. Equation (3.13) can be expressed more explicitly as

$$\frac{dl^\alpha}{d\sigma} + \Gamma^\alpha_{\beta\gamma} l^\beta l^\gamma = 0 \,, \tag{3.14}$$

where σ is an affine parameter along the light-ray trajectory, and

$$\Gamma^\alpha_{\beta\gamma} = \frac{1}{2} g^{\alpha\mu} \left(\partial_\gamma g_{\mu\beta} + \partial_\beta g_{\mu\gamma} - \partial_\mu g_{\beta\gamma} \right) \,, \tag{3.15}$$

are the Christoffel symbols.

Finally, contracting equation (3.8) with l^γ, and using (3.7), (3.9), and (3.10) along with (3.12) we can show that in the first approximation

$$l^\gamma \nabla_\gamma a_{\alpha\beta} + \vartheta a_{\alpha\beta} = 0 \,, \tag{3.16}$$

where

$$\vartheta \equiv (1/2) \nabla_\alpha l^\alpha \,, \tag{3.17}$$

is the expansion of the light-ray congruence defined at each point of spacetime by the derivative of the wave vector l^α.

Equation (3.16) represents the law of propagation of the tensor amplitude of electromagnetic wave along the light ray. In the most general general case, when the expansion $\vartheta \neq 0$, the tensor amplitude of the electromagnetic wave is not parallel-transported along the light rays. It can be shown that the expansion ϑ of the light-ray congruence is defined only by the stationary components of the gravitational field of the isolated astronomical system determined by its mass $\mathcal{M}$, and spin $\mathcal{S}^i$, but it does not depend on the higher order multipole moments. It means that gravitational waves do not contribute to the expansion of the light-ray congruence in the linearized approximation of general relativity and their impact on ϑ is postponed to the terms of the second order of magnitude with respect to the universal gravitational constant G.

3.3 Electromagnetic eikonal and light-ray geodesics

The unperturbed congruence of light rays

We have assumed that geometric optics approximation is valid and electromagnetic waves propagate in vacuum. We also assume that each electromagnetic wave has a wavelength $\lambda_{\rm em}$ much smaller than the characteristic wavelength $\lambda_{\rm gw}$ of gravitational waves emitted by the isolated astronomical system. Physical speed of light in vacuum, measured locally, is equal to the speed of propagation of gravitational waves, and is equal to the fundamantal speed c in tangent Minkowski spacetime. We have neglected all relativistic effects associated with the curvature tensor of spacetime in equations of light propagation. In accordance with the consideration given in previous Section 3.2, a kinematic description of propagation of each electromagnetic wave can be given by tracking position of its phase φ, which is a null hypersurface in spacetime, as a function of time or by following the congruence of light rays that are orthogonal to the phase. Quantum electrodynamics tells us that the light rays are tracks of massless particles of the quantized electromagnetic field (photons) which are moving along light-ray geodesics defined by equation (3.14).

Particular solution of these equations can be found after imposing the initial-boundary conditions

$$\boldsymbol{x}(t_0) = \boldsymbol{x}_0, \qquad \frac{d\boldsymbol{x}(-\infty)}{dt} = \boldsymbol{k}\,. \tag{3.18}$$

These conditions determine the spatial position, $\boldsymbol{x}_0$, of an electromagnetic signal (a photon) at the time of its emission, t_0, and the initial direction of its propagation given by the unit vector, $\boldsymbol{k}$, at the past null infinity, that is at the infinite spatial distance and at the infinite past [2, 3] where the spacetime is assumed to be flat (see Figure 1). We imply that vector $\boldsymbol{k}$ is directed toward observer. Notice that the initial-boundary conditions (3.18) have been chosen as a matter of convenience only. Instead of them, we could chose two boundary conditions when both the point of emission and that of observation of the electromagnetic signal are fixed in time and space. It is always possible to convert solution of equations of the null geodesics given in terms of the initial-boundary conditions to that given in terms of the boundary conditions. We discuss it in Section 5.3.

In the next sections, we will derive an explicit form of equations of null geodesics and solve them by iterations with the initial boundary conditions (3.18). At the first iteration we can neglect relativistic perturbation of photon's motion and approximate it by a straight line

$$x^i = x^i_N(t) \equiv x^i_0 + k^i(t - t_0)\,, \tag{3.19}$$

where t_0 is the time of emission of electromagnetic signal, x^i_0 are spatial coordinates of the source of the electromagnetic signal taken at the time t_0, and $k^i = \boldsymbol{k}$ is the unit vector along the trajectory of photon's motion defined in (3.18).

The bundle of light rays makes 2+1 split of space by projecting any point in space onto the plane being orthogonal to the bundle (see Figure 2). This allows to make a transformation to new independent variables, τ and ξ^i, defined as follows:

$$\tau = k_i x^i_N\,, \quad \xi^i = P^i{}_j x^j_N\,, \tag{3.20}$$

where

$$P^i{}_j = \delta^i_j - k^i k_j\,, \tag{3.21}$$

is the operator of projection on the plane being orthogonal to $\boldsymbol{k}$. It is easy to see that the parameter τ is equivalent to time

$$\tau \equiv \boldsymbol{k} \cdot \boldsymbol{x} = t - t^*\,, \tag{3.22}$$

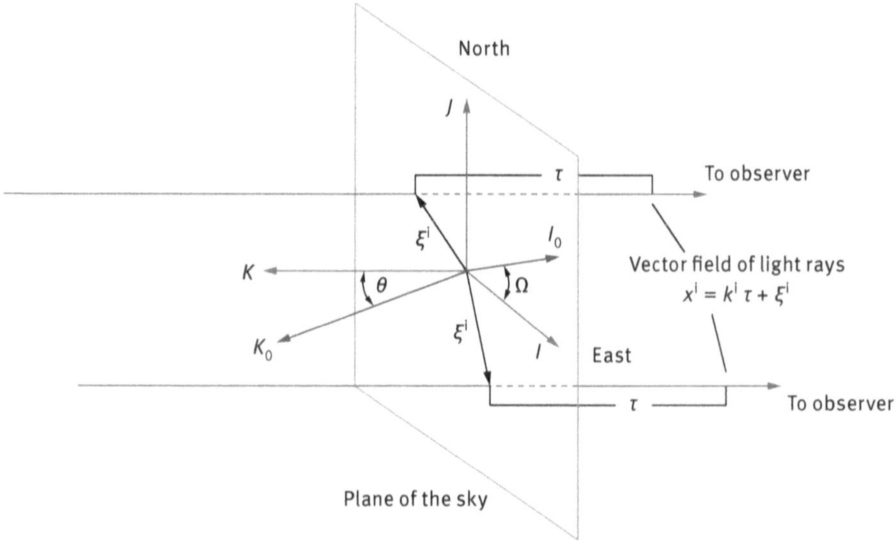

Fig. 2. Astronomical coordinates used in calculation of light propagation. The origin of the coordinates is at the center-of-mass of the source of gravitational waves. The bundle of light rays is defined by a vector field k^i. Vector $K^i = -k^i + O(c^{-2})$ is directed from observer toward the source of light. Vector K_0^i is directed from the observer toward the source of gravitational waves. We define $K_0^i = -N^i = -x^i/r$, where x^i are the coordinates of observer with respect to the source of gravitational waves, and $r = |\boldsymbol{x}|$. The picture shows the plane of the sky being orthogonal to vector K^i.

where

$$t^* \equiv \boldsymbol{k} \cdot \boldsymbol{x}_0 - t_0 \,, \tag{3.23}$$

is the time of the closest approach of the electromagnetic signal to the origin of the spatial coordinates which is taken, in our case, coinciding with the center of mass of the isolated astronomical system. Because for each light ray the time t^* is fixed, we conclude that the time differential $d\tau = dt$ on the light ray. The reader may expect that the results of our calculation of observable quantities are to depend on the parameters ξ^i and t^*. This is, however, not true since ξ^i and t^* depend on the choice of the origin of the coordinates and direction of its spatial axes that is they are coordinate dependent. The observable quantities have nothing to do with the choice of coordinates and, thus, ξ^i and t^* cannot enter the expressions for observable quantities. Inspection of the resulting equations in the sections which follow, shows that parameters ξ^i and t^* do vanish from the observed quantities.

The unperturbed light-ray trajectory (3.19) written in terms of the new variables (3.20) reads

$$x_N^i(\tau) = k^i \tau + \xi^i \,, \tag{3.24}$$

so that the new variable $\xi^i \equiv \boldsymbol{\xi} = \boldsymbol{k} \times (\boldsymbol{x} \times \boldsymbol{k})$ should be understood as a vector drawn from the origin of the coordinate system toward the point of the closest approach of the ray to the origin. For vectors k^i and ξ^i are orthogonal, the unperturbed distance $r_N = \sqrt{x_i x^i}$ between the photon and the origin of the coordinate system

$$r_N = \sqrt{\tau^2 + d^2}\,, \tag{3.25}$$

where $d = |\boldsymbol{\xi}|$ is the impact parameter of the unperturbed light-ray trajectory with respect to the coordinate origin.

We introduce two other operators of partial derivatives with respect to τ and ξ^i determined for any smooth function taken on the congruence of light rays. These operators will be denoted with a hat above them and are defined as

$$\hat{\partial}_\tau \equiv \frac{\partial}{\partial \tau}\,, \quad \hat{\partial}_i \equiv P_i^{\,j}\frac{\partial}{\partial \xi^j}\,, \quad \hat{\partial}_{t^*} \equiv \frac{\partial}{\partial t^*} \tag{3.26}$$

so that, for example,

$$k^i \hat{\partial}_i = 0\,. \tag{3.27}$$

An important consequence of the projective structure of the bundle of the light rays is that for any smooth function $F(t, \boldsymbol{x})$ defined on the light-ray trajectories, one has

$$\left[\left(\frac{\partial}{\partial x^i} + k_i \frac{\partial}{\partial t}\right) F(t\,,\,\boldsymbol{x})\right]_{\boldsymbol{x}=\boldsymbol{x}_0+\boldsymbol{k}(t-t_0)} = \left(\frac{\partial}{\partial \xi^i} + k_i \frac{\partial}{\partial \tau}\right) F(t^* + \tau\,,\,\boldsymbol{\xi} + \boldsymbol{k}\tau)\,, \tag{3.28}$$

$$\left[\left(\frac{\partial}{\partial t} + k^i \frac{\partial}{\partial x^i}\right) F(t\,,\,\boldsymbol{x})\right]_{\boldsymbol{x}=\boldsymbol{x}_0+\boldsymbol{k}(t-t_0)} = \frac{d}{d\tau} F(t^* + \tau\,,\,\boldsymbol{\xi} + \boldsymbol{k}\tau)\,, \tag{3.29}$$

$$\left[\frac{\partial}{\partial t} F(t\,,\,\boldsymbol{x})\right]_{\boldsymbol{x}=\boldsymbol{x}_0+\boldsymbol{k}(t-t_0)} = \frac{\partial}{\partial t^*} F(t^* + \tau\,,\,\boldsymbol{\xi} + \boldsymbol{k}\tau)\,. \tag{3.30}$$

Here, on the left-hand sides of Equations (3.28)–(3.30) one must, first, calculate the partial derivatives and only after that substitute the unperturbed trajectory of the light ray, $\boldsymbol{x} = \boldsymbol{x}_0 + \boldsymbol{k}(t - t_0)$, while in the right side of these equations one, first, substitute the unperturbed trajectory parameterized by the variables τ and ξ^i and, then, differentiate. Equations (3.28)–(3.30) define the commutation rule of interchanging the operations of substitution of the light ray trajectory to a function defined on spacetime manifold and the calculation of the partial derivatives from the function. It turns out to be very effective for analytic integration of the light-ray geodesic equations.

It is worth noting that (3.30) allows us to rewrite (3.28) as follows:

$$\left[\frac{\partial F(t, \boldsymbol{x})}{\partial x^i}\right]_{\boldsymbol{x}=\boldsymbol{x}_0+\boldsymbol{k}(t-t_0)} = \left(\frac{\partial}{\partial \xi^i} + k_i \frac{\partial}{\partial \tau} - k_i \frac{\partial}{\partial t^*}\right) F(t^* + \tau\,,\,\boldsymbol{\xi} + \boldsymbol{k}\tau)\,. \tag{3.31}$$

This equation will be used later for decomposing STF spatial derivatives from the potentials of gravitational field depending on retarded time.

The eikonal equation

The eikonal φ is related to the wave vector l^α of the electromagnetic wave as $l_\alpha = \partial_\alpha \varphi$. This definition along with equation (3.11) immediately gives us a differential Hamilton–Jacobi equation for the eikonal propagation [1]

$$g^{\alpha\beta}\frac{\partial\varphi}{\partial x^\alpha}\frac{\partial\varphi}{\partial x^\beta} = 0\,. \tag{3.32}$$

The unperturbed solution of this equation is a plane electromagnetic wave

$$\varphi_N = \varphi_0 + \omega k_\alpha x^\alpha\,, \tag{3.33}$$

where φ_0 is a constant, $\omega = 2\pi\nu_\infty$, ν_∞ is a constant frequency of the electromagnetic wave at infinity, and k_α is the unperturbed direction of the co-vector l_α. Equation (3.32) assumes that k_α is a null co-vector with respect to the Minkowski metric in the sense that

$$\eta^{\alpha\beta}k_\alpha k_\beta = 0\,. \tag{3.34}$$

We postulate that the co-vector $k_\alpha = (-1, \boldsymbol{k})$ where the unit Euclidean vector $\boldsymbol{k}$ is defined at past null infinity by equation (3.18).

In the linearized approximation of general relativity the eikonal can be decomposed in a linear combination of unperturbed, φ_N, and perturbed, ψ, parts

$$\varphi = \varphi_N + \omega\psi\,, \tag{3.35}$$

so that the wave co-vector

$$l_\alpha = \omega\left(k_\alpha + \frac{\partial\psi}{\partial x^\alpha}\right)\,, \tag{3.36}$$

Making use of equations (2.2) and (3.32)–(3.36) yield a partial differential equation of the first order for the perturbed part of the eikonal

$$k^\alpha\frac{\partial\psi}{\partial x^\alpha} = \frac{1}{2}h_{\alpha\beta}k^\alpha k^\beta\,. \tag{3.37}$$

This equation can be solved in all space by the method of characteristics [**?**] which, in the case under consideration, are the unperturbed light-ray geodesics given by Equation (3.24). Hence, after making use of relation (3.29) one gets an ordinary differential equation for finding the eikonal perturbation

$$\frac{d\psi}{d\tau} = \frac{1}{2}h^{\rm can.}_{\alpha\beta}(\tau,\xi)k^\alpha k^\beta + \hat{\partial}_\tau\left(k^i w^i - w^0\right)\,, \tag{3.38}$$

where both the gauge functions w^α and the canonical metric tensor perturbation $h^{\rm can.}_{\alpha\beta}$ are taken on the unperturbed light-ray trajectory. In particular, the components of the metric tensor perturbation have the following form:

$$h^{\rm can.}_{00} = \frac{2\mathcal{M}}{r} + 2\sum_{l=2}^{\infty}\sum_{p=0}^{l}\sum_{q=0}^{p}\underset{(M)}{\mathrm{h}}{}^{lpq}_{00}(t^*,\tau,\xi)\,, \tag{3.39}$$

$$h_{0i}^{\text{can.}} = -\frac{2\epsilon_{iba}\mathcal{S}^b N^a}{r^2} + 4\sum_{l=2}^{\infty}\sum_{p=0}^{l-1}\sum_{q=0}^{p}\left[\underset{(S)}{\mathrm{h}}{}_{0i}^{lpq}(t^*,\tau,\boldsymbol{\xi}) + \underset{(M)}{\mathrm{h}}{}_{0i}^{lpq}(t^*,\tau,\boldsymbol{\xi})\right], \tag{3.40}$$

$$h_{ij}^{\text{can.}} = \delta_{ij}h_{00}^{\text{can.}} + q_{ij}^{\text{can.}}, \tag{3.41}$$

$$q_{ij}^{\text{can.}} = 4\sum_{l=2}^{\infty}\sum_{p=0}^{l-2}\sum_{q=0}^{p}\underset{(M)}{\mathrm{q}}{}_{ij}^{lpq}(t^*,\tau,\boldsymbol{\xi}) - 8\sum_{l=2}^{\infty}\sum_{p=0}^{l-2}\sum_{q=0}^{p}\underset{(S)}{\mathrm{q}}{}_{ij}^{lpq}(t^*,\tau,\boldsymbol{\xi}), \tag{3.42}$$

where

$$\underset{(M)}{\mathrm{h}}{}_{00}^{lpq}(t^*,\tau,\boldsymbol{\xi}) = \frac{(-1)^{l+p-q}}{l!}C_l(l-p,p-q,q)k_{<a_1\dots a_p}\hat{\partial}_{a_{p+1}\dots a_l>}\hat{\partial}_\tau^q\left[\frac{\mathcal{I}_{A_l}^{(p-q)}(t-r)}{r}\right], \tag{3.43}$$

$$\underset{(M)}{\mathrm{h}}{}_{0i}^{lpq}(t^*,\tau,\boldsymbol{\xi}) = \frac{(-1)^{l+p-q}}{l!}C_{l-1}(l-p-1,p-q,q) \times k_{<a_1\dots a_p}\hat{\partial}_{a_{p+1}\dots a_{l-1}>}\hat{\partial}_\tau^q\left[\frac{\mathcal{I}_{iA_{l-1}}^{(p-q+1)}(t-r)}{r}\right], \tag{3.44}$$

$$\underset{(S)}{\mathrm{h}}{}_{0i}^{lpq}(t^*,\tau,\boldsymbol{\xi}) = \frac{(-1)^{l+p-q}\,l}{(l+1)!}C_{l-1}(l-p-1,p-q,q) \times(\hat{\partial}_a + k_a\hat{\partial}_\tau - k_a\hat{\partial}_{t^*})\hat{\partial}_\tau^q\left[\frac{\epsilon_{iab}\mathcal{S}_{bA_{l-1}}^{(p-q)}(t-r)}{r}\right], \tag{3.45}$$

$$\underset{(M)}{\mathrm{q}}{}_{ij}^{lpq}(t^*,\tau,\boldsymbol{\xi}) = \frac{(-1)^{l+p-q}}{l!}C_{l-2}(l-p-2,p-q,q) \times k_{<a_1\dots a_p}\hat{\partial}_{a_{p+1}\dots a_{l-2}>}\hat{\partial}_\tau^q\left[\frac{\mathcal{I}_{ijA_{l-2}}^{(p-q+2)}(t-r)}{r}\right], \tag{3.46}$$

$$\underset{(S)}{\mathrm{q}}{}_{ij}^{lpq}(t^*,\tau,\boldsymbol{\xi}) = \frac{(-1)^{l+p-q}l}{(l+1)!}C_{l-2}(l-p-2,p-q,q)(\hat{\partial}_a + k_a\hat{\partial}_\tau - k_a\hat{\partial}_{t^*}) \times k_{<a_1\dots a_p}\hat{\partial}_{a_{p+1}\dots a_{l-2}>}\hat{\partial}_\tau^q\left[\frac{\epsilon_{ba(i}\mathcal{S}_{j)bA_{l-2}}^{(p-q+1)}(t-r)}{r}\right]. \tag{3.47}$$

All quantities on the right-hand side of (3.43)–(3.47), which are explicitly shown as functions of x^i, $r = |\boldsymbol{x}|$ and t, must be understood as taken on the unperturbed light-ray trajectory and expressed in terms of ξ^i, $d = |\boldsymbol{\xi}|$, τ and t^* in accordance with equations (3.22) and (3.25). For example, the ratio $\mathcal{I}_{A_l}^{(p-q)}(t-r)/r$ in equation (3.43) must be understood as

$$\frac{\mathcal{I}_{A_l}^{(p-q)}(t-r)}{r} \equiv \frac{\mathcal{I}_{A_l}^{(p-q)}(t^* + \tau - \sqrt{\tau^2+d^2})}{\sqrt{\tau^2+d^2}}, \tag{3.48}$$

and the same replacement rule is applied to the other equations. After accounting for (3.39)–(3.47), equation (3.38) can be solved analytically with the mathematical technique shown in Section 4.

Light-ray geodesics

The geodesic equation for light rays is given in (3.14). It is reduced to a more explicit form after making use of the linearized post-Minkowskian expressions for the Christoffel symbols

$$\Gamma^0_{00} = -\frac{1}{2}\partial_t h_{00}(t, \boldsymbol{x}) , \tag{3.49}$$

$$\Gamma^0_{0i} = -\frac{1}{2}\partial_i h_{00}(t, \boldsymbol{x}) , \tag{3.50}$$

$$\Gamma^0_{ij} = -\frac{1}{2}\left[\partial_i h_{0j}(t, \boldsymbol{x}) + \partial_j h_{0i}(t, \boldsymbol{x}) - \partial_t h_{ij}(t, \boldsymbol{x})\right] , \tag{3.51}$$

$$\Gamma^i_{00} = \partial_t h_{0i}(t, \boldsymbol{x}) - \frac{1}{2}\partial_i h_{00}(t, \boldsymbol{x}) , \tag{3.52}$$

$$\Gamma^i_{0j} = \frac{1}{2}\left[\partial_j h_{0i}(t, \boldsymbol{x}) - \partial_i h_{0j}(t, \boldsymbol{x}) + \partial_t h_{ij}(t, \boldsymbol{x})\right] , \tag{3.53}$$

$$\Gamma^i_{jp} = \frac{1}{2}\left[\partial_j h_{ip}(t, \boldsymbol{x}) + \partial_p h_{ij}(t, \boldsymbol{x}) - \partial_i h_{jp}(t, \boldsymbol{x})\right] . \tag{3.54}$$

The affine parameter σ in this equation is an implicit function of the coordinate time t. Relation between σ and t is derived from the time component of (3.14)

$$\frac{d^2t}{d\sigma^2} + \left(\Gamma^0_{00} + \Gamma^0_{0p}\dot{x}^p + \Gamma^0_{pq}\dot{x}^p\dot{x}^q\right)\left(\frac{dt}{d\sigma}\right)^2 = 0 , \tag{3.55}$$

where the dot above coordinates denote a derivative with respect to the coordinate time, $\dot{x}^p \equiv dx^p/dt$. Effectively, there is no need to solve (3.55) explicitly as we are not interested in the parameter σ because the coordinate time t is more practical parameter which can be measured with clocks. Therefore, we express the spatial components of the geodesic equation (3.14) in terms of the Christoffel symbols (3.49)–(3.54), and replace differentiation with respect to the canonical parameter σ by differentiation with respect to the coordinate time t. With the help of equation (3.55) the spatial components of the geodesic equation for light ray propagation becomes

$$\ddot{x}^i(t) = \frac{1}{2}h_{00,i} - h_{0i,0} - \frac{1}{2}h_{00,0}\dot{x}^i - h_{ik,0}\dot{x}^k - (h_{0i,k} - h_{0k,i})\dot{x}^k - h_{00,k}\dot{x}^k\dot{x}^i - \left(h_{ik,j} - \frac{1}{2}h_{kj,i}\right)\dot{x}^k\dot{x}^j + \left(\frac{1}{2}h_{kj,0} - h_{0k,j}\right)\dot{x}^k\dot{x}^j\dot{x}^i , \tag{3.56}$$

where h_{00}, h_{0i}, h_{ij} are components of the metric tensor taken on the unperturbed light-ray trajectory as shown in equations (3.39)–(3.47), that is $h_{\alpha\beta} \equiv h_{\alpha\beta}(t, \boldsymbol{x}_N(t))$.

Equation (3.56) can be further simplified after substituting the unperturbed light-ray trajectory (3.24) to the right-hand side of Equation (3.56) and making use of equation (3.28). Working in arbitrary coordinates one obtains

$$\frac{d^2x^i(\tau)}{d\tau^2} = \frac{1}{2}k^\alpha k^\beta \hat{\partial}_i h^{\rm can.}_{\alpha\beta} - \hat{\partial}_\tau\left(k^\alpha h^{\rm can.}_{i\alpha} - \frac{1}{2}k^i k^j k^p q^{\rm can.}_{jp}\right) - \hat{\partial}_{\tau\tau}(w^i - k^i w^0)\,. \quad (3.57)$$

where all functions in equation (3.57) are taken (before any differentiation) on the unperturbed light-ray trajectory given by equation (3.24) and the gauge functions w^α (they are explained in (2.3)) have not yet been specified which means that equations (3.57) are gauge invariant. We discuss the choice of the gauge functions in the next subsection.

The main advantage of the form (3.57) to which we have reduced the light ray propagation equation (3.14) is the convenience of its analytic integration. Indeed, when we integrate along the light-ray path the following rules, applied to any smooth function $F(\tau,\xi)$, can be used

$$\int \frac{\partial}{\partial\tau}F(\tau,\xi)\,d\tau = F(\tau,\xi) + C(\xi)\,, \quad (3.58)$$

$$\int \frac{\partial}{\partial\xi^i}F(\tau,\xi)\,d\tau = \frac{\partial}{\partial\xi^i}\int F(\tau,\xi)\,d\tau\,. \quad (3.59)$$

This means that terms which are represented as partial derivatives with respect to the time parameter τ can be immediately integrated out by making use of (3.58). At the same time (3.59) shows that one can change the order of integration and differentiation with respect to the parameter ξ^i. It allows us to calculate the integral along the light ray from a more simple scalar expression instead of integrating a vector function. This technique will be demonstrated explicitly in the next sections.

Equation (3.57) is linear with respect to the perturbation of the metric tensor, $h_{\alpha\beta}$. Hence, it can be linearly decomposed in the equation for perturbations of the light-ray trajectory caused separately by mass and spin multipole moments. Substitution of the metric tensor (2.5)–(2.8) to Equation (3.57) and replacement of spatial derivatives with respect to x^i with those with respect to parameters ξ^i and τ by making use of (3.30) yield the following linear superposition:

$$\ddot{x}^i = \ddot{x}^i_{(G)} + \ddot{x}^i_{(M)} + \ddot{x}^i_{(S)}\,, \quad (3.60)$$

where $\ddot{x}^i_{(M)}$ and $\ddot{x}^i_{(S)}$ are the components of photon's coordinate acceleration caused by mass and spin multipoles of the metric tensor, respectively, and $\ddot{x}^i_{(G)}$ is the gauge-dependent acceleration. These components read

$$\ddot{x}^i_{(G)} = \hat{\partial}_{\tau\tau}\left[(k^i\phi^0 - \phi^i) + (k^i w^0 - w^i)\right]\,, \quad (3.61)$$

$$\ddot{x}^i_{(M)} = 2(\hat{\partial}_i - k_i\hat{\partial}_\tau)\frac{\mathcal{M}}{r} \quad (3.62)$$

$$+ 2\hat{\partial}_i \sum_{l=2}^{\infty}\sum_{p=0}^{l}\sum_{q=0}^{p} \frac{(-1)^{l+p-q}}{l!} C_l(l-p, p-q, q) H(2-q)$$

$$\times \left(1 - \frac{p-q}{l}\right)\left(1 - \frac{p-q}{l-1}\right) k_{<a_1 \ldots a_p} \hat{\partial}_{a_{p+1} \ldots a_l>} \hat{\partial}_{t^*}^{p-q} \hat{\partial}_{\tau}^{q} \left[\frac{\mathcal{I}_{A_l}(t-r)}{r}\right]$$

$$- 2\hat{\partial}_\tau \sum_{l=2}^{\infty}\sum_{p=0}^{l} \frac{(-1)^{l+p}}{l!} C_l(l-p, p) \left(1 - \frac{p}{l}\right)$$

$$\times \left\{\left(1 + \frac{p}{l-1}\right) k_{i<a_1 \ldots a_p} \hat{\partial}_{a_{p+1} \ldots a_l>} \hat{\partial}_{t^*}^{p} \left[\frac{\mathcal{I}_{A_l}(t-r)}{r}\right]\right.$$

$$\left. - \frac{2p}{l-1} k_{<a_1 \ldots a_{p-1}} \hat{\partial}_{a_p \ldots a_{l-1}>} \hat{\partial}_{t^*}^{p} \left[\frac{\mathcal{I}_{iA_{l-1}}(t-r)}{r}\right]\right\},$$

and

$$\ddot{x}^i_{(S)} = 2\left(k_j \hat{\partial}_{ia} - \delta_{ij} \hat{\partial}_{a\tau}\right) \frac{\epsilon_{jba} \mathcal{S}_b}{r} \tag{3.63}$$

$$- 4 k_j \hat{\partial}_{ia} \sum_{l=2}^{\infty}\sum_{p=0}^{l-1}\sum_{q=0}^{p} \frac{(-1)^{l+p-q} l}{(l+1)!} C_{l-1}(l-p-1, p-q, q) H(2-q)$$

$$\times \left(1 - \frac{p-q}{l-1}\right) k_{<a_1 \ldots a_p} \hat{\partial}_{a_{p+1} \ldots a_{l-1}>} \hat{\partial}_{t^*}^{p-q} \hat{\partial}_{\tau}^{q} \left[\frac{\epsilon_{jba} \mathcal{S}_{bA_{l-1}}(t-r)}{r}\right]$$

$$+ 4\left(\hat{\partial}_a - k_a \hat{\partial}_{t^*}\right) \hat{\partial}_\tau \sum_{l=2}^{\infty}\sum_{p=0}^{l-1} \frac{(-1)^{l+p} l}{(l+1)!} C_{l-1}(l-p-1, p)$$

$$\times \left(1 - \frac{p}{l-1}\right) k_{<a_1 \ldots a_p} \hat{\partial}_{a_{p+1} \ldots a_{l-1}>} \hat{\partial}_{t^*}^{p} \left[\frac{\epsilon_{iba} \mathcal{S}_{bA_{l-1}}(t-r)}{r}\right]$$

$$+ 4 k_j \hat{\partial}_\tau \sum_{l=2}^{\infty}\sum_{p=0}^{l-1} \frac{(-1)^{l+p} l}{(l+1)!} C_{l-1}(l-p-1, p) \tag{3.64}$$

$$\times k_{<a_1 \ldots a_p} \hat{\partial}_{a_{p+1} \ldots a_{l-1}>} \hat{\partial}_{t^*}^{p} \left[\frac{\epsilon_{jba_{l-1}} \dot{\mathcal{S}}_{ibA_{l-2}}(t-r)}{r}\right],$$

where a dot above any function denotes a partial time derivative with respect to the parameter t^* [113], w^α are the gauge functions coming from (2.3), and ϕ^α are the gauge functions which appear as a result of integration of the light ray propagation equation. These terms enter the equations of motion in the form of combination $k^i\phi^0 - \phi^i$ given in the next section. It is worth emphasizing that we do not intend to separate $k^i\phi^0 - \phi^i$ in two functions, ϕ^i and ϕ^0 because such a separation is not unique while the linear combination $k^i\phi^0 - \phi^i$ is unambiguous. We have not combined functions ϕ^α with the gauge functions w^α for two reasons:

(1) to indicate that the solution of equations of light-ray geodesic, performed in one specific coordinate system, leads to generation of terms which can be eliminated by gauge transformation,
(2) to simplify the final form of the result of the integration as all terms with second and higher order time derivatives are immediately integrated in accordance with (3.58).

Gauge functions w^α are still arbitrary which makes our equations gauge invariant. However, for the sake of physical interpretation of the result of integration of equations of light-ray geodesic, we shall choose a specific form of functions w^α to make our coordinate system both harmonic and ADM which makes the coordinate description of motion of free-falling particles in these coordinates simple. Specific form of the gauge functions w^α at arbitrary field point is shown in equations (2.17) and (2.18) and their form at any point on the light-ray trajectory is given in equations (3.66), (3.67).

It is important to notice that all terms depending on mass-type multipoles of the order l on the right-hand sides of (3.62) have a numerical factor $1 - p/l$ where p is the summation index. Such terms vanish when $p = l$. It means that (3.62) does not contain terms with the time derivatives of the order l from the multipoles of the lth order which, actually, describe gravitational wave emission from the astrophysical system because they decay slowly as $1/r$. The same is true with regard to the spin-type multipoles of the order $l - 1$ in (3.63) – the time derivatives of the order of $l - 1$ from the spin-type multipoles (which decay as $1/r$) vanish on the right-hand side of (3.63). Explicit analytic integration of such terms would be impossible but they simply do not present in the solution of general relativistic equations of light propagation for the reason mentioned above. It is this property of the null geodesic in general relativity which prevents the amplification of the gravitational wave perturbation for a light ray propagating closely to an astrophysical system emitting multipolar gravitational radiation (binary star, etc.). This fact was established in [100] in a quadrupole approximation and extended to any multipole in [103]. This chapter confirms this result. The reader should notice that the cancellation of these terms occurs only in general relativity.

Gauge freedom of equations of propagation of light

Gauge functions w^α, generating the coordinate transformation from the canonical harmonic coordinates to the ADM-harmonic ones, are given by equations (2.19) and (2.20). They transform the metric tensor as follows:

$$h^{\text{can.}}_{\alpha\beta} = h_{\alpha\beta} - \partial_\alpha w_\beta - \partial_\beta w_\alpha \,, \tag{3.65}$$

where $h^{\text{can.}}_{\alpha\beta}$ is the canonical form of the metric tensor in harmonic coordinates given by equations (2.5)–(2.8) and $h_{\alpha\beta}$ is the metric tensor given in the ADM-harmonic coordinates by equations (2.21)–(2.24).

The gauge functions taken on the light-ray trajectory and expressed in terms of the variables ξ and τ can be written down in the next form

$$w^0 = \sum_{l=2}^{\infty}\sum_{p=0}^{l}\sum_{q=0}^{p}\int_{-\infty}^{\tau+t^*} du\, \underset{(M)}{\mathrm{h}}{}_{00}^{lpq}(u,\tau,\xi)\,, \tag{3.66}$$

$$w^i = (\hat{\partial}_i + k_i\hat{\partial}_\tau - k_i\hat{\partial}_{t^*})\sum_{l=2}^{\infty}\sum_{p=0}^{l}\sum_{q=0}^{p}\int_{-\infty}^{\tau+t^*} dv \int_{-\infty}^{v} du\, \underset{(M)}{\mathrm{h}}{}_{00}^{lpq}(u,\tau,\xi) \tag{3.67}$$
$$- 4\sum_{l=2}^{\infty}\sum_{p=0}^{l-1}\sum_{q=0}^{p}\int_{-\infty}^{\tau+t^*} du\left[\underset{(M)}{\mathrm{h}}{}_{0i}^{lpq}(u,\tau,\xi) + \underset{(S)}{\mathrm{h}}{}_{0i}^{lpq}(u,\tau,\xi)\right],$$

where $\underset{(M)}{\mathrm{h}}{}_{00}^{lpq}(u,\tau,\xi)$, $\underset{(M)}{\mathrm{h}}{}_{0i}^{lpq}(u,\tau,\xi)$ and $\underset{(S)}{\mathrm{h}}{}_{0i}^{lpq}(u,\tau,\xi)$ are defined by the Equations (3.43), (3.44), and (3.45) after making use of the substitution $t^* \to u$. It is worth noting the following relationships:

$$\frac{\partial w^0}{\partial t^*} = \sum_{l=2}^{\infty}\sum_{p=0}^{l}\sum_{q=0}^{p} \underset{(M)}{\mathrm{h}}{}_{00}^{lpq}(t^*,\tau,\xi)\,, \tag{3.68}$$

$$\frac{\partial w^i}{\partial t^*} = \sum_{l=2}^{\infty}\sum_{p=0}^{l}\sum_{q=0}^{p}\int_{-\infty}^{\tau+t^*} du(\hat{\partial}_i + k_i\hat{\partial}_\tau)\, \underset{(M)}{\mathrm{h}}{}_{00}^{lpq}(u,\tau,\xi) \tag{3.69}$$
$$- 4\sum_{l=2}^{\infty}\sum_{p=0}^{l-1}\sum_{q=0}^{p}\left[\underset{(M)}{\mathrm{h}}{}_{0i}^{lpq}(t^*,\tau,\xi) + \underset{(S)}{\mathrm{h}}{}_{0i}^{lpq}(t^*,\tau,\xi)\right],$$

and

$$k^i\frac{\partial w^i}{\partial t^*} - \frac{\partial w^0}{\partial t^*} = \sum_{l=2}^{\infty}\sum_{p=0}^{l}\sum_{q=0}^{p}\left[\int_{-\infty}^{\tau+t^*} du\hat{\partial}_\tau\, \underset{(M)}{\mathrm{h}}{}_{00}^{lpq}(u,\tau,\xi) - \underset{(M)}{\mathrm{h}}{}_{00}^{lpq}(t^*,\tau,\xi)\right] \tag{3.70}$$
$$- 4\sum_{l=2}^{\infty}\sum_{p=0}^{l-1}\sum_{q=0}^{p}\left[k^i\, \underset{(M)}{\mathrm{h}}{}_{0i}^{lpq}(t^*,\tau,\xi) + k^i\, \underset{(S)}{\mathrm{h}}{}_{0i}^{lpq}(t^*,\tau,\xi)\right],$$

which are helpful in calculation of the gravitational shift of the frequency of light.

A linear combination, $k^i\phi^0 - \phi^i$, of the gauge-dependent functions ϕ^α that appear in (3.61), is given by the expressions

$$k^i\phi^0 - \phi^i = (k^i\phi^0_{(M)} - \phi^i_{(M)}) + (k^i\phi^0_{(S)} - \phi^i_{(S)})\,, \tag{3.71}$$

$$k^i\phi^0_{(M)} - \phi^i_{(M)} = 2\hat{\partial}_i\sum_{l=2}^{\infty}\sum_{p=2}^{l}\sum_{q=2}^{p}\frac{(-1)^{l+p-q}}{l!}C_l(l-p,p-q,q) \tag{3.72}$$
$$\times\left(1-\frac{p-q}{l}\right)\left(1-\frac{p-q}{l-1}\right)k_{<a_1\dots a_p}\hat{\partial}_{a_{p+1}\dots a_l>}\hat{\partial}_\tau^{q-2}\left[\frac{\mathcal{I}_{A_l}^{(p-q)}(t-r)}{r}\right]$$

$$+2\sum_{l=2}^{\infty}\sum_{p=1}^{l}\sum_{q=1}^{p}\frac{(-1)^{l+p-q}}{l!}C_l(l-p,p-q,q)$$
$$\times\left(1-\frac{p-q}{l}\right)\left\{\left(1+\frac{p-q}{l-1}\right)k_{i<a_1...a_p}\hat{\partial}_{a_{p+1}...a_l>}\hat{\partial}_{\tau}^{q-1}\left[\frac{\mathcal{I}^{(p-q)}_{A_l}(t-r)}{r}\right]\right.$$
$$\left.-2\frac{p-q}{l-1}k_{<a_1...a_{p-1}}\hat{\partial}_{a_p...a_{l-1}>}\hat{\partial}_{\tau}^{q-1}\left[\frac{\mathcal{I}^{(p-q)}_{iA_{l-1}}(t-r)}{r}\right]\right\},$$

$$k^i\phi^0_{(S)}-\phi^i_{(S)}=2\frac{\epsilon_{iab}k^a\mathcal{S}^b}{r} \tag{3.73}$$
$$+4k_j\hat{\partial}_{ia}\sum_{l=3}^{\infty}\sum_{p=2}^{l-1}\sum_{q=2}^{p}\frac{(-1)^{l+p-q}\,l}{(l+1)!}C_{l-1}(l-p-1,p-q,q)$$
$$\times\left(1-\frac{p-q}{l-1}\right)k_{<a_1...a_p}\hat{\partial}_{a_{p+1}...a_{l-1}>}\hat{\partial}_{\tau}^{q-2}\left[\frac{\epsilon_{jab}\mathcal{S}^{(p-q)}_{bA_{l-1}}(t-r)}{r}\right]$$
$$-4\left(\hat{\partial}_a-k_a\hat{\partial}_{t^*}\right)\sum_{l=2}^{\infty}\sum_{p=1}^{l-1}\sum_{q=1}^{p}\frac{(-1)^{l+p-q}l}{(l+1)!}C_{l-1}(l-p-1,p-q,q)$$
$$\times\left(1-\frac{p-q}{l-1}\right)k_{<a_1...a_p}\hat{\partial}_{a_{p+1}...a_{l-1}>}\hat{\partial}_{\tau}^{q-1}\left[\frac{\epsilon_{iab}\mathcal{S}^{(p-q)}_{bA_{l-1}}(t-r)}{r}\right]$$
$$-4k_a\sum_{l=2}^{\infty}\sum_{p=0}^{l-1}\sum_{q=0}^{p}\frac{(-1)^{l+p-q}l}{(l+1)!}C_{l-1}(l-p-1,p-q,q)$$
$$\times\left(1-\frac{p-q}{l-1}\right)k_{<a_1...a_p}\hat{\partial}_{a_{p+1}...a_{l-1}>}\hat{\partial}_{\tau}^{q}\left[\frac{\epsilon_{iab}\mathcal{S}^{(p-q)}_{bA_{l-1}}(t-r)}{r}\right]$$
$$+4k_j\sum_{l=2}^{\infty}\sum_{p=1}^{l-1}\sum_{q=1}^{p}\frac{(-1)^{l+p-q}l}{(l+1)!}C_{l-1}(l-p-1,p-q,q)$$
$$\times k_{<a_1...a_p}\hat{\partial}_{a_{p+1}...a_{l-1}>}\hat{\partial}_{\tau}^{q-1}\left[\frac{\epsilon_{jba_{l-1}}\mathcal{S}^{(p-q+1)}_{ibA_{l-2}}(t-r)}{r}\right]$$

3.4 Polarization of light and the Stokes parameters

Reference tetrad

Propagation of electromagnetic fields in vacuum and evolution of their physical parameters in curved spacetime can be studied with various mathematical techniques.

One of the most convenient techniques was worked by Newman and Penrose and is called the Newman–Penrose formalism [114, 115]. This formalism introduces at each point of spacetime a null tetrad of four vectors associated with the bundle of light rays defined by the electromagnetic wave vector field l^α. The Newman–Penrose tetrad consists of two real and two complex null vectors – $(l^\alpha, n^\alpha, m^\alpha, \overline{m}^\alpha)$ – where the bar above function indicates complex conjugation. The null tetrad vectors are normalized in such a way that $n_\alpha l^\alpha = -1$ and $m_\alpha \overline{m}^\alpha = +1$ are the only nonvanishing products among the four vectors of the tetrad.

The vectors of the null tetrad are not uniquely determined by specifying l^α. Indeed, for a fixed direction l^α the normalization conditions for the tetrad vectors are preserved under the linear transformations (null rotation) [110, 114]

$$l'^\alpha = A l^\alpha \,, \tag{3.74}$$

$$n'^\alpha = A^{-1}\left(n^\alpha + \overline{B} m^\alpha + B\overline{m}^\mu + B\overline{B} l^\mu\right) , \tag{3.75}$$

$$m'^\alpha = e^{-i\Theta}\left(m^\alpha + \overline{B} l^\alpha\right) , \tag{3.76}$$

$$\overline{m}'^\alpha = e^{i\Theta}\left(\overline{m}^\alpha + B l^\alpha\right) , \tag{3.77}$$

where A, Θ are real scalars and $B = B_1 + iB_2$ is a complex scalar. These transformations form a four-parameter (A, B_1, B_2, Θ) subgroup of the homogeneous Lorentz group which is equivalent to the point-like Lorentz transformations [109].

For doing mathematical analysis of the intensity and polarization of electromagnetic waves it is useful to introduce a local orthonormal reference frame of observer moving with a 4-velocity u^α who is seeing the electromagnetic wave traveling in the positive direction of z axis of the reference frame. It means that at each point of spacetime the observer uses a tetrad frame $e^\alpha{}_{(\beta)} = \{e^\alpha{}_{(0)}, e^\alpha{}_{(1)}, e^\alpha{}_{(2)}, e^\alpha{}_{(3)}\}$ defined in such a way that

$$e^\alpha{}_{(0)} = u^\alpha \,, \quad e^\alpha{}_{(3)} = (-l_\alpha u^\alpha)^{-1}\left[l^\alpha + (l_\beta u^\beta) u^\alpha\right] , \tag{3.78}$$

and two other vectors of the observer's tetrad, $e^\alpha{}_{(1)}$ and $e^\alpha{}_{(2)}$, are the unit space-like vectors being orthogonal to each other as well as to $e^\alpha{}_{(0)}$ and $e^\alpha{}_{(3)}$. In other words, vectors of the observer tetrad are subject to the following normalization conditions:

$$g_{\alpha\beta} e^\alpha{}_{(\mu)} e^\beta{}_{(\nu)} = \eta_{\mu\nu} \,, \quad \eta^{\mu\nu} e^\alpha{}_{(\mu)} e^\beta{}_{(\nu)} = g^{\alpha\beta} \,. \tag{3.79}$$

It is worth noticing that the observer's tetrad e^α_β has two group of indices. The indices without round brackets run from 0 to 3 and are associated with time and space coordinates. The indices enclosed in the round brackets numerate vectors of the tetrad and also run from 0 to 3. The coordinate-type indices of the tetrad have no relation to the tetrad indices. If one changes spacetime coordinates (passive coordinate transformation) it does not affect the tetrad indices while the coordinate indices of the tetrad change in accordance with the transformation law for vectors. On the other hand, one can change the tetrad vectors at each point in spacetime by doing the Lorentz transformation (active coordinate transformation) without changing the coordinate chart [115].

Let us define at each point of spacetime a coordinate basis of static observers

$$E^{\alpha}_{\ (0)} = \left(1 + \frac{1}{2}h_{00}\,,\ 0\,,\ 0\,,\ 0\right)\,, \tag{3.80}$$

$$E^{\alpha}_{\ (1)} = \left(h_{0j}a^j\,,\ a^i - \frac{1}{2}h_{ij}a^j\right)\,, \tag{3.81}$$

$$E^{\alpha}_{\ (2)} = \left(h_{0j}b^j\,,\ b^i - \frac{1}{2}h_{ij}b^j\right)\,, \tag{3.82}$$

$$E^{\alpha}_{\ (3)} = \left(h_{0j}k^j\,,\ k^i - \frac{1}{2}h_{ij}k^j\right)\,, \tag{3.83}$$

which is written down for the case of weak gravitational field, $g_{\alpha\beta} = \eta_{\alpha\beta} + h_{\alpha\beta}$. Here the unit spatial vectors $\boldsymbol{a} = (a^1, a^2, a^3)$, $\boldsymbol{b} = (b^1, b^2, b^3)$, and $\boldsymbol{k} = (k^1, k^2, k^3)$ are orthonormal in the Euclidean sense ($\delta_{ij}a^ib^j = \delta_{ij}a^ik^j = \delta_{ij}b^ik^j = 0$ and $\delta_{ij}a^ia^j = \delta_{ij}b^ib^j = \delta_{ij}k^ik^j = 1$) with vector $\boldsymbol{k}$ pointing to the direction of propagation of the light ray at infinity as given in (3.18). These basis vectors are convenient to track the changes in the parameters of the electromagnetic wave as it travels from the point of emission of light to the point of its observation.

The local tetrad $e^{\alpha}_{\ (\beta)}$ of observers moving with 4-velocity u^{α} with respect to the static observers relates to the tetrad $E^{\alpha}_{\ (\beta)}$ by means of the Lorentz transformation

$$e^{\alpha}_{\ (\beta)} = \Lambda^{\gamma}_{\ \beta}E^{\alpha}_{\ (\gamma)}\,, \qquad E^{\alpha}_{\ (\beta)} = \lambda^{\gamma}_{\ \beta}e^{\alpha}_{\ (\gamma)}\,, \tag{3.84}$$

where the matrix of the Lorentz transformation is [2]

$$\Lambda^0_{\ 0} = u^0 \equiv \gamma \tag{3.85}$$

$$\Lambda^i_{\ 0} = \Lambda^0_{\ i} = u^0\,, \tag{3.86}$$

$$\Lambda^i_{\ j} = \delta^{ij} + \frac{u^iu^j}{1+\gamma}\,, \tag{3.87}$$

and the inverse matrix of the Lorentz transformation $\lambda^{\alpha}_{\ \beta}$ is obtained from $\Lambda^{\alpha}_{\ \beta}$ by replacing $u^i \to -u^i$ that complies with the definition of the inverse matrix $\Lambda^{\alpha}_{\ \beta}\lambda^{\beta}_{\ \gamma} = \delta^{\alpha}_{\gamma}$.

The connection between the null tetrad and the observer's tetrad frame, $e^{\alpha}_{\ (\beta)}$, is given by equations

$$l^{\alpha} = -(l_{\gamma}u^{\gamma})\left[e^{\alpha}_{\ (0)} + e^{\alpha}_{\ (3)}\right]\,, \tag{3.88}$$

$$n^{\alpha} = -\frac{1}{2}(l_{\gamma}u^{\gamma})\left[e^{\alpha}_{\ (0)} - e^{\alpha}_{\ (3)}\right]\,, \tag{3.89}$$

$$m^{\alpha} = \frac{1}{\sqrt{2}}\left[e^{\alpha}_{\ (1)} + ie^{\alpha}_{\ (2)}\right]\,, \tag{3.90}$$

$$\overline{m}^{\alpha} = \frac{1}{\sqrt{2}}\left[e^{\alpha}_{\ (1)} - ie^{\alpha}_{\ (2)}\right]\,. \tag{3.91}$$

Vector pairs $e^\alpha_{(0)}$, $e^\alpha_{(3)}$ and $e^\alpha_{(1)}$, $e^\alpha_{(2)}$ split spacetime at the point in two subspaces. In particular, vectors $e^\alpha_{(1)}$, $e^\alpha_{(2)}$ defines the plane of polarization in spacetime. If vectors $e^\alpha_{(0)}$ and $e^\alpha_{(3)}$ are fixed, then, vectors $e^\alpha_{(1)}$ and $e^\alpha_{(2)}$ are defined up to an arbitrary rotation in the plane of polarization. Transformations (3.76) and (3.77) with $B = 0$ yield

$$e'^\alpha_{(1)} = \cos\Theta\, e^\alpha_{(1)} + \sin\Theta\, e^\alpha_{(2)}\,, \tag{3.92}$$

$$e'^\alpha_{(2)} = -\sin\Theta\, e^\alpha_{(1)} + \cos\Theta\, e^\alpha_{(2)}\,, \tag{3.93}$$

where Θ is the rotation angle of the vectors in the plane of polarization. We notice that, since vectors $e^\alpha_{(1)}$, $e^\alpha_{(2)}$ are orthogonal to $e^\alpha_{(0)} = u^\alpha$, the two null vectors, m^α and $\overline{m}^\alpha$, are also orthogonal to the 4-velocity,

$$m_\alpha u^\alpha = 0\,, \quad \overline{m}_\alpha u^\alpha = 0\,. \tag{3.94}$$

The null vector n^α is also orthogonal to u^α, $n_\alpha u^\alpha = 0$. On the other hand, the scalar product of l^α with 4-velocity yields the angular frequency of electromagnetic wave, $\omega \equiv -l_\alpha u^\alpha$.

Propagation laws for the reference tetrad

Discussion of the rotation of the polarization plane and the change of the Stokes parameters of electromagnetic radiation is inconceivable without understanding of how the local reference frame propagates along the light-ray geodesic from the point of emission of light to the point of its observation. To this end we construct the reference tetrad frame of observer at the point of observation of light and render a parallel transport of it backward in time along the light-ray geodesic. By definition, vectors of the tetrad frame of the observer, $e^\alpha_{(\beta)}$, and those of the null tetrad $(l^\alpha, n^\alpha, m^\alpha, \overline{m}^\alpha)$ do not change in a covariant sense as they are parallel transported along the light ray. The propagation equation for the tetrad vectors are, thus, obtained by applying the operator $D \equiv l^\alpha \nabla_\alpha$ of the parallel transport along the null vector l^α. Explicit form of the parallel transport of the reference tetrad is

$$\frac{de^\alpha_{(\mu)}}{d\sigma} + \Gamma^\alpha_{\beta\gamma}\, l^\beta\, e^\gamma_{(\mu)} = 0\,, \tag{3.95}$$

where σ is an affine parameter along the light ray. Using definition of the Christoffel symbols (3.15) and changing parameter σ to the proper time τ of the observer we recast (3.95) to

$$\frac{d}{d\tau}\left[e^\alpha_{(\mu)} + \frac{1}{2}h^\alpha{}_\beta e^\beta_{(\mu)}\right] = \frac{1}{2}\eta^{\alpha\nu}\left(\partial_\nu h_{\gamma\beta} - \partial_\gamma h_{\nu\beta}\right) k^\beta e^\gamma_{(\mu)}\,. \tag{3.96}$$

The propagation of the null vectors m^α and $\overline{m}^\alpha$ along the direction of the null vector l^α is given by

$$\frac{dm^\alpha}{d\lambda} + \Gamma^\alpha_{\beta\gamma}\, l^\beta\, m^\gamma = 0\,, \tag{3.97}$$

$$\frac{d\overline{m}^\alpha}{d\lambda} + \Gamma^\alpha_{\beta\gamma}\, l^\beta\, \overline{m}^\gamma = 0\,, \tag{3.98}$$

and the same laws are valid for n^α and l^α (see, for example, (3.14)). Equations (3.96)–(3.98) are the main equations for the discussion of the rotation of the plane of polarization and variation of the Stokes parameters.

Relativistic description of polarized electromagnetic radiation

We consider propagation of a bundle of plane electromagnetic waves from the point of emission to the point of observation. Each of these waves have an electromagnetic tensor $F_{\alpha\beta} = \mathcal{F}_{\alpha\beta} + O(\varepsilon)$ defined in the first approximation by equation [110, 111]

$$\mathcal{F}_{\alpha\beta} = a_{\alpha\beta} \exp\left(\frac{i\varphi}{\varepsilon}\right)\,, \tag{3.99}$$

$$a_{\alpha\beta} = \Phi\, m_{[\alpha} l_{\beta]} + \bar{\Phi}\, \overline{m}_{[\alpha} l_{\beta]} \tag{3.100}$$

where Φ is a complex scalar amplitude of the wave with a real and imaginary components which are independent of each other in the most general case of incoherent radiation. In the proper frame of the observer with 4-velocity u^α the components of the electric and magnetic field vectors are defined, respectively, as $E^\alpha = -F^{\alpha\beta} u_\beta$ and $H^\alpha = (-1/2\sqrt{-g})\epsilon^{\alpha\beta\gamma\delta} F_{\gamma\delta} u_\beta$ [3]. The electric field is a product of slowly changing amplitude $\mathcal{E}_\alpha = -a_{\alpha\beta} u^\beta$ and fast-oscillating phase exponent

$$E_\alpha = \mathcal{E}_\alpha \exp\left(\frac{i\varphi}{\varepsilon}\right)\,. \tag{3.101}$$

The polarization properties of electromagnetic radiation consisting of an ensemble of the waves with equal frequencies but different phases are defined by the components of the electric field measured by observer. In the rest frame of the observer with 4-velocity u^α, the intensity and polarization properties of the electromagnetic radiation are described in terms of the *polarization* tensor [1, 107]

$$J_{\alpha\beta} = \langle E_\alpha \overline{E}_\beta \rangle = \langle \mathcal{E}_\alpha \overline{\mathcal{E}}_\beta \rangle\,, \tag{3.102}$$

where the angular brackets represent an average with respect to an ensemble of the electromagnetic waves with randomly distributed phases. This averaging eliminates all fast-oscillating terms from $J_{\alpha\beta}$. One has to notice [1] that the polarization tensor $J_{\alpha\beta}$ is symmetric only for a linearly polarized radiation. In all other cases, the polarization tensor is not symmetric. The polarization tensor $J_{\alpha\beta}$ is purely spatial at the point of observation which means it is orthogonal to the 4-velocity of observer, $J_{\alpha\beta}u^{\beta}$. Furthermore, because the polarization tensor is defined in the subspace of the polarization plane, it is orthogonal to the wave vector $J_{\alpha\beta}l^{\beta} = 0$. This equality follows directly from its definition (3.102) and (3.9).

The vector amplitude $\mathcal{E}_{\alpha}$ of the electric field can be decomposed in two independent components in the plane of polarization. Two vectors of the null tetrad, $m^{\alpha}, \overline{m}^{\alpha}$, form the circular-polarization basis, and vectors $e^{\alpha}_{\ (1)}, e^{\alpha}_{\ (2)}$ form a linear polarization basis. The decomposition reads

$$\mathcal{E}^{\alpha} = \mathcal{E}_L m^{\alpha} + \mathcal{E}_R \overline{m}^{\alpha} , \tag{3.103}$$

$$\mathcal{E}^{\alpha} = \mathcal{E}_{(1)} e^{\alpha}_{\ (1)} + \mathcal{E}_{(2)} e^{\alpha}_{\ (2)} , \tag{3.104}$$

where

$$\mathcal{E}_L = \frac{1}{2}\omega\Phi , \quad \mathcal{E}_R = \frac{1}{2}\omega\bar{\Phi} , \tag{3.105}$$

are left and right circularly polarized components of the electric field,

$$\mathcal{E}_{(1)} = \frac{1}{\sqrt{2}}\left(\mathcal{E}_L + \mathcal{E}_R\right) , \quad \mathcal{E}_{(2)} = \frac{i}{\sqrt{2}}\left(\mathcal{E}_L - \mathcal{E}_R\right) , \tag{3.106}$$

are linearly polarized components, $\omega = -l_{\alpha}u^{\alpha}$ is the angular frequency of the electromagnetic wave, and we have taken into account the condition of orthogonality (3.94).

There are for electromagnetic Stokes parameters $S_{\alpha} = (S_0, S_1, S_2, S_3)$. They are defined by projecting the polarization tensor $J_{\alpha\beta}$ on four independent combination of the tensor products of the two vectors, $e^{\alpha}_{\ (1)}, e^{\alpha}_{\ (2)}$ making up the polarization plane. More specifically [107, 116]

$$S_0 = J_{\alpha\beta}\left[e^{\alpha}_{\ (1)}e^{\beta}_{\ (1)} + e^{\alpha}_{\ (2)}e^{\beta}_{\ (2)}\right] , \tag{3.107}$$

$$S_1 = J_{\alpha\beta}\left[e^{\alpha}_{\ (1)}e^{\beta}_{\ (1)} - e^{\alpha}_{\ (2)}e^{\beta}_{\ (2)}\right] , \tag{3.108}$$

$$S_2 = J_{\alpha\beta}\left[e^{\alpha}_{\ (1)}e^{\beta}_{\ (2)} + e^{\alpha}_{\ (2)}e^{\beta}_{\ (1)}\right] , \tag{3.109}$$

$$S_3 = iJ_{\alpha\beta}\left[e^{\alpha}_{\ (1)}e^{\beta}_{\ (2)} - e^{\alpha}_{\ (2)}e^{\beta}_{\ (1)}\right] , \tag{3.110}$$

where $S_0 \equiv I$ is the intensity, $S_1 \equiv Q$ and $S_2 \equiv U$ characterize the degree of a linear polarization, and $S_3 \equiv V$ is the degree of a circular polarization of the electromagnetic wave.

Making use of equation (3.102) in (3.107)–(3.110) allows us to represent the Stokes parameters in the linear polarization basis as follows [107]:

$$S_0 = \langle|\mathcal{E}_{(1)}|^2 + |\mathcal{E}_{(2)}|^2\rangle , \tag{3.111}$$

$$S_1 = \langle|\mathcal{E}_{(1)}|^2 - |\mathcal{E}_{(2)}|^2\rangle , \tag{3.112}$$

$$S_2 = \langle\mathcal{E}_{(1)}\overline{\mathcal{E}}_{(2)} + \overline{\mathcal{E}}_{(1)}\mathcal{E}_{(2)}\rangle , \tag{3.113}$$

$$S_3 = i\langle\mathcal{E}_{(1)}\overline{\mathcal{E}}_{(2)} - \overline{\mathcal{E}}_{(1)}\mathcal{E}_{(2)}\rangle , \tag{3.114}$$

where $\mathcal{E}_{(n)} = \mathcal{E}_{\alpha}e^{\alpha}_{\ (n)}$ for $n = 1, 2$.

It is important to emphasize that though the Stokes parameters have four components, they do not form a four-dimensional spacetime vector because they do not behave like a vector under transformations of the Lorentz group [1, 107]. Indeed, if we perform a pure Lorentz boost all four Stokes parameters remain invariant [1]. However, for a constant rotation of angle Θ in the polarization plane, the Stokes parameters transform as [1]

$$S_0' = S_0 , \tag{3.115}$$

$$S_1' = S_1 \cos 2\Theta + S_2 \sin 2\Theta , \tag{3.116}$$

$$S_2' = S_1 \cos 2\Theta - S_2 \sin 2\Theta , \tag{3.117}$$

$$S_3' = S_3 . \tag{3.118}$$

This is what would be expected for a spin-1/2 field. That is, under a duality rotation of $\Theta = \pi/4$, one linear polarization state turns into the other, while the circular polarization state remains the same. The transformation properties (3.115)–(3.118) of the Stokes parameters point out that the Stokes parameters S_1, S_2 represent a linearly polarized components, and S_3 represents a circularly polarized component.

The polarization vector $\boldsymbol{P} = (P_1, P_2, P_3)$ and the degree of polarization $P = |\boldsymbol{P}|$ of the electromagnetic radiation can be defined in terms of the normalized Stokes parameters by $\boldsymbol{P} = (S_1/I, S_2/I, S_3/I)$. Any partially polarized wave may be thought of as an incoherent superposition of a completely polarized wave with the degree of polarization P and the polarization vector $\boldsymbol{P}$, and a completely unpolarized wave with the degree of polarization $1 - P$ and nil polarization vector, $\boldsymbol{P} = 0$, so that for an arbitrary polarized radiation one has: $(S_0, S_1, S_2, S_3) = I(P, P_1, P_2, P_3) + I(1 - P, 0, 0, 0)$. For completely polarized waves, vector $\boldsymbol{P}$ describes the surface of the unit sphere introduced by Poincaré [1]. The center of the Poincaré sphere corresponds to an unpolarized radiation and the interior to a partially polarized radiation. Orthogonally polarized waves represent any two conjugate points on the Poincaré sphere. In particular, $(P_1 = \pm 1, P_2 = 0, P_3 = 0)$, and $(P_1 = 0, P_2 = 0, P_3 = \pm 1)$ represent orthogonally polarized waves corresponding to the linear and circular polarization bases, respectively.

Propagation law of the Stokes parameters

Taking definition (3.100) of the electromagnetic tensor and accounting for the parallel transport of the null vectors $l^\alpha, m^\alpha, \overline{m}^\alpha$ along the light ray and the laws of propagation of the electromagnetic tensor given by equations (3.16), yield the law of propagation of the complex scalar functions Φ and $\overline{\Phi}$

$$\frac{d\Phi}{d\sigma} + \vartheta\Phi = 0\,, \tag{3.119}$$

$$\frac{d\overline{\Phi}}{d\sigma} + \vartheta\overline{\Phi} = 0\,, \tag{3.120}$$

where σ is an affine parameter along the ray and ϑ is the expansion of the light-ray congruence defined in (3.17).

Let us consider a sufficiently small, two-dimensional area $\mathcal{A}$ in the cross-section of the congruence of light rays lying on a null hypersurface of constant phase φ that is in the polarization plane. The law of transportation of the cross-sectional area is [2, 71, 110]

$$\frac{d\mathcal{A}}{d\sigma} - 2\vartheta\mathcal{A} = 0\,. \tag{3.121}$$

Thus, the product, $\mathcal{A}|\Phi|^2 = \mathcal{A}\Phi\bar{\Phi}$, remains constant along the congruence of light rays:

$$l^\alpha\nabla_\alpha\left(\mathcal{A}|\Phi|^2\right) = 0\,. \tag{3.122}$$

This law of propagation for the product $\mathcal{A}|\Phi|^2$ corresponds to the conservation of photon's flux [2, 71].

The law of conservation of the number of photons propagating along the light ray, corresponds to the propagation law of vector $|\Phi|^2 l^\alpha$. Indeed, taking covariant divergence of this quantity and making use of equations (3.119) and (3.120) along with definition (3.17) for the expansion ϑ of the bundle of light rays, yields [2, 71]

$$\nabla_\alpha\left(|\Phi|^2 l^\alpha\right) = 0\,. \tag{3.123}$$

This equation assumes that the scalar amplitude Φ of the electromagnetic wave can be interpreted in terms of the number density, $\mathcal{N}$, of photons in phase space and the energy of one photon, $E_\omega = \hbar\omega$, as follows:

$$|\Phi| = \sqrt{8\pi\hbar}\left(\frac{\mathcal{N}}{E_\omega}\right)^{1/2}\,, \tag{3.124}$$

where the reduced Planck constant $\hbar = h/2\pi$ and the normalizing factor were introduced for consistency between classical and quantum definitions of the energy of an electromagnetic wave [2].

Each of the Stokes parameter is proportional to the square of frequency of light, $\omega = -u^\alpha l_\alpha$, as directly follows from equations (3.111)–(3.114) and (3.106). Therefore, the variation of the Stokes parameters along the light ray can be obtained directly from their definitions (3.107)–(3.110) along the light ray and making use of the laws of propagation (3.119) and (3.120). However, the set of the Stokes parameters S_α $(\alpha = 0, 1, 2, 3)$ is not directly observed in astronomy and we do not discuss their laws of propagation.

Instead the set of four other polarization parameters $(F_\omega, P_1, P_2, P_3)$ is practically measured [71, 117] and we focus on the discussion of the laws of propagation for these parameters. Here $\boldsymbol{P} = (P_1, P_2, P_3)$ is the polarization vector as defined at the end of the preceding section, and F_ω is the *specific flux* of radiation (also known as the *monochromatic flux* of a light source [117]) entering a telescope from a given source. The specific flux is defined as an integral of the *specific intensity* (also known as the *surface brightness* [117]) of the radiation, $I_\omega \equiv I_\omega(\omega, \boldsymbol{l})$, over the total solid angle (assumed $\ll 4\pi$) subtended by the source on the observer's sky:

$$F_\omega = \int I_\omega(\omega, \boldsymbol{l}) d\Omega(\hat{\boldsymbol{l}}) , \tag{3.125}$$

where $\hat{\boldsymbol{l}} = (\sin\hat{\theta}\cos\hat{\phi}, \sin\hat{\theta}\sin\hat{\phi}, \cos\hat{\theta})$ is the unit vector in the direction of the radiation flow and $d\hat{\Omega}(\boldsymbol{l}) = \sin\hat{\theta} d\hat{\theta} d\hat{\phi}$ is the element of the solid angle formed by light rays from the source and measured in the observer's local Lorentz frame.

The *specific intensity* I_ω of radiation at a given frequency $\omega = 2\pi\nu$, flowing in a given direction, $\hat{\boldsymbol{l}}$, as measured in a specific local Lorentz frame, is defined by

$$I_\omega = \frac{d(\text{energy})}{d(\text{time})d(\text{area})d(\text{frequency})d(\text{solid angle})} . \tag{3.126}$$

A simple calculation (see, for instance, the problem 5.10 in [118]), reveals that

$$\mathcal{N} = \frac{8\pi^3}{h^4}\frac{I_\omega}{\omega^3} , \tag{3.127}$$

where h is the Planck's constant. The number density $\mathcal{N}$ is invariant along the light ray and does not change under the Lorentz transformation. Invariance of $\mathcal{N}$ is a consequence of the kinetic equation for photons (radiative transfer equation) which in the case of gravitational field and without any other scattering processes, assumes the following form [2]:

$$\frac{d\mathcal{N}}{d\lambda} = 0 . \tag{3.128}$$

Equations (3.127) and (3.128) tell us that the ratio I_ω/ω^3 is invariant along the light-ray trajectory, that is

$$\frac{I_\omega}{I_{\omega_0}} = \left(\frac{\omega}{\omega_0}\right)^3 , \tag{3.129}$$

where ω_0, ω are frequency of light at the point of emission and observation, respectively, I_{ω_0} is the *surface brightness* of the source of light at the point of emission, and I_ω is the *surface brightness* of the source of light at the point of observation.

Equations (3.119)–(3.129) make it evident that in the geometric optics approximation the gravitational field does not mix up the linear and circular polarizations of the electromagnetic radiation but can change its *surface brightness* I_ω due to gravitational (and Doppler) shift of the light frequency caused by the time-dependent part of the gravitational field of the isolated system emitting gravitational waves. Furthermore, the *monochromatic flux* from the source of radiation changes due to the distortion of the domain of integration in equation (3.125) caused by the gravitational light-bending effect. Taking into account that the gravitationally unperturbed solid angle $d\Omega(\boldsymbol{k}) = \sin\theta d\theta d\phi$, and introducing the Jacobian, $J(\theta,\phi)$, of transformation between the spherical coordinates $(\hat{\theta},\hat{\phi})$ and (θ,ϕ) at the point of observation, one obtains that the measured monochromatic flux is

$$F_\omega = \int d\phi \int I_{\omega_0}(\theta,\phi)J(\theta,\phi)(\omega/\omega_0)^3 \sin\theta d\theta\,. \tag{3.130}$$

Equation (3.130) tells us that the monochromatic flux of the source of light can vary due to:

(1) the gravitational Doppler shift of the electromagnetic frequency of light when it travels from the point of emission to the point of observation;
(2) the change in the solid angle at the point of observation caused by the gravitational light deflection. The "magnification" matrix is the Jacobian of the transformation of the null directions on the celestial sphere generated by the bending of the light-ray trajectories by the gravitational field of the isolated system.

Spatial orientation of two components, P_1 and P_2, of the polarization vector $\boldsymbol{P}$ at the point of observation differs from that taken at the point of emission of the electromagnetic wave due to the parallel transport of the polarization vector. The reference tetrad with resepct to which the orientation of the polarization vector is measured is subject to the same law of the parallel transport. Therefore, in order to understand the change in the orientation of the polarization vector it is sufficient to consider the change in the orientation of the reference tetrad as it propagates from the point of emission of light to the point of observation. The change in the polarization can be easily extracted then from the law of transformation of the Stokes parameters under the change of the reference tetrad. As we will see later in Section 6.4, only two vectors of the tetrad, $e^\alpha_{(1)}$ and $e^\alpha_{(2)}$, will undergo the change in their orientation as they propagate along the light ray. Consequently, only two components of the polarization vector, P_1 and P_2 will change. This can be seen after taking into account equations (3.92) and (3.93) where the rotational angle Θ is determined as a solution of the parallel transport equation (3.96) for the reference tetrad (see Section 6.4). Thus, the gravitational field changes the tilt angle of the polarization ellipse as light propagates along the light-ray trajectory. This gravitationally induced rotation of the polarization ellipse is called the Skrotskii effect [119, 120]. It was also predicted and discussed independently in [121, 122]. Its observation would play a significant role for detection of gravitational waves of cosmological origin by CMBR-radiometry space missions [38, 123]. The third component of the polarization vector, P_3, remains the same along the light-ray trajectory because it represents the circularly polarized component of the radiation and is not affected by the rotation of the reference tetrad as it propagates along the light-ray path.

4 Mathematical technique for analytic integration of light-ray equations

This section provides mathematical technique for performing integration of equations of propagation of various characteristics of electromagnetic wave from the point of emission of light $\boldsymbol{x}_0$ to the point of its observation $\boldsymbol{x}$. It has been worked out in [6, 100, 103, 104]. The basic function, which is to be integrated, is the metric tensor perturbation, $h_{\alpha\beta}(t, \boldsymbol{x})$, generated by the localized astronomical system and taken at the points lying on the light-ray trajectory. The metric tensor depends on the multipole moments and/or their derivatives taken at the retarded instant of time $s = t - r$ divided by the radial distance from the system, r, taken to some power. The most difficult integrals are those taken from the functions like this, $F(s)/r$, with $F(s)$ being a multipole moment of the gravitating system depending on the retarded time s. Calculation of the integrals goes slightly differently for stationary and nonstationary components of the metric tensor and these issues are treated in the next two subsections.

4.1 Monopole and dipole light-ray integrals

The monopole and dipole parts of the metric tensor perturbation (2.5)–(2.8) are formed by terms being proportional to the mass $\mathcal{M}$ and spin $\mathcal{S}^i$ of the isolated system. These terms are given in the solution of the light-ray equations by integrals $[\mathcal{M}(s)/r]^{[-1]}$, $[\mathcal{M}(s)/r]^{[-2]}$ and $[\mathcal{S}^i(s)/r]^{[-1]}$, $[\mathcal{S}^i(s)/r]^{[-2]}$. In this section we shall assume for simplicity that mass $\mathcal{M}$ and spin $\mathcal{S}^i$ of the isolated system are constant during the time of propagation of light from the source of light to observer. The assumption of constancy of the mass $\mathcal{M}$ and spin $\mathcal{S}^i$ of the isolated system is valid as long as one neglects the energy emitted in the form of the gravitational waves by the isolated system under consideration. For light ray propagating from the sources at the edge of our visible universe (quasars) the characteristic interval of time of emission of gravitational waves by an isolated system is comparable with the interval of time the light takes to travel from the point of emission to observer. In this case the time evolution of the mass-monopole and spin-dipole is to be taken into account for correct calculation of the perturbations of the trajectory of light ray (see Section 5.1 for more detail). Mass and spin of the isolated system can also change due to a catastrophic disruption of the isolated system resulting in supernova explosion. Specific details of how this process affects the light-ray propagation are not considered in this chapter but, perhaps, are worthwhile to study.

In the case when mass and spin are constant, the integrals that we need to carry out are reduced to $[1/r]^{[-1]}$ and $[1/r]^{[-2]}$ which are formally divergent at the lower limit of integration at the past null infinity when time $t \to -\infty$. However, one must bear in mind that these integrals do not enter equations of light-ray geodesics (3.57) alone but appear in these equations after taking at least one partial derivative with respect to either ξ^i or τ parameters. This differentiation effectively eliminates divergent parts of the integrals from the final result. Hence, in what follows, we drop out the formally divergent terms so that the integrals under discussion assume the following form:

$$\left[\frac{1}{r}\right]^{[-1]} \equiv \int_{-\infty}^{t} \frac{d\tau}{r} = \int \frac{d\tau}{\sqrt{d^2+\tau^2}} = -\ln\left[\frac{r(\tau)-\tau}{r_{\rm E}}\right], \tag{4.1}$$

$$\left[\frac{1}{r}\right]^{[-2]} \equiv \int_{-\infty}^{t} \left[\frac{1}{r}\right]^{[-1]} d\tau = -\tau\ln\left[\frac{r(\tau)-\tau}{r_{\rm E}}\right] - r(\tau), \tag{4.2}$$

where $\tau = t - t^*$ and we used equation (3.25) explicitly while expressing the distance $r = r_N(\tau)$ as a function of time τ. The constant distance $r_{\rm E}$ was introduced to make the argument of the logarithmic function dimensionless. This constant is not important for calculations as it always cancel out in final formulas. However, in the case of gravitational lensing it is convenient to identify the scale constant $r_{\rm E}$ with the radius of the Einstein ring [71–73, 124]

$$r_{\rm E} = \left(\frac{4G\mathcal{M}}{c^2}\frac{rr_0}{R}\right)^{1/2}, \tag{4.3}$$

where r, r_0 are the distances from the isolated system (the deflector of light) to observer and to the source of light, respectively, and $R = \tau + \tau_0$. The radius of the Einstein ring is a characteristic distance separating naturally the case of weak gravitational lensing ($d > r_{\rm E}$) from the strong lensing ($d < r_{\rm E}$). The Einstein ring at the observer's point has the angular size given by

$$\theta_{\rm E} = \frac{r_{\rm E}}{r} = \left(\frac{4G\mathcal{M}}{c^2}\frac{r_0}{Rr}\right)^{1/2}. \tag{4.4}$$

The angular radius $\theta_{\rm E}$ defines the angular scale for a lensing situation. For example, in cosmology typical mass of an isolated system is $M \simeq 10^{12}M_\odot$ and distances r, r_0, R are of the order of 1 Gpc (Gigaparsec). Consequently, the angular Einstein radius $\theta_{\rm E}$ is of the order of one arcsecond, and the linear radius $r_{\rm E}$ is of the order of 1 Kpc (Kiloparsec). A typical star within our galaxy has mass $M \simeq M_\odot$ and distances r, r_0, R are of the order 10 Kpc. It yields an angular Einstein radius $\theta_{\rm E}$ of the order of one 1μas (one milli-arcsecond) and the linear radius $r_{\rm E}$ is of the order of 10 au (10 astronomical units).

4.2 Light-ray integrals from quadrupole and higher order multipoles

Time-dependent terms in the metric tensor (2.5)–(2.8) result from the multipole moments which can be either periodic (a binary system) or aperiodic (a supernova explosion) functions of time. The most straightforward way to calculate the impact of the gravitational field of such a source on the propagation of light would be to decompose its multipole moments $F(s) := \{\mathcal{I}_L(s), \mathcal{S}_L(s)\}$ in the Fourier series [100]

$$F(t-r) = \frac{1}{\sqrt{2\pi}} \int\limits_{-\infty}^{+\infty} \tilde{F}(\tilde{\omega}) e^{i\tilde{\omega}(t-r)} d\tilde{\omega} \,, \tag{4.5}$$

where $\tilde{\omega}$ is a Fourier frequency, then to substitute this decomposition to the light-ray propagation equations and to integrate them term by term. This method makes an impression that in order to obtain the final result of the integration the (complex-valued) Fourier image $\tilde{F}(\tilde{\omega})$ of the multipole moments of the isolated system must be specified explicitly, otherwise neither explicit integration nor the convolution of the integrated Fourier series will be possible. However, this impression is misleading, at least in general relativity. We shall show below that the explicit structure of $\tilde{F}(\tilde{\omega})$ is irrelevant for general-relativistic calculation of the light-ray integrals. This may be understood better, if one recollects that in general relativity gravitational field propagates with the same speed as light in vacuum. In alternative theories of gravity the speed of gravity and light may be different [125], so it can lead to the appearance of terms being proportional to the difference between the speed of gravity and the speed of light that will drastically complicates calculations which will do require to know the explicit form of the Fourier images of gravitational-wave sources as functions of $\tilde{\omega}$. All such terms, which might depend on the difference between the speed of gravity and the speed of light are canceled out in general relativity, thus, making integration of the light ray propagation equations manageable and applicable for any source of gravitational waves without particular specification of its temporal behavior. It may be worth mentioning that the experimental limit on the difference between the speed of gravity and the speed of light in general relativity can be measured in the solar system experiments with major planets as predicted in [126]. It was experimentally tested with the precision 20% in the jovian light-ray deflection experiment with VLBI network [127]. No deviation from general relativity was detected.

We notice that the integration of the light-ray equations is effectively reduced to the calculation of only two types of integrals along the light ray: $[F(s)/r]^{[-1]}$ and $[F(s)/r]^{[-2]}$, where $F(s) = F(t-r)$ denotes any type of the time-dependent multipole moments of the gravitational field of the localized astronomical system. These integrals can be performed after introducing a new variable [100]

$$y \equiv s - t^* = \tau - r(\tau) = \tau - \sqrt{d^2 + \tau^2} \,, \tag{4.6}$$

which is the interval of time between two spacetime events: position of photon $x^\alpha = (\tau, \boldsymbol{x})$ on the light ray and that of the center of mass of the isolated system $z^\alpha = (y, 0)$ taken at the retarded time y[1]. Equation (4.6) is a retarded solution of the null cone equation in flat spacetime

$$\eta_{\alpha\beta}(x^\alpha - z^\alpha)(x^\beta - z^\beta) = 0\,, \tag{4.7}$$

giving the time of propagation of gravity from the isolated system to the photon along the null cone characteristic of Einstein's equations. This retardation of gravity effect presents in the time argument of solution (2.5)–(2.8) of the linearized Einstein equations. The replacement of the time argument τ with the retarded time $y = \tau - r(\tau)$ allows us to perform integration of equations of light-ray geodesics completely without making specific assumptions about the time dependence of the multipole moments. It is worth noting that while the parameter τ runs from $-\infty$ to $+\infty$, the retarded time y runs from $-\infty$ to 0, that is, y is always negative, $y \le 0$.

Equation (4.6) leads to other useful transformations

$$\tau = \frac{y^2 - d^2}{2y}\,, \qquad \sqrt{d^2 + \tau^2} = -\frac{1}{2}\frac{d^2 + y^2}{y}\,, \tag{4.8}$$

and

$$d\tau = \frac{1}{2}\frac{d^2 + y^2}{y^2}dy\,. \tag{4.9}$$

Making use of the new variable y and relationships (4.8) and (4.9) the integrals under discussion can be explicitly displayed as follows:

$$\left[\frac{F(t-r)}{r}\right]^{[-1]} = -\int_{-\infty}^{y} \frac{F(t^* + \zeta)}{\zeta}d\zeta\,, \tag{4.10}$$

$$\left[\frac{F(t-r)}{r}\right]^{[-2]} = -\frac{1}{2}\int_{-\infty}^{y}\int_{-\infty}^{\eta} \frac{F(t^* + \zeta)}{\zeta}d\zeta d\eta - \frac{d^2}{2}\int_{-\infty}^{y}\frac{1}{\eta^2}\int_{-\infty}^{\eta} \frac{F(t^* + \zeta)}{\zeta}d\zeta d\eta\,, \tag{4.11}$$

where ζ, η are the dummy variables of the integration replacing the integration along the light-ray trajectory by that along a null characteristic of the gravitational field, and t^* is the time of the closest approach of photon to the origin of the coordinate chart coinciding with the center of mass of the isolated gravitational system. The time t^* has no physical meaning in the most general case and appears in calculations as an auxiliary (constant) parameter which vanishes in the final result. The reason is simple, by choosing the origin of the coordinate chart at a different point we change the numerical value of t^* but it cannot change physical observables – the angle of light-ray

1 The retarded time y should not be confused with the Cartesian coordinate y.

deflection, frequency shift, etc. The time of the closest approach t^* can be used as an approximation of the retarded time $s = t-r$ in the case of a small impact parameter d of the light ray with respect to the isolated astronomical system, that is $s \simeq t^* + O(d^2/cr)$ (see equation (7.6)). This approximation makes an erroneous impression that the physical effect of propagation of gravity is unimportant for calculation of observable effects caused by variable gravitational field of the isolated system. This misinterpretation of the role of gravity's propagation in the interpretation of the observable astronomical effects did happen in some papers [53, 128] which used the simplifying approximation $s = t^*$ from the very beginning of calculations but it precludes to spatially disentangle the null cone characteristics of gravity and light. The only case when the time t^* makes physical sense is when observer is located in infinity. Indeed, in this case the radial distance $r = \infty$, and the retarded time $s = t^*$ exactly, because the residual terms $O(d^2/r) = 0$ irrespectively of the choice of the coordinate origin [99].

A remarkable property of the integrals on the right-hand side of equations (4.10) and (4.11) is that they depend on the parameters ξ^i and τ only through either the upper limit of the integration, which is the variable $y = \tau - \sqrt{d^2 + \tau^2}$, or the square of the impact parameter, $d^2 = (\boldsymbol{\xi} \cdot \boldsymbol{\xi})$ standing in front of the second integral on the right-hand side of equation (4.11). For this reason, differentiation of the integrals in the left part of equations (4.10) and (4.11) with respect to either ξ^i or τ will effectively eliminate the integration along the light ray trajectory. For example,

$$\hat{\partial}_i \left\{ \left[\frac{F(t-r)}{r} \right]^{[-1]} \right\} = -\frac{F(t^* + y)}{y} \hat{\partial}_i y = \frac{\xi^i}{yr} F(t-r) . \tag{4.12}$$

Similar calculation can be easily performed in case of differentiation of integral $[F(t-r)/r]^{[-1]}$ with respect to τ. It results in

$$\hat{\partial}_\tau \left\{ \left[\frac{F(t-r)}{r} \right]^{[-1]} \right\} = -\frac{F(t^* + y)}{y} \hat{\partial}_\tau y = -\frac{F(t^* + y)}{y} \left(1 - \frac{\tau}{r} \right) = \frac{F(t-r)}{r} , \tag{4.13}$$

as it could be expected because the partial differentiation with respect to τ keeps ξ^i and t^* fixed and, hence, is equivalent to taking a total time derivative with respect to t along the light ray. Since all terms in the solution of the light geodesic equation are represented as partial derivatives from the integrals $[F(t-r)/r]^{[-1]}$ and $[F(t-r)/r]^{[-2]}$ with respect to the parameters ξ^i and/or τ, it is clear from equations (4.12) and (4.13) that the solution will not contain any single integral like $[F(t-r)/r]^{[-1]}$ at all – only the derivatives of this integral will appear which are ordinary functions.

Dealing with the double integrals of type $[F(t-r)/r]^{[-2]}$ is more sophisticated. We notice that taking the first and second derivatives from the double integral $[F(t-r)/r]^{[-2]}$ do not eliminate the integration along the light ray trajectory

$$\hat{\partial}_k \left\{ \left[\frac{F(t-r)}{r} \right]^{[-2]} \right\} = \xi^k \left\{ \frac{1}{y} \left[\frac{F(t-r)}{r} \right]^{[-1]} - \int_{-\infty}^{y} \frac{1}{\eta^2} \int_{-\infty}^{\eta} \frac{F(t^* + \zeta)}{\zeta} d\zeta d\eta \right\} , \tag{4.14}$$

$$\hat{\partial}_{jk}\left\{\left[\frac{F(t-r)}{r}\right]^{[-2]}\right\} = \frac{\xi^k\xi^j}{y^2}\frac{F(t-r)}{r} + P^{jk}\left\{\frac{1}{y}\left[\frac{F(t-r)}{r}\right]^{[-1]} - \int_{-\infty}^{y}\frac{1}{\eta^2}\int_{-\infty}^{\eta}\frac{F(t^*+\zeta)}{\zeta}d\zeta d\eta\right\}, \tag{4.15}$$

However, taking one more (a third) derivative eliminates all integrals from equation (4.15) completely. More specifically, we have

$$\hat{\partial}_{ijk}\left\{\left[\frac{F(t-r)}{r}\right]^{[-2]}\right\} = \frac{1}{y}\left\{\left(P^{ij}+\frac{\xi^{ij}}{yr}\right)\hat{\partial}_k + P^{jk}\hat{\partial}_i + \xi^j\hat{\partial}_{ik}\right\}\left[\frac{F(t-r)}{r}\right]^{[-1]}, \tag{4.16}$$

and making use of equations (4.12) and (4.15) for explicit calculation of partial derivatives on the right-hand side of equation (4.16) proves that all integrations disappear, that is

$$\hat{\partial}_{ijk}\left\{\left[\frac{F(t-r)}{r}\right]^{[-2]}\right\} = \frac{P^{ij}\xi^k + P^{jk}\xi^i + P^{ik}\xi^j}{y^2}\frac{F(t-r)}{r} + \frac{\xi^i\xi^j\xi^k}{y^2r^2}\left[\left(\frac{2}{y}-\frac{1}{r}\right)F(t-r) - \dot{F}(t-r)\right], \tag{4.17}$$

where $\dot{F}(t-r) \equiv \partial_t F(t-r)$. The same kind of reasoning works for the third derivatives with respect to the parameter τ and to the mixed derivatives taken with respect to both ξ^i and τ.

We shall obtain solution of the light geodesic in terms of (STF) partial derivatives taken with respect to the impact parameter vector, ξ^i, of the light ray acting on the single and double integrals having a symbolic form $\hat{\partial}_{<a_1...a_k>}[F(t-r)/r]^{[-1]}$ and $\hat{\partial}_{\langle a_1...a_k\rangle}[F(t-r)/r]^{[-2]}$ where the angular brackets around the spatial indices indicate the STF symmetry. Explicit expression for this STF derivative can be obtained directly by applying the differentiation rules shown in equations (4.14)–(4.16). In what follows, we shall see that the solution of the equations of propagation of light rays will always contain three or more derivatives acting on the integrals from the higher order multipole moments (with the multipole index $l \geq 2$) which have the same structure as that shown in equations (4.10) and (4.11). It means that the final result of the integration of the light-ray propagation equations depending on the higher order multipoles of the localized astronomical system can be expressed completely solely in terms of these multipoles taken at the retarded instant of time $s = t - r$. In other words, the observed effects of light deflection, etc. does not depend in *general relativity* directly on the past history of photon's propagation that is the effects of temporal variations of the gravitational field (gravitational waves) do not accumulate. On the other hand, the past history of the isolated system can affect the propagation of light ray, at least in principle, through the tails of the gravitational waves which contribute to the present value of multipole moments of the system as shown in equations (2.12), (2.13).

We would like to point out that if an electromagentic wave propagated through a dispersive medium, the physical speed of light would be different from the speed of gravity. In such a case the effects of the temporal variations of the gravitational field can do accumulate and the motion of the electromagnetic signal depends on its past history. This effect was studied by Bertotti and Catenacci [129] in order to set an upper limit on the light scintillations caused by the stochastic gravitational wave background. They concluded that such a limit is not very interesting but it might be a good time now to reconsider this conclusion in application to the more advanced level of observational technologies.

5 Gravitational perturbations of the light ray

5.1 Relativistic perturbation of the electromagnetic eikonal

Perturbation of eikonal (phase) of electromagnetic signal propagating from the point $\boldsymbol{x}_0$ to the point $\boldsymbol{x}$ are obtained by solving equation (3.38). This solution is found by integrating the metric tensor perturbation along the light-ray trajectory and can be written down in the linearized approximation as an algebraic sum of three separate terms

$$\psi = \psi_{(G)} + \psi_{(M)} + \psi_{(S)} , \tag{5.1}$$

where $\psi_{(G)}$ represents the gauge-dependent part of the eikonal, and $\psi_{(M)}$, $\psi_{(S)}$ are eikonal's perturbations caused by the mass and spin multipole moments correspondingly. Their explicit expressions are as follows:

$$\psi_{(G)} = (k^i\phi^i - \phi^0) + (k^i w^i - w^0) , \tag{5.2}$$

$$\begin{aligned}\psi_{(M)} = {} & 2\left[\frac{\mathcal{M}(t-r)}{r}\right]^{[-1]} + 2\sum_{l=2}^{\infty}\sum_{p=0}^{l-1}\frac{(-1)^{l+p}}{l!}C_l(l-p,p)\left(1-\frac{p}{l}\right) \\ & \times\left\{\left(1+\frac{p}{l-1}\right)k_{<a_1\ldots a_p}\hat{\partial}_{a_{p+1}\ldots a_l>}\left[\frac{\mathcal{J}^{(p)}_{A_l}(t-r)}{r}\right]^{[-1]}\right. \\ & \left. -\frac{2p}{l-1}k_{<a_1\ldots a_{p-1}}\hat{\partial}_{a_p\ldots a_{l-1}>}\left[\frac{k^i\mathcal{J}^{(p)}_{iA_{l-1}}(t-r)}{r}\right]^{[-1]}\right\}\end{aligned} \tag{5.3}$$

$$\begin{aligned}\psi_{(S)} = {} & 2\epsilon_{abi}k^a\mathcal{S}^b\hat{\partial}_i\ln(r-\tau) + 4\hat{\partial}_a\sum_{l=2}^{\infty}\sum_{p=0}^{l-1}\frac{(-1)^{l+p}l}{(l+1)!}C_{l-1}(l-p-1,p) \\ & \times\left(1-\frac{p}{l-1}\right)k_{<a_1\ldots a_p}\hat{\partial}_{a_{p+1}\ldots a_{l-1}>}\left[\frac{k^i\epsilon_{iba}\mathcal{S}^{(p)}_{bA_{l-1}}(t-r)}{r}\right]^{[-1]} ,\end{aligned} \tag{5.4}$$

where $H(p-q)$ is the Heaviside function defined by expression (1.21), $C_l(p,q)$ are the polynomial coefficient given by equation (1.20) and $F^{(p)}(t-r)$ denotes $\partial^p F(t-r)/\partial t^p$ where r is considered as constant.

We note that the gauge-dependent part of the eikonal, $\psi_{(G)}$, contains combination of terms $k^i\phi^i - \phi^0$ defined by equations (3.71)–(3.73) which can be, in principle, eliminated with an appropriate choice of the gravitational field gauge functions w^0 and w^i. However, such a procedure will introduce a reference frame in the sky with a coordinate grid being dependent on the direction to the source of light rays, that is, to the direction of the unit vector k^i. The coordinate frame obtained in this way will have the direction-dependent distortions which can change as time goes on because of the proper motion of stars. For this reason the elimination of functions ϕ^0 and ϕ^i from equation (5.2) may be not practically justified. The ADM-harmonic coordinate chart admits a more straightforward treatment of observable relativistic effects. Thus, we leave the gauge functions ϕ^0 and ϕ^i in equation (5.2) along with the gauge functions w^0 and w^i that are defined by formulas (3.66), (3.67).

Expressions (5.3) and (5.4) for the eikonal contain derivatives from the retarded integrals of the mass- and spin-multipoles. After taking the derivatives one can prove that the integrals from all high-order multipoles ($l \geq 2$) are eliminated. Indeed, scrutiny inspection of equations (5.3) and (5.4) elucidates that all the integrals from the multipole moments enter the equation in combination with at least one derivative with respect to the impact parameter vector ξ^i. Hence, the differentiation rule (4.12) is applied which eliminates the integration.

The only integral which must be performed, is that from the mass monopole in equation (5.3). It can be taken by parts as follows:

$$\left[\frac{\mathcal{M}(t-r)}{r}\right]^{[-1]} = -\mathcal{M}(t-r)\ln(r-\tau) + \int_{-\infty}^{y} \dot{\mathcal{M}}(t^* + \zeta)\ln\zeta d\zeta\,, \tag{5.5}$$

where $\dot{\mathcal{M}} \equiv \partial\mathcal{M}/\partial t^*$. If one assumes that the mass $\mathcal{M}$ is constant, then the second term on the right-hand side of equation (5.5) vanishes and the eikonal does not contain any integral dependence on the past history of the light propagation. This assumption is usually implied, for example, in the theory of gravitational lensing [71–73] and other applications of the relativistic theory of light propagation. Here, we extend our approach to take into account the case of time-dependent mass of the isolated system, $\dot{\mathcal{M}} \neq 0$. The mass may change because stars are losing mass in the form of the stellar wind. We shall not consider this case but focus on another process of the lost of mass by the isolated system due to the emission of gravitational radiation. The rate of the mass loss is, then, given by [1, 2]

$$\dot{\mathcal{M}}(t) = -\frac{1}{c^7}\mathcal{I}^{(3)}_{ij}(t)\mathcal{I}^{(3)}_{ij}(t) + O\left(c^{-9}\right)\,, \tag{5.6}$$

where $\mathcal{I}^{(3)}_{ij}$ represents a third time derivative from the mass quadrupole moment and terms of order $O(c^{-9})$ describe contribution of higher order mass and spin multipoles which we neglect here. Making use of equation (5.6) in (5.5) yields

$$\left[\frac{\mathcal{M}(t-r)}{r}\right]^{[-1]} = -\mathcal{M}(t-r)\ln(r-\tau) - \frac{1}{c^7}\int_{-\infty}^{y} \mathcal{I}^{(3)}_{ij}(t^*+\zeta)\mathcal{I}^{(3)}_{ij}(t^*+\zeta)\ln\zeta d\zeta + O\left(c^{-9}\right) . \tag{5.7}$$

It is instructive to evaluate contribution of the second term on the right-hand side of equation (5.5) in the case of light propagating in gravitational field of a binary system consisting of two stars with masses m_1 and m_2 orbiting each other on a circular orbit. The total mass $\mathcal{M}$ of the system is defined in accordance with equation (2.10) (for $l = 0$) which takes into account the first post-Newtonian correction

$$\mathcal{M}(t) = m_1 + m_2 - \frac{1}{2c^2}\frac{\mu\mathcal{M}}{a(t)} + O\left(c^{-4}\right) , \tag{5.8}$$

where $\mu = m_1 m_2/\mathcal{M}$ is the reduced mass of the system and $a = a(t)$ is the orbital radius of the system. Assuming that the lost of the orbital energy is due to the emission of gravitational waves from the system in accordance with the quadrupole formula approximation (5.6), the time evolution of the orbital radius is defined by equation [1, 2]

$$\dot{a} = -\frac{64}{5c^5}\frac{\mu\mathcal{M}^2}{a^3} . \tag{5.9}$$

It has a simple solution [2]

$$a(t) = a_0\left(1-\frac{\tau}{T}\right)^{1/4} , \tag{5.10}$$

$$T = \frac{5c^5}{256}\frac{a_0^4}{\mu\mathcal{M}^2} , \tag{5.11}$$

where $\tau = t-t^*$, t^* is the time of the closest approach of light ray to the binary system, $a_0 = a(t^*)$ is the orbital radius of the binary system at the time of the closest approach, and T is the spiral time of the binary. The lost of the total mass due to the emission of the gravitational waves is

$$\dot{\mathcal{M}} = -\frac{1}{8c^2}\frac{\mu\mathcal{M}}{a_0 T}\left(1-\frac{\tau}{T}\right)^{-5/4} . \tag{5.12}$$

Substituting equation (5.12) to the right-hand side of equation (5.5) and performing integration yields

$$\int_{-\infty}^{y} \dot{\mathcal{M}}(t^*+\zeta)\ln\zeta d\zeta = -\frac{1}{2}\frac{\mu\mathcal{M}}{c^2 a(s)}\ln(r-\tau) + \frac{1}{2}\frac{m_1 m_2}{c^2 a_0}\left\{\ln\left[\frac{a(s)-a_0}{a(s)+a_0}\right] + 2\arctan\left[\frac{a(s)}{a_0}\right]\right\} , \tag{5.13}$$

where $a(s) \equiv a(t-r)$ denotes the orbital radius of the binary system taken at the retarded time $s = t - r$. Putting all terms in equations (5.5), (5.8), and (5.13) together yields

$$\left[\frac{\mathcal{M}(t-r)}{r}\right]^{[-1]} = -(m_1 + m_2)\ln(r-\tau) + \frac{1}{2}\frac{m_1 m_2}{c^2 a_0}\left\{\ln\left[\frac{a(s)-a_0}{a(s)+a_0}\right] + 2\arctan\left[\frac{a(s)}{a_0}\right]\right\} , \tag{5.14}$$

where the first term on the right-hand side of this equation is the standard Shapiro time delay in the gravitational field of the binary system with constant total mass $m_1 + m_2$, and the second term represents relativistic correction due to the emission of gravitational waves by the system causing the overall loss of its orbital energy.

Eikonal describes propagation of a wave front of the electromagnetic wave. Light rays are orthogonal to the wave front and their trajectories can be easily calculated as soon as the eikonal is known. In this chapter we do not use this technique and obtain solution for the light rays directly from the light-ray geodesic equations that gives identical results.

5.2 Relativistic perturbation of the coordinate velocity of light

Integration of the light-ray propagation equations (3.60)–(3.63) is fairly straightforward. Performing one integration of these equations with respect to time yields

$$\dot{x}^i(\tau) = k^i + \dot{\Xi}^i(\tau,\xi) , \tag{5.15}$$

$$\dot{\Xi}^i(\tau,\xi) = \dot{\Xi}^i_{(G)}(\tau,\xi) + \dot{\Xi}^i_{(M)}(\tau,\xi) + \dot{\Xi}^i_{(S)}(\tau,\xi) , \tag{5.16}$$

where the relativistic perturbations of photon's trajectory are given by

$$\dot{\Xi}^i_{(G)}(\tau,\xi) = \hat{\partial}_\tau\left[(\phi^i - k^i\phi^0) + (w^i - k^iw^0)\right], \tag{5.17}$$

$$\begin{aligned}\dot{\Xi}^i_{(M)}(\tau,\xi) &= 2(\hat{\partial}_i - k_i\hat{\partial}_\tau)\left[\frac{\mathcal{M}}{r}\right]^{[-1]} \\ &+ 2\hat{\partial}_i\sum_{l=2}^{\infty}\sum_{p=0}^{l}\sum_{q=0}^{p}\frac{(-1)^{l+p-q}}{l!}C_l(l-p,p-q,q)H(2-q) \\ &\times\left(1-\frac{p-q}{l}\right)\left(1-\frac{p-q}{l-1}\right)k_{<a_1\ldots a_p}\hat{\partial}_{a_{p+1}\ldots a_l>}\hat{\partial}^q_\tau\left[\frac{\mathcal{I}^{(p-q)}_{A_l}(t-r)}{r}\right]^{[-1]} \\ &- 2\sum_{l=2}^{\infty}\sum_{p=0}^{l-1}\frac{(-1)^{l+p}}{l!}C_l(l-p,p)\left(1-\frac{p}{l}\right) \\ &\times\left\{\left(1+\frac{p}{l-1}\right)k_{i<a_1\ldots a_p}\hat{\partial}_{a_{p+1}\ldots a_l>}\left[\frac{\mathcal{I}^{(p)}_{A_l}(t-r)}{r}\right]\right. \\ &\left.-\frac{2p}{l-1}k_{<a_1\ldots a_{p-1}}\hat{\partial}_{a_p\ldots a_{l-1}>}\left[\frac{\mathcal{I}^{(p)}_{iA_{l-1}}(t-r)}{r}\right]\right\}, \end{aligned} \tag{5.18}$$

$$\begin{aligned}\dot{\Xi}^i_{(S)}(\tau,\xi) &= 2k_j\hat{\partial}_{ia}\left[\frac{\epsilon_{jba}\mathcal{S}_b}{r}\right]^{[-1]} - 2\hat{\partial}_a\frac{\epsilon_{iba}\mathcal{S}_b}{r} \\ &- 4k_j\hat{\partial}_{ia}\sum_{l=2}^{\infty}\sum_{p=0}^{l-1}\sum_{q=0}^{p}\frac{(-1)^{l+p-q}l}{(l+1)!}C_{l-1}(l-p-1,p-q,q)H(2-q) \\ &\times\left(1-\frac{p-q}{l-1}\right)k_{<a_1\ldots a_p}\hat{\partial}_{a_{p+1}\ldots a_{l-1}>}\hat{\partial}^q_\tau\left[\frac{\epsilon_{jba}\mathcal{S}^{(p-q)}_{bA_{l-1}}(t-r)}{r}\right]^{[-1]} \\ &+ 4\left(\hat{\partial}_a - k_a\hat{\partial}_{t^*}\right)\sum_{l=2}^{\infty}\sum_{p=0}^{l-1}\frac{(-1)^{l+p}l}{(l+1)!}C_{l-1}(l-p-1,p) \\ &\times\left(1-\frac{p}{l-1}\right)k_{<a_1\ldots a_p}\hat{\partial}_{a_{p+1}\ldots a_{l-1}>}\left[\frac{\epsilon_{iba}\mathcal{S}^{(p)}_{bA_{l-1}}(t-r)}{r}\right] \\ &+ 4k_j\sum_{l=2}^{\infty}\sum_{p=0}^{l-1}\frac{(-1)^{l+p}l}{(l+1)!}C_{l-1}(l-p-1,p) \\ &\times k_{<a_1\ldots a_p}\hat{\partial}_{a_{p+1}\ldots a_{l-1}>}\left[\frac{\epsilon_{jba_{l-1}}\mathcal{S}^{(p+1)}_{ibA_{l-2}}(t-r)}{r}\right]. \end{aligned} \tag{5.19}$$

Here $H(p-q)$ is a Heaviside function defined by expression (1.21) and $C_l(p,q)$ are the polynomial coefficients (1.20). The gauge functions are given in equations (3.66)–(3.73).

Mass monopole and spin dipole terms are written down in equations (5.18) and (5.19) in symbolic form and after taking derivatives are simplified

$$2(\hat{\partial}_i - k_i\hat{\partial}_\tau)\left[\frac{\mathcal{M}}{r}\right]^{[-1]} = \frac{2\mathcal{M}}{r}\left(\frac{\xi^i}{y} - k^i\right), \tag{5.20}$$

$$2k_j\hat{\partial}_{ia}\left[\frac{\epsilon_{jba}\mathcal{S}_b}{r}\right]^{[-1]} = 2k^j\epsilon_{jba}\hat{\partial}_a\left(\frac{\mathcal{S}_b\xi^i}{yr}\right). \tag{5.21}$$

Remaining integrals shown on the right-hand side of equations (5.18) and (5.19) are convenient for presentation of the result of integration in the symbolic form. The integration is actually not required because the integrals are always appear in combination with, at least, one operator of a partial derivative $\hat{\partial}_i$ with respect to the impact parameter vector of the light ray. The derivative operator acts on the integrals in accordance with equation (4.12) converting the integrals into functions of the retarded time $y = \tau - r(\tau)$

$$\hat{\partial}_i\left[\frac{\mathcal{I}^{(p-q)}_{A_l}(t-r)}{r}\right]^{[-1]} = \mathcal{I}^{(p-q)}_{A_l}(t^* + y)\hat{\partial}_i\ln(-y), \tag{5.22}$$

$$\hat{\partial}_i\left[\frac{\epsilon_{jba}\mathcal{S}^{(p-q)}_{bA_{l-1}}(t-r)}{r}\right]^{[-1]} = \epsilon_{jba}\mathcal{S}^{(p-q)}_{bA_{l-1}}(t^* + y)\hat{\partial}_i\ln(-y). \tag{5.23}$$

We conclude that at each point of the wave front of the electromagnetic wave the relativistic perturbation of the direction of propagation of light ray (wave vector) caused by the time-dependent gravitational field of the isolated system depends only on the value of its multipole moments taken at the retarded instant of time and it does not depend on the past history of the light propagation.

5.3 Perturbation of the light-ray trajectory

Integration of equation (5.15) with respect to time yields the relativistic perturbation of the trajectory of the light ray

$$x^i(\tau) = x^i_N + \Delta\underset{(G)}{\Xi^i} + \Delta\underset{(M)}{\Xi^i} + \Delta\underset{(S)}{\Xi^i}, \tag{5.24}$$

where

$$\Delta\underset{(G)}{\Xi^i} \equiv \underset{(G)}{\Xi^i}(\tau,\xi) - \underset{(G)}{\Xi^i}(\tau_0,\xi), \quad \Delta\underset{(M)}{\Xi^i} \equiv \underset{(M)}{\Xi^i}(\tau,\xi) - \underset{(M)}{\Xi^i}(\tau_0,\xi),$$
$$\Delta\underset{(S)}{\Xi^i} \equiv \underset{(S)}{\Xi^i}(\tau,\xi) - \underset{(S)}{\Xi^i}(\tau_0,\xi). \tag{5.25}$$

Here, the term

$$\underset{(G)}{\Xi^i}(\tau,\xi) = (\phi^i - k^i\phi^0) + (w^i - k^i w^0)\,, \tag{5.26}$$

is the gauge-dependent part of the trajectory's perturbation, and the physically meaningful perturbations due to the mass and spin multipoles

$$\begin{aligned}\underset{(M)}{\Xi^i}(\tau,\xi) &= 2(\hat{\partial}_i - k_i\hat{\partial}_\tau)\left[\frac{\mathcal{M}}{r}\right]^{[-2]} \\ &+ 2\hat{\partial}_i \sum_{l=2}^{\infty}\sum_{p=0}^{l}\sum_{q=0}^{p}\frac{(-1)^{l+p-q}}{l!}C_l(l-p,p-q,q)H(2-q) \\ &\times\left(1-\frac{p-q}{l}\right)\left(1-\frac{p-q}{l-1}\right)k_{<a_1\dots a_p}\hat{\partial}_{a_{p+1}\dots a_l>}\hat{\partial}^q_\tau\left[\frac{\mathcal{I}^{(p-q)}_{A_l}(t-r)}{r}\right]^{[-2]} \\ &-2\sum_{l=2}^{\infty}\sum_{p=0}^{l-1}\frac{(-1)^{l+p}}{l!}C_l(l-p,p)\left(1-\frac{p}{l}\right) \\ &\times\left\{\left(1+\frac{p}{l-1}\right)k_{i<a_1\dots a_p}\hat{\partial}_{a_{p+1}\dots a_l>}\left[\frac{\mathcal{I}^{(p)}_{A_l}(t-r)}{r}\right]^{[-1]}\right. \\ &\left.-\frac{2p}{l-1}k_{<a_1\dots a_{p-1}}\hat{\partial}_{a_p\dots a_{l-1}>}\left[\frac{\mathcal{I}^{(p)}_{iA_{l-1}}(t-r)}{r}\right]^{[-1]}\right\}\,,\end{aligned} \tag{5.27}$$

$$\begin{aligned}\underset{(S)}{\Xi^i}(\tau,\xi) &= 2k_j\hat{\partial}_{ia}\left[\frac{\epsilon_{jba}\mathcal{S}_b}{r}\right]^{[-2]} - 2\hat{\partial}_a\left[\frac{\epsilon_{iba}\mathcal{S}_b}{r}\right]^{[-1]} \\ &-4k_j\hat{\partial}_{ia}\sum_{l=2}^{\infty}\sum_{p=0}^{l-1}\sum_{q=0}^{p}\frac{(-1)^{l+p-q}l}{(l+1)!}C_{l-1}\left(l-p-1,p-q,q\right)H(2-q) \\ &\times\left(1-\frac{p-q}{l-1}\right)k_{<a_1\dots a_p}\hat{\partial}_{a_{p+1}\dots a_{l-1}>}\hat{\partial}^q_\tau\left[\frac{\epsilon_{jba}\mathcal{S}^{(p-q)}_{bA_{l-1}}(t-r)}{r}\right]^{[-2]} \\ &+4\left(\hat{\partial}_a - k_a\hat{\partial}_{t^*}\right)\sum_{l=2}^{\infty}\sum_{p=0}^{l-1}\frac{(-1)^{l+p}l}{(l+1)!}C_{l-1}\left(l-p-1,p\right) \\ &\times\left(1-\frac{p}{l-1}\right)k_{<a_1\dots a_p}\hat{\partial}_{a_{p+1}\dots a_{l-1}>}\left[\frac{\epsilon_{iba}\mathcal{S}^{(p)}_{bA_{l-1}}(t-r)}{r}\right]^{[-1]}\end{aligned} \tag{5.28}$$

$$
+\,4k_j\sum_{l=2}^{\infty}\sum_{p=0}^{l-1}\frac{(-1)^{l+p}l}{(l+1)!}C_{l-1}(l-p-1,p)
$$

$$
\times\, k_{<a_1\ldots a_p}\hat{\partial}_{a_{p+1}\ldots a_{l-1}>}\left[\frac{\epsilon_{jba_{l-1}}\mathcal{S}^{(p+1)}_{ibA_{l-2}}(t-r)}{r}\right]^{[-1]}.
$$

Here, again $H(p-q)$ is a Heaviside function defined by equation (1.21) and $C_l(p,q)$ are the polynomial coefficients (1.20).

Relativistic perturbations (5.27) and (5.28) contain two types of integrals along the light ray which has symbolic form as $[F(s)/r]^{[-1]}$ and $[F(s)/r]^{[-2]}$, where $s=t-r$ is the retarded time. In Section 4, we have shown that taking a derivative from an integral $[F(s)/r]^{[-1]}$ reduces the integral to an ordinary function of the retarded time s with no integral dependence on the past history of the propagation of light ray. In order to eliminate the integration in integrals $[F(s)/r]^{[-2]}$ one needs to take three or more derivatives. One can notice that in (5.27) and (5.28) there are terms with the summation index $p=l-1$ or $p=l$ in which the number of the derivatives is less than three that is not sufficient to eliminate the integrals. However, all such integrals are multiplied with a numerical coefficients like $1-p/l$, etc. which gets nil for those values of the summation index p. It effectively eliminates all the integrals in (5.27) and (5.28) in which the integration cannot be eliminated by the partial derivatives. As an example let us examine equation (5.27) for $\underset{(M)}{\Xi^i}(\tau,\xi)$. In this expression the term with the double integral $[\mathcal{I}^{(p-q)}_{A_l}(s)/r]^{[-2]}$ has $l-p+1$ derivatives $\hat{\partial}_a$ and q derivatives $\hat{\partial}_\tau$. After performing the differentiation, the expression $\hat{\partial}_{\langle a_1\ldots a_{l-p+1}\rangle}\hat{\partial}^q_\tau[\mathcal{I}^{(p-q)}_{A_l}(s)/r]^{[-2]}$ can contain the explicit integrals in the following cases: $q=0$ and $p=l-1$; $q=0$ and $p=l$; $q=1$ and $p=l$. However, these integrals are multiplied with $[1-(p-q)/l][1-(p-q)/(l-1)]$ and in any case of p and q values mentioned above either the factor $1-(p-q)/l$ or $1-(p-q)/(l-1)$ will vanish. Similar reasoning is applied to the spin-dependent perturbation $\underset{(S)}{\Xi^i}(\tau,\xi)$ which shows that there is no need to explicitly integrate the spin multipole moments in order to calculate the perturbation of light ray trajectory.

The only integral dependence of the light-ray trajectory perturbation depending on the past history of the propagating photon remains in equation (5.26) which contains the gauge functions ϕ^α and w^α used later in this chapter for interpretation of observable effects caused by the gravitational waves from the isolated system. The past-history dependence of the light-ray trajectory may come into play also through the integrals from mass monopole and spin-dipole terms: $[\mathcal{M}/r]^{[-2]}$ and $[\mathcal{S}/r]^{[-2]}$, if mass and/or spin of the isolated system are not conserved and change as time passes on. The nonconservation can be caused, for example, by emission of gravitational waves carrying away the orbital energy and angular momentum of the system. The past-history contribution of these integrals can be calculated similarly to the eikonal perturbation considered in equations (5.6)–(5.14).

Solution of the light-ray equation with the initial-boundary conditions (3.18) depends on the unit vector $\boldsymbol{k}$ defining direction of the light-ray propagation extrapolated backward in time to the past null infinity. In real practice the light ray is emitted at the point $\boldsymbol{x}_0$ where the source of light is located at time t_0, and it arrives to observer at time t to the point $\boldsymbol{x}$ separated from the source of light by a finite coordinate distance $R = |\boldsymbol{x} - \boldsymbol{x}_0|$. Therefore, solution of the light-ray equations must be expressed in this case in terms of the integrals of the boundary-value problem. It is formulated in terms of the coordinates of the initial, $\boldsymbol{x}_0$, and the final, $\boldsymbol{x}$, positions of the photon

$$\boldsymbol{x}(t) = \boldsymbol{x}\,, \qquad \boldsymbol{x}(t_0) = \boldsymbol{x}_0\,, \tag{5.29}$$

and a unit vector

$$K^i = -\frac{x^i - x_0^i}{|\boldsymbol{x} - \boldsymbol{x}_0|}\,, \tag{5.30}$$

that defines a coordinate direction from the observer toward the source of light and can be interpreted as a unit vector in the Euclidean space in the sense that $\delta_{ij}K^iK^j = 1$.

In what follows it is convenient to make use of the astronomical coordinates $\boldsymbol{x} \equiv x^i = (x^1, x^2, x^3)$ based on a triad of the Euclidean unit vectors $(\boldsymbol{I}_0, \boldsymbol{J}_0, \boldsymbol{K}_0)$ as shown in Figure 2. Vector $\boldsymbol{K}_0$ points from observer toward the isolated system emitting gravitational waves and deflecting light rays. Vectors $\boldsymbol{I}_0$ and $\boldsymbol{J}_0$ lie in the plane being orthogonal to vector $\boldsymbol{K}_0$. The unit vector $\boldsymbol{I}_0$ is directed to the east, and that $\boldsymbol{J}_0$ points toward the north celestial pole. The origin of the coordinate system is chosen to lie at the center of mass of the isolated system.

We will need in our discussion an another reference frame based on a triad of the Euclidean unit vectors $(\boldsymbol{I}, \boldsymbol{J}, \boldsymbol{K})$ that are turned at some angle with respect to vectors $(\boldsymbol{I}_0, \boldsymbol{J}_0, \boldsymbol{K}_0)$. Vector $\boldsymbol{K}$ points from the observer toward the source of light, and vectors $\boldsymbol{I}$ and $\boldsymbol{J}$ lie in the *plane of the sky* defined to be orthogonal to vector $\boldsymbol{K}$. Mutual orientation of the triads is defined by the orthogonal transformation (rigid rotation)

$$\boldsymbol{I}_0 = \boldsymbol{I}\cos\Omega + \boldsymbol{J}\sin\Omega\,, \tag{5.31}$$

$$\boldsymbol{J}_0 = -\boldsymbol{I}\cos\theta\sin\Omega + \boldsymbol{J}\cos\theta\cos\Omega + \boldsymbol{K}\sin\theta\,, \tag{5.32}$$

$$\boldsymbol{K}_0 = \boldsymbol{I}\sin\theta\sin\Omega - \boldsymbol{J}\sin\theta\cos\Omega + \boldsymbol{K}\cos\theta\,, \tag{5.33}$$

where the rotational angles, Ω and θ, are constant.

It is rather straightforward to obtain solution of the boundary value problem (5.29) for light propagation in terms of the unit vector $\boldsymbol{K}$ instead of the unit vector $\boldsymbol{k}$ of the initial-boundary value problem. All what we need, is to convert the unit vector $\boldsymbol{k}$ to $\boldsymbol{K}$ written in terms of spatial coordinates of the points of emission, $\boldsymbol{x}_0$, and observation, $\boldsymbol{x}$, of the light ray. From formula (5.24) one has

$$k^i = -K^i - \beta^i(\tau, \xi)\,, \tag{5.34}$$

where the relativistic correction $\beta^i(\tau, \xi)$ to the vector K^i is derived from the solution of the initial-boundary value problem, It is explicitly defined as follows:

$$\beta^i(\tau, \xi) = \frac{P^i{}_j \left[\underset{(G)}{\Delta \Xi^j} + \underset{(M)}{\Delta \Xi^j} + \underset{(S)}{\Delta \Xi^j} \right]}{|\boldsymbol{x} - \boldsymbol{x}_0|} , \tag{5.35}$$

with $P^i{}_j = \delta^i{}_j - k^i k_j$ being the operator of projection on the plane being orthogonal to vector k^i. Denominator of equation (5.35) contains the distance $R = |\boldsymbol{x} - \boldsymbol{x}_0|$ between observer and source of light which can be very large, thus, making an impression that the relativistic correction $\beta^i(\tau, \xi)$ is probably negligibly small. However, the difference $\Delta\Xi^j(\tau, \xi)$ in the numerator of (5.35) is proportional either to the distance $r = |\boldsymbol{x}|$ between the observer and the isolated system or to the distance $r_0 = |\boldsymbol{x}_0|$ between the source of light and the isolated system. Either one of them or both distances can be comparable with R so that the relativistic correction $\beta^i(\tau, \xi)$ cannot be neglected in general case, and must be taken into account for calculation of relativistic perturbations of light-ray trajectory. Only in a case where observer and source of light reside at extremely large distances on opposite sides of the source of gravitational waves can the relativistic correction β^i be neglected.

6 Observable relativistic effects

In this section, we derive general expressions for four observable relativistic effects – the time delay, the light bending, the frequency shift, and the rotation of the polarization plane.

6.1 Gravitational time delay of light

Relativistic time delayfor light propagating through time-dependent gravitational field can be obtained either directly from expression (5.24) or from the electromagnetic eikonal (3.35) and (5.1) by observing that the eikonal is constant not only on the null hypersurface of the wave front of electromagnetic wave but also along the light rays citepfrolov. Both derivations lead, of course, to the same result

$$t - t_0 = |\boldsymbol{x} - \boldsymbol{x}_0| + \Delta(\tau, \tau_0) , \tag{6.1}$$

$$\Delta(\tau, \tau_0) = \underset{(G)}{\Delta}(\tau, \tau_0) + \underset{(M)}{\Delta}(\tau, \tau_0) + \underset{(S)}{\Delta}(\tau, \tau_0) , \tag{6.2}$$

where $x_0^\alpha = (t_0, \boldsymbol{x}_0)$ are four-coordinates of the point of emission of light, $x^\alpha = (t, \boldsymbol{x})$ are four-coordinates of the point of observation, $\underset{(G)}{\Delta}(\tau, \tau_0)$, $\underset{(M)}{\Delta}(\tau, \tau_0)$ and $\underset{(S)}{\Delta}(\tau, \tau_0)$ are functions describing correspondingly the delay of the electromagnetic signal due to the gauge freedom, mass and spin multipoles of the gravitational field of the isolated system. These functions are expressed as follows:

$$\underset{(G)}{\Delta}\left(\tau,\tau_0\right) = -k_i\left[\underset{(G)}{\Xi^i}(\tau,\xi) - \underset{(G)}{\Xi^i}\left(\tau_0,\xi\right)\right], \tag{6.3}$$

$$\underset{(M)}{\Delta}\left(\tau,\tau_0\right) = -k_i\left[\underset{(M)}{\Xi^i}(\tau,\xi) - \underset{(M)}{\Xi^i}\left(\tau_0,\xi\right)\right], \tag{6.4}$$

$$\underset{(S)}{\Delta}\left(\tau,\tau_0\right) = -k_i\left[\underset{(S)}{\Xi^i}(\tau,\xi) - \underset{(S)}{\Xi^i}\left(\tau_0,\xi\right)\right]. \tag{6.5}$$

There exists a relation between the relativistic perturbations of the eikonal and the light-ray trajectory

$$\psi(\tau,\xi) = -k_i\Xi^i(\tau,\xi), \tag{6.6}$$

where the eikonal perturbation, ψ, reads off equations (5.1)–(5.4). From equations (6.3)–(6.5) one can infer that in the linear approximation the functions describing the time delay are just the projections of vector functions describing the coordinate perturbation of the light-ray trajectory onto the unperturbed direction k^i from the source of light to the observer.

Equation (6.1) defines the light time delay effect in the global time t of the ADM-harmonic reference frame. In order to convert it to observable proper time T of the observer, we assume for simplicity that the observer is in a state of free fall and moves with velocity V^i with respect to the reference frame of the isolated system. Transformation from the ADM-harmonic coordinate time t to the proper time T is made with the help of the standard formula [1, 2]

$$dT^2 = -g_{\alpha\beta}dx^\alpha dx^\beta. \tag{6.7}$$

Substituting the metric tensor expansion $g_{\alpha\beta} = \eta_{\alpha\beta} + h_{\alpha\beta}$, where $h_{\alpha\beta}$ is given by equations (2.21)–(2.24), to equation (6.7) yields

$$\frac{dT}{dt} = \sqrt{1 - V^2 - h_{00}(1+V^2) - 2h_{0i}V^i - h_{ij}^{TT}V^iV^j}. \tag{6.8}$$

In the most simple case, when observer is at rest ($V = 0$) with respect to the ADM-harmonic reference frame equation (6.8) is drastically simplified and depends only on h_{00} component of the metric tensor. Implementation of formula (2.21) for h_{00} and subsequent integration of (6.8) with respect to time then yields

$$T = \left(1 - \frac{\mathcal{M}}{r}\right)(t - t_i), \tag{6.9}$$

where t_i is the initial epoch of observation. Another simple case of equation (6.8) is obtained for observer located at the distance r so large that one can neglect h_{00} and h_{0i} quasi-static perturbations of the metric tensor. Then, the difference between the observer's proper time T and coordinate time t is

$$\frac{dT}{dt} = \sqrt{1 - V^2 - h_{ij}^{TT}V^iV^j}, \tag{6.10}$$

leading in the case of a small velocity to

$$T = \sqrt{1 - V^2}(t - t_i) - \frac{V^i V^j}{\sqrt{1 - V^2}} \int_{t_i}^{t} h_{ij}^{TT} dt . \quad (6.11)$$

The last term on the right-hand side of this equation describes the effect of the gravitational waves on the rate of the proper time of observer which may be important under some specific circumstances but is enormously small in real practice and, as a rule, can be neglected. Nonetheless, it would be interesting to study this effect in more detail as a possible tool to detect gravitational waves emitted, for example, in galactic supernova explosions or by powerful gamma-ray bursts (GRB) in distant galaxies.

6.2 Gravitational deflection of light

The coordinate direction to a source of light measured at the point of observation $\boldsymbol{x}$ is defined by the spatial components of four-vector of photon l^α incoming to an observer from the source of light. This vector depends on the energy (frequency) of photon which is irrelevant for the present section. Normalization of l^α to its frequency eliminates the frequency dependence and brings about a four-vector $p^\alpha = (1, p^i)$ where $p^i = -\dot{x}^i$ and the dot denotes derivative with respect to coordinate time t. Vector Φ^a is null, $p_\alpha p^\alpha = 0$, and is only direction dependent. The spatial part of this vector can be presented as a small deviation from the direction k^i of the unperturbed photon's trajectory

$$p^i = -k^i - \dot{\Xi}^i , \quad (6.12)$$

where the minus sign indicates that the tangent vector p^i is directed from the observer to the source of light.

The coordinate direction p^i is not a directly observable quantity as it is defined with respect to the chosen coordinate grid on the curved spacetime manifold. A real observable vector toward the source of light, $s^\alpha = (1, s^i)$, is defined with respect to the local inertial frame co-moving with the observer [130]. In this frame $s^i = -dX^i/dT$, where T is the observer's proper time, and X^i are the spatial coordinates of the local inertial frame with the observer at its origin. We shall assume for simplicity that the observer is at rest with respect to the global ADM-harmonic coordinates (t, x^i). The case of an observer moving with respect to the ADM–harmonic system with velocity V^i can be treated with the help of the Lorentz transformation which is a straightforward procedure so that we do not discuss it. In case of a static observer, transformation from the global ADM-harmonic coordinates (t, x^i) to the local coordinates (T, X^i) is given by the formulas

$$dT = \Lambda^0{}_0\, dt + \Lambda^0{}_j\, dx^j , \quad dX^i = \Lambda^i{}_0\, dt + \Lambda^i{}_j\, dx^j , \quad (6.13)$$

where the matrix of transformation Λ^{α}_{β} is defined by the requirements of orthonormality

$$g_{\alpha\beta} = \eta_{\mu\nu}\Lambda^{\mu}{}_{\alpha}\Lambda^{\nu}{}_{\beta} \,. \tag{6.14}$$

In particular, the orthonormality condition (6.14) assumes that the spatial angles and lengths at the point of observations are measured with the Euclidean metric δ_{ij}. Because the vector s^{α} is null ($s_{\alpha}s^{\alpha} = 0$) with respect to the Minkowski metric $\eta_{\alpha\beta}$, we conclude that the Euclidean length, $|\boldsymbol{s}|$, of vector s^i is equal to 1. Indeed, one has

$$\eta_{\alpha\beta}s^{\alpha}s^{\beta} = -1 + \boldsymbol{s}^2 = 0 \,. \tag{6.15}$$

Hence, $|\boldsymbol{s}| = 1$.

In the linear approximation with respect to the universal gravitational constant G, the matrix of the transformation is as follows [69, 70]:

$$\begin{aligned}
\Lambda^{0}{}_{0} &= 1 - \frac{1}{2}h_{00}(t, \boldsymbol{x}) \,, \\
\Lambda^{0}{}_{i} &= -h_{0i}(t, \boldsymbol{x}) \,, \\
\Lambda^{i}{}_{0} &= 0 \,, \\
\Lambda^{i}{}_{j} &= \left[1 + \frac{1}{2}h_{00}(t, \boldsymbol{x})\right]\delta_{ij} + \frac{1}{2}h^{TT}_{ij}(t, \boldsymbol{x}) \,.
\end{aligned} \tag{6.16}$$

Using transformation (6.13) we obtain relationship between the observable vector s^i and the coordinate direction p^i

$$s^i = -\frac{\Lambda^{i}{}_{0} - \Lambda^{i}{}_{j}\, p^j}{\Lambda^{0}{}_{0} - \Lambda^{0}{}_{j}\, p^j} \,. \tag{6.17}$$

In the linear approximation this takes the form

$$s^i = \left(1 + h_{00} - h_{0j}p^j\right)p^i + \frac{1}{2}h^{TT}_{ij}\, p^j \,. \tag{6.18}$$

Remembering that vector $|\boldsymbol{s}| = 1$, we find the Euclidean norm of the vector p^i from the relationship

$$|\boldsymbol{p}| = 1 - h_{00} + h_{0j}p^j - \frac{1}{2}h^{TT}_{ij}\, p^i p^j \,, \tag{6.19}$$

which brings equation (6.18) to the form

$$s^i = m^i + \frac{1}{2}P^{ij}m^q h^{TT}_{jq}(t, \boldsymbol{x}) \,, \tag{6.20}$$

where $P^{ij} = \delta^{ij} - k^i k^j$ is the operator of projection onto the plane being orthogonal to k^i, and the Euclidean unit vector $m^i = p^i/|\boldsymbol{p}|$.

Let us now denote by α^i the dimensionless vector describing the total angle of deflection of the light ray measured at the point of observation, and calculated with respect to vector k^i given at past null infinity. It is defined according to [74]

$$\alpha^i(\tau,\xi) = k^i \boldsymbol{k}\cdot\dot{\boldsymbol{\Xi}}(\tau,\xi) - \dot{\Xi}^i(\tau,\xi)\,, \tag{6.21}$$

or

$$\alpha^i(\tau,\xi) = -\,P^i{}_j\,\dot{\Xi}^j(\tau,\xi)\,. \tag{6.22}$$

As a consequence of definitions (6.12) and (6.22), we conclude that

$$m^i = -k^i + \alpha^i(\tau,\xi)\,. \tag{6.23}$$

Accounting for expressions (6.20), (6.23), and (5.34) we obtain the observed direction to the source of light

$$s^i(\tau,\xi) = K^i + \alpha^i(\tau,\xi) + \beta^i(\tau,\xi) + \gamma^i(\tau,\xi)\,, \tag{6.24}$$

where the unit vector K^i is given by equation (5.30), relativistic correction β^i is defined by equation (5.35), and the perturbation

$$\gamma^i(\tau,\xi) = -\frac{1}{2}P^{ij}k^q h^{TT}_{jq}(t,\boldsymbol{x}) \tag{6.25}$$

describes a deformation of the local coordinates of observer with respect to the global ADM-harmonic frame caused by the transverse-traceless part of the gravitational field of the isolated system at the point of observation.

Let two sources of light (quasars, stars, etc.) be observed along the directions $\boldsymbol{s}_1$ and $\boldsymbol{s}_2$ corresponding to two different light rays passing near the isolated gravitating system at the minimal distances corresponding to the (vector) impact parameters, ξ_1 and ξ_2. The angle Ψ between them, measured in the local inertial frame is

$$\cos\Psi = \boldsymbol{s}_1\cdot\boldsymbol{s}_2\,, \tag{6.26}$$

where the dot between the two vectors denotes the usual Euclidean scalar product. It is worth emphasizing that the observed direction s^i to each source of light includes relativistic deflection of the light ray in the form of three perturbations. Two of them, α^i and γ^i, depend only on the quantities taken at the point of observation, but β^i, according to equation (5.35), is also sensitive to the strength of the gravitational field taken at the point of emission of light. This remark reveals that according to relation (6.24) a single gravitational wave signal may cause different angular displacements and/or time delays for different sources of light located at different distances from the source of gravitational waves even if the directions to the light sources are the same.

Without going into further details of the observational procedure we give an explicit expression for the total angle of the light deflection α^i. We have

$$\alpha^i(\tau,\xi) = \underset{(G)}{\alpha^i}(\tau,\xi) + \underset{(M)}{\alpha^i}(\tau,\xi) + \underset{(S)}{\alpha^i}(\tau,\xi)\,, \tag{6.27}$$

where

$$\underset{(G)}{\alpha^i}(\tau,\xi) = -P^{ij}\hat{\partial}_\tau\left(\phi^j + w^j\right), \tag{6.28}$$

$$\begin{aligned}\underset{(M)}{\alpha^i}(\tau,\xi) = &-\frac{2\mathcal{M}}{r}\frac{\xi^i}{y} \\ &- 2\hat{\partial}_i\sum_{l=2}^{\infty}\sum_{p=0}^{l}\sum_{q=0}^{p}\frac{(-1)^{l+p-q}}{l!}C_l(l-p,p-q,q)H(2-q) \\ &\times\left(1-\frac{p-q}{l}\right)\left(1-\frac{p-q}{l-1}\right)k_{<a_1\dots a_p}\hat{\partial}_{a_{p+1}\dots a_l>}\hat{\partial}^q_\tau\left[\frac{\mathcal{I}^{(p-q)}_{A_l}(t-r)}{r}\right]^{[-1]} \\ &- 4P^{ij}\sum_{l=2}^{\infty}\sum_{p=0}^{l-1}\frac{(-1)^{l+p}}{l!}C_{l-2}(l-p-1,p-1) \\ &\times k_{<a_1\dots a_{p-1}}\hat{\partial}_{a_p\dots a_{l-1}>}\left[\frac{\mathcal{I}^{(p)}_{jA_{l-1}}(t-r)}{r}\right],\end{aligned} \tag{6.29}$$

$$\begin{aligned}\underset{(S)}{\alpha^i}(\tau,\xi) = &-2k^j\epsilon_{jba}\hat{\partial}_a\left(\frac{\mathcal{S}_b\xi^i}{yr}\right) + 2P^{ij}\hat{\partial}_a\frac{\epsilon_{jba}\mathcal{S}_b}{r} \\ &+ 4\hat{\partial}_{ia}\sum_{l=2}^{\infty}\sum_{p=0}^{l-1}\sum_{q=0}^{p}\frac{(-1)^{l+p-q}l}{(l+1)!}C_{l-1}(l-p-1,p-q,q)\,H(2-q) \\ &\times\left(1-\frac{p-q}{l-1}\right)k_{<a_1\dots a_p}\hat{\partial}_{a_{p+1}\dots a_{l-1}>}\hat{\partial}^q_\tau\left[\frac{k^j\epsilon_{jba}\mathcal{S}^{(p-q)}_{bA_{l-1}}(t-r)}{r}\right]^{[-1]} \\ &- 4\left(\hat{\partial}_a - k_a\hat{\partial}_{t^*}\right)\sum_{l=2}^{\infty}\sum_{p=0}^{l-1}\frac{(-1)^{l+p}l}{(l+1)!}C_{l-2}(l-p-2,p) \\ &\times k_{<a_1\dots a_p}\hat{\partial}_{a_{p+1}\dots a_{l-1}>}\left[P^{ij}\frac{\epsilon_{jba}\mathcal{S}^{(p)}_{bA_{l-1}}(t-r)}{r}\right] \\ &- 4\sum_{l=2}^{\infty}\sum_{p=0}^{l-1}\frac{(-1)^{l+p}l}{(l+1)!}C_{l-1}(l-p-1,p) \\ &\times k_{<a_1\dots a_p}\hat{\partial}_{a_{p+1}\dots a_{l-1}>}\left[\frac{P^{iq}k^j\epsilon_{jba_{l-1}}\mathcal{S}^{(p+1)}_{qbA_{l-2}}(t-r)}{r}\right].\end{aligned} \tag{6.30}$$

These expressions do not contain any explicit integration along the light ray trajectory because all explicit integrals are either eliminated after taking partial derivatives with respect to the upper limit of the integrals or they are vanish because the numerical coefficient in front of them become nil.

6.3 Gravitational shift of frequency

Exact calculation of the gravitational shift of frequency of electromagnetic wave traveling from the point of emission to observer plays a crucial role for the adequate interpretation of spectral astronomical investigations of high resolution including the astronomical measurements of radial velocities of stars [131], anisotropy of cosmic microwave background radiation (CMBR), and others. In the last decade, the technique for measuring the radial velocity of stars had reached an unprecedented precision of 1 m/sec [132, 133]. At this level the post-Newtonian effects in the orbital motion of spectroscopic binary stars can be measured [134, 135]. Gravitational shift of frequency of light affect the apparent brightness of the observed sources according to equation (3.125). Therefore, it can be important in highly accurate photometric measurements of faint radio sources with large radio telescope like SKA [136].

Let a source of light move with respect to the ADM-harmonic coordinate frame (t, x^i) with velocity $\boldsymbol{V}_0 = \dot{\boldsymbol{x}}_0(t_0)$ and emit continuous electromagnetic radiation at frequency $\nu_0 = 1/(\delta T_0)$, where t_0 and T_0 are the coordinate and proper time of the source of light, respectively. We denote by $\nu = 1/(\delta T)$ the observed frequency of the electromagnetic radiation measured at the point of observation by an observer moving with velocity $\boldsymbol{V} = \dot{\boldsymbol{x}}(t)$ with respect to the ADM-harmonic coordinate frame. In the geometric optics approximation we can consider the increments of the proper time, δT_0 and δT, as infinitesimally small which allows us to operate with them as with differentials [69, 70]. Time delay equation (6.1) can be considered as an implicit function of the emission time $t_0 = t_0(t)$ having the time of observation t as its argument. Because the coordinate and proper time at the points of emission of light and its observation are connected through the metric tensor we conclude that the observed gravitational shift of frequency $1 + z = \nu/\nu_0$ can be defined through the consecutive differentiation of the proper time of the source of light, T_0, with respect to the proper time of the observer, T, [69, 70]

$$1 + z = \frac{dT_0}{dT} = \frac{dT_0}{dt_0}\frac{dt_0}{dt}\frac{dt}{dT} \,. \tag{6.31}$$

Synge calls relation (6.31) the *Doppler effect in terms of frequency* [137, p. 122]. It is fully consistent with the definition of the *Doppler shift in terms of energy* [137, p. 231] when one compares the energy of photon at the points of emission and observation of light. The *Doppler shift in terms of energy* is given by

$$1 + z = \frac{\nu}{\nu_0} = \frac{l_\alpha u^\alpha}{l_{0\alpha} u_0^\alpha} \,, \tag{6.32}$$

where u_0^α, u^α and $l_{0\alpha}$, l_α are 4-velocities of the source of light and observer and four momenta of photon at the points of emission and observation, respectively. It is quite easy to see that both mentioned formulations of the Doppler shift effect are equivalent. Indeed, taking into account that $u^\alpha = dx^\alpha/dT$ and $l_\alpha = \partial\varphi/\partial x^\alpha$, where φ is the phase of the electromagnetic wave (eikonal), we obtain $l_\alpha u^\alpha = d\varphi/dT$, and $l_{0\alpha}u_0^\alpha = d\varphi_0/dT_0$, respectively. Thus, equation (6.32) yields

$$1 + z = \frac{d\varphi}{d\varphi_0}\frac{dT_0}{dT} \, . \tag{6.33}$$

The phase of electromagnetic wave remains constant along the light ray trajectory. For this reason, $d\varphi/d\varphi_0 = 1$ and, hence, equation (6.33) is reduced to equation (6.31) as expected on the ground of physical intuition. Detailed comparison of the two definitions of the Doppler shift and the proof of their identity in general theory of relativity is thoroughly discussed in [54, 138].

We emphasize that in equation (6.31) the time derivative

$$\frac{dT_0}{dt_0} = \left[1 - V_0^2 - (1 + V_0^2)h_{00}(t_0, \boldsymbol{x}_0) - 2V_0^i h_{0i}(t_0, \boldsymbol{x}_0) - V_0^i V_0^j h_{ij}^{TT}(t_0, \boldsymbol{x}_0)\right]^{1/2} \, , \tag{6.34}$$

is taken at the time t_0 at the point of emission of light $\boldsymbol{x}_0$ along the world line of the emitter of light while the time derivative

$$\frac{dt}{dT} = \left[1 - V^2 - (1 + V^2)h_{00}(t, \boldsymbol{x}) - 2V^i h_{0i}(t, \boldsymbol{x}) - V^i V^j h_{ij}^{TT}(t, \boldsymbol{x})\right]^{-1/2} \, , \tag{6.35}$$

is calculated at the time of observation t at the position of observer $\boldsymbol{x}$ along the world line of the observer.

The time derivative dt_0/dt is taken along the light-ray trajectory and calculated from the time delay equation (6.1) where we have to take into account that function $\Delta(\tau, \tau_0)$ depends on times t_0 and t indirectly through the retarded times $s_0 = t_0 - r_0$ and $s = t - r$ that are arguments of the multipole moments of the isolated system and wherein $r_0 \equiv |\boldsymbol{x}_0| = |\boldsymbol{x}(t_0)|$, $r = |\boldsymbol{x}| = |\boldsymbol{x}(t)|$ are functions of time t_0 and t, respectively. Function $\Delta(\tau, \tau_0)$ also depends on the time of the closest approach of light ray, t^*, through variables $\tau = t - t^*$, $\tau_0 = t_0 - t^*$, and on the unit vector $\boldsymbol{k}$. Both t^* and $\boldsymbol{k}$ should be considered as parameters depending on t_0 and t because of the relative motion of the observer with respect to the source of light which causes variation in the relative position of the source of light and observer and, consequently, to the corresponding change in the parameters characterizing trajectory of the light ray, that is in t^* and $\boldsymbol{k}$. Therefore, the function $\Delta(\tau, \tau_0)$ must be viewed as parametrically dependent on four variables $\Delta = \Delta(s, s_0, t^*, \boldsymbol{k})$. Accounting for these remarks the derivative along the light ray reads as follows:

$$\frac{dt_0}{dt} = \frac{1 + \boldsymbol{K}\cdot\boldsymbol{V} - \left\{\frac{\partial s}{\partial t}\frac{\partial}{\partial s} + \frac{\partial s_0}{\partial t}\frac{\partial}{\partial s_0} + \frac{\partial t^*}{\partial t}\frac{\partial}{\partial t^*} + \frac{\partial k^i}{\partial t}\frac{\partial}{\partial k^i}\right\}\Delta(s, s_0, t^*, \boldsymbol{k})}{1 + \boldsymbol{K}\cdot\boldsymbol{V}_0 + \left\{\frac{\partial s}{\partial t_0}\frac{\partial}{\partial s} + \frac{\partial s_0}{\partial t_0}\frac{\partial}{\partial s_0} + \frac{\partial t^*}{\partial t_0}\frac{\partial}{\partial t^*} + \frac{\partial k^i}{\partial t_0}\frac{\partial}{\partial k^i}\right\}\Delta(s, s_0, t^*, \boldsymbol{k})} \, , \tag{6.36}$$

where the unit vector $\boldsymbol{K}$ is defined in (5.30) and where we explicitly show the dependence of function $\Delta(\tau, \tau_0)$ on all parameters which implicitly depend on time.

The time derivative of vector $\boldsymbol{k}$ is calculated using the approximation $\boldsymbol{k} = -\boldsymbol{K}$ and formula (5.30) where the coordinates of the source of light, $\boldsymbol{x}_0(t_0)$, and of the observer, $\boldsymbol{x}(t)$, are considered as functions of time. These derivatives are

$$\frac{\partial k^i}{\partial t} = \frac{(\mathbf{k}\times(\boldsymbol{V}\times\mathbf{k}))^i}{R}\,, \qquad \frac{\partial k^i}{\partial t_0} = -\frac{(\mathbf{k}\times(\boldsymbol{V}_0\times\mathbf{k}))^i}{R}\,, \tag{6.37}$$

where $R = |\boldsymbol{x} - \boldsymbol{x}_0|$ is the coordinate distance between the observer and the source of light.

Time derivatives of the retarded times s and s_0 with respect to t and t_0 are calculated from their definitions, $s = t - r$ and $s_0 = t_0 - r_0$, where we have to take into account that the spatial position of the point of observation is connected to the point of emission of light by the unperturbed trajectory of light, $\boldsymbol{x} = \boldsymbol{x}_0 + \boldsymbol{k}\,(t - t_0)$. More explicitly, we use for the calculations the following relations:

$$s = t - |\boldsymbol{x}_0(t_0) + \boldsymbol{k}\,(t - t_0)|\,, \tag{6.38}$$

$$s_0 = t_0 - |\boldsymbol{x}_0(t_0)|\,, \tag{6.39}$$

where the unit vector $\boldsymbol{k} = \boldsymbol{k}(t, t_0)$ must be also considered as a function of two arguments t, t_0 with its derivatives given by equation (6.37). It is instructive to notice that relation (6.38) combines two characteristics of the null cone – the first one is related to the propagation of gravitational field from the isolated system to an observer, and the second one is related to the propagation of light from the source of light to the same observer. Equation (6.39) describes a null cone characteristic corresponding to the propagation of gravitational field from the isolated system to the point of emission of light. Calculation of the infinitesimal variations of equations (6.38), (6.39) immediately gives a set of partial derivatives

$$\frac{\partial s}{\partial t} = 1 - \boldsymbol{k}\cdot\boldsymbol{N} - (\boldsymbol{k}\times\boldsymbol{V})\cdot(\boldsymbol{k}\times\boldsymbol{N})\,, \tag{6.40}$$

$$\frac{\partial s}{\partial t_0} = (1 - \boldsymbol{k}\cdot\boldsymbol{V}_0)(\boldsymbol{k}\cdot\boldsymbol{N})\,, \tag{6.41}$$

$$\frac{\partial s_0}{\partial t_0} = 1 - \boldsymbol{V}_0\cdot\boldsymbol{N}_0\,, \tag{6.42}$$

$$\frac{\partial s_0}{\partial t} = 0\,, \tag{6.43}$$

where $N^i = x^i/r$ and $N_0^i = x_0^i/r_0$ are the unit vectors directed from the isolated system to the observer and to the source of light, respectively.

Time derivatives of the parameter $t^* = t_0 - \boldsymbol{k}\cdot\boldsymbol{x}_0(t_0)$, where again $\boldsymbol{k} = \boldsymbol{k}(t, t_0)$, read

$$\frac{\partial t^*}{\partial t_0} = 1 - \boldsymbol{k}\cdot\boldsymbol{V}_0 + \frac{\boldsymbol{V}_0\cdot\boldsymbol{\xi}}{R}\,, \qquad \frac{\partial t^*}{\partial t} = -\frac{\boldsymbol{V}\cdot\boldsymbol{\xi}}{R}\,. \tag{6.44}$$

Terms of the order $|\xi|/R$ in both formulas are produced by the time derivatives of vector $\boldsymbol{k}$.

In what follows we shall restrict ourselves with a static case of an observer and a source of light, that is we shall assume velocities $V = V_0 = 0$. Taking into account this restriction in equations (6.37)–(6.44) we expand denominator of (6.36) leaving only the linear with respect to the universal gravitational constant G terms. Reducing, then, similar terms allows us to simplify (6.36) to

$$\frac{dt_0}{dt} = 1 - \left\{\frac{\partial}{\partial s} + \frac{\partial}{\partial s_0} + \frac{\partial}{\partial t^*}\right\}\Delta(s, s_0, t^*, \boldsymbol{k})\,. \tag{6.45}$$

Function $\Delta = \Delta(s, s_0, t^*, \boldsymbol{k})$ is defined by (6.2) which (in the case of the static observer and the source of light) depends on the retarded times s, s_0 and the time of the closest approach t^* through the arguments (see (4.6)) $y = s - t^*$, $y_0 = s_0 - t^*$, and t^*, that is

$$\Delta(s, s_0, t^*, \boldsymbol{k}) \equiv \Delta(y, y_0, t^*) = \psi(y, t^*) - \psi(y_0, t^*)\,, \tag{6.46}$$

where $\psi = -k^i \Xi$ is the relativistic perturbation of the eikonal defined in (6.6). This particular dependence of ψ on its arguments means that the partial derivative of $\psi(y, t^*)$ with respect to y and and that of $\psi(y_0, t^*)$ with respect to y_0 vanish in (6.45) which is reduced to a simpler form

$$\frac{dt_0}{dt} = 1 - \frac{\partial\psi(y, t^*)}{\partial t^*} + \frac{\partial\psi(y_0, t^*)}{\partial t^*}\,. \tag{6.47}$$

The partial time derivative from $\psi(y, t^*)$ is found by differentiating relations (5.2)–(5.4)

$$\frac{\partial\psi(y, t^*)}{\partial t^*} = \frac{\partial\psi_{(G)}(y, t^*)}{\partial t^*} + \frac{\partial\psi_{(M)}(y, t^*)}{\partial t^*} + \frac{\partial\psi_{(S)}(y, t^*)}{\partial t^*} \tag{6.48}$$

where

$$\frac{\partial\psi_{(G)}(y, t^*)}{\partial t^*} = (k^i\frac{\partial\phi^i}{\partial t^*} - \frac{\partial\phi^0}{\partial t^*}) + \left(k^i\frac{\partial w^i}{\partial t^*} - \frac{\partial w^0}{\partial t^*}\right)\,, \tag{6.49}$$

$$\frac{\partial\psi_{(M)}(y, t^*)}{\partial t^*} = 2\sum_{l=2}^{\infty}\sum_{p=0}^{l-1}\frac{(-1)^{l+p}}{l!}C_l(l-p, p)\left(1 - \frac{p}{l}\right) \tag{6.50}$$
$$\times\left\{\left(1 + \frac{p}{l-1}\right)k_{<a_1\ldots a_p}\hat{\partial}_{a_{p+1}\ldots a_l>}\left[\frac{\mathcal{I}^{(p+1)}_{A_l}(t-r)}{r}\right]^{[-1]}\right.$$
$$\left. - \frac{2p}{l-1}k_{<a_1\ldots a_{p-1}}\hat{\partial}_{a_p\ldots a_{l-1}>}\left[\frac{k^i\mathcal{I}^{(p+1)}_{iA_{l-1}}(t-r)}{r}\right]^{[-1]}\right\}$$

$$\frac{\partial\psi_{(S)}(y,t^*)}{\partial t^*}=4\hat{\partial}_a\sum_{l=2}^{\infty}\sum_{p=0}^{l-1}\frac{(-1)^{l+p}l}{(l+1)!}C_{l-1}(l-p-1,p)\times\left(1-\frac{p}{l-1}\right)k_{<a_1\dots a_p}\hat{\partial}_{a_{p+1}\dots a_{l-1}>}\left[\frac{k^i\epsilon_{iba}\mathcal{S}^{(p+1)}_{bA_{l-1}}(t-r)}{r}\right]^{[-1]}. \tag{6.51}$$

When deriving these equations we assumed that the time evolution of both the mass $\mathcal{M}$ and the angular momentum $\mathcal{S}^i$ of the isolated system can be neglected. In the opposite case, the right side of equations (6.50) and (6.51) would contain also time derivatives from the mass and the angular momentum. Partial derivative from function $\psi(y_0,t^*)$ is obtained from equations (6.48)–(6.51) after replacements $y\to y_0$, $t\to t_0$, and $r\to r_0$. Equations (6.49)–(6.51) do not contain explicit integrals along the light ray trajectory because all of the explicit integrals vanish after taking the partial derivatives in these equations.

Frequency shift, z, is given by equation (6.31). In the case when both observer and source of light are at rest with respect to the ADM-harmonic reference frame, the frequency (energy) of a photon propagating through the gravitational potential of the isolated astronomical system changes in accordance with

$$\frac{\nu}{\nu_0}=\sqrt{\frac{1-h_{00}(t_0,\boldsymbol{x}_0)}{1-h_{00}(t,\boldsymbol{x})}}\left[1-\frac{\partial\psi(y,t^*)}{\partial t^*}+\frac{\partial\psi(y_0,t^*)}{\partial t^*}\right]. \tag{6.52}$$

In the linear with respect to G approximation, equation (6.52) is simplified

$$\frac{\delta\nu}{\nu_0}=\frac{1}{2}h_{00}(t,\boldsymbol{x})-\frac{1}{2}h_{00}(t_0,\boldsymbol{x}_0)-\frac{\partial\psi(y,t^*)}{\partial t^*}+\frac{\partial\psi(y_0,t^*)}{\partial t^*}, \tag{6.53}$$

where $\delta\nu\equiv\nu-\nu_0$. The time of the closest approach t^* enters equation (6.53) explicitly. At the first glance, the reader may think that it must be known in order to calculate the gravitational shift of frequency. However, the partial derivative with respect to t^* have to be understood in the sense of equation (3.30) which makes it evident that the time derivative with respect to t^* taken on the light-ray path is, in fact, the time derivative with respect to time t taken before the using of the light-ray trajectory substitution. Thus, the gravitational shift of frequency is actually not sensitive to the time of the closest approach t^*, and can be recast to the form which is more suitable for practical applications,

$$\frac{\delta\nu}{\nu_0}=\frac{1}{2}h_{00}(t,\boldsymbol{x})-\frac{1}{2}h_{00}(t_0,\boldsymbol{x}_0)-\frac{\partial\psi(t,\boldsymbol{x})}{\partial t}+\frac{\partial\psi(t_0,\boldsymbol{x}_0)}{\partial t_0}, \tag{6.54}$$

where $\psi(t,\boldsymbol{x})$ and $\psi(t_0,\boldsymbol{x}_0)$ are relativistic perturbations of the electromagnetic eikonal taken at the points of observation and emission of light, respectively. These time derivatives are given by the same equations (6.48)–(6.51) after making use of (3.30).

Equation (6.54) has been derived by making use of the definition (6.31). It is straightforward to prove that the definition (6.32) brings about the same result. Indeed, in the case of the static observer and the source of light their 4 velocities are, $u^\alpha = (dt/dT, 0, 0, 0)$ and $u_0^\alpha = (dt_0/dT_0, 0, 0, 0)$, and the wave vector $l_\alpha = \omega(k_\alpha + \partial_\alpha \psi)$ (see equation (3.36)) where ω is a constant frequency of light wave which is conserved along the light ray because of the equation of the parallel transport (3.14). Substituting these relations to (6.32) leads immediately to equation (6.54) as expected.

Physical interpretation of the relativistic frequency shift given by equation (6.54) is straightforward although the calculation of different components of the time derivatives from the eikonal are tedious. The first two terms on the left-hand side of (6.54) reads

$$\frac{1}{2}h_{00}(t, \boldsymbol{x}) - \frac{1}{2}h_{00}(t_0, \boldsymbol{x}_0) = \frac{G\mathcal{M}}{r} - \frac{G\mathcal{M}}{r_0}, \tag{6.55}$$

and represents the difference between the values of the spherically symmetric part of the Newtonian gravitational potential of the isolated system taken at the point of observation and emission of light. The time derivatives of the eikonal depending on the mass and spin multipole moments of the isolated system are given in (6.50) and (6.51). Scrutiny examination of these equations reveal that these components of the frequency shift depend on the first and higher order time derivatives of the multipole moments which vanish in the stationary case. The gravitational frequency shift contains the gauge-dependent contribution as well. This contribution is given by equation (6.49) and can be calculated by making use of equations (3.66)–(3.73). The result is as follows:

$$k^i \frac{\partial w^i}{\partial t^*} - \frac{\partial w^0}{\partial t^*} = k^i \nabla_i \sum_{l=2}^{\infty} \frac{(-1)^l}{l!} \left[\frac{\mathcal{I}_{A_l}^{(-1)}(t-r)}{r}\right]_{,A_l} - \sum_{l=2}^{\infty} \frac{(-1)^l}{l!} \left[\frac{\mathcal{I}_{A_l}(t-r)}{r}\right]_{,A_l} \tag{6.56}$$

$$- 4\sum_{l=2}^{\infty} \frac{(-1)^l}{l!} \left[\frac{k^i \dot{\mathcal{I}}_{iA_{l-1}}(t-r)}{r}\right]_{,A_{l-1}} \tag{6.57}$$

$$+ 4\sum_{l=2}^{\infty} \frac{(-1)^l l}{(l+1)!} \left[\frac{k^i \epsilon_{iba} \mathcal{S}_{bA_{l-1}}(t-r)}{r}\right]_{,aA_{l-1}},$$

$$k^i \frac{\partial \phi^i}{\partial t^*} - \frac{\partial \phi^0}{\partial t^*} = -2\sum_{l=2}^{\infty}\sum_{p=1}^{l}\sum_{q=1}^{p} \frac{(-1)^{l+p-q}}{l!} C_l(l-p, p-q, q)\left(1 - \frac{p-q}{l}\right) \tag{6.58}$$

$$\times \left\{\left(1 + \frac{p-q}{l-1}\right) k_{<a_1 \ldots a_p} \hat{\partial}_{a_{p+1} \ldots a_l>} \hat{\partial}_\tau^{q-1} \left[\frac{\mathcal{I}_{A_l}^{(p-q+1)}(t-r)}{r}\right]\right.$$

$$\left. - 2\frac{p-q}{l-1} k_{<a_1 \ldots a_{p-1}} \hat{\partial}_{a_p \ldots a_{l-1}>} \hat{\partial}_\tau^{q-1} \left[\frac{k^i \mathcal{I}_{iA_{l-1}}^{(p-q+1)}(t-r)}{r}\right]\right\}$$

$$
\begin{aligned}
& - 4\hat{\partial}_a \sum_{l=2}^{\infty} \sum_{p=1}^{l-1} \sum_{q=1}^{p} \frac{(-1)^{l+p-q} l}{(l+1)!} C_{l-1}(l-p-1, p-q, q) \\
& \times \left(1 - \frac{p-q}{l-1}\right) k_{<a_1 \ldots a_p} \hat{\partial}_{a_{p+1} \ldots a_{l-1}>} \hat{\partial}_\tau^{q-1} \left[\frac{k^i \epsilon_{iab} \mathcal{S}^{(p-q+1)}_{bA_{l-1}}(t-r)}{r} \right] .
\end{aligned}
$$

Time derivatives $\hat{\partial}_{t_0^*}(k^i w^i - w^0)$ and $\hat{\partial}_{t_0^*}(k^i \varphi^i - \varphi^0)$ can be obtained from equations (6.56) and (6.58) after making replacements $t \to t_0$, $r \to r_0$, and $\tau \to \tau_0$.

6.4 Gravity-induced rotation of the plane of polarization of light

Any kind of axisymmetric gravitational field induces a relativistic effect of the rotation of the polarization plane of an electromagnetic wave propagating through this field. To some extent this effect is similar to Faraday's effect in electrodynamics [108]. The Faraday effect is caused by the presence of magnetic field along the trajectory of propagation of electromagnetic wave while the gravity-induced rotation of the plane of polarization of light is caused by the presence of the, so-called, gravitomagnetic field associated with the angular momentum and spin-type multipoles of the isolated system [139, 140]. This gravitomagnetic effect was first discussed by Skrotskii ([119, 120]) and a number of other researches [121, 122, 141–145]. Recently, we have studied the Skrotskii effect caused by a spinning body moving arbitrarily fast and derived the Lorentz-invariant expression for this effect [146]. In this chapter we further generalize the Skrotskii effect to the case of an isolated system emitting gravitational waves of arbitrary multipolarity.

We consider the parallel transport of the reference polarization tetrad $e^\alpha_{\ (\beta)}$ defined by equations (3.84) along the light ray. We assume for simplicity that at the past null infinity the spatial vectors of the tetrad, $e^0_{\ (i)}$, coincide with the spatial unit vectors of the coordinate tetrad defined by equation (3.80). The parallel transport of the tetrad along the light ray is defined by equation (3.96). We are interested only in solving equation (3.96) for vectors $e^\alpha_{\ (n)}$ $(n = 1, 2)$ that are used in description of polarization of light. The propagation equation for the spatial components $e^i_{\ (n)}$ can be written in the following form:

$$
\frac{d}{d\tau}\left[e^i_{\ (n)} + \frac{1}{2} h_{ij} e^p_{\ (n)} \right] + \epsilon_{ijp} e^j_{\ (n)} \Omega^p = 0 \,, \quad (n = 1, 2) \tag{6.59}
$$

where the angular velocity vector

$$
\Omega^i = -\frac{1}{2} \epsilon_{ijp} \partial_j \left(h_{p\alpha} k^\alpha \right) , \tag{6.60}
$$

describes the rate of the rotation of the plane of polarization of electromagnetic wave caused by the presence of the gravitomagnetic field. As soon as equation (6.59) is solved, the time component $e^0_{(n)}$ of the tetrad is obtained from the orthogonality condition, $l_\alpha e^\alpha_{(n)} = 0$, which implies that

$$e^0_{(n)} = k_i e^i_{(n)} + h_{0i} e^i_{(n)} + h_{ij} k^i e^j_{(n)} + \delta_{ij} \dot{\Xi}^i e^j_{(n)} , \tag{6.61}$$

where the relativistic perturbation $\dot{\Xi}^i$ of the light-ray trajectory is given in equation (5.15).

Let us decompose vector Ω^i into three components that are parallel and perpendicular to the unit vector k^i. We can use in the first approximation the well-known decomposition of the Kroneker symbol in the orthonormal basis

$$\delta^{ij} = a^i a^j + b^i b^j + k^i k^j , \tag{6.62}$$

where $(\boldsymbol{a}, \boldsymbol{b}, \boldsymbol{k})$ are three orthogonal unit vectors of the reference tetrad at infinity. Decomposition of Ω^i is, then, given by

$$\Omega^i = (\mathbf{a} \cdot \boldsymbol{\Omega}) a^i + (\mathbf{b} \cdot \boldsymbol{\Omega}) b^i + (\boldsymbol{k} \cdot \boldsymbol{\Omega}) k^i . \tag{6.63}$$

Taking into account that at any point on the light-ray trajectory, vectors $e^i_{(1)} = a^i + O(h_{\alpha\beta})$, $e^i_{(2)} = b^i + O(h_{\alpha\beta})$, one obtains equations of the parallel transport of these vectors (6.59) in the following form:

$$\frac{d}{d\tau}\left[e^i_{(1)} + \frac{1}{2} h_{ij} e^j_{(1)}\right] - (\boldsymbol{k} \cdot \boldsymbol{\Omega}) e^i_{(2)} = 0 , \tag{6.64}$$

$$\frac{d}{d\tau}\left[e^i_{(2)} + \frac{1}{2} h_{ij} e^j_{(2)}\right] + (\boldsymbol{k} \cdot \boldsymbol{\Omega}) e^i_{(1)} = 0 , \tag{6.65}$$

where equalities $\varepsilon_{ijl} e^i_{(1)} k^l = -e^i_{(2)} + O(h_{\alpha\beta})$, and $\varepsilon_{ijl} e^i_{(2)} k^l = e^i_{(1)} + O(h_{\alpha\beta})$ have been used.

Solutions of equations (6.64) and (6.65) in the linear approximation with respect to the (post-Newtonian) angular velocity Ω^i read

$$e^i_{(1)} = a^i - \frac{1}{2} h_{ij} a^j + \left(\int_{-\infty}^{\tau} \boldsymbol{k} \cdot \boldsymbol{\Omega}(\sigma)\, d\sigma \right) b^i , \tag{6.66}$$

$$e^i_{(2)} = b^i - \frac{1}{2} h_{ij} b^j - \left(\int_{-\infty}^{\tau} \boldsymbol{k} \cdot \boldsymbol{\Omega}(\sigma)\, d\sigma \right) a^i , \tag{6.67}$$

where the second term on the right-hand side of (6.66) and (6.67) preserve orthogonality of vectors $e^i_{(1)}$ and $e^i_{(2)}$ in the presence of gravitational field while the last term describes the Skrotskii effect which is a small rotation of each of the vectors at the angle

$$\Phi(\tau) = \int_{-\infty}^{\tau} \boldsymbol{k} \cdot \boldsymbol{\Omega}\, d\sigma \tag{6.68}$$

about the direction of light propagation, $\boldsymbol{k}$, in the local plane of vectors $\boldsymbol{e}_{(1)}$ and $\boldsymbol{e}_{(2)}$. It is worth noting that the Euclidean dot product $\boldsymbol{k}\cdot\boldsymbol{\Omega}$ can be expressed in terms of partial differentiation with respect to the vector ξ^i of the impact parameter. This can be done by making use of equation (3.31) and noting that $\varepsilon_{ijp}k^jk^p\equiv 0$, so that

$$\boldsymbol{k}\cdot\boldsymbol{\Omega}=\frac{1}{2}k^\alpha k^j\varepsilon_{jpq}\hat{\partial}_q h_{\alpha p}\,. \tag{6.69}$$

Hence, the transport equation for the angle Φ assumes the following form:

$$\frac{d\Phi}{d\tau}=\frac{1}{2}k^\alpha k^j\epsilon_{jpq}\hat{\partial}_q h_{\alpha p}\,. \tag{6.70}$$

This equation can be split in three, linearly independent parts corresponding to the contributions of the gauge, $\Phi_{(G)}$, the mass, $\Phi_{(M)}$, and the spin, $\Phi_{(S)}$, multipoles of the gravitational field to the Skrotskii effect. Specifically, we have

$$\Phi=\Phi_{(G)}+\Phi_{(M)}+\Phi_{(S)}+\Phi_0\,, \tag{6.71}$$

where Φ_0 is a constant angle characterizing the initial orientation of the polarization ellipse of the electromagnetic wave in the reference plane formed by the $\boldsymbol{e}_{(1)}$ and $\boldsymbol{e}_{(2)}$ vectors.

The gauge-dependent part of the Skrotskii effect is easily integrated so that we obtain

$$\Phi_{(G)}=\frac{1}{2}k^j\epsilon_{jpq}\hat{\partial}_q\left(w^p+\chi^p\right)\,, \tag{6.72}$$

where the gauge vector functions $w^i=w^i_{(M)}+w^i_{(S)}$ are given in equations (2.20). The gauge functions χ^i appear in the process of integration of the equation of the parallel transport. They can be linearly decomposed in two parts corresponding to the mass and spin multipoles:

$$\chi^i=\chi^i_{(M)}+\chi^i_{(S)}\,, \tag{6.73}$$

where

$$\chi^i_{(M)}=4\sum_{l=2}^{\infty}\sum_{p=1}^{l-1}\sum_{q=1}^{p}\frac{(-1)^{l+p-q}}{l!}C_{l-1}(l-p-1,p-q,q)\left(1-\frac{p-q}{l-1}\right)$$
$$\times\, k_{<a_1\dots a_p}\hat{\partial}_{a_{p+1}\dots a_{l-1}>}\hat{\partial}_\tau^{q-1}\left[\frac{\mathcal{I}^{(p-q+1)}_{iA_{l-1}}(t-r)}{r}\right]\,, \tag{6.74}$$

$$\chi^i_{(S)}=-4\sum_{l=1}^{\infty}\sum_{p=0}^{l-1}\sum_{q=0}^{p}\frac{(-1)^{l+p-q}l}{(l+1)!}C_{l-1}(l-p-1,p-q,q)\left[1-\frac{p-q}{l-1}H(l-1)\right]$$
$$\times\left[H(q)(\hat{\partial}_a-k_a\hat{\partial}_{t^*})+k_a\hat{\partial}_\tau\right]k_{<a_1\dots a_p}\hat{\partial}_{a_{p+1}\dots a_{l-1}>}\hat{\partial}_\tau^{q-1}\left[\frac{\epsilon_{iba}\mathcal{S}^{(p-q)}_{bA_{l-1}}(t-r)}{r}\right] \tag{6.75}$$

$$+4\sum_{l=3}^{\infty}\sum_{p=1}^{l-1}\sum_{q=1}^{p}\frac{(-1)^{l+p-q}l}{(l+1)!}C_{l-1}(l-p-1,p-q,q)\left(1-\frac{p}{l}\right)\left(1-\frac{q}{p}\right)$$
$$\times k_{<a_1\dots a_p}\hat{\partial}_{a_{p+1}\dots a_{l-1}>a}\hat{\partial}_{\tau}^{q-1}\left[\frac{\epsilon_{baa_{l-1}}\mathcal{S}^{(p-q)}_{ibA_{l-2}}(t-r)}{r}\right].$$

The gauge-dependent equation (6.72) has the following, more explicit form:

$$\frac{1}{2}k^j\epsilon_{jmn}\hat{\partial}_n\chi^m_{(M)}=2\sum_{l=2}^{\infty}\sum_{p=1}^{l}\sum_{q=1}^{p}\frac{(-1)^{l+p-q}}{l!}\frac{l-p+q}{(l-1)}\frac{p-q}{p}C_l(l-p,p-q,q)\tag{6.76}$$
$$\times k_{<a_1\dots a_p}\hat{\partial}_{a_{p+1}\dots a_l>j}\hat{\partial}_{\tau}^{q-1}\left[\frac{\epsilon_{jba_l}\mathcal{J}^{(p-q)}_{bL-1}(t-r)}{r}\right]^{[-1]},$$

$$\frac{1}{2}k^j\epsilon_{jmn}\hat{\partial}_n\chi^m_{(S)}=2\sum_{l=1}^{\infty}\sum_{p=0}^{l}\sum_{q=0}^{p}\frac{(-1)^{l+p-q}l}{(l+1)!}C_l(l-p,p-q,q)\tag{6.77}$$
$$\times\left\{H(q)\left(1-\frac{p}{l}\right)\left[1+H(l-1)\left(\frac{l-p-1}{l-1}-\frac{p-q}{l-1}\right)\right]\hat{\partial}^2_{t^*}\right.$$
$$-\left[\frac{l-p}{l}+2\frac{p-q}{l}-H(l-1)\frac{p-q}{l}\left(\frac{l-p}{l-1}+2\frac{p-q-1}{l-1}\right)\right]\hat{\partial}_{t^*\tau}$$
$$\left.+\frac{p-q}{l}\left[1-H(l-1)\frac{p-q-1}{l-1}\right]\hat{\partial}^2_{\tau}\right\}$$
$$\times k_{<a_1\dots a_p}\hat{\partial}_{a_{p+1}\dots a_l>}\hat{\partial}_{\tau}^{q-1}\left[\frac{\mathcal{S}^{(p-q-1)}_{L}(t-r)}{r}\right]^{[-1]},$$

$$\frac{1}{2}k^j\epsilon_{jmn}\hat{\partial}_n w^m_{(M)}=-2\sum_{l=2}^{\infty}\sum_{p=0}^{l}\sum_{q=0}^{p}\frac{(-1)^{l+p-q}}{l!}\frac{p-q}{p}C_l(l-p,p-q,q)\tag{6.78}$$
$$\times k_{<a_1\dots a_p}\hat{\partial}_{a_{p+1}\dots a_l>j}\hat{\partial}_{\tau}^{q}\left[\frac{\epsilon_{jba_l}\mathcal{J}^{(p-q-1)}_{bL-1}(t-r)}{r}\right],$$

$$\frac{1}{2}k^j\epsilon_{jmn}\hat{\partial}_n w^m_{(S)}=-2\sum_{l=1}^{\infty}\sum_{p=0}^{l}\sum_{q=0}^{p}\frac{(-1)^{l+p-q}l}{(l+1)!}C_l(l-p,p-q,q)\tag{6.79}$$
$$\times\left\{\left(1-\frac{p}{l}\right)\hat{\partial}^2_{t^*}-\left(\frac{l-p}{l}+2\frac{p-q}{l}\right)\hat{\partial}_{t^*\tau}+\frac{p-q}{l}\hat{\partial}^2_{\tau}\right\}$$
$$\times k_{<a_1\dots a_p}\hat{\partial}_{a_{p+1}\dots a_l>}\hat{\partial}_{\tau}^{q}\left[\frac{\mathcal{S}^{(p-q-2)}_{L}(t-r)}{r}\right].$$

The reader can notice the presence of integrals on the right-hand side of equations (6.76) and (6.77). The integrals are actually not supposed to be calculated explicitly since there is sufficient number of partial derivatives in front of them which cancels the integration in correspondence with the rules of differentiation of such integrals which have been explained in Section 4.

The Skrotskii effect due to the mass-type multipoles of the isolated system is given by

$$\Phi_{(M)}(\tau) = 2\sum_{l=2}^{\infty}\sum_{p=0}^{l}\frac{(-1)^{l+p}}{l!}C_l(l-p,p)\frac{l-p}{l-1}k_{<a_1...a_p}\hat{\partial}_{a_{p+1}...a_l>j}\left[\frac{\epsilon_{jba_l}\mathcal{I}^{(p)}_{bA_{l-1}}(t-r)}{r}\right]^{[-1]}. \tag{6.80}$$

The gravitational field of the spin-type multipoles rotates the polarization plane of the electromagnetic wave at the following angle:

$$\Phi_{(S)}(\tau) = 2\sum_{l=1}^{\infty}\sum_{p=0}^{l}\frac{(-1)^{l+p}l}{(l+1)!}C_l(l-p,p)\left(1-\frac{p}{l}\right) \tag{6.81}$$
$$\times\left[1+H(l-1)\left(1-\frac{2p}{l-1}\right)\right]k_{<a_1...a_p}\hat{\partial}_{a_{p+1}...a_l>}\left[\frac{\mathcal{S}^{(p+1)}_{A_l}(t-r)}{r}\right]^{[-1]}$$

Integrals in equations (6.80) and (6.81) are eliminated after taking at least one partial derivative so that we do not need to integrate.

7 Light propagation through the field of gravitational lens

This section considers propagation of light in a special case of gravitational lens approximation when the impact parameter d of the light ray with respect to an isolated system, is much smaller than both the distance r_0 from the isolated system to the

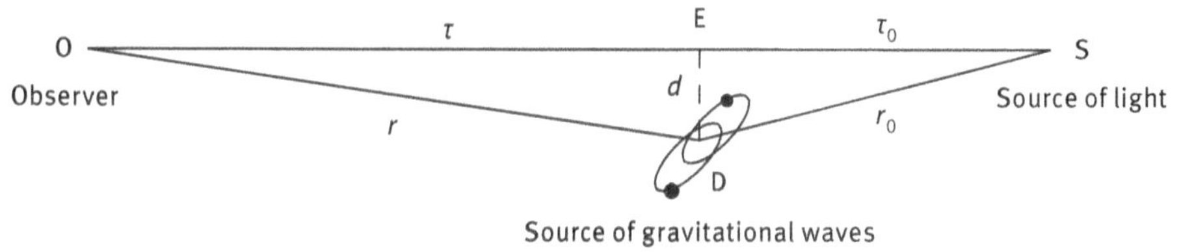

Fig. 3. Relative configuration of an observer (O), a source of light (S), and a localized system emitting gravitational waves (D). Static part of gravitational field of the localized system and gravitational waves deflect light rays which are emitted at the moment t_0 at the point S and received at the moment t at the point O. The point E on the line OS corresponds to the moment t^* of the closest approach of a light ray to the deflector D. We denoted the distances as follows: $OS = R$, $DO = r$, $DS = r_0$, the impact parameter $DE = d$, $OE = \tau > 0$, $ES = \tau_0 = \tau - R < 0$. The impact parameter d is much smaller as compared with all other distances.

source of light and distance r from the isolated system to observer, as illustrated in Figure 3.

7.1 Small parameters and asymptotic expansions

In the case of a small impact parameter of the light ray with respect to the isolated system the near-zone gravitational field of the system strongly affects propagation of the light ray only when the light particle (photon) moves in the close proximity to the system where the effects on the light ray propagation caused by gravitational waves are suppressed [99, 100]. In what follows, we assume that the impact parameter d of the light ray is small as compared with both distances r and r_0, that is, $d \ll \min[r, r_0]$ (see Figure 3). This assumption allows us to introduce two small parameters: $\varepsilon \equiv d/r$ and $\varepsilon_0 \equiv d/r_0$. If the light ray is propagated through the near-zone of the isolated system one more small parameter can be introduced, $\varepsilon_\lambda \equiv d/\lambda$, where λ is the characteristic wavelength of the gravitational radiation emitted by the system proportional to the product of the speed of gravity ($= c$ in general relativity) and the characteristic period of matter oscillations in the isolated system. Parameters ε, ε_0, and ε_λ are not physically correlated. Parameter ε_λ is used for the post-Newtonian expansion of the equations describing the observed effects – deflection of light, time delay, etc. This expansion can be also viewed as a Taylor expansion with respect to the parameter v/c, where v is the characteristic speed of motion of matter composing the isolated system and c is the speed of propagation of gravity. If the light ray does not enter the near zone of the system the parameter ε_λ is not small and the post-Newtonian expansion cannot be performed.

The small-impact-parameter expansions for the retarded time variables y and y_0 yield:

$$y = \sqrt{r^2 - d^2} - r = -d\left(\frac{\varepsilon}{2} - \sum_{k=2}^{\infty} \mathcal{C}_k \varepsilon^{2k-1}\right), \tag{7.1}$$

$$y_0 = -\sqrt{r_0^2 - d^2} - r_0 = -2r_0 + d\left(\frac{\varepsilon_0}{2} - \sum_{k=2}^{\infty} \mathcal{C}_k \varepsilon_0^{2k-1}\right), \tag{7.2}$$

and

$$\frac{1}{yr} = -\frac{1}{d^2}\left(2 + \sum_{k=1}^{\infty} \mathcal{C}_k \varepsilon^{2k}\right), \tag{7.3}$$

$$\frac{1}{y_0 r_0} = \frac{1}{d^2} \sum_{k=1}^{\infty} \mathcal{C}_k \varepsilon_0^{2k}, \tag{7.4}$$

where the numerical coefficients entering the expansions are

$$\mathcal{C}_k = \frac{(-1)^k}{k!}\frac{1}{2}\left(\frac{1}{2} - 1\right) \cdot \ldots \cdot \left(\frac{1}{2} - k + 1\right) = \frac{(2k-1)!}{(2k)!!}. \tag{7.5}$$

Retarded time variables, $s = t - r$ and $s_0 = t_0 - r_0$, are expanded as follows:

$$s = t^* - d\left(\frac{\varepsilon}{2} - \sum_{k=2}^{\infty} \mathcal{C}_k \varepsilon^{2k-1}\right), \tag{7.6}$$

$$s_0 = t^* - 2r_0 + d\left(\frac{\varepsilon_0}{2} - \sum_{k=2}^{\infty} \mathcal{C}_k \varepsilon_0^{2k-1}\right), \tag{7.7}$$

where $\mathcal{C}_k$ is given by equation (7.5) and t^* is the time of the closest approach of the light ray to the barycenter of the isolated system.

Using (7.6) and (7.7) we can write down the post-Newtonian expansions for functions of the retarded time $t - r$ as follows:

$$F(t-r) = \sum_{k=0}^{\infty} \frac{(-1)^k d^k}{k!}\left(\frac{\varepsilon}{2} - \sum_{k=2}^{\infty} \mathcal{C}_k \varepsilon^{2k-1}\right)^k F^{(k)}(t^*) \tag{7.8}$$
$$= F(t^*) - \varepsilon\frac{d}{2}\dot{F}(t^*) + O\left(\varepsilon^2\varepsilon_\lambda^2\right),$$

$$F(t_0 - r_0) = \sum_{k=0}^{\infty} \frac{d^k}{k!}\left(\frac{\varepsilon_0}{2} - \sum_{k=2}^{\infty} \mathcal{C}_k \varepsilon_0^{2k-1}\right)^k F^{(k)}(t^* - 2r_0) \tag{7.9}$$
$$= F(t^* - 2r_0) + \varepsilon_0\frac{d}{2}\dot{F}(t^* - 2r_0) + O\left(\varepsilon_0^2\varepsilon_\lambda^2\right),$$

where the dot above functions denote the total time derivative. We notice that convergence of the post-Newtonian time series for light propagation depends, in fact, not just on a single parameter, $\varepsilon_\lambda \sim v/c$, that is typical in the post-Newtonian celestial mechanics of extended bodies [54, 69, 70], but on the product of two parameters. Therefore, the convergence requires satisfaction of two conditions

$$\varepsilon\varepsilon_\lambda \ll 1, \tag{7.10}$$

$$\varepsilon_0\varepsilon_\lambda \ll 1, \tag{7.11}$$

where the numerical value of the parameter ε_λ must be taken for the smallest wavelength, $\lambda_{\min}$, in the spectrum of the gravitational radiation emitted by the isolated system. Conditions (7.10) and (7.11) ensure convergence of the post-Newtonian series in (7.8) and (7.9), respectively. If the source of light rays and observer are at infinite distances from the isolated system then $\varepsilon \simeq 0$ and $\varepsilon_0 \simeq 0$, and the requirements (7.10), (7.11) are satisfied automatically, irrespective of the structure of the Fourier spectrum (4.5) of the gravitational radiation emitted by the isolated system. In a real astronomical practice such an assumption may not be always satisfied. In such cases, it is more natural to avoid the post-Newtonian expansions of the metric tensor and/or observable effects and operate with the functions of the retarded time. It is important to notice that the retarded time $s = t - r$ which enters the result of the calculation of the light propagation, is a characteristic of the Einstein equations of gravitational field, not the

Maxwell equations. Therefore, measuring the effect of the gravitational deflection of light caused by time-dependent gravitational field allows us to measure the speed of gravity with respect to the speed of light. This type of experiments have been proposed in our paper [126] and successfully performed in 2003 [127]. There were several publications (see review [147]) arguing that the speed of gravity is irrelevant in the light-ray deflection experiments by moving bodies. Unfortunately, all the authors of those publications operated with the post-Newtonian expansion of the gravitational field which replaces the retarded time $s = t - r$ of the gravitational field with the time t^* of the closest approach of light to the light-ray deflecting body like in equation (7.6). This explains why the effect of the retardation of gravity was confused with the retardation of light in [147].

If we assume that the mass and angular momentum of the isolated system are conserved the asymptotic expansions of integrals (4.1) and (4.2) of the stationary part of the metric tensor have the following form:

$$\begin{aligned}\left[\frac{1}{r}\right]^{[-1]} &= -\ln\left(\frac{r}{2r_{\mathrm{E}}}\right) - 2\ln\varepsilon - \sum_{k=1}^{\infty}\frac{(2k-1)!}{2^{2k}(k!)^2}\varepsilon^{2k} \\ &= -2\ln d + \ln r + \ln(2r_{\mathrm{E}}) + O(\varepsilon^2)\,,\end{aligned} \tag{7.12}$$

$$\left[\frac{1}{r_0}\right]^{[-1]} = -\ln\left(\frac{2r_0}{r_{\mathrm{E}}}\right) - \sum_{k=1}^{\infty}\frac{(2k-1)!}{2^{2k}(k!)^2}\varepsilon_0^{2k} = -\ln\left(\frac{2r_0}{r_{\mathrm{E}}}\right) + O(\varepsilon_0^2)\,, \tag{7.13}$$

$$\begin{aligned}\left[\frac{1}{r}\right]^{[-2]} &= -r\left\{1 + \left(1 + \sum_{k=1}^{\infty}\mathfrak{C}_k\varepsilon^{2k}\right)\left[\ln\left(\frac{\varepsilon^2 r}{2r_{\mathrm{E}}}\right) + \sum_{k=1}^{\infty}\frac{(2k-1)!}{2^{2k}(k!)^2}\varepsilon^{2k}\right]\right\} \\ &= -r - 2r\ln d + r\ln(2rr_{\mathrm{E}}) - \varepsilon\frac{d}{2}\left[\frac{1}{2} - \ln\left(\frac{d^2}{2rr_{\mathrm{E}}}\right)\right] + O(\varepsilon^2)\,,\end{aligned} \tag{7.14}$$

$$\begin{aligned}\left[\frac{1}{r_0}\right]^{[-2]} &= -r_0\left\{1 - \left(1 + \sum_{k=1}^{\infty}\mathfrak{C}_k\varepsilon_0^{2k}\right)\left[\ln\left(\frac{2r_0}{r_{\mathrm{E}}}\right) + \sum_{k=1}^{\infty}\frac{(2k-1)!}{2^{2k}(k!)^2}\varepsilon_0^{2k}\right]\right\} \\ &= -r_0 + r_0\ln\left(\frac{2r_0}{r_{\mathrm{E}}}\right) - \varepsilon_0\frac{d}{2}\left[\frac{1}{2} + \ln\left(\frac{2r_0}{r_{\mathrm{E}}}\right)\right] + O(\varepsilon_0^2)\,.\end{aligned} \tag{7.15}$$

Next several equations yield the estimates for the partial derivatives with respect to the parameters ξ^i and τ from functions of the retarded time (of gravity). In these estimates the numbers m and n depend on l; we give the estimates from below for m and n for $l \geq 1$.

$$\hat{\partial}_{\langle a_1 \dots a_l\rangle}\frac{F(t-r)}{r} = O\left(\varepsilon^2\frac{F}{d^{l+1}}\right)\,, \tag{7.16}$$

$$\hat{\partial}_{\langle a_1 \dots a_l\rangle}\frac{F(t_0-r_0)}{r_0} = O\left(\varepsilon^2\frac{F}{d^{l+1}}\right)\,, \tag{7.17}$$

$$\hat{\partial}^l_\tau\left[\frac{F(t-r)}{r}\right] = O\left(\varepsilon^2\frac{F}{d^{l+1}}\right)\,, \tag{7.18}$$

$$\hat{\partial}^l_\tau \left[\frac{F(t_0 - r_0)}{r_0}\right] = O\left(\varepsilon \frac{F}{d^{l+1}}\right) , \tag{7.19}$$

$$\hat{\partial}_{\langle a_1 \dots a_l \rangle} \left[\frac{F(t_0 - r_0)}{r_0}\right]^{[-1]} = O\left(\varepsilon^2 \frac{F}{d^l}\right) , \tag{7.20}$$

$$\hat{\partial}_{\langle a_1 \dots a_l \rangle} \left[\frac{F(t_0 - r_0)}{r_0}\right]^{[-2]} = O\left(\varepsilon^3 \frac{F}{d^{l-1}}\right) . \tag{7.21}$$

Two asymptotic expansions will be also useful.

$$\hat{\partial}_{\langle a_1 \dots a_l \rangle} \left[\frac{F(t - r)}{r}\right]^{[-1]} = -2F(t-r)\hat{\partial}_{\langle a_1 \dots a_l \rangle} \ln d + O\left(\varepsilon\varepsilon_\lambda \frac{F}{d^l}\right) , \tag{7.22}$$

$$\hat{\partial}_{\langle a_1 \dots a_l \rangle} \left[\frac{F(t - r)}{r}\right]^{[-2]} = -2rF(t-r)\hat{\partial}_{\langle a_1 \dots a_l \rangle} \ln d + O\left(\varepsilon_\lambda \frac{F}{d^{l-1}}\right) . \tag{7.23}$$

These estimates and asymptotic expansions will be used for obtaining the observable relativistic effects in the gravitational lens approximation.

7.2 Asymptotic expressions for observable effects

This section provides the reader with the asymptotic expressions for potentially observable relativistic effects by taking into account only the leading terms and neglecting all terms which are proportional to the parameter ε.

The relativistic time delay is given by

$$\Delta(\tau, \tau_0) = \underset{(M)}{\Delta}(\tau, \tau_0) + \underset{(S)}{\Delta}(\tau, \tau_0) , \tag{7.24}$$

where

$$\underset{(M)}{\Delta}(\tau, \tau_0) = -4\mathcal{M} \ln d + 2\mathcal{M} \ln(4rr_0) \tag{7.25}$$
$$- 4 \sum_{l=2}^{\infty} \sum_{p=0}^{l-2} \frac{(-1)^{l+p}}{l!} C_l(l-p, p) \left(1 - \frac{p}{l}\right)\left(1 - \frac{p}{l-1}\right)$$
$$\times \hat{\partial}^p_{t^*} \mathcal{I}_{A_l}(t - r) k_{<a_1 \dots a_p} \hat{\partial}_{a_{p+1} \dots a_l>} \ln d ,$$

$$\underset{(S)}{\Delta}(\tau, \tau_0) = -4\epsilon_{iba} k_i \mathcal{S}_b \hat{\partial}_a \ln d \tag{7.26}$$
$$+ 8\epsilon_{iba} k_i \hat{\partial}_a \sum_{l=2}^{\infty} \sum_{p=0}^{l-1} \frac{(-1)^{l+p} l}{(l+1)!} C_{l-1}(l-p-1, p) \left(1 - \frac{p}{l-1}\right)$$
$$\times \hat{\partial}^p_{t^*} \mathcal{S}_{bA_{l-1}}(t - r) k_{<a_1 \dots a_p} \hat{\partial}_{a_{p+1} \dots a_{l-1}>} \ln d .$$

The observable unit vector in the direction from the observer to the source of light is given by the following expression:

$$s^i(\tau, \xi) = K^i + \alpha^i(\tau, \xi) + \beta^i(\tau, \xi) , \tag{7.27}$$

where we have dropped off the quantities $\beta^i(\tau_0, \xi)$ and $\gamma^i(\tau, \xi)$ as being negligibly small. Vector $\alpha^i(\tau, \xi)$, characterizing the deflection of light, is given by expression

$$\alpha^i(\tau, \xi) = \alpha^i_{(M)}(\tau, \xi) + \alpha^i_{(S)}(\tau, \xi) , \tag{7.28}$$

where

$$\begin{aligned}\alpha^i_{(M)}(\tau, \xi) = {} & 4\mathcal{M}\hat{\partial}_i \ln d \\ & + 4\hat{\partial}_i \sum_{l=2}^{\infty}\sum_{p=0}^{l-2} \frac{(-1)^{l+p}}{l!} C_l(l-p, p)\left(1-\frac{p}{l}\right)\left(1-\frac{p}{l-1}\right) \\ & \times \hat{\partial}^p_{t^*} \mathcal{I}_{A_l}(t-r) k_{<a_1 \dots a_p} \hat{\partial}_{a_{p+1} \dots a_l>} \ln d ,\end{aligned} \tag{7.29}$$

$$\begin{aligned}\alpha^i_{(S)}(\tau, \xi) = {} & 4\epsilon_{jba} k_j \mathcal{S}_b \hat{\partial}_{ia} \ln d \\ & - 8\epsilon_{iba} k_i \hat{\partial}_{ia} \sum_{l=2}^{\infty}\sum_{p=0}^{l-1} \frac{(-1)^{l+p} l}{(l+1)!} C_{l-1}(l-p-1, p)\left(1-\frac{p}{l-1}\right) \\ & \times \hat{\partial}^p_{t^*} \mathcal{S}_{bA_{l-1}}(t-r) k_{<a_1 \dots a_p} \hat{\partial}_{a_{p+1} \dots a_{l-1}>} \ln d .\end{aligned} \tag{7.30}$$

The corresponding relativistic correction to the light-ray deflection is

$$\beta^i(\tau, \xi) = \beta^i_{(M)}(\tau, \xi) + \beta^i_{(S)}(\tau, \xi) , \tag{7.31}$$

where

$$\beta^i_{(M)}(\tau, \xi) = -\frac{r}{R}\alpha^i_{(M)} , \tag{7.32}$$

$$\beta^i_{(S)}(\tau, \xi) = -\frac{r}{R}\alpha^i_{(S)} . \tag{7.33}$$

We can use (7.8) and (7.9) to rewrite functions of the retarded time $s = t - r$ in (7.25), (7.26), (7.29), and (7.30) as functions taken at the moment of the closest approach t^* of photon to the gravitating system. This is accomplished by formal replacing $t - r \rightarrow t^*$ because the corrections will be of the higher order with respect to the parameter ε (see Equation (7.8)). It should not confuse the reader about the physical meaning of the retardation, which is due to the finite speed of propagation of gravity.

The time delay (7.24) and the light-ray deflection (7.28) can be written in a very short and concise form by making use of the gravitational lens potential, ψ, which is just the eikonal perturbation. More specifically,

$$\Delta(\tau, \tau_0) = -4\psi + 2\mathcal{M}\ln(4rr_0) , \tag{7.34}$$

$$\alpha^i(\tau, \xi) = 4\hat{\partial}_i\psi , \tag{7.35}$$

where

$$\psi = \psi_{(M)} + \psi_{(S)} , \tag{7.36}$$

and

$$\psi_{(M)} = \mathcal{M}\ln d + \sum_{l=2}^{\infty}\sum_{p=0}^{l-2}\frac{(-1)^{l+p}}{l!}C_l(l-p,p)\left(1-\frac{p}{l}\right)\left(1-\frac{p}{l-1}\right) \times \hat{\partial}^p_{t^*}\mathcal{I}_{A_l}(t^*)k_{<a_1\dots a_p}\hat{\partial}_{a_{p+1}\dots a_l>}\ln d\,, \tag{7.37}$$

$$\psi_{(S)} = \epsilon_{jba}k_j\mathcal{S}_b\hat{\partial}_a\ln d - 2\epsilon_{jba}k_j\hat{\partial}_a\sum_{l=2}^{\infty}\sum_{p=0}^{l-1}\frac{(-1)^{l+p}l}{(l+1)!}C_{l-1}(l-p-1,p)\left(1-\frac{p}{l-1}\right) \times \hat{\partial}^p_{t^*}\mathcal{S}_{bA_{l-1}}(t^*)k_{<a_1\dots a_p}\hat{\partial}_{a_{p+1}\dots a_{l-1}>}\ln d\,. \tag{7.38}$$

This expression takes into account the multipole moments of the isolated gravitating system of all orders and presents a generalization of our previous results obtained in [6] (stationary gravitational field) and [100] (the quadrupolar gravitational field).

The angle of rotation of the polarization plane of electromagnetic wave is simplified if one uses the approximation of gravitational lens. More specifically, the total angle of the rotation is given by

$$\Delta\Phi = \Delta\Phi_{(M)} + \Delta\Phi_{(S)}\,, \tag{7.39}$$

where

$$\Delta\Phi_{(M)} = -4\sum_{l=2}^{\infty}\sum_{p=0}^{l}\frac{(-1)^{l+p}}{l!}C_l(l-p,p)\frac{l-p}{l-1} \times \hat{\partial}^p_{t^*}\epsilon_{jba_l}\mathcal{I}_{bA_{l-1}}(t^*)k_{<a_1\dots a_p}\hat{\partial}_{a_{p+1}\dots a_l>j}\ln d\,, \tag{7.40}$$

$$\Delta\Phi_{(S)} = -4\sum_{l=1}^{\infty}\sum_{p=0}^{l}\frac{(-1)^{l+p}l}{(l+1)!}C_l(l-p,p)\left(1-\frac{p}{l}\right)\left[1+H(l-1)\left(1-\frac{2p}{l-1}\right)\right] \times \hat{\partial}^{p+1}_{t^*}\mathcal{S}_{A_l}(t^*)k_{<a_1\dots a_p}\hat{\partial}_{a_{p+1}\dots a_l>}\ln d\,. \tag{7.41}$$

8 Light propagation through the field of plane gravitational waves

8.1 Plane-wave asymptotic expansions

We consider the source of light and observer located, respectively, at radial distances $r_0 = |\boldsymbol{r}_0|$ and $r = |\boldsymbol{r}|$ from the isolated system emitting gravitational waves. The distance between the source of light and observer is $R = |\boldsymbol{r} - \boldsymbol{r}_0|$. In the plane gravitational-wave approximation we assume that

(a) the characteristic wavelength of gravitational waves, $\lambda \ll \min[r, r_0]$,
(b) the distance between the source of light and observer, $R \ll \min[r, r_0]$.

Condition (a) tells us that both the source of light and observer are lying in the wave zone of the isolated system. Condition (b) allows us to consider the gravitational waves emitted by the isolated system as plane waves when they propagate from the source of light to observer. This approximation is vizualized in Figure 4.

We introduce small parameters $\delta_\lambda = \max[\lambda/r, \lambda/r_0]$, $\delta = R/r$ and $\delta_0 = R/r_0$ The coordinate relation between r, r_0 and $R = |\boldsymbol{r} - \boldsymbol{r}_0|$ can be written as follows:

$$r_0^2 = r^2 - 2rR\cos\theta + R^2 = r^2(1 - 2\delta\cos\theta + \delta^2), \tag{8.1}$$

where θ – is the angle between the directions "observer – the source of light" and "observer – the gravitating system" (see Figure 4). From (8.1) it follows that

$$r_0 = r(1 - \delta\cos\theta) + O(\delta^2), \tag{8.2}$$

$$\frac{1}{r_0} = \frac{1}{r}(1 + \delta\cos\theta) + O(\delta^2). \tag{8.3}$$

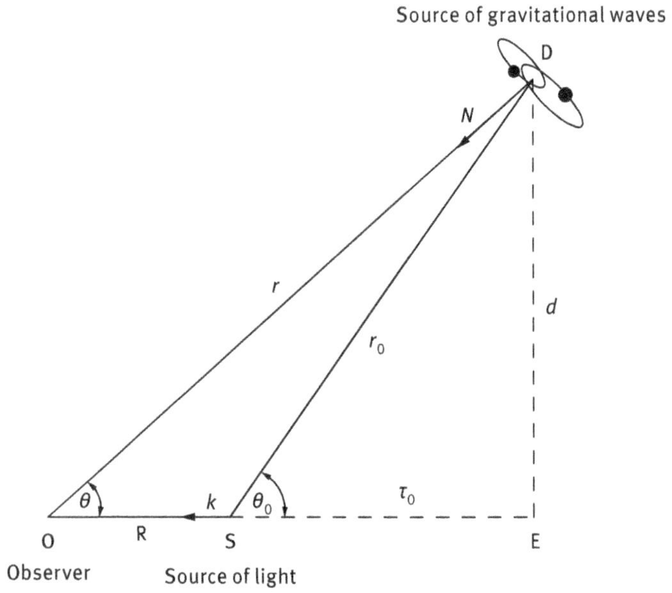

Fig. 4. Relative configuration of observer (O), a source of light (S), and a localized system emitting of gravitational waves (D). The gravitational waves deflect light rays which are emitted at the moment t_0 at the point S and received at the moment t at the point O. The point E on the line OS corresponds to the moment of the closest approach of light ray to the system D. Notations for distances are $OS = R$, $DO = r$, $DS = r_0$, $DE = d$ – the impact parameter, $OE = \tau = r\cos\theta$, $ES = \tau_0 = \tau - R$. The distance R is much smaller than both r and r_0. There is no limitation on the impact parameter d which can be or may be not small as compared to all other distances.

For the variables τ and τ_0 we have exact relations,

$$\tau = r\cos\theta, \qquad \tau_0 = \tau - R = r\cos\theta - R\,. \tag{8.4}$$

Quantities d and y are given in terms of the distance r and the angle τ as follows:

$$d = r\sin\theta\,, \tag{8.5}$$

$$y = \tau - r = -r(1-\cos\theta)\,. \tag{8.6}$$

The retarded instants of time $s = t - r$ and $s_0 = t_0 - r_0$ are related to each other as follows:

$$t_0 - r_0 = t - r - R(1-\cos\theta)\,. \tag{8.7}$$

Using this expression we can write the Taylor expansion for the functions of the retarded time $s = t - r$

$$F(t_0 - r_0) = F(t-r) - R(1-\cos\theta)\dot{F}(t-r) + O\left(R^2/\lambda^2\right)\,. \tag{8.8}$$

The impact parameter vector, ξ^i, can be decomposed as follows:

$$\xi^i = r\left(N^i - k^i\cos\theta\right)\,. \tag{8.9}$$

This form of the impact parameter vector ξ^i is useful in subsequent approximations.

Let us now write out the asymptotic expressions for the derivatives with respect to ξ^i and τ of functions depending on the retarded time $s = t - r$. We have

$$\hat{\partial}^n_\tau\left[\frac{F(t-r)}{r}\right] = (1-\cos\theta)^n\hat{\partial}^n_{t^*}\left[\frac{F(t-r)}{r}\right] + O(\delta^2)\,, \tag{8.10}$$

$$\hat{\partial}_{a_1\ldots a_n}\left[\frac{F(t-r)}{r}\right] = (-1)^n\frac{\xi_{a_1}\ldots\xi_{a_n}}{r^n}\hat{\partial}^n_{t^*}\left[\frac{F(t-r)}{r}\right] + O(\delta^2)\,, \tag{8.11}$$

$$\hat{\partial}^n_\tau\left[\frac{F(t-r)}{r}\right]^{[-1]} = (1-\cos\theta)^{n-1}\hat{\partial}^{n-1}_{t^*}\left[\frac{F(t-r)}{r}\right] + O(\delta^2)\,, \tag{8.12}$$

$$\hat{\partial}_{a_1\ldots a_n}\left[\frac{F(t-r)}{r}\right]^{[-1]} = \frac{(-1)^n}{1-\cos\theta}\frac{\xi_{a_1}\ldots\xi_{a_n}}{r^n}\hat{\partial}^{n-1}_{t^*}\left[\frac{F(t-r)}{r}\right] + O(\delta^2)\,. \tag{8.13}$$

One more asymptotic expression is given for the integral taken from an STF partial derivative

$$\left\{\left[\frac{F(t-r)}{r}\right]_{,\langle A_l\rangle}\right\}^{[-1]} = \sum_{p=0}^{l}\sum_{q=0}^{p}(-1)^{p-q}C_l(l-p,p-q,q) \times k_{<a_1\ldots a_p}\hat{\partial}_{a_{p+1}\ldots a_l>}\hat{\partial}^{p-q}_{t^*}\hat{\partial}^q_\tau\left[\frac{F(t-r)}{r}\right]^{[-1]}\,. \tag{8.14}$$

Taking into account only the leading-order terms in (8.15), we obtain the asymptotic expansion

$$\left\{\left[\frac{F(t-r)}{r}\right]_{,\langle A_l\rangle}\right\}^{[-1]} = \frac{1}{1-\cos\theta}\left[\frac{{}^{(-1)}F(t-r)}{r}\right]_{,\langle A_l\rangle} - \frac{(-1)^l}{1-\cos\theta}k_{\langle A_l\rangle}\hat{\partial}^{l-1}_{t^*}\frac{F(t-r)}{r} + (-1)^l k_{\langle A_l\rangle}\hat{\partial}^{l}_{t^*}\left[\frac{F(t-r)}{r}\right]^{[-1]} + O(\delta_\lambda)\,, \tag{8.15}$$

where $N^i = x^i/r$. Equations (8.10)–(8.15) can be checked by induction.

Corresponding expressions for the functions taken at the retarded instant of time $s_0 = t_0 - r_0$ can be obtained from (8.10)–(8.15) by replacing the arguments t, r and θ with t_0, r_0, and θ_0, respectively.

8.2 Asymptotic expressions for observable effects

In this section, we give expressions for the relativistic effects of the time delay, bending of light and the Skrotskii effect in the gravitational plane-wave approximation. In this approximation we neglect all terms of the order of δ^2, δ_0^2 and δ_λ^2, and higher. For the time delay we have

$$\Delta = \underset{(M)}{\Delta}(\tau,\tau_0) + \underset{(S)}{\Delta}(\tau,\tau_0)\,, \tag{8.16}$$

where

$$\underset{(M)}{\Delta}(\tau,\tau_0) = 2\mathcal{M}\frac{R}{r} + \frac{2k_ik_j}{1-\cos\theta}\sum_{l=2}^{\infty}\frac{(-1)^l}{l!}\left\{\left[\frac{\dot{\mathcal{I}}_{ijA_{l-2}}(t-r)}{r}\right]^{TT}_{,A_{l-2}} - \left[\frac{\dot{\mathcal{I}}_{ijA_{l-2}}(t_0-r_0)}{r_0}\right]^{TT}_{,A_{l-2}}\right\}, \tag{8.17}$$

$$\underset{(S)}{\Delta}(\tau,\tau_0) = -2\frac{\epsilon_{iba}k_i\xi_a\mathcal{S}_b}{1-\cos\theta}\left(\frac{1}{r}-\frac{1}{r_0}\right) - \frac{4k_ik_j}{1-\cos\theta} \times\sum_{l=2}^{\infty}\frac{(-1)^l l}{(l+1)!}\left\{\left[\frac{\epsilon_{ba(i}\dot{\mathcal{S}}_{j)bA_{l-2}}(t-r)}{r}\right]^{TT}_{,aA_{l-2}} - \left[\frac{\epsilon_{ba(i}\dot{\mathcal{S}}_{j)bA_{l-2}}(t_0-r_0)}{r_0}\right]^{TT}_{,aA_{l-2}}\right\}, \tag{8.18}$$

Here the transverse-traceless (TT) part of the tensors depending on the multipole moments of the isolated astronomical system is taken with respect to the direction N^i. Taking into account expression (2.8) for the components of the metric tensor h_{ij} we can rewrite expressions (8.16)–(8.18) for the time delay as follows:

$$\Delta = 2\mathcal{M}\frac{R}{r} - 2\frac{\epsilon_{iba}k_i\xi_a\mathcal{S}_b}{1-\cos\theta}\left(\frac{1}{r}-\frac{1}{r_0}\right) - \frac{k_ik_j}{2(1-\cos\theta)}\left[\int_{-\infty}^{t} h_{ij}^{TT}(\tau,\boldsymbol{x})d\tau - \int_{-\infty}^{t_0} h_{ij}^{TT}(\tau,\boldsymbol{x}_0)d\tau\right]. \tag{8.19}$$

The observed astrometric direction from observer to the source of light in the plane gravitational-wave approximation is

$$s^i(\tau,\xi) = K^i + \alpha^i(\tau,\xi) + \gamma^i(\tau,\xi)\,, \tag{8.20}$$

where

$$\alpha^i(\tau,\xi) = \frac{1}{2}\frac{k_pk_q}{1-\cos\theta}\left[(\cos\theta-2)k^i + N^i\right]h_{pq}^{TT}(t,\boldsymbol{x}) + k^q h_{ip}^{TT}(t,\boldsymbol{x})\,, \tag{8.21}$$
$$\gamma^i(\tau,\xi) = -\frac{1}{2}P^{ij}k^q h_{jq}^{TT}(t,\boldsymbol{x})\,.$$

In expression (8.20) we have dropped off the quantities $\beta^i(\tau,\xi)$ which are negligibly small. Truncated expressions (8.19)–(8.21) were obtained in [100] in the spin-dipole, mass-quadrupole approximation. Current expressions (8.19)–(8.21) include all multipole moments of the arbitrary order.

In the case when the distance between observer and the source of light is much smaller than the wavelength of the gravitational waves $R \ll \lambda$, expression (8.19) is reduced to a well known result [2] for gravitational wave detectors located in a wave-zone of an isolated system

$$\frac{\Delta R}{R} = \frac{1}{2}k_{ij}h_{ij}^{TT}(t,\boldsymbol{x}) + O(\delta^2) + O\left(\frac{R}{\lambda}\right)\,, \tag{8.22}$$

where $\Delta R = c\,\Delta(\tau,\tau_0)$.

Relativistic rotation of the polarization plane of light (the Skrotskii effect) in the gravitational plane-wave approximation assumes the next form

$$\Delta\Phi = \frac{1}{2}\frac{(\boldsymbol{k}\times\mathbf{N})^ik^j}{1-\boldsymbol{k}\cdot\mathbf{N}}h_{ij}^{TT}(t,\boldsymbol{x}) - \frac{1}{2}\frac{(\boldsymbol{k}\times\mathbf{N}_0)^ik^j}{1-\boldsymbol{k}\cdot\mathbf{N}_0}h_{ij}^{TT}(t_0,\boldsymbol{x}_0)\,. \tag{8.23}$$

References

[1] L. D. Landau and E. M. Lifshitz, *The Classical Theory of Fields*, Pergamon Press: London, 1962.
[2] C. W. Misner, K. S. Thorne and J. A. Wheeler, *Gravitation*, W. H. Freeman and Company: San Francisco, 1973.
[3] R. M. Wald *General Relativity*, The University of Chicago Press: Chicago, 1984.
[4] V. A. Fock, *The Theory of Space, Time and Gravitation*, Pergamon Press: London, 1959.
[5] K. S. Thorne, Review of Modern Physics, **52**, 299, 1980.
[6] S. M. Kopeikin, J. Math. Phys., **38**, 2587, 1997.
[7] D. A. Brown et al., Classical and Quantum Gravity, **21**, 1625, 2004.
[8] Y. Tsunesada and the TAMA Collaboration, Classical and Quantum Gravity, **21**, 703, 2004.
[9] P. Astone et al., Classical and Quantum Gravity, **21**, 759, 2004.
[10] L. Cadonati, E. Katsavounidis and the LIGO Scientific Collaboration's Bursts Working Group, Classical and Quantum Gravity, **20**, 633, 2003.
[11] M. Cerdonio, L. Baggio and G. A. Prodi, Proceedings of the SPIE, **4856**, 189, 2003.
[12] L. Blanchet, T. Damour, G. Esposito-Farese and B. R. Iyer , Physics Review Letters, **93**, 091101, 2004.
[13] T Damour, A. Gopakumar and B. R. Iyer , Physical Review D, **70**, id. 064028 (23 pages), 2004.
[14] A. Buonanno, Y. Chen, Y. Pan and M. Vallisneri , Physical Review D, **70**, 104003, 2004.
[15] A. Buonanno, Y. Chen and M. Vallisneri , Physical Review D, **67**, 024016, 2003.
[16] T. Damour, B. R. Iyer and B. S. Sathyaprakash, Physical Review D, **63**, 044023, 2001.
[17] T. Damour, B. R. Iyer and B. S. Sathyaprakash, Physical Review D, **66**, 027502, 2002.
[18] B. J. Owen and B. S. Sathyaprakash, Physical Review D, **60**, 022002, 1999.
[19] C. Ungarelli and A. Vecchio, Classical and Quantum Gravity, **21**, 857, 2004.
[20] K. Riles, APS Meeting Abstracts, 9001, 2004.
[21] D. Sigg, Classical and Quantum Gravity, **21**, 409, 2004.
[22] F. Acernese et al., Astroparticle Physics, **21**, 1, 2004.
[23] M. Fujimoto and TAMA Collaboration, Advances in Space Research, **25**, 1161, 2000.
[24] B. Willke et al., Proceedings of the SPIE, **4856**, 238, 2003.
[25] J. Weber, *General Relativity and Gravitational Waves*, Interscience Publishers: New York, 1961.
[26] E. Coccia, Physica B Condensed Matter, **280**, 525, 2000.
[27] L. Conti, de Rosa, M., F. Marin, L. Taffarello and M. Cerdonio, Journal of Applied Physics, **93**, 3589, 2003.
[28] O. D. Aguiar, L. A. Andrade, J. J. Barroso, L. Camargo Filho, L. A. Carneiro, C. S. Castro, P. J. Castro, C. A. Costa, K. M. F. Costa, J. C. N. de Araujo, A. U. de Lucena, W. de Paula, E. C. de Rey Neto, S. T. de Souza, A. C. Fauth, C. Frajuca, G. Frossati, S. R. Furtado, L. C. Lima, N. S. Magalhães, R. M. Marinho Jr., E. S. Matos, J. L. Melo, O. D. Miranda, N. F. Oliveira Jr., B. W. Paleo, M. Remy, K. L. Ribeiro, C. Stellati, W. F. Velloso Jr. and J. Weber , Classical and Quantum Gravity, **21**, 457, 2004.
[29] K. Danzmann, Advances in Space Research, **32**, 1233, 2003.
[30] P. Amaro-Seoane, S. Aoudia, S. Babak et al., GW Notes, **6**, 4-110, 6, 4, 2013.
[31] T. Appourchaux, R. Burston, Y. Chen et al., Experimental Astronomy, **23**, 491, 2009.
[32] W.-T. Ni, International Journal of Modern Physics D, **22**, 41004, 2013.
[33] K. S. Thorne 1995, "Gravitational Waves", in: Proc. of the Snowmass 95 Summer Study on Particle and Nuclear Astrophysics and Cosmology, E. W. Kolb and R. Peccei (eds.), World Scientific: Singapore, p. 398

[34] B. Schutz 1997, in: Proc. of the Alpbach Summer School 1997, *Fundamental Physics in Space*, A. Wilson (ed.), ESA Publications Division, Noordwijk: The Netherlands, SP-420, p. 229
[35] K. Riles, Progress in Particle and Nuclear Physics, **68**, 1, 2013.
[36] M. Kramer, 38th COSPAR Scientific Assembly, **38**, 4273, 2010.
[37] P. Coles, R. A. Frewin and A. G. Polnarev, “CMBR Polarization from Gravitational Waves”, in: Proceedings of the International Workshop Held in Rome, Italy, 18–21 May 1994, F. Occhionero (ed.), Springer-Verlag Berlin Heidelberg New York, 1995. Also Lecture Notes in Physics, **455**, 273–276, 1995.
[38] A. G. Polnarev, N. J. Miller and B. G. Keating, Monthly Notices of the Royal Astronomical Society, **386**, 1053–1063, 2008.
[39] B. S. Sathyaprakash, *Comptes Rendus Physique*, **14** (4), 272–287, 2013.
[40] B. Warner, Ap&SS, **2525**, 249, 1995.
[41] D. Hils, P. Bender and R. F. Webbink, Gravitational Radiation from the Galaxy, The Astrophysical Journal, **360**, 75–94, 1990.
[42] M. Rajagopal and R. W. Romani, Ultra–Low-Frequency Gravitational Radiation from Massive Black Hole Binaries, The Astrophysical Journal, **446**, 543, 1995.
[43] L. P. Grishchuk, Amplification of Gravitational Waves in an Isotropic Universe, Sov. Phys. JETP, **40**, 409, 1975.
[44] L. P. Grishchuk, Lecture Notes in Physics, Springer: Berlin, **562**, 167, 2001.
[45] B. Allen, Stochastic Gravity-Wave Background in Inflationary-Universe Models, Physical Review D, **37**, 2078–2085, 1988.
[46] B. Allen, “The Stochastic Gravity-Wave Background: Sources and Detection”, in: Proc. Spanish Relativity Meeting 1996, J. A. Miralles, J. A. Morales and D. Séz (eds.), Editions Frontiers: Paris, pp. 3–49, 1997.
[47] X. Martin and A. Vilenkin, Gravitational Wave Background From Hybrid Topological Defects, Physics Review Letters, **77**, 2879–2882, 1996.
[48] C. M. Will, *Theory and Experiment in Gravitational Physics*, (Cambridge University Press: Cambridge), 1993.
[49] S. Kopeikin and I. Vlasov, Physics Reports, **400**, 209–318, 2004.
[50] S. M. Kopeikin, The Astrophysical Journal Letters, **556**, L1–L5, 2001.
[51] E. B. Fomalont and S. M. Kopeikin, The Astrophysical Journal, **598**, 704–711, 2003.
[52] E. Fomalont, S. Kopeikin, D. Jones, M. Honma and O. Titov, Recent VLBA/VERA/IVS tests of general relativity, in: *Relativity in Fundamental Astronomy: Dynamics, Reference Frames, and Data Analysis*, Proceedings of the International Astronomical Union, IAU Symposium 261, pp. 291–295, 2010.
[53] C. M. Will, The Astrophysical Journal, **590**, 683–690, 2003.
[54] S. Kopeikin, M. Efroimsky and G. Kaplan, *Relativistic Celestial Mechanics of the Solar System*, Wiley-VCH: Berlin, 2011.
[55] S. M. Kopeikin and E. B. Fomalont, Foundations of Physics, **36**, 1244–1285, 2006.
[56] H. Stephani, D. Kramer, M. MacCallum, C. Hoenselaers and E. Herlt, *Exact solutions of Einstein’s field equations*, Cambridge University Press: Cambridge, 2003.
[57] H. Bondi, M. G. J. van der Burg and A. W. K. Metzner, Proc. Roy. Soc. Lond., **A269**, 21, 1962.
[58] R. K. Sachs, Proc. Roy. Soc. Lond., **A270**, 103, 1962.
[59] R. Penrose, Proc. Roy. Soc. Lond., **A284**, 159, 1965.
[60] R. P. Geroch, J. Math. Phys., **13**, 956, 1972.
[61] R. Arnowitt, S. Deser and C. W. Misner, The Dynamics of General Relativity, in: *Gravitation: An Introduction to Current Research*, L. Witten (ed.), John Wiley: New York, p. 227, 1962.
[62] A. Ashtekar, Asymptotic Structure of the Gravitational Field at Spatial Infinity, in: *General Relativity and Gravitation*, Vol. 2, A. Held (ed.), Plenum Press: New York.

[63] R. O. Hansen, J. Math. Phys., **15**, 46–52, 1974.
[64] L. Blanchet and T. Damour, Philos. Trans. R. Soc. London, **A320**, 379, 1986.
[65] L. Blanchet and T. Damour, Ann. Inst. H. Poincare, **50**, 337, 1989.
[66] L. Blanchet and T. Damour, Physical Review D, **46**, 4304, 1992.
[67] L. Blanchet, Physical Review D, **51**, 2559, 1995.
[68] T. Damour and B. R. Iyer, Physical Review D, **43**, 3259, 1991.
[69] V. A. Brumberg, *Relativistic Celestial Mechanics*, Nauka: Moscow (in Russian), 1972.
[70] V. A. Brumberg, *Essential Relativistic Celestial Mechanics* Adam Hilger: Bristol, 1991.
[71] P. Schneider, J. Ehlers and E. E. Falco„ *Gravitational Lenses*, Springer: Berlin, 1992.
[72] A. F. Zakharov, *Gravitational Lenses and Microlenses*, Janus-K: Moscow (in Russian), 1997.
[73] F. Courbin and D. Minniti (eds.), *Gravitational Lensing: An Astrophysical Tool*, Lecture Notes in Physics, **608**, Springer: Berlin, 2002.
[74] S. A. Klioner and S. M. Kopeikin, The Astronomical Journal, **104**, 897–914, 1992.
[75] J. H. Taylor, Classical and Quantum Gravity, **10**, 167, 1993.
[76] J. M. Weisberg and J. H. Taylor, Radio Pulsars, in: ASP Conf. Ser. **302**, M. Bailes, D. J. Nice and S. E. Thorsett (eds.), Astronomical Society of the Pacific: San Francisco, pp. 93–108, 2003.
[77] M. V. Sazhin, Sov. Astron., **22**, 36, 1978.
[78] S. Detweiler, The Astrophysical Journal, **234**, 1100, 1979.
[79] M. V. Sazhin and M. V. Saphonova, Astron. Astrophys. Space Sci., **208**, 93, 1993.
[80] M. Tinto, Physics Review Letters, **106**, 191101, 2011.
[81] K. J. Lee, N. Wex, M. Kramer et al., Monthly Notices of the Royal Astronomical Society, **414**, 3251–3264, 2011.
[82] D. R. B. Yardley, W. A. Coles, G. B. Hobbs et al., Monthly Notices of the Royal Astronomical Society, **414**, 1777–1787, 2011.
[83] H. Wahlquist, Gen. Rel. Grav., **19**, 1101, 1987.
[84] F. B. Estabrook and H. D. Wahlquist, GRG, **6**, 439, 1975.
[85] M. Tinto 1996, Physical Review D, **53**, 5354
[86] B. Bertotti, A. Vecchio and L. Iess, Physical Review D, **59**, id. 082001, 1999.
[87] J. W. Armstrong, Low-Frequency Gravitational Wave Searches Using Spacecraft Doppler Tracking, Living Reviews in Relativity, vol. 9, issue 1, 2006, URL: http://www.livingreviews.org/lrr-2006-1 (cited on November 19, 2013).
[88] V. B. Braginsky, N. S. Kardashev, A. G. Polnarev and I. D. Novikov, Nuovo Cim. B, **105**, 1141, 1990
[89] V. B. Braginsky and L. P. Grishchuk, Soviet Physics – JETP, **62**, 427, 1985.
[90] N. Kaiser and A. Jaffe, The Astrophysical Journal, **484**, 545, 1997.
[91] T. Pyne, C. R. Gwinn, M. Birkinshaw, T. M. Eubanks and D. N. Matsakis, The Astrophysical Journal, **465**, 566, 1996.
[92] C. R. Gwinn, T. M. Eubanks, T. Pyne, M. Birkinshaw and D. N. Matsakis, The Astrophysical Journal, **485**, 87, 1997.
[93] E. Montanari, Class. Quant. Grav., **15**, 2493, 1998.
[94] R. Fakir, The Astrophysical Journal, **426**, 74, 1994.
[95] R. Fakir, Physical Review D, **50**, 3795, 1994.
[96] R. Durrer, Physics Review Letters, **72**, 3301, 1994.
[97] A. Labeyrie, A&A, **268**, 823, 1993.
[98] S. Pogrebenko, M. Mingaliev, S. Montebugnoli, S. Neizvestny, N. Borisov and V. Stolyarov, GPS QSO 2022+171 – A Possible Target for VLBI Detection of Gravitational Waves, in: Proc. 2nd EVN/JIVE Symposium, Toruń, p. 33, 1994.
[99] T. Damour and G. Esposito-Farèse, Physical Review D, **58**, 042001, 1998.

[100] S. M. Kopeikin, G. Schäfer, C. R. Gwinn and T. M. Eubanks, Physical Review D, 59, 084023, 1999.
[101] R. Moenchmeyer, G. Schäfer, E. Mueller and R. E. Kates, Astronomy and Astrophysics, **246**, 417, 1991.
[102] J. D. Bekenstein, The Astrophysical Journal, **183**, 657, 1973.
[103] S. M. Kopeikin, P. Korobkov and A. G. Polnarev, Classical and Quantum Gravity, **23**, 4299–4322, 2006.
[104] S. M. Kopeikin and V. V. Makarov, Physical Review D, **75**, id. 062002 (22 pages), 2007.
[105] L. Blanchet, Gravitational Radiation from Post-Newtonian Sources and Inspiralling Compact Binaries, Living Rev. Relativity **5**, 3, 2002. [Online article]: http://www.livingreviews.org/lrr-2002-3 (cited on November 21, 2013).
[106] L. Blanchet, S. Kopeikin and G. Schäfer, *Gravitational Radiation Theory and Light Propagation*, Lecture Notes in Physics **562**, Springer-Verlag: Berlin, pp. 141–166, 2001.
[107] J. D. Jackson, *Classical Electrodynamics*, John Wiley: New York, 1998.
[108] L. D. Landau, E. M. Lifshitz and L. P. Pitaevskii, *Electrodynamics of Continuous Media*, Elsevier Butterworth-Heinemann: Oxford 2004 (2nd edition), 1982.
[109] R. Penrose and W. Rindler, *Spinors and Spacetime*, Cambridge University Press: Cambridge, 1988.
[110] V. P. Frolov, Problems in the General Theory of Relativity and Theory of Group Representations, in: Proc. of the P. N. Lebedev Physics Institute **96**, N. G. Basov (ed.), 73, 1977.
[111] J. Ehlers, Survey of General Relativity Theory, in: *Relativity, Astrophysics and Cosmology*, W. Israel (ed.), 1973.
[112] Synge [137] had developed an elegant theory of propagation of null rays in arbitrary transparent medium and gravitational field. This theory makes analysis of the problem under consideration in this chapter much more sophisticated and will not be considered here.
[113] That is, $\dot{F}(t-r) = \hat{\partial}_{t^*} F(t^* + \tau - \sqrt{\tau^2 + d^2})$.
[114] E. T. Newman and R. Penrose, J. Math. Phys., **3**, 566, 1962.
[115] R. Penrose, Structure of spacetime, in: *Battelle Recontres: 1967 Lectures in Mathematics and Physics*, C. M. DeWitt and J. A. Wheeler (eds.), Benjamin: New York, 1968.
[116] A. M. Anile and R. A. Breuer, Astrophys. J., **189**, 39, 1974.
[117] A. O. Petters, H. Levine and J. Wambsganss, *Singularity Theory and Gravitational Lensing*, Birkhäuser: Boston, 2001.
[118] A. P. Lightman, W. H. Press, R. H. Price and S. A. Teukolsky, *Problem Book in Relativity and Gravitation*, Princeton Univ. Press: Princeton, 1975.
[119] G. V. Skrotskii, Doklady Akademii Nauk SSSR, **114**, 73–76 (in Russian), 1957.
[120] G. V. Skrotskii and A. A. Izmest'ev, Doklady Akademii Nauk SSSR, **178**, 77 (in Russian), 1968.
[121] N. L. Balaz, Physical Review D, **110**, 236, 1958.
[122] J. Plebanski, Physical Review, **118**, 1396, 1960.
[123] B. G. Keating, A. G. Polnarev, N. J. Miller and D. Baskaran, The Polarization of the Cosmic Microwave Background due to Primordial Gravitational Waves, International Journal of Modern Physics A, **21**, 2459–2479, 2006.
[124] A. F. Zakharov and M. V. Sazhin, Gravitational microlensing, Soviet Physics Uspekhi, **41**, 945–982, 1998.
[125] S. M. Kopeikin, Classical and Quantum Gravity, **21**, 3251–3286, 2004.
[126] S. M. Kopeikin, Astrophys. J. Lett., **556**, L1–L5, 2002.
[127] E. B. Fomalont and S. M. Kopeikin, The Astrophysical Journal, **598**, 704, 2003.
[128] H. Asada, The Light Cone Effect on the Shapiro Time Delay, The Astrophysical Journal Letters, **574**, L69–L70, 2002.
[129] B. Bertotti and R. Catenacci, Gen. Rel. Grav., **6**, 329, 1975.

[130] S. Kopeikin, Relativistic Astrometry, Scholarpedia, **6** (8), id. 11382, 2011.

[131] L. Lindegren and D. Dravins, The fundamental definition of "radial velocity", Astronomy & Astrophysics, **401**, 1185–1201, 2003.

[132] M. Konacki, M. W. Muterspaugh, S. R. Kulkarni and K. G. Helminiak, High-precision Orbital and Physical Parameters of Double-lined Spectroscopic Binary Stars – HD78418, HD123999, HD160922, HD200077, and HD210027, The Astrophysical Journal, **719**, 1293–1314, 2010.

[133] D. A. Fischer, G. Laughlin, G. W. Marcy et al., The N2K Consortium. III. Short-Period Planets Orbiting HD 149143 and HD 109749, The Astrophysical Journal, **637**, 1094–1101, 2006.

[134] S. M. Kopeikin and L. M. Ozernoy, Post-Newtonian Theory for Precision Doppler Measurements of Binary Star Orbits, The Astrophysical Journal, **523**, 771–785, 1999.

[135] S. Zucker and T. Alexander, Spectroscopic Binary Mass Determination Using Relativity, The Astrophysical Journal Letters, **654**, L83–L86, 2007.

[136] M. Kramer, The HTRA potential of the SKA and its relation to observations with large optical telescopes, in: Proceedings of High Time Resolution Astrophysics – The Era of Extremely Large Telescopes (HTRA-IV). May 5–7, 2010, Agios Nikolaos, Crete Greece, 2010. Published online at http://pos.sissa.it/cgi-bin/reader/conf.cgi?confid=108, id. 38.

[137] J. L. Synge, *Relativity: The General Theory*, North-Holland: Amsterdam, 1960.

[138] S. M. Kopeikin and G. Schäfer, Physical Review D, **60**, id. 124002, 1999.

[139] V. B. Braginskii, C. M. Caves and K. S. Thorne, *Phys. Rev D*, **15**, 2047, 1977.

[140] S. M. Kopeikin, Gravitomagnetism and the Speed of Gravity, International Journal of Modern Physics D, **15**, 305–320, 2006.

[141] B. Mashhoon, Ann. Phys. (NY), **89**, 254, 1975.

[142] T. Piran and P. N. Safier, Nature, **318**, 271, 1985.

[143] F. S. Su and R. L. Mallett, The Astrophysical Journal, **238**, 1111, 1980.

[144] I. Ciufolini, S. Kopeikin, B. Mashhoon and F. Ricci, Physics Letters A, **308**, 101, 2003.

[145] M. Sereno, Physical Review D, **69**, id. 087501, 2004.

[146] S. M. Kopeikin and B. Mashhoon, Physical Review D, **65**, id. 064025, 2002.

[147] C. M. Will, The Confrontation between General Relativity and Experiment, Living Rev. Relativity **9**, 3, 2006. URL: http://www.livingreviews.org/lrr-2006-3 (cited on December 1, 2013).

Toshifumi Futamase

On the backreaction problem in cosmology

1 Introduction

The standard Friedman–Lemaitre–Robertson–Walker (FLRW) universe model assuming exact homogeneity and isotropy have had tremendous success in describing the early universe. Even at the present epoch, our universe where there exist local inhomogeneities in various scales, recent observations of the cosmic microwave background radiation [1, 2] and large scale galaxy surveys, such as SDSS [3], shows that our universe is remarkably isotropic and homogeneous over scales larger than 100 Mpc. Then one can ask the following question. Let us suppose there are two universes. One has a completely smooth distribution of matter and the other has a very inhomogeneous distribution of matter but with the same averaged density. Is these two universes expands exactly same way or not? One would think that there is some kind of gravitational potential energy in the inhomogeneous case and thus there will be some difference in the expansion from the homogeneous case. This question leads to another question as well. The distribution of matter is nowhere homogeneous and isotropic in real universe, and thus how can one make sense of homogeneous and isotropic "background" metric? Physically one can think of global homogeneity and isotropy as a result of some sort of averaging. However, the problem is not simple because of nonlinear nature of general relativity. The solution of the Einstein equation with averaged homogeneous matter distribution does not solve the Einstein equation with realistic matter distribution. Thus it is naturally conjectured that if we average somehow a locally inhomogeneous universe, the expansion of the averaged spacetime will be affected by the local inhomogeneities. We call this effect as the backreaction due to local inhomogeneities.

Neglecting nonlinear nature of general relativity in cosmology was questioned by Ellis [4] and the effect was identified as the backreaction in the 1980s by the present author using the averaged Friedman equation in the framework of cosmological post-Newtonian approximation realizing that the metric perturbation remains small even if the density contrast is highly nonlinear. It was found that the backreaction acts as a curvature term in the averaged Friedman equation [5, 6]. An attempt to make a rigorous statement for averaging Einstein equation has been developed by Zalaletdinov [7, 8], but no explicit treatment based on the scheme appears in realistic cosmological situation. The Issacson averaging [9, 10] was employed for this problem in Ref. [11] where the Hubble parameter is defined by the trace of the second fundamental form K and

Toshifumi Futamase: Astronomical Institute, Tohoku University, Sendai 980-8578, Japan

the Friedmann equations with the backreaction term are derived by the averaging the Hamiltonian constraint and the evolution equation for K within the framework of the cosmological post-Newtonian approximation. Later an averaging scheme without assuming post-Newtonian type metric is developed in the synchronous comoving coordinate by Buchert [12, 13]. There have been many studies motivated by the work of Buchert [14–16]. Recently the averaging scheme using one-parameter family of spacetimes developed by Burnett [17, 18] is applied in the cosmological situation [19].

On the other hand, a relativistic treatment has been developed by generalizing Zeldovich approximation [20, 21]. Although it is not easy to compare these expressions because of different gauge choices and create some confusion. Both approaches coincide that the effect is of the order of 10^{-6} compared with the critical density and is sufficiently negligible. Recently relativistic higher order study of Lagrangian approximation is developed by Buchert and his collaborators [22, 23]. There is a series of study on this problem by using Lagrange approach which claims the effect is not negligible [12]. There have been renewed interest on the averaging problem in the 2000s by the discovery of accelerated expansion of the present universe [24, 25]. The hope is to explain the cosmic acceleration by the backreaction and to avoid the introduction of the cosmological constant or more generally the dark energy [26, 27]. A possible large effect on the cosmic expansion by the backreaction in some of the works comes from the contribution from superhorizon perturbation, but such perturbations should be regarded as a part of background. Thus the redefinition of the background will eliminate such a contribution, and keep the backreaction small [37]. Perturbation treatment on the backreacion problem is also discussed in detail [29, 30], and conclude that the backreaction cannot explain the observed cosmic acceleration.

There is another approach to explain the observed acceleration using the voids of matter distribution. The SDSS commissioning data indicates the existence of a void of scales of the order of 250/h Mpc [31]. It is not totally impossible to assume that we are living inside of such a void. The void expands more rapidly compared with the surrounding region and thus it is natural to expect that the distance–redshift relation in this situation will change dramatically compared with that in the homogeneous universe. There have been some studies of the luminosity distance in such a situation [32, 33]. Moreover, there exists a model using the Lemaitre–Tolman–Bondi (LTB) spherical symmetric dust solution which gives a good fit to the Type Ia supernova observation [34]. There is an arbitrary function in the LTB solution which can be chosen to fit the observational data [35, 36]. Although such models are very useful to see the general nature of the backreaction, but it seems to the present author that they are too special to apply for realistic universes like ours.

Here we shall investigate the backreaction problem and clarifies some of confusion on the nature of backreaction based on our works [37, 38]. Because there is no unique choice of the averaged spacetime in inhomogeneous universe, the averaging should be defined in such a way that it corresponds to the actual definition of the background geometry from the observation using theory. Namely the background FLRW spacetime is defined as the geometry generated by the conserved mass density in the comoving space. However any averaging of the mass distribution does not conserve in the comoving space once the backreaction is taken into account. Thus it is natural to defined the conserved averaged quantity in the comoving space for the effective mass term in the Friedman equations. This is what we will do in this chapter.

There are obviously many other approaches and references which the author unfortunately cannot cover here and does not even realize. The reader may consult with the recent review article and different opinion on this problem by Clarkson et al. [39] and by Buchert and Rasanen [40].

This paper is organized as follows. In Section 2, we present the basic equation in 3+1 formalism in order to formulate the problem as generally as possible, and the averaging is introduced. In Section 3, we introduce an averaging and the scale factor to rewrite Einstein equations using these variables. Then the conserved effective density for the matter in the comoving space to make a direct comparison with the observation, and the averaged Friedman equations are write down in Section 4. There we also discuss the nature of backreaction in some detail. Finally, we discuss observational effect of the backreaction and give conclusion in Section 5.

2 Formulation and averaging

In this section, we would like to set the problem as general as possible. In order to do so, we shall employ the 3+1 formalism in general relativity to write down our basic equations. The line element may be written in 3+1 formalism as follows:

$$ds^2 = -(Ndt)^2 + \gamma_{ij}\left(dx^i + N^i dt\right)\left(dx^j + N^j dt\right), \tag{2.1}$$

where N, N^i are the lapse function and shift vector, respectively. Without any loss of generality, we can set $N^i = 0$. The unit vector normal to a hypersurface foliated by t = const. is defined by

$$n^\mu = \left(\frac{1}{N}, 0, 0, 0\right). \tag{2.2}$$

Then extrinsic curvature is defined as

$$K^i{}_j \equiv \frac{1}{2N}\gamma^{ik}\dot{\gamma}_{kj}, \tag{2.3}$$

where the dot denotes differentiation with respect to time coordinate t. Then the basic equations are obtained by projecting the Einstein equation

$$G_{\mu\nu} + \Lambda g_{\mu\nu} = 8\pi T_{\mu\nu} \tag{2.4}$$

onto the normal and tangential direction to the hypersurface

$$^{(3)}R + \left(K^i_{\ i}\right)^2 - K^i_{\ j}K^j_{\ i} = 16\pi GE + 2\Lambda \tag{2.5}$$

$$K^i_{\ j|i} - K^j_{\ i|j} = 8\pi GJ_i \tag{2.6}$$

$$\dot{K}^i_{\ i} + NK^i_{\ j}K^j_{\ i} - N^{|i}_{\ \ |i} = -4\pi GN\,(E + S) + N\Lambda\,, \tag{2.7}$$

where $^{(3)}R$ is a three-dimensional Ricci scalar curvature, and | denotes the three-dimensional covariant derivative. The other symbols are as follows:

$$E = T_{\mu\nu}n^\mu n^\nu = \frac{1}{N^2}T_{00} \tag{2.8}$$

$$J_i = -T_{\nu i}n^\nu = \frac{1}{N}T_{0i} \tag{2.9}$$

$$S = T_{ij}\gamma^{ij}\,. \tag{2.10}$$

The scale factor may be introduced by the following consideration. Let us define a one-parameter family $\gamma_s(t)$ of timelike geodesic (world line of galaxies), and let η^μ be the orthogonal deviation vector from γ_0 which is defined to be the world line of our galaxy. Thus η^μ represents a spatial displacement from us to a neighboring galaxy. We define the spatial distance by

$$\delta\ell = \left(h_{\mu\nu}\eta^\mu\eta^\nu\right)^{1/2}\,, \tag{2.11}$$

where $h_{\mu\nu} = g_{\mu\nu} + n_\mu n_\nu$. Then the rate of change of the distance is calculated to be

$$\frac{d}{d\tau}\,(\delta\ell) = u^\mu\,(\delta\ell)_{;\mu} = \left(\frac{1}{3}K + \sigma_{ij}e^ie^j\right)\delta\ell\,, \tag{2.12}$$

where $e^i = \eta^i/\delta\ell$ with $\eta^i = h^i_{\ \mu}\eta^\mu$, and σ_{ij} is the trace free part of the extrinsic curvature:

$$K_{ij} = \sigma_{ij} + \frac{1}{3}K\gamma_{ij}\,. \tag{2.13}$$

The above equation may be interpreted as describing the cosmic expansion where the first and second terms represent isotropic and anisotropic expansion, respectively. Therefore, one would like to introduce the scale factor proportional to K. Thus introduced scale factor is, however, a function of event. In the split of 3+1 formalism, we would like to have the scale factor which is independent of position on t = const. hyprersurfaces. This implies, we need to introduce a spatial averaging over sufficiently large compact domain D. We define the spatial averaging as follows:

$$\langle A\rangle \equiv \frac{1}{V}\int_D A\sqrt{\gamma}\,d^3x\,. \tag{2.14}$$

This is consistent with the definition of the volume of the domain D.

$$V_D = \int_D \sqrt{\gamma}\,d^3x\,, \tag{2.15}$$

where $\gamma = \det(\gamma_{ij})$. By averaging over the domain D of Equation (2.12) we have

$$\frac{d}{dt}(\delta\ell) = \left\langle N\left(\frac{1}{3}K + \sigma_{ij}e^i e^j\right)\right\rangle \delta\ell\,. \tag{2.16}$$

Thus, we can define the scale factor which depends on the domain D as follows:

$$\frac{\dot{a}_D}{a_D} = \frac{1}{3}\langle NK\rangle\,. \tag{2.17}$$

One can see that this is consistent with the definition of the volume as well. In fact,

$$3\frac{\dot{a}}{a} \equiv \frac{\dot{V}}{V} = \frac{1}{V}\int_D \frac{1}{2}\gamma^{ij}\dot{\gamma}_{ij}\sqrt{\gamma}\,d^3x\,. \tag{2.18}$$

If we remember Equation (2.3), one can see that the last expression is same with Equation (2.13). Then we can define the deviation from the uniform Hubble flow as

$$V^i_j \equiv NK^i{}_j - \frac{\dot{a}_D}{a_D}\delta^i{}_j = 3N\sigma^i{}_j\,, \tag{2.19}$$

and this satisfies $\langle V^i{}_i\rangle = 0$.

Using the scale factor the averaged Einstein equations are written as follows:

$$\begin{aligned}\left(\frac{\dot{a}}{a}\right)^2 &= \frac{8\pi G}{3}\langle N^2 E\rangle - \frac{1}{6}\langle N^{2\,(3)}R\rangle + \frac{\Lambda}{3}\langle N^2\rangle \\ &\quad - \frac{1}{6}\langle (V^i_i)^2 - V^i_j V^j_i\rangle,\end{aligned} \tag{2.20}$$

$$\begin{aligned}\frac{\ddot{a}}{a} &= -\frac{4\pi G}{3}\langle N^2(E+S)\rangle + \frac{1}{3}\langle (V^i_i)^2 - V^i_j V^j_i\rangle \\ &\quad + \frac{1}{3}\langle NN^{|i}{}_{|i} + \dot{N}K\rangle + \frac{\Lambda}{3}\langle N^2\rangle\,.\end{aligned} \tag{2.21}$$

Up to this point the treatment is completely general.

3 Calculation in the Newtonian gauge

In this section, we give more detailed and practical treatment for the backreaction. This requires explicit expression for the matter and for the inhomogeneous metric. For the matter, we use irrotational dust with the following stress energy tensor:

$$T^{\mu\nu} = \rho u^\mu u^\nu\,, \tag{3.1}$$

where u^μ is the 4-velocity of the fluid flow. This is reasonable because we are interested in the backreaction due to the collective effect of clumpy matter distribution much later than decoupling.

For the metric, we specify a particular form motivated by the post-Newtonian approximation in the cosmological circumstance [41–43] since by doing so physical meaning of the backreaction is easy to interpret and the equations are closed in perturbative sense. This choice is reasonable since the metric perturbation remains small even when the density contrast is highly nonlinear [5]. As seen below this assumption allows us to solve the Friedman equations iteratively. To see the validity of this approximation, we shall introduce two smallness parameters ϵ and κ. The parameter ϵ characterizes the order of the gravitational potential ϕ of the material clumps, $\phi \simeq \epsilon^2$. The parameter κ is the ratio between the horizon scale, $1/H$, and the scale, ℓ, of the density fluctuation, $\kappa = \ell/H$. The relative size of ϵ and κ depends on the system we have in mind. The metric perturbation is generated by the density contrast $\delta = \delta\rho/\delta_b$ (where the choice of the background is not relevant in the discussion of the order of magnitude) via the Poisson equation and may be evaluated as

$$\delta \simeq \frac{\Delta\phi}{G\rho_b} \simeq \frac{\epsilon^2}{\kappa^2}, \tag{3.2}$$

where ρ_b is the averaged density and ϕ is the Newtonian potential generated by the density contrast. Thus the linear and nonlinear regions may be characterized by the conditions $\epsilon \ll \kappa$ and $\epsilon \gg \kappa$, respectively. Thus we are interested in the parameter range $\kappa \ll \epsilon \ll 1$ which will be assumed in the below equation. It has been shown that the use of the post-Newtonian approximation is guaranteed by the following parameter range [38]:

$$\epsilon^3 \ll \kappa\,. \tag{3.3}$$

This condition is satisfied for almost all practical situations since the metric perturbation is of the order of 10^{-3} at most except very special places such as the vicinity of a compact stars.

Neglecting higher order terms the cosmological post-Newtonian, the metric may be written in the following form:

$$ds^2 = -(1+2\phi)\,dt^2 + a^2(1-2\phi)\,\delta_{ij}dx^i dx^j\,, \tag{3.4}$$

where δ_{ij} denotes the Kronecker delta, and we assumed the totally flat universe. Since we restrict ourselves to inhomogeneities with scales much smaller than the horizon scale $\kappa \ll 1$, we can safely ignore the time dependence of the potential. Actually the potential begins to decay when the cosmological constant begins to dominate the cosmic expansion. However, as the time goes on, small-scale structures drop out from the expansion and larger structures are homogenized. Then the time dependence will be neglected. Only period we cannot ignore the time dependence would be during nonlinear structure formation. Although we do not consider here, it would be interesting to see the effect of backreaction in this period.

One may wonder the relation between the volume dependent scale factor a_D and the volume independent scale factor a introduced in Equation (3.4). The difference may be calculate as follows. First Einstein equations give

$$\frac{1}{a^2}\phi_{i,i} = \frac{3}{2}\frac{\dot{a}^2}{a^2}(2\phi + \delta) - \frac{1}{2}\Lambda\delta \tag{3.5}$$

$$\delta_{ij}v^j = -2\frac{\dot{a}}{a}\left(3\dot{a}^2 - a^2\Lambda\right)^{-1}\phi_{,i}\,, \tag{3.6}$$

where $\delta = (\rho-\rho_b)/\rho_b$, and $\rho_b = \langle T_{00}\rangle$ are the averaged density. Then one can calculate V^i_j as follows:

$$V^i_j = NK^i_j - \frac{\dot{a}_D}{a_D} = \left(\frac{\dot{a}}{a} - \frac{\dot{a}_D}{a_D}\right)\delta^i_j - \frac{\dot{a}}{a}\phi^2\delta^i_j + \mathcal{O}(\phi^2)\,. \tag{3.7}$$

Remembering $\langle V^i_i\rangle = 0$, we find

$$\left(\frac{\dot{a}}{a} - \frac{\dot{a}_D}{a_D}\right) = \frac{\dot{a}}{a}\phi^2 + \mathcal{O}(\phi^2)\,. \tag{3.8}$$

We also note

$$\begin{aligned}\langle \phi'^i{}_{,i}\rangle &= \frac{1}{V}\int \phi'^i{}_{,i}\sqrt{\gamma}\,d^3x \\ &= \frac{1}{V}\left(\int \phi'^{|i}{}_{|i}\sqrt{\gamma}\,d^3x + 3\int \phi'^i\phi_{,i}\sqrt{\gamma}\,d^3x\right) \\ &= 3\langle \phi'^i\phi_{,i}\rangle\,,\end{aligned} \tag{3.9}$$

where we used periodic boundary condition. Using the above equations, we finally get the averaged Einstein equations as follows:

$$\left(\frac{\dot{a}}{a}\right)^2 = \frac{8\pi G}{3}\langle T_{00}\rangle + \frac{1}{a^2}\langle \phi_{,i}\phi_{i}\rangle + \frac{\Lambda}{3}, \tag{3.10}$$

$$\frac{\ddot{a}}{a} = -\frac{4\pi G}{3}\langle T_{00} + \rho_b a^2 v^2\rangle - \frac{1}{3a^2}\langle \phi_{,i}\phi_{,i}\rangle + \frac{\Lambda}{3}\,, \tag{3.11}$$

where $v^2 = \delta_{ij}v^i v^j$. The potential is calculated by Equation (3.5) and thus the above equations are closed and solved iteratively at least.

Although the effect of local inhomogeneities is described mathematically in the above equations, it is not straightforward to connect these equations with observation. In order to do so, we need to consider how we compare the observation and theory which will treated in the next section.

4 Definition of the background

Naively one might think that the above Equations (3.10) and (3.11) indicate that the nonlinear backreaction give a contribute to positive acceleration. However we should remember how we compare the observation with theory. The cosmological parameters are determined, at least in principle, by comparing the observed expansion behavior against the standard Friedman equation with the conserved matter energy density in the comoving space.

This suggests to use a newly defined conserved energy density $\bar{\rho}$ instead of the naive averaged density ρ_b, which satisfies the usual conservation law

$$\dot{\bar{\rho}} + 3\frac{\dot{a}}{a}\bar{\rho} = 0 . \tag{4.1}$$

In order to find $\bar{\rho}$, we set

$$\bar{\rho} = \langle T_{00}\rangle + a^{-3}A(a) . \tag{4.2}$$

Then one finds

$$\dot{A}(a) = -a^3\langle T^{i0}\phi_{,i}\rangle + a^4\dot{a}\rho\langle v^2\rangle . \tag{4.3}$$

From this expression and Equation (4.1), it is easy to find that

$$\bar{\rho} = \langle T_{00}\rangle + \left(a^{-2}\frac{5}{12\pi G} + a\frac{\Omega_\Lambda}{24\pi G\Omega_m}\right)\langle\phi_{,i}\phi_{,j}\delta^{ij}\rangle , \tag{4.4}$$

where we have used Equation (3.7) to rewrite the term containing $\langle v^2\rangle$ in terms of $\langle\phi_{,i}\phi_{,i}\rangle$. The density parameters Ω_m, Ω_Λ are defined as usual using the critical density $\rho_{cr} = H_0^2/8\pi G$, H_0 being the Hubble parameter.

Now we rewrite Equations (3.10) and (3.11) in terms of the conserved density $\bar{\rho}$

$$\left(\frac{\dot{a}}{a}\right)^2 = \frac{8\pi G}{3}\bar{\rho} + \frac{\Lambda}{3} - \frac{1}{9}\left(\frac{1}{a^2} + \frac{\Lambda_\Lambda}{\Omega_m}a\right)\langle\phi_{,i}\phi_{,i}\rangle \tag{4.5}$$

$$\frac{\ddot{a}}{a} = -\frac{4\pi G}{3}\bar{\rho} + \frac{\Lambda}{3} - a\frac{\Omega_\Lambda}{6\Omega_m}\langle\phi_{,i}\phi_{,i}\rangle . \tag{4.6}$$

These are the Friedmann equations for the averaged FLRW model of a locally inhomogeneous universe, and they should be used to interpret the effect of the local inhomogeneities on the background expansion. The third terms on the right-hand side of these equations are interpreted as the nonlinear backreaction due to local inhomogeneities. As seen from the above expressions, there are two types of backreaction. One is proportional to a^{-2}, and thus behaves as a positive curvature. The other is proportional to a and the cosmological constant. Thus if there is no cosmological constant, the backreaction does not appear in the equation for the cosmic acceleration. This means one cannot explain the cosmic acceleration by the backreaction. With the cosmological constant, the backreaction does affect the cosmic acceleration, but it tends to decelerate the cosmic expansion.

It is interesting to see the effective density and pressure giving to backreaction due to the existence of the cosmological constant only. We find

$$\rho_\Lambda^{\rm BR} = -a\frac{\Omega_\Lambda}{24\pi G\Omega_m}\langle\phi_{,i}\phi_{i}\rangle \quad (4.7)$$

$$P_\Lambda^{\rm BR} = a\frac{\Omega_\Lambda}{18\pi G\Omega_m}\langle\phi_{,i}\phi_{i}\rangle\,. \quad (4.8)$$

Thus, the equation of state is $P_\Lambda^{\rm br} = -\frac{4}{3}\rho_\Lambda^{\rm br}$, that looks like the equation of state of a phantom energy but the effective energy density is negative. It satisfies the null energy condition, but violates the weak energy condition.

We note that the cosmological constant begins to dominate the expansion, the potential starts decaying. Thus the backreaction obtained above also decays. As mentioned above, we have ignored the time dependence of the potential. This is allowed as long as we consider perturbations with scales much smaller than the horizon scale. This may be not a bad assumption for our universe because density contrast becomes small for larger scales. In fact, we can evaluate the backreaction using the power spectrum $P^\phi(k)$ of gravitational potential as

$$\langle\phi_{,i}\phi_{,i}\rangle = \int\frac{d^3k}{(2\pi)^3}k^2P^\phi(k)\,, \quad (4.9)$$

where

$$P^\phi(k) = \frac{9\Omega_m H_0^2}{4a^2k^4}P_\delta(k)\,, \quad (4.10)$$

P_δ is the matter power spectrum. Employing a realistic Λ CDM model for the power spectrum, we find that $\langle\phi_{,i}\phi_{,i}\rangle$ changes less than 0.3% when the lower bound of the integral region changes from H_0 to $0.1H_0$.

Finally, we mention the averaging issues and gauge issues in the backreaction problem. One may wonder if the results here would change by employing different averaging and other gauge. Changes induced by using a different averaging scheme occur simultaneously only inside $\bar{\rho}$ of Equations (4.5) and (4.6). Furthermore, the nonlinear backreaction term is invariant to second order with respect to the choice of the averaging procedure. Concerning gauge issue, one may wonder the result here would change under a different gauge condition. So far only the comoving synchronous gauge (CS) except post-Newtonian gauge is used to investigate the backreaction problem [44, 45], and the result in CS gauge obtained equations similar to Equations (3.9) and (4.4). showing no change in the cosmic acceleration in the case of no cosmological constant [45]. Although there is no general proof available at present, this strongly suggests that the above conclusion is valid for any choice of the gauge.

5 Conclusions

We have studied the backreaction problem in cosmology based on our previous works. We find that the backreaction on the global cosmic expansion due to local inhomogeneities does exist and acts as a curvature term in the Friedman equations, but the effect is of the order of 10^{-6} or so in the critical density and thus negligibly small at least for the present observations. In fact, recent PLANCK observation with the addition of BAO data sets the constraint of the total density parameter as $|\Omega_{\text{total}} - 1| \leq 10^{-3}$ [2]. More explicitly, we can calculate the change due to the backreaction obtained above in the luminosity distance from the present to $z = 1000$ is less than 0.001% from the standard distance in the ΛCDM model with $\Omega_m = 0.3$ and $\Omega_\Lambda = 0.7$. However, the progress of the cosmological observation would reach such a accuracy sometime in future.

Although the effect is small, its nature is very interesting. First, we find that no cosmic acceleration occurs as a result of the backreaction of a Newtonian perturbed FLRW metric in the case of no cosmological constant. This is in clear contrast with other works which show only the smallness of the backreaction [29, 30]. Second, the cosmological constant does induce the backreaction on the cosmic acceleration, but in a negative direction. Namely the development of the structure tends to decrease the cosmic acceleration. If we write the Friedman equations as follows:

$$\left(\frac{\dot{a}}{a}\right)^2 = \frac{8\pi G}{3}(\bar{\rho} + \rho_{\text{eff}}) \tag{5.1}$$

$$\frac{\ddot{a}}{a} = -\frac{4\pi G}{3}(\bar{\rho} + \rho_{\text{eff}} + 3P_{\text{eff}}) , \tag{5.2}$$

and

$$P_{\text{eff}} = \omega_{\text{eff}}\rho_{\text{eff}} , \tag{5.3}$$

then, $\omega_{\text{eff}} + 1$, changes sign from negative to positive at around $z \simeq 0.25$ for $\Omega_\Lambda = 0.7$ flat cosmology although the deviation from $\omega_\Lambda = -1$ is of the order of 10^{-6}.

This will set the fundamental limit of the observation to determine the nature of the dark energy if it is the cosmological constant or a time dependent vacuum energy.

Acknowledgement: I would like to thank Prof. S. Kopeikin for giving me a chance to contribute this chapter to a volume celebrating Prof. V. A. Brumberg on his 80th birthday. I would also like to thank Prof. M. Kasai and Prof. H. Asada for fruitful collaborations and useful discussions. I apologize not to cite and mention many relevant works on the problem of averaging and strongly recommend reader to read the review articles mentioned in the text. This work is supported by a Grant-in-Aid for Scientific Research from JSPS (Nos. 18072001, 20540245).

References

[1] G. Hinshaw et al., (2012) arXiv:1212.5226.
[2] P. A. R. Ade et al., arXiv:1303.5075.
[3] M. Tegmark et al., Astrophysics. Journals 606 (2004), 702.
[4] G. F. R. Ellis, in *General Relativity and Gravitation*, edited by B. Bertotti, F. de Felice and A. Pascolini (D. Reidel Publishing Company, 1984) pp. 215.
[5] T. Futamase, Monthly Notices of Royal Astronomical Society 237 (1989), 187.
[6] T. Futamase, Physical Review Letters 61 (1988), 2175.
[7] X. Zalaletdinov, General Relativity and Gravitation 24 (1992), 1015.
[8] X. Zalatetdinov, General Relativity and Gravitation 25 (1993), 673.
[9] R. A. Issacson, Physical Review 166 (1968), 1263.
[10] Physical Review 166 (1968), 1272.
[11] T. Futamase, Physical Review D53 (1996), 2330.
[12] T. Buchert, General Relativity and Gravitation 32 (2000), 105.
[13] T. Buchert, General Relativity and Gravitation 33 (2001), 1381.
[14] D. L. Wiltshire, Physical Review Letters 99 (2008), 251101.
[15] S. Rasanen, Journal of Cosmology and Astroparticle Physics 0611 (2006), 003.
[16] R. A. Sussman and L. Garcia-Trujillo, Classcal and Quantum Gravity 19 (2002), 2897.
[17] G. A. Burnett, Journal of Mathematical Physics 30 (1989), 90.
[18] see also T. Futamase and P. A. Hogan, Journal of Mathematical Physics 34 (1992), 3.
[19] S. R. Green and R. M. Wald, Phys. Rev. D83 (2011), 084020; see also arXive:1304.2318.
[20] M. Kasai, Physical Review Letters 69, (1992), 681.
[21] M. Kasai, Physical Review D52, (1995), 5605.
[22] T. Buchert and M. Ostermann, Physical Review D86 (2012), 023520.
[23] T. Buchert and C. Nayet and A. Wiegand, Physical Review D87 (2013), 123503.
[24] A. Riese et al., Astronomy Journal 116 (1998) 1009.
[25] S. Perlmutter et al., Nature 391 (1998), 391.
[26] E. W. Kolb, S. Matarrese, A. Notari and A. Riotto, Physical Review D71 (2005), 023524.
[27] V. Marra, E. W. Kolb and S. Matarrese, Physical Review D 77 (2008), 023003.
[28] T. Buchert, Astronomy and Astrophysics 454 (2006), 415.
[29] A. Ishibashi and R. M. Wald, Classical and Quantum Gravity 23 (2006), 235.
[30] M. Hirata and U. Seljal, Physical Review D72 (2005), 083501.
[31] M. R. Blanton et al., Astronomy Journal 121 (2001), 2358.
[32] K. Tomita, Astrophysical Journal 529 (2000) 26.
[33] K. Tomita, Progress of Theoretical Physics 106 (2001), 26.
[34] H. Alnes, M. Amarzguioui and O. Gron, Physical Review D73 (2006), 08519.
[35] T. Kai, H. Kozao, N. Nakao, K. Nambu and C. M. Yoo, Progress of Theoretical Physics 117 (2007), 229.
[36] C. M. Yoo, T. Kai and K. i. Nakao, Progress of Theoretical Physics 120 (2008) 937.
[37] M. Kasai, H. Asada and T. Futamase, Progress of Theoretical Physics 115 (2006), 827.
[38] H. Tanaka and T. Futamase, Progress of Theoretical Physics 117 (2007), 183.
[39] C. Clarkson. G. F. L. Ellis, J. Larena and O. Umeh, Report on Progress of Physics 74 (2011), 112901.
[40] T. Buchert and S. Rasanen, Annual Review of Nuclear and Particle Science 62 (2012), 57.
[41] H. Nariai and Y. Ueno, Progress of Theoretical Physics 23 (1960), 241.

[42] W. M. Irvine, Annual Physics 32 (1965), 322.
[43] K. Tomita, Progress of Theoretical Physics 79 (1988), 322.
[44] M. Kasai, Physical Review D47 (1993), 2330.
[45] H. Russ, M. H. Soffel, M. Kasai and G. Boerner, Physical Review D56 (1997), 2044.

Alexander Petrov and Sergei Kopeikin

Post-Newtonian approximations in cosmology

1 Introduction

Post-Newtonian celestial mechanics is a branch of fundamental gravitational physics [18, 19, 66, 104] that deals with the theoretical concepts and experimental methods of measuring gravitational fields and testing general relativity both in the solar system and beyond [20, 118]. In particular, the relativistic celestial mechanics of binary pulsars (see [82], and references therein) was instrumental in providing conclusive evidences for the existence of gravitational radiation as predicted by Einstein's theory of relativity [105, 115].

Over the last few decades, various groups within the International Astronomical Union (IAU) have been active in exploring the application of general relativity to the modeling and interpretation of high-accuracy astrometric observations in the solar system. A Working Group on Relativity in Celestial Mechanics and Astrometry was formed in 1994 to define and implement a relativistic theory of reference frames and time scales. This task was successfully completed with the adoption of a series of resolutions on astronomical reference systems, time scales, and Earth rotation models by 24th General Assembly of the IAU, held in Manchester, UK, in 2000. The IAU resolutions are based on the first post-Newtonian approximation of general relativity which is a conceptual basis of the fundamental astronomy in the solar system [103].

The mathematical formalism of the post-Newtonian approximations is getting progressively complicated as one goes from the Newtonian to higher orders [31, 99]. For this reason, the theory has been primarily developed for an isolated astronomical systems with a matter distribution having a compact support and under simplifying assumptions that gravitational field perturbation is weak everywhere, decays rapidly enough at infinity, and the background spacetime is asymptotically flat. Mathematically, it means that the full spacetime metric, $\gamma_{\alpha\beta}$, is decomposed around the background Minkowskian metric, $\eta_{\alpha\beta} = \mathrm{diag}(-1, 1, 1, 1)$, into a linear combination

$$g_{\alpha\beta} = \eta_{\alpha\beta} + h_{\alpha\beta} , \tag{1.1}$$

Alexander Petrov: Sternberg Astronomical Institute, Moscow Lomonosov State University, Universitetskiy Prospect 13, Moscow 119992, Russia
Sergei Kopeikin: Department of Physics and Astronomy, University of Missouri, 322 Physics Bldg., Columbia, Missouri 65211, USA

where the perturbation $h_{\alpha\beta}$ is represented as the post-Minkowskian[1] series decomposition with respect to the powers of the universal gravitational constant,

$$h_{\alpha\beta} = G\underset{(1)}{h_{\alpha\beta}} + G^2\underset{(2)}{h_{\alpha\beta}} + G^3\underset{(3)}{h_{\alpha\beta}} + \cdots , \tag{1.2}$$

where each term, $\underset{(k)}{h_{\alpha\beta}}$, $(k = 1, 2, 3, \ldots)$ of the post-Minkowskian series is decomposed into the post-Newtonian series

$$\underset{(k)}{h_{\alpha\beta}} = c^{-2}\underset{(k)}{h^{[2]}_{\alpha\beta}} + c^{-3}\underset{(k)}{h^{[3]}_{\alpha\beta}} + c^{-4}\underset{(k)}{h^{[4]}_{\alpha\beta}} + \cdots , \tag{1.3}$$

with respect to the powers of $1/c$, where c is the speed of gravity in general relativity[2]. Post-Minkowskian series (1.2) is analytic with respect to the parameter G while the post-Newtonian series (1.3) loses analyticity at higher order approximations where the backreaction of gravitational radiation becomes important [13].

Post-Newtonian approximations suggest that there exists a method to determine $h_{\alpha\beta}$ by doing successive iterations of Einstein's field equations with the tensor of energy–momentum of matter field Φ of the localized astronomical system, $\mathfrak{T}_{\alpha\beta}(\Phi, g_{\alpha\beta})$, taken as a source of the gravitational field perturbation $h_{\alpha\beta}$. The iterations start from $h_{\alpha\beta} = 0$ which is inserted to the expression for $\mathfrak{T}_{\alpha\beta}$ which becomes a well-defined function of the matter variables Φ. Einstein's equations are solved at the first iteration yielding $\underset{(1)}{h_{\alpha\beta}}$. This solution is substituted back to the tensor $\mathfrak{T}_{\alpha\beta}$ which is used to find $\underset{(2)}{h_{\alpha\beta}}$, and so on. The post-Minkowskian solution for the metric perturbations $\underset{(k)}{h_{\alpha\beta}}$ naturally depend on the retarded time $s = t - r/c$ which accounts for the finite speed of propagation of gravity passing the distance r from the mass emitting gravitational radiation. The post-Newtonian decomposition (1.3) of the metric tensor perturbation represents an additional expansion of the retarded functions around the time event t. Thus, the post-Newtonian expansion assumes $r/c \ll T$ or $r \ll \lambda$, where λ is a characteristic wavelength of gravitational radiation. It means that the post-Newtonian series (1.3) is valid only in the near zone of the isolated astronomical system.

The solution of the field equations and the equations of motion of the astronomical bodies are derived in some coordinates $r^\alpha = \{ct, \boldsymbol{r}\}$ where t is the coordinate time,

---

1 The term "post-Minkowskian" was introduced by Damour and Blanchet [14] to emphasize that the metric tensor $g_{\alpha\beta}$ is built as a perturbative series around the Minkowski metric $\eta_{\alpha\beta}$, and it does not assume any limitation on the velocity of matter generating gravitational field.

2 A common convention is to call c the speed of light irrespectively of the nature of the fundamental interaction under consideration [118] but it may lead to confusion and misinterpretation of gravitational experiments and astronomical observations [42, 68].

and $\boldsymbol{r} = \{x, y, z\}$ are spatial coordinates. The post-Newtonian theory in asymptotically flat spacetime has a well-defined Newtonian limit determined by

(1) solution of Poisson's equation for the Newtonian potential, $U \equiv \underset{(1)}{h}{}^{[2]}_{00}/2$,

$$U(t, \boldsymbol{r}) = \int\limits_{\mathcal{V}} \frac{\rho(t, \boldsymbol{r}')d^3r'}{|\boldsymbol{r} - \boldsymbol{r}'|} , \tag{1.4}$$

where $\rho = c^{-2}\mathfrak{T}_{00}$, is the mass density of matter producing the gravitational field,

(2) equation of motion for massive particles

$$\ddot{\boldsymbol{r}} = \nabla U , \tag{1.5}$$

where $\nabla = \{\partial_x, \partial_y, \partial_z\}$ is the operator of gradient, $\boldsymbol{r} = \boldsymbol{r}(t)$ is time-dependent position of a particle (world line of the particle), and the dot denotes a total derivative with respect to time t,

(3) equation of motion for light (massless particles)

$$\ddot{\boldsymbol{r}} = 0 . \tag{1.6}$$

These equations are foundational for creation of astronomical ephemerides of celestial bodies in the solar system [19, 66] and in any other localized system of self-gravitating bodies like a binary pulsar [82]. In all practical cases they have to be extended to take into account the post-Newtonian corrections sometimes up to the 3D post-Newtonian order of magnitude [119]. It is important to notice that in the Newtonian limit the coordinate time t of the gravitational equations of motion (1.5) and (1.6) coincides with the proper time of observer τ that is practically measured with an atomic clock.

So far, the post-Newtonian theory was mathematically successful and passed through numerous experimental tests with a flying color. Nevertheless, it hides several pitfalls. The first one is the problem of convergence of the post-Newtonian series and regularization of divergent integrals that appear in the post-Newtonian calculations at higher post-Newtonian orders [99]. The second problem is that the background manifold is not asymptotically flat Minkowskian spacetime but the FLRW metric, $\bar{g}_{\alpha\beta}$. We live in the expanding universe which rate of expansion is determined by the present value H_0 of the Hubble parameter $H = H(t)$ depending on time. Therefore, the right thing would be to replace the post-Newtonian decomposition (1.1) with a more adequate post-Friedmannian series [109]

$$g_{\alpha\beta} = \bar{g}_{\alpha\beta} + \varkappa_{\alpha\beta} , \tag{1.7}$$

where

$$\varkappa_{\alpha\beta} = \varkappa^{\{0\}}_{\alpha\beta} + H\varkappa^{\{1\}}_{\alpha\beta} + H^2\varkappa^{\{2\}}_{\alpha\beta} + \cdots , \tag{1.8}$$

is the metric perturbation around the cosmological background represented as a series with respect to the Hubble parameter, H. Each term of the series has its own expansion into the post Minkowskian/Newtonian series like (1.2) and (1.3). For example, $\varkappa^{\{0\}}_{\alpha\beta} = h_{\alpha\beta}$, and there is no asymptotically flat spacetime analog to $\varkappa^{\{1\}}_{\alpha\beta}$, $\varkappa^{\{2\}}_{\alpha\beta}$, etc. Generalization of the theory of post-Newtonian approximations from the Minkowski spacetime to that of the expanding universe is important for extending the applicability of the post-Newtonian celestial dynamics to testing cosmological effects, for more deep understanding of the process of formation of large and small scale structure in the universe and gravitational interaction between pairs of galaxies and their clusters.

Whether cosmological expansion affects gravitational dynamics of bodies inside a localized astronomical system was a matter of considerable efforts of many researchers [17, 24, 37, 38, 71, 86, 87, 100]. Most of the previous works on celestial dynamics in cosmology assumed spherical symmetry of matter distribution and gravitational field which allowed to use exact spherically symmetric solutions of Einstein's equations approximating the Schwarzschild solution near the body and a cosmological solution far outside of it. Matching of the two solutions in the intermediate zone was achieved in several different ways but all of them suggest some kind of a fine tuning of the size of the matching zone to the cosmological parameters and the mass of the central body. This fine tuning is physically unrealistic. Furthermore, real astronomical systems in cosmology (galaxies, clusters, filaments, etc.) have no spherical symmetry. McVittie's solution [87] is perhaps the most successful mathematically among the spherically symmetric approaches but yet lacks a clear physical interpretation [24].

Cosmological observations are now performed so accurately that we need a precise mathematical formulation of the post-Newtonian theory for interpretation of these observations. This theory is not to be limited by the assumption of the spherical symmetry of the isolated astronomical system which must be coupled to the time-dependent background geometry through the gravitational interaction. Theoretical description of the post-Newtonian dynamics of a localized astronomical system in expanding universe should correspond in the limit of vanishing H to the post-Newtonian dynamics in the asymptotically flat spacetime. Such a description will allow us to directly compare the equations of the standard post-Newtonian celestial dynamics with its cosmological counterpart. Therefore, the task is to derive a set of the post-Newtonian equations in cosmology in some coordinates introduced on the background manifold, and to map them onto the set of the Newtonian equations (1.4)–(1.6) in asymptotically flat spacetime. The post-Newtonian celestial dynamics would be of a paramount importance for extending the tools of experimental gravitational physics to the field of cosmology, for example, to properly formulate the cosmological extension of the PPN formalism [117]. This chapter discusses the main ideas and principal results of such a theoretical approach in the linearized approximation with respect to the gravitational perturbations of the cosmological background caused by the presence of a localized astronomical system. The formalism of this chapter has been employed in [69] to check the theoretical consistency of equations (1.4)–(1.6) on expanding cosmological background and to analyze the outcome of some experiments like the excessive Doppler effect discovered by Anderson et al. [3, 4] in the hyperbolic motion of Pioneer 10 and 11 space probes in the solar system.

The original goal in developing the theory of cosmological perturbations was to relate the physics of the early universe to CMB anisotropy and to explain the formation and growth of large-scale structure from a primordial spectrum. The ultimate goal of this theory is to establish a mathematical link between the fundamental physical laws at the Planck epoch and the output of the gravitational wave detectors which are the only experimental devices being capable to measure the parameters and the state of the universe at that time [64]. Originally, the two basic approximation schemes for calculating cosmological perturbations have been invented by Lifshitz with his collaborators [78, 79] and, later on, by Bardeen [8]. Lifshitz [78] worked out a coordinate-dependent theory of cosmological perturbations in a synchronous gauge while Bardeen [8] concentrated on finding the gauge-invariant combinations for perturbed quantities and derivation of a perturbation technique based on gauge-invariant field equations. At the same time, Lukash [83] had suggested an original approach for deriving the gauge-invariant scalar equations based on the thermodynamic theory of the Clebsch potential [102] also known in cosmology as the scalar velocity potential [73, 102] or the Taub potential [108]. It turns out that the variational principle with a Lagrangian of cosmological matter formulated in terms of the Clebsch potential, is the most useful mathematical device for developing the theory of relativistic celestial dynamics of localized astronomical systems embedded in expanding cosmological manifold [70].

In the years that followed, the gauge-invariant formalism was refined and improved by Durrer and Straumann [34, 35], Ellis et al. [39–41] and, especially, by Mukhanov et al. [91, 93]. Irrespectively of the approach a specific gauge must be fixed in order to solve equations for cosmological perturbations. Any gauge is allowed and its particular choice is simply a matter of mathematical convenience. Imposing a gauge condition eliminates four degrees of freedom in the cosmological metric perturbations and brings the differential equations for them to a solvable form. Nonetheless, the residual gauge freedom associated with the tensor nature of the gravitational field remains. This residual gauge freedom leads to appearance of spurious perturbations which must be disentangled from the physical modes. Lifshitz's theory of cosmological perturbations [78, 79] is worked out in a synchronous gauge and contains the spurious modes but they are easily isolated from the physical perturbations and suppressed [49]. The other gauges are described in Bardeen's article [8] and used in cosmological perturbation theory as well. Among them, the longitudinal (conformal or Newtonian) gauge is one of the most common. This gauge is advocated by Mukhanov [91] because it removes spurious coordinate degrees of freedom in scalar perturbations. Detailed comparison of the cosmological perturbation theory in the synchronous and conformal gauges was given by Ma and Bertschinger [84].

Unfortunately, none of the previously known cosmological gauges can be applied for analysis of the cosmological perturbations caused by localized matter distributions like an isolated astronomical system which can be a single star, a planetary system, a galaxy, or even a cluster of galaxies. The reason is that the synchronous gauge has no Newtonian limit and is applicable only for freely falling test particles while the longitudinal gauge separates the scalar, vector, and tensor modes in the metric tensor perturbation in the way that is incompatible with the technique of the post-Newtonian approximation schemes having been worked out in asymptotically flat spacetime [66]. We also notice that standard cosmological perturbation technique often operates with harmonic (Fourier) decomposition of both the metric tensor and matter perturbations when one is interested in statistical statements based on the cosmological principle. This technique is unsuitable and must be avoided in subhorizon approximation for working out the post-Newtonian celestial dynamics of self-gravitating isolated systems.

Current paradigm is that the cosmological generalization of the Newtonian field equations of an isolated gravitating system like the solar system or a galaxy or a cluster of galaxies can be easily obtained by just making use of the linear principle of superposition with a simple algebraic addition of the local system to the tensor of energy momentum of the background matter. It is assumed that the superposition procedure is equivalent to operating with the Newtonian equations of motion derived in asymptotically flat spacetime and adding to them ("by hands") the tidal force due to the presence of the external universe (see, for example, [86]). Though such a procedure may look pretty obvious it lacks a rigorous mathematical analysis of the perturbations induced on the background cosmological manifold by the local system. This analysis should be done in the way that embeds cosmological variables to the field equations of standard post-Newtonian approximations not by "hands" but by precise mathematical technique which is the goal of this chapter. The variational calculus on manifolds is the most convenient for joining the standard theory of cosmological perturbations with the post-Newtonian approximations in asymptotically flat spacetime. It allows us to track down the rich interplay between the perturbations of the background manifold with the dynamic variables of the local system which cause these perturbations. The output is the system of the post-Newtonian field equations with the cosmological effects incorporated to them in a physically transparent and mathematically rigorous way. This system can be used to solve a variety of physical problems starting from celestial dynamics of localized systems in cosmology to gravitational wave astronomy in expanding universe that can be useful for deeper exploration on scientific capability of such missions as LISA and Big Bang Observer (BBO) [30].

In fact, the problem of whether the cosmological expansion affects the long-term evolution of an isolated N-body system (galaxy, solar system, binary system, etc.) had a long controversial history. The reason is that there was no adequate mathematical formalism for describing the cosmological perturbations caused by an isolated system so that different authors have arrived to opposite opinions. It seems that McVittie [87] was first who had considered the influence of the expansion of the universe on the dynamics of test particles orbiting around a massive point-like body immersed to the cosmological background. He found an exact solution of the Einstein equations in his model which assumed that the mass of the central body is not constant but decreases as the universe expands. Einstein and Straus [37, 38] suggested a different approach to discuss motion of particles in gravitationally self-interacting systems residing on the expanding background. They showed that a Schwarzschild solution could be smoothly matched to the Friedman universe on a spherical surface separating the two solutions. Inside the surface ("vacuole") the motion of the test particles is totally unaffected by the expansion. Thus, Einstein and Straus [37, 38] concluded that the cosmic expansion is irrelevant for the Solar system. Bonnor [17] generalized the Einstein–Straus vacuole and matched the Schwarzschild region to the inhomogeneous Lemaître–Tolman–Bondi model thus, making the average energy density inside the vacuole be independent of the exterior energy density while in the Einstein–Straus model they must be equal. Bonnor [17] concluded that the local systems expand but at a rate which is negligible compared with the general cosmic expansion. Similar conclusion was reached by Mashhoon et al. [86] who analyzed the tidal dynamics of test particles in the Fermi coordinates.

The vacuole solutions are not appropriate for adequate physical solution of the N-body problem in the expanding universe. There are several reasons for it. First, the vacuole is spherically symmetric while majority of real astronomical systems are not. Second, the vacuole solution imposes physically unrealistic boundary conditions on the matching surface that relates the central mass to the size of the surface and to the cosmic energy density. Third, the vacuole is unstable against small perturbations. In order to overcome these difficulties a realistic approach based on the approximate analytic solution of the Einstein equations for the N-body problem immersed to the cosmological background, is required. In the case of a flat spacetime there are two the most advanced techniques for finding approximate solution of the Einstein equations describing gravitational field of an isolated astronomical system – the post-Newtonian and post-Minkowskian approximations [31] that have been briefly discussed in Section 1. The post-Newtonian approximation technique is applicable to the systems with weak gravitational field and slow motion of matter. The post-Minkowskian approximations also assume that the field is weak but does not imply any limitation on the speed of matter. The post-Newtonian iterations are based on solving the elliptic-type Poisson equations while the post-Minkowskain approach operates with the hyperbolic-type (wave) D'Alembert equations. The post-Minkowskian approximations naturally include description of the gravitational radiation emitted by the isolated system while the post-Newtonian scheme has to use additional mathematical methods to describe generation of the gravitational waves [25]. In this chapter, we concentrate on the development of a generic scheme for calculation of cosmological perturbations caused by a localized distribution of matter (small-scale structure) which preserves many advantages of the post-Minkowskian approximation scheme. The cosmological post-Newtonian approximations are derived from the post-Minkowskian perturbation scheme by making use of the slow-motion expansion with respect to a small parameter v/c, where v is the characteristic velocity of matter in the N-body system and c is the fundamental speed.

There were several attempts to work out a physically adequate and mathematically rigorous approximation schemes in general relativity in order to construct and to adequately describe dynamics of small-scale structures in the universe. The most notable work in this direction has been done by Kurskov and Ozernoy [72], Futamase et al. [12, 45, 46, 107] (see also Chapter 6 of this book), Buchert and Ehlers [22, 36], Mukhanov et al. [1, 91–93], Zalaletdinov [120]. These approximation schemes have been designed to track the temporal evolution of the cosmological perturbations from a very large down to a small scale up to the epoch when the perturbation becomes isolated from the expanding cosmological background. These approaches looked hardly connected between each other until recent works by Clarkson et al. [28, 29], Li and Schwarz [76, 77], Räsänen [98], Buchert and Räsänen [23] and Wiegand and Schwarz [116]. In particular, Wiegand and Schwarz [116] have shown that the idea of cosmic variance (that is a standard way of thinking) is closely related to the cosmic averages defined by Buchert and Ehlers [22, 36]. All researchers agree that the post-Newtonian approximations are important to understand the backreaction of the cosmological perturbations on the expansion rate of the universe [1, 45, 56, 57, 93, 120]).

Development of observational cosmology and gravitational wave astronomy demands to extend the linearized theory of cosmological perturbations to second and higher order of approximation. A fair number of works have been devoted to solving this problem. Nonlinear perturbations of the metric tensor and matter affect evolution of the universe and this backreaction of the perturbations should be taken into account. This requires derivation of the effective stress–energy tensor for cosmological perturbations formed by freely propagating gravitational waves and scalar field [1, 91–93]. The laws of conservation for the effective stress–energy tensor are important for derivation of the post-Newtonian equations of motion of the isolated astronomical system.

In this chapter, we construct a nonlinear theory of successive cosmological perturbations for isolated systems which generalizes the post-Minkowskian approximation scheme in asymptotically flat spacetime. We implement the Lagrangian-based theory of dynamical perturbations of gravitational field on a curved background manifold which has been worked out in [50, 96] (see also [6]). This theory has a number of specific advantages over other perturbation methods among which the most important are:

- Lagrangian-based approach is covariant and can be implemented for any curved background spacetime that is a solution of the Einstein gravity field equations;
- the system of the partial differential equations describing dynamics of the perturbations is determined by a dynamic Lagrangian $\mathcal{L}^D$ which is derived from the total Lagrangian $\mathcal{L}$ by making use of its Taylor expansion with respect to the perturbations and accounting for the background field equations. The dynamic La-

grangian $\mathcal{L}^D$ defines the conserved quantities for the perturbations (energy, angular momentum, etc.) that depend on the symmetries of the background manifold;
- the dynamic Lagrangian $\mathcal{L}^D$ and the corresponding field equations for the perturbations are gauge-invariant in any order of the perturbation theory. Gauge transformations map the background manifold onto itself and are associated with arbitrary (analytic) coordinate transformations on the background spacetime;
- the entire perturbation theory is self-reproductive and is extended to the next perturbative order out of a previous iteration by making use of the same equations with a corresponding substitution of quantities from the previous iteration. The linearized approximation is the basic starting point of the theory.

Perhaps, it would be more appropriate to call the perturbative technique explained in this chapter as the post-Friedmannian approximations – the term proposed by Tegmark [109]. However, we shall continue to use the conventional name of post-Newtonian approximations to emphasize that it is applicable not only to large-scale perturbations but also to the discussion of formation and dynamics of small-scale structures in cosmology – the topic being intimately related to relativistic celestial mechanics.

The chapter is organized as follows. In Section 2, we describe the variational and Lie derivatives on manifold. These derivatives are crucial for understanding the mathematical technique of this chapter. Section 3 introduces the Lagrangian of gravitational field and matter of the background cosmological model as well as the Lagrangian of an isolated astronomical system which perturbs the background cosmological manifold. Section 4 describes the geometric structure of the background spacetime manifold of the cosmological model and the corresponding equations of motion of the matter and field variables. Section 5 introduces the reader to the theory of the Lagrangian perturbations of the cosmological manifold and the dynamic variables. Section 6 makes use of the preceding sections in order to derive the field equations in the gauge-invariant form. Beginning from Section 7, we focus on the spatially flat universe in order to derive the post-Newtonian field equations that generalize the post-Newtonian equations in the asymptotically flat spacetime. These equations are coupled in the scalar sector of the proposed theory. Therefore, we consider in Section 8 a few particular cases when the equations can be fully decoupled one from another, and solved in terms of the retarded potentials. This section also provides a proof of the Lorentz invariance of the retarded potentials for the wave equations describing propagation of weak gravitational and sound waves on the background cosmological manifold.

We use G to denote the universal gravitational constant and c for the ultimate speed in the Minkowski spacetime. Every time, when there is no confusion about the system of units, we shall choose a geometrized system of units such that $G = c = 1$. We put a bar over any function that belongs to the background manifold of the FLRW cosmological model. Any function without such a bar belongs to the perturbed manifold.

The other notations used in this chapter are as follows:

- Greek indices $\alpha, \beta, \gamma, \ldots$ run through values 0, 1, 2, 3, and Roman indices $i, j, k, \ldots$ take values 1, 2, 3,
- Einstein summation rule is applied for repeated (dummy) indices, for example, $P^\alpha Q_\alpha \equiv P^0 Q_0 + P^1 Q_1 + P^2 Q_2 + P^3 Q_3$, and $P^i Q_i \equiv P^1 Q_1 + P^2 Q_2 + P^3 Q_3$,
- $g_{\alpha\beta}$ is a full metric on the cosmological spacetime manifold,
- $\mathfrak{g}^{\alpha\beta} \equiv \sqrt{-g} g^{\alpha\beta}$,
- $\bar{g}_{\alpha\beta}$ is the FLRW metric on the background spacetime manifold,
- $\mathfrak{f}_{\alpha\beta}$ is the metric on the conformal spacetime manifold,
- $\eta_{\alpha\beta} \equiv \mathrm{diag}\{-1, +1, +1, +1\}$ is the Minkowski metric,
- T and $X^i \equiv \{X, Y, Z\}$ are the coordinate time and isotropic spatial coordinates on the background manifold,
- $X^\alpha \equiv \{X^0, X^i\} = \{c\eta, X^i\}$ are the conformal coordinates with η being a conformal time,
- $x^\alpha \equiv \{x^0, x^i\} = \{ct, x^i\}$ is an arbitrary coordinate chart on the background manifold,
- a bar, $\bar{F}$ above a geometric object F, denotes the unperturbed value of F on the background manifold,
- a prime $F' \equiv dF/d\eta$ denotes a total derivative with respect to the conformal time η,
- a dot $\dot{F} \equiv dF/dT$ denotes a total derivative with respect to the cosmic time T,
- $\partial_\alpha \equiv \partial/\partial x^\alpha$ is a partial derivative with respect to the coordinate x^α,
- a comma with a following index $F_{,\alpha} \equiv \partial_\alpha F$ is another designation of a partial derivative with respect to a coordinate x^α,
- a vertical bar, $F_{|\alpha}$ denotes a covariant derivative of a geometric object F (a scalar, a vector, a tensor) with respect to the background metric $\bar{g}_{\alpha\beta}$,
- a semicolon, $F_{;\alpha}$ denotes a covariant derivative of a geometric object F (a scalar, a vector, a tensor) with respect to the conformal metric $\mathfrak{f}_{\alpha\beta}$,
- the tensor indices of geometric objects on the background manifold are raised and lowered with the background metric $\bar{g}_{\alpha\beta}$,
- the tensor indices of geometric objects on the conformal spacetime are raised and lowered with the conformal metric $\mathfrak{f}_{\alpha\beta}$,
- the scale factor of the FLRW metric is denoted as $R \equiv R(T)$, or as $a \equiv a(\eta) = R[T(\eta)]$,
- the Hubble parameter, $H \equiv \dot{R}/R$, and the conformal Hubble parameter, $\mathcal{H} = a'/a$.

Other notations will be introduced and explained as they appear in the text.

2 Derivatives on the geometric manifold

2.1 Variational derivative

Theory of perturbations of physical fields on manifolds rely on the principle of the least action of a functional S called action. Variational derivative arises in the problem of finding solutions of the gravitational field equation that extremize the action

$$S = \int \mathcal{F} d^4x \,, \tag{2.1}$$

where $\mathcal{F} \equiv f\sqrt{-g}$, is a scalar density of weight $+1$. Let the set of $\mathcal{F} = \mathcal{F}(Q, Q_\alpha, Q_{\alpha\beta})$ depends on the field variable Q, its first – $Q_\alpha \equiv Q_{,\alpha}$ – and second – $Q_{\alpha\beta} \equiv Q_{,\alpha\beta}$ – partial derivatives that play here a similar role as velocity and acceleration in the Lagrangian mechanics of point-like particles. The field variable Q can be a tensor field of an arbitrary type with the covariant and/or contravariant indices. For the time being, we suppress the tensor indices of Q as it may not lead to a confusion. Function $\mathcal{F}$ depends on the determinant g of the metric tensor and can also depend on its derivatives. We shall discuss this case in the sections that follow.

A certain care should be taken in choosing the dynamic variables of the Lagrangian formalism in the case when the variable Q is a tensor field. For example, if we choose a covariant vector field A_μ as an independent variable, the corresponding "velocity" and "acceleration" variables must be chosen as $A_{\mu,\alpha}$ and $A_{\mu,\alpha\beta}$, respectively. On the other hand, if the independent variable is chosen as a contravariant vector A^μ, the corresponding "velocity" and "acceleration" variables must be chosen as $A^\mu{}_{,\alpha}$ and $A^\mu{}_{,\alpha\beta}$. The same remark is applied to any other tensor field. The reason behind is that A_μ and A^μ are interrelated via the metric tensor, $A^\mu = g^{\mu\nu} A_\nu$. Therefore, derivative of A^μ differs from that of A_μ by an additional term involving the derivative of the metric tensor which, if being improperly introduced, can bring about spurious terms to the field equations derived from the principle of the least action.

Variational derivative, $\delta\mathcal{F}/\delta Q$, taken with respect to the variable Q relates a change, δS, in the functional S to a change, $\delta\mathcal{F}$, in the function $\mathcal{F}$ that the functional depends on,

$$\delta S = \int \delta\mathcal{F} d^4x \,, \tag{2.2}$$

where

$$\delta\mathcal{F} = \frac{\partial\mathcal{F}}{\partial Q}\delta Q + \frac{\partial\mathcal{F}}{\partial Q_\alpha}\delta Q_\alpha + \frac{\partial\mathcal{F}}{\partial Q_{\alpha\beta}}\delta Q_{\alpha\beta} \,. \tag{2.3}$$

This is a functional increment of $\mathcal{F}$. The variational derivative is obtained after we single out a total divergence on the right-hand side of (2.3) by making use of the commutation relations, $\delta Q_\alpha = (\delta Q)_{,\alpha}$ and $\delta Q_{\alpha\beta} = (\delta Q)_{,\alpha\beta}$. The total divergence is reduced to a surface term in the integral (2.2) which vanishes on the boundary of the volume of integration. Thus, the variation of S with respect to Q is given by

$$\delta S = \int \frac{\delta \mathcal{F}}{\delta Q} \delta Q d^4 x \,, \tag{2.4}$$

where

$$\frac{\delta \mathcal{F}}{\delta Q} \equiv \frac{\partial \mathcal{F}}{\partial Q} - \frac{\partial}{\partial x^\alpha} \frac{\partial \mathcal{F}}{\partial Q_\alpha} + \frac{\partial^2}{\partial x^\alpha \partial x^\beta} \frac{\partial \mathcal{F}}{\partial Q_{\alpha\beta}} \,. \tag{2.5}$$

Similar procedure can be applied to S by varying it with respect to Q_α and $Q_{\alpha\beta}$. In such a case, we get the variational derivatives of $\mathcal{F}$ with respect to Q_α

$$\frac{\delta \mathcal{F}}{\delta Q_\alpha} \equiv \frac{\partial \mathcal{F}}{\partial Q_\alpha} - \frac{\partial}{\partial x^\beta} \frac{\partial \mathcal{F}}{\partial Q_{\alpha\beta}} \,, \tag{2.6}$$

and that of $\mathcal{F}$ with respect to $Q_{\alpha\beta}$,

$$\frac{\delta \mathcal{F}}{\delta Q_{\alpha\beta}} \equiv \frac{\partial \mathcal{F}}{\partial Q_{\alpha\beta}} \,. \tag{2.7}$$

Let us assume that there is another geometric object, $\mathcal{T}(Q, Q_\alpha, Q_{\alpha\beta})$, which differs from the original one $\mathcal{F}(Q, Q_\alpha, Q_{\alpha\beta})$ by a total divergence

$$\mathcal{T}\left(Q, Q_\alpha, Q_{\alpha\beta}\right) = \mathcal{F}\left(Q, Q_\alpha, Q_{\alpha\beta}\right) + \partial_\beta H^\beta \left(Q, Q_\alpha\right) \,. \tag{2.8}$$

It is well-known [53, 90] that taking the variational derivative (2.5) from $\mathcal{T}$ and $\mathcal{F}$ yields the same result

$$\frac{\delta \mathcal{T}}{\delta Q} \equiv \frac{\delta \mathcal{F}}{\delta Q} \,, \tag{2.9}$$

because the variational derivative from the divergence is zero identically. In fact, it is straightforward to prove that the variational derivative (2.5), after it applies to a partial derivative of an arbitrary smooth function, vanishes

$$\frac{\delta}{\delta Q} \left(\frac{\partial \mathcal{F}}{\partial x^\alpha} \right) \equiv 0 \,. \tag{2.10}$$

However, this property does not hold on for a covariant derivative in the most general case [90].

The variational derivatives are covariant geometric object, that is, they do not depend on the choice of a particular coordinates on manifold [66, 90]. In case, when the dynamic variable Q is not a metric tensor, this statement can be proved by taking the first, $\mathcal{Q}_\alpha \equiv Q_{;\alpha}$, and second, $\mathcal{Q}_{\alpha\beta} \equiv Q_{;\alpha\beta}$, covariant derivatives of Q as independent dynamic variables instead of its partial derivatives, Q_α and $Q_{\alpha\beta}$. In this case, the procedure of derivation of variational derivatives (2.5) and (2.6) remains the same and the result is

$$\frac{\delta \mathcal{F}}{\delta Q} = \frac{\partial \mathcal{F}}{\partial Q} - \left[\frac{\partial \mathcal{F}}{\partial \mathcal{Q}_\alpha} \right]_{;\alpha} + \left[\frac{\partial \mathcal{F}}{\partial \mathcal{Q}_{\alpha\beta}} \right]_{;\beta\alpha} \,. \tag{2.11}$$

The order, in which the covariant derivatives are taken, is imposed by the procedure of the extracting the total divergence from the variation of the action in (2.2). The order of the derivatives is important because the covariant derivatives do not commute.

Variational derivative of $\mathcal{F}$ with respect to the metric tensor $g_{\mu\nu}$ is defined by the same equations (2.5)–(2.7) where we identify $Q \equiv g_{\mu\nu}$, $Q_\alpha \equiv g_{\mu\nu,\alpha}$, and $Q_{\mu\nu} \equiv g_{\mu\nu,\alpha\beta}$. It yields

$$\frac{\delta\mathcal{F}}{\delta g_{\mu\nu}} \equiv \frac{\partial\mathcal{F}}{\partial g_{\mu\nu}} - \frac{\partial}{\partial x^\alpha}\frac{\partial\mathcal{F}}{\partial g_{\mu\nu,\alpha}} + \frac{\partial^2}{\partial x^\alpha \partial x^\beta}\frac{\partial\mathcal{F}}{\partial g_{\mu\nu,\alpha\beta}} \tag{2.12}$$

$$\frac{\delta\mathcal{F}}{\delta g_{\mu\nu,\alpha}} \equiv \frac{\partial\mathcal{F}}{\partial g_{\mu\nu,\alpha}} - \frac{\partial}{\partial x^\beta}\frac{\partial\mathcal{F}}{\partial g_{\mu\nu,\alpha\beta}} , \tag{2.13}$$

$$\frac{\delta\mathcal{F}}{\delta g_{\mu\nu,\alpha\beta}} \equiv \frac{\partial\mathcal{F}}{\partial g_{\mu\nu,\alpha\beta}} . \tag{2.14}$$

Covariant generalization of (2.12)–(2.14) is not quite straightforward because the covariant derivative of the metric tensor $g_{\mu\nu;\alpha} = 0$, and we cannot use it as a dynamic variable. In this case, we consider the set of the metric tensor, $g_{\mu\nu}$, the Christoffel symbols $\Gamma^\alpha{}_{\mu\nu}$, and the Riemann tensor $R^\alpha{}_{\beta\mu\nu}$ as a set of independent dynamic variables. The action is given by (2.1) where $\mathcal{F} \equiv \sqrt{-g}f(g_{\mu\nu}, \Gamma^\alpha{}_{\mu\nu}, R^\alpha{}_{\beta\mu\nu})$ is a scalar density of weight $+1$. Variation of $\mathcal{F}$ is

$$\delta\mathcal{F} = \frac{\partial\mathcal{F}}{\partial g_{\mu\nu}}\delta g_{\mu\nu} + \frac{\partial\mathcal{F}}{\partial\Gamma^\alpha{}_{\mu\nu}}\delta\Gamma^\alpha{}_{\mu\nu} + \frac{\partial\mathcal{F}}{\partial R^\alpha{}_{\beta\mu\nu}}\delta R^\alpha{}_{\beta\mu\nu} , \tag{2.15}$$

where variations of the Christoffel symbols and the Riemann tensor are tensors that can be expressed in terms of the variation $\delta g_{\mu\nu}$ of the metric tensor [112]

$$\delta\Gamma^\alpha{}_{\mu\nu} = \frac{1}{2}g^{\alpha\sigma}\left[(\delta g_{\sigma\mu})_{;\nu} + (\delta g_{\sigma\nu})_{;\mu} - (\delta g_{\mu\nu})_{;\sigma}\right] , \tag{2.16}$$

$$\delta R^\alpha{}_{\beta\mu\nu} = (\delta\Gamma^\alpha{}_{\beta\nu})_{;\mu} - (\delta\Gamma^\alpha{}_{\beta\mu})_{;\nu} . \tag{2.17}$$

Now, we replace (2.16) and (2.17) in (2.15) and single out a total divergence[3]. It yields

$$\delta\mathcal{F} = \frac{\delta\mathcal{F}}{\delta g_{\mu\nu}}\delta g_{\mu\nu} + \mathcal{B}^\alpha{}_{,\alpha} , \tag{2.18}$$

3 The fact that $\mathcal{F}$ is a scalar density is essential for the transformation of covariant derivatives to the total divergence. The total divergences can be converted to surface integrals which vanish on the boundary of integration and, hence, can be dropped off the calculations.

where the total divergence vanishes on the boundary of integration of the action, and the covariant variational derivative is

$$\frac{\delta \mathcal{F}}{\delta g_{\mu\nu}} = \frac{\partial \mathcal{F}}{\partial g_{\mu\nu}} - \frac{1}{2}\left(g^{\sigma\mu}\frac{\partial \mathcal{F}}{\partial \Gamma^{\sigma}{}_{\nu\alpha}} + g^{\sigma\nu}\frac{\partial \mathcal{F}}{\partial \Gamma^{\sigma}{}_{\mu\alpha}} - g^{\sigma\alpha}\frac{\partial \mathcal{F}}{\partial \Gamma^{\sigma}{}_{\mu\nu}}\right)_{;\alpha} + \left(g^{\sigma\mu}\frac{\partial \mathcal{F}}{\partial R^{\sigma}{}_{\alpha\beta\nu}} + g^{\sigma\nu}\frac{\partial \mathcal{F}}{\partial R^{\sigma}{}_{\mu\beta\alpha}} - g^{\sigma\alpha}\frac{\partial \mathcal{F}}{\partial R^{\sigma}{}_{\mu\beta\nu}}\right)_{;\beta\alpha} . \tag{2.19}$$

Variational derivative with respect to the contravariant metric tensor is

$$\frac{\delta \mathcal{F}}{\delta g^{\mu\nu}} = \frac{\partial g_{\alpha\beta}}{\partial g^{\mu\nu}}\frac{\delta \mathcal{F}}{\delta g_{\alpha\beta}} = -g_{\alpha\mu}g_{\beta\nu}\frac{\delta \mathcal{F}}{\delta g_{\alpha\beta}} . \tag{2.20}$$

The variational derivatives are not linear operators. For example, they do not obey Leibniz's rule [48, Section 2.3]. More specifically, for any geometric object, $H = \mathcal{F}\mathcal{T}$, that is a corresponding product of two other geometric objects, $\mathcal{F} = \mathcal{F}(Q, Q_{\alpha}, Q_{\alpha\beta})$ and $\mathcal{T} = \mathcal{T}(Q, Q_{\alpha}, Q_{\alpha\beta})$, the variational derivative

$$\frac{\delta(\mathcal{F}\mathcal{T})}{\delta Q} \neq \frac{\delta \mathcal{F}}{\delta Q}\mathcal{T} + \mathcal{F}\frac{\delta \mathcal{T}}{\delta Q} , \tag{2.21}$$

in the most general case. The chain rule with regard to the variational derivative is preserved in a limited sense. More specifically, let us consider a geometric object $\mathcal{F} = \mathcal{F}(Q, Q_{\alpha}, Q_{\alpha\beta})$ where Q is a function of a variable P, that is, $Q = Q(P)$. Then, the variational derivative

$$\frac{\delta \mathcal{F}}{\delta P} = \frac{\delta \mathcal{F}}{\delta Q}\frac{\partial Q}{\partial P} , \tag{2.22}$$

that can be confirmed by inspection [96]. On the other hand, if we have a singled-valued function $H = H(Q)$, and $Q = Q(P, P_{\alpha}, P_{\alpha\beta})$, the chain rule

$$\frac{\delta H}{\delta P} = \frac{\partial \mathcal{F}}{\partial Q}\frac{\delta Q}{\delta P} , \tag{2.23}$$

is also valid. The chain rule (2.23) will be often used in calculations of this chapter.

2.2 Lie derivative

Lie derivative on the manifold can be viewed as being induced by a diffeomorphism

$$x'^{\alpha} = x^{\alpha} + \xi^{\alpha}(x) , \tag{2.24}$$

such that a vector field ξ^{α} has no self-intersections; thus, defining a congruence of curves which provides a natural mapping of the manifold into itself. Lie derivative of a geometric object $\mathcal{F}$ is denoted as $£_{\xi}\mathcal{F}$. It is defined by a standard rule

$$£_{\xi}\mathcal{F} = \mathcal{F}'(x) - \mathcal{F}(x) , \tag{2.25}$$

where $\mathcal{F}'$ is calculated by doing its coordinate transformation induced by the change of the coordinates (2.24) with subsequent pulling back the transformed object from the point x'^{α} to x^{α} along the congruence ξ^{α} [66]. In particular, for any tensor density $\mathcal{F} = \mathcal{F}^{\mu_1 \ldots \mu_p}_{\nu_1 \ldots \nu_q}$ of type (p, q) and weight m one has

$$\begin{aligned} \pounds_{\xi} \mathcal{F}^{\mu_1 \ldots \mu_p}_{\nu_1 \ldots \nu_q} &= \xi^{\alpha} \mathcal{F}^{\mu_1 \ldots \mu_p}_{\nu_1 \ldots \nu_q ,\alpha} + m \xi^{\alpha}{}_{,\alpha} \mathcal{F}^{\mu_1 \ldots \mu_p}_{\nu_1 \ldots \nu_q} \\ &+ \mathcal{F}^{\mu_1 \ldots \mu_p}_{\alpha \ldots \nu_q} \xi^{\alpha}{}_{,\nu_1} + \cdots + \mathcal{F}^{\mu_1 \ldots \mu_p}_{\nu_1 \ldots \alpha} \xi^{\alpha}{}_{,\nu_q} - \mathcal{F}^{\alpha \ldots \mu_p}_{\nu_1 \ldots \nu_q} \xi^{\mu_1}{}_{,\alpha} - \cdots - \mathcal{F}^{\mu_1 \ldots \alpha}_{\nu_1 \ldots \nu_q} \xi^{\mu_p}{}_{,\alpha} \,. \end{aligned} \tag{2.26}$$

We notice that *all* partial derivatives on the right-hand side of equation (2.26) can be simultaneously replaced with the covariant derivatives because the terms containing the Christoffel symbols cancel each other.

The Lie derivative commutes with a partial (but not a covariant) derivative

$$\partial_{\alpha} \left(\pounds_{\xi} \mathcal{F} \right) = \pounds_{\xi} \left(\partial_{\alpha} \mathcal{F} \right) \,. \tag{2.27}$$

This property allows us to prove that a Lie derivative from a geometric object $\mathcal{F}(Q, Q_{\alpha}, Q_{\alpha\beta})$ can be calculated in terms of its variational derivative. Indeed,

$$\pounds_{\xi} \mathcal{F} = \frac{\partial \mathcal{F}}{\partial Q} \pounds_{\xi} Q + \frac{\partial \mathcal{F}}{\partial Q_{\alpha}} \pounds_{\xi} Q_{\alpha} + \frac{\partial \mathcal{F}}{\partial Q_{\alpha\beta}} \pounds_{\xi} Q_{\alpha\beta} \,. \tag{2.28}$$

Now, after using the commutation property (2.27) and changing the order of partial derivatives in $\pounds_{\xi} Q_{\alpha}$ and $\pounds_{\xi} Q_{\alpha\beta}$, one can express (2.28) as an algebraic sum of the variational derivative and a total divergence

$$\pounds_{\xi} \mathcal{F} = \frac{\delta \mathcal{F}}{\delta Q} \pounds_{\xi} Q + \frac{\partial}{\partial x^{\alpha}} \left(\frac{\delta \mathcal{F}}{\delta Q_{\alpha}} \pounds_{\xi} Q + \frac{\delta \mathcal{F}}{\delta Q_{\alpha\beta}} \pounds_{\xi} Q_{\beta} \right) \,. \tag{2.29}$$

This property of the Lie derivative indicates its close relation to the variational derivative on the manifold and will be used in the calculations that follow. It is also worth pointing out that (2.29) is used for derivation of Noether's theorem of conservation of the canonical stress–energy tensor of the field Q in the case when $\mathcal{F} = \mathcal{L}$ is the Lagrangian density of the field for which the variational derivative vanishes on-shell, $\delta\mathcal{F}/\delta Q = \delta\mathcal{L}/\delta Q = 0$, and $\pounds_{\xi}\mathcal{L} = \partial_{\alpha}(\xi^{\alpha}\mathcal{L})$.

3 Lagrangian and field variables

We accept the Einstein's theory of general relativity and consider a universe filled up with matter consisting of three components. The first two components are: (1) an ideal fluid composed of particles of one type with transmutations excluded; (2) a scalar field; and (3) a matter of the localized astronomical system. The ideal fluid consists of baryons and cold dark matter (CDM), while the scalar field describes dark energy [2].

We assume that these two components do not interact with each other directly, and are the source of the *Friedmann–Lemître–Robertson–Walker* (FLRW) geometry. There is no dissipation in the ideal fluid and in the scalar field so that they can only interact through the gravitational field. It means that the equations of motion for the fluid and the scalar field are decoupled in the main approximation, and we can calculate their evolution separately. Mathematically, it means that the Lagrangian of the ideal fluid and that of the scalar field depend only on their own field variables and the metric tensor.

The tensor of energy–momentum of matter of the localized astronomical system is not specified in agreement with the approach adopted in the post-Newtonian approximation scheme developed in the asymptotically flat spacetime [31, 67]. This allows us to generate all possible types of cosmological perturbations: scalar, vector, and tensor modes. We are mostly interested in developing our formalism for application to the astronomical system of massive bodies bound together by intrinsic gravitational forces such as the solar system, galaxy, or a cluster of galaxies. It means that our approach admits a large density contrast between the background matter and the matter of the localized system. The localized system perturbs the background matter and gravitational field of FLRW metric locally but it is not included to the matter source of the background geometry, at least, in the approximation being linearized with respect to the metric tensor perturbation. Our goal is to study how the perturbations of the background matter and gravitational field are incorporated to the gravitational field perturbations of the standard post-Newtonian theory of relativistic celestial dynamics.

Let us now consider the action functional and the Lagrangian of each component.

3.1 Action functional

We shall consider a theory with the action functional

$$S = \int_{\mathcal{M}} \mathcal{L} d^4x\,, \tag{3.1}$$

where the integration is performed over the entire spacetime manifold $\mathcal{M}$. The Lagrangian $\mathcal{L}$ is comprised of four terms

$$\mathcal{L} = \mathcal{L}^{\mathrm{g}} + \mathcal{L}^{\mathrm{m}} + \mathcal{L}^{\mathrm{q}} + \mathcal{L}^{\mathrm{p}}\,, \tag{3.2}$$

where $\mathcal{L}^{\mathrm{g}}$, $\mathcal{L}^{\mathrm{m}}$, $\mathcal{L}^{\mathrm{q}}$ are the Lagrangians of gravitational field, the dark matter, the scalar field that governs the accelerated expansion of the universe [47], and $\mathcal{L}^{\mathrm{p}}$ is the Lagrangian describing the source of the cosmological perturbations. Gravitational field Lagrangian is

$$\mathcal{L}^{\mathrm{g}} = -\frac{1}{16\pi}\sqrt{-g}R\,, \tag{3.3}$$

where R is the Ricci scalar built of the metric $g_{\alpha\beta}$ and its first and second derivatives [89]. Other Lagrangians depend on the metric and the matter variables. Correct choice of the matter variables is a key element in the development of the Lagrangian theory of the post-Newtonian perturbations of the cosmological manifold caused by a localized astronomical system.

3.2 Lagrangian of the ideal fluid

The ideal fluid is characterized by the following thermodynamic parameters: the rest-mass density $\rho_{\rm m}$, the specific internal energy $\Pi_{\rm m}$ (per unit of mass), pressure $p_{\rm m}$, and entropy $s_{\rm m}$ where the subindex "m" stands for "matter." We shall assume that the entropy of the ideal fluid remains constant, that excludes it from further consideration. The standard approach to the theory of cosmological perturbations presumes that the constant entropy excludes rotational (vector) perturbations of the fluid component from the start, and only scalar (adiabatic) perturbations are generated [2, 91, 113, 114]. However, this chapter deals with the cosmological perturbations that are generated by a localized astronomical system which is described by its own Lagrangian (see Section 3.4) which is left as general as possible. This leads to the tensor of energy–momentum of the matter of the localized system that incorporates the rotational motion of matter which is the source of the rotational perturbations of the background ideal fluid. This extrapolates the concept of the gravitomagnetic field of the post-Newtonian dynamics of localized systems in the asymptotically flat spacetime [19, 27, 66] to cosmology. Further details regarding the vector perturbations are given in Section 6.5 of this chapter.

The total energy density of the fluid is

$$\epsilon_{\rm m} = \rho_{\rm m}(1 + \Pi_{\rm m}) \,. \tag{3.4}$$

One more thermodynamic parameter is the specific enthalpy of the fluid defined as

$$\mu_{\rm m} = \frac{\epsilon_{\rm m} + p_{\rm m}}{\rho_{\rm m}} = 1 + \Pi_{\rm m} + \frac{p_{\rm m}}{\rho_{\rm m}} \,. \tag{3.5}$$

In the most general case, the thermodynamic equation of state of the fluid is given by equation $p_{\rm m} = p_{\rm m}(\rho_{\rm m}, \Pi_{\rm m})$, where the specific internal energy $\Pi_{\rm m}$ is related to pressure by the first law of thermodynamics.

Since the entropy has been assumed to be constant, the first law of thermodynamics reads

$$d\Pi_{\rm m} + p_{\rm m} d\left(\frac{1}{\rho_{\rm m}}\right) = 0 \,. \tag{3.6}$$

It can be used to derive the following thermodynamic relationships:

$$dp_{\rm m} = \rho_{\rm m} d\mu_{\rm m} \,, \tag{3.7}$$

$$d\epsilon_{\rm m} = \mu_{\rm m} d\rho_{\rm m} \,, \tag{3.8}$$

which means that all thermodynamic quantities are solely functions of the specific enthalpy μ_{m}, for example, $\rho_{\mathrm{m}} = \rho_{\mathrm{m}}(\mu_{\mathrm{m}})$, $\Pi_{\mathrm{m}} = \Pi_{\mathrm{m}}(\mu_{\mathrm{m}})$, etc. The equation of state is also a function of the variable μ_{m}, that is,

$$p_{\mathrm{m}} = p_{\mathrm{m}}(\mu_{\mathrm{m}}) . \tag{3.9}$$

Derivatives of the thermodynamic quantities with respect to μ_{m} can be calculated by making use of Equations (3.7) and (3.8), and the definition of the (adiabatic) speed of sound v_{s} of the fluid

$$\frac{\partial p_{\mathrm{m}}}{\partial \epsilon_{\mathrm{m}}} = \frac{v_{\mathrm{s}}^2}{c^2} , \tag{3.10}$$

where the partial derivative is taken under a condition that the entropy, s_{m}, of the fluid does not change. Then, the derivatives of the thermodynamic quantities take on the following form:

$$\frac{\partial p_{\mathrm{m}}}{\partial \mu_{\mathrm{m}}} = \rho_{\mathrm{m}} , \qquad \frac{\partial \epsilon_{\mathrm{m}}}{\partial \mu_{\mathrm{m}}} = \frac{c^2}{v_{\mathrm{s}}^2} \rho_{\mathrm{m}} , \qquad \frac{\partial \rho_{\mathrm{m}}}{\partial \mu_{\mathrm{m}}} = \frac{c^2}{v_{\mathrm{s}}^2} \frac{\rho_{\mathrm{m}}}{\mu_{\mathrm{m}}} , \tag{3.11}$$

where all partial derivatives are performed under the same condition of constant entropy.

The Lagrangian of the ideal fluid is usually taken in the form of the total energy density, $L^{\mathrm{m}} = \sqrt{-g}\epsilon_{\mathrm{m}}$ [89]. However, this form is less convenient for applying the variational calculus on manifolds. The above thermodynamic relationships and the integration by parts of the action (3.1) allow us to recast the Lagrangian $L^{\mathrm{m}} = \sqrt{-g}\epsilon_{\mathrm{m}}$ to the form of pressure, $L^{\mathrm{m}} = -\sqrt{-g}p_{\mathrm{m}}$, so that the Lagrangian density becomes (see [66, pp. 334–335] for more details)

$$\mathcal{L}^{\mathrm{m}} = -\sqrt{-g}p_{\mathrm{m}} = \sqrt{-g}\left(\epsilon_{\mathrm{m}} - \rho_{\mathrm{m}}\mu_{\mathrm{m}}\right) . \tag{3.12}$$

Theoretical description of the ideal fluid as a dynamic system on spacetime manifold is given the most conveniently in terms of the Clebsch potential, Φ which is also called the velocity potential [102]. In the case of a single-component ideal fluid the Clebsch potential is introduced by the following relationship:

$$\mu_{\mathrm{m}} w_\alpha = -\Phi_{,\alpha} . \tag{3.13}$$

In fact, equation (3.13) is a solution of relativistic equations of motion of the ideal fluid [73].

The Clebsch potential is a primary field variable in the Lagrangian description of the isentropic ideal fluid. The 4-velocity is normalized to $w^\alpha w_\alpha = g_{\alpha\beta} w^\alpha w^\beta = -1$, so that the specific enthalpy can be expressed in the following form:

$$\mu_{\mathrm{m}} = \sqrt{-g^{\alpha\beta}\Phi_{,\alpha}\Phi_{,\beta}} . \tag{3.14}$$

One may also notice that

$$\mu_{\mathrm{m}} = w^\alpha \Phi_{,\alpha} . \tag{3.15}$$

It is important to notice that the Clebsch potential Φ has no direct physical meaning as it can be changed to another value $\Phi \to \Phi' = \Phi + \tilde{\Phi}$ such that the gauge function, $\tilde{\Phi}$, is constant along the world lines of the fluid: $w^\alpha \tilde{\Phi}_{,\alpha} = 0$.

In terms of the Clebsch potential the Lagrangian (3.12) of the ideal fluid is

$$\mathcal{L}^{\mathrm{m}} = \sqrt{-g}\left(\epsilon_{\mathrm{m}} - \rho_{\mathrm{m}}\sqrt{-g^{\alpha\beta}\Phi_{,\alpha}\Phi_{,\beta}}\right) . \tag{3.16}$$

Metrical tensor of energy–momentum of the ideal fluid is obtained by taking a variational derivative of the Lagrangian (3.16) with respect to the metric tensor,

$$T^{\mathrm{m}}_{\alpha\beta} = \frac{2}{\sqrt{-g}}\frac{\delta\mathcal{L}^{\mathrm{m}}}{\delta g^{\alpha\beta}} . \tag{3.17}$$

Calculation yields

$$T^{\mathrm{m}}_{\alpha\beta} = \left(\epsilon_{\mathrm{m}} + p_{\mathrm{m}}\right) w_\alpha w_\beta + p_{\mathrm{m}} g_{\alpha\beta} , \tag{3.18}$$

where $w^\alpha = dx^\alpha/d\tau$ is the 4-velocity of the fluid, and τ is the proper time of the fluid element taken along its world line. This is a standard form of the tensor of energy–momentum of the ideal fluid [89]. Because the Lagrangian (3.16) is expressed in terms of the dynamical variable Φ, the Noether approach based on taking the variational derivative of the Lagrangian with respect to the field variable, can be applied to derive the canonical tensor of the energy–momentum of the ideal fluid. This calculation has been done in [66, pp. 334–335] and it leads to the expression (3.18). It could be expected because we assumed that the ideal fluid consists of bosons. The metrical and canonical tensors of energy–momentum for the liquid differ, if and only if, the liquid's particles are fermions (see [66, pp. 331–332] for more detail). We do not consider the fermionic liquids in this chapter.

3.3 Lagrangian of scalar field

The Lagrangian of the scalar field Ψ is given by

$$\mathcal{L}^{\mathrm{q}} = \sqrt{-g}\left(\frac{1}{2}g^{\alpha\beta}\partial_\alpha\Psi\partial_\beta\Psi + W\right) , \tag{3.19}$$

where $W \equiv W(\Psi)$ is a potential of the scalar field. We assume that there is no direct coupling between the scalar field and the matter of the ideal fluid. They can interact only through the gravitational field. Many different potentials of the scalar field are used in cosmology [2]. At this step, we do not chose a specific form of the potential which will be selected later.

Metrical tensor of energy–momentum of the scalar field is obtained by taking a variational derivative

$$T^{\mathrm{q}}_{\alpha\beta} = \frac{2}{\sqrt{-g}}\frac{\delta\mathcal{L}^{\mathrm{q}}}{\delta g^{\alpha\beta}} , \tag{3.20}$$

that yields

$$T^{\mathrm{q}}_{\alpha\beta} = \partial_\alpha \Psi \partial_\beta \Psi - g_{\alpha\beta}\left[\frac{1}{2}g^{\mu\nu}\partial_\mu \Psi \partial_\nu \Psi + W(\Psi)\right] . \tag{3.21}$$

The canonical tensor of energy–momentum of the scalar field is obtained by applying the Neother theorem and leads to the same expression (3.21).

One can *formally* reduce the tensor (3.21) to the form similar to that of the ideal fluid by making use of the following procedure. First, we define the analog of the specific enthalpy of the scalar field "fluid"

$$\mu_{\mathrm{q}} = \sqrt{-g^{\sigma\nu}\partial_\sigma \Psi \partial_\nu \Psi} , \tag{3.22}$$

and the effective 4-velocity, v^α, of the "fluid"

$$\mu_{\mathrm{q}} v_\alpha = -\partial_\alpha \Psi . \tag{3.23}$$

The 4-velocity v^α is normalized to $v_\alpha v^\alpha = -1$. Therefore, the scalar field enthalpy μ_{q} can be expressed in terms of the partial derivative from the scalar field

$$\mu_{\mathrm{q}} = v^\alpha \partial_\alpha \Psi . \tag{3.24}$$

Then, we introduce the analog of the rest mass density ρ_{q} of the scalar field "fluid" by defining,

$$\rho_{\mathrm{q}} \equiv \mu_{\mathrm{q}} = v^\alpha \partial_\alpha \Psi = \sqrt{-g^{\sigma\nu}\partial_\sigma \Psi \partial_\nu \Psi} . \tag{3.25}$$

As a consequence of the above definitions, the energy density, ϵ_{q} and pressure p_{q} of the scalar field "fluid" can be introduced as follows:

$$\epsilon_{\mathrm{q}} \equiv -\frac{1}{2}g^{\sigma\nu}\partial_\sigma \Psi \partial_\nu \Psi + W(\Psi) = \frac{1}{2}\rho_{\mathrm{q}}\mu_{\mathrm{q}} + W(\Psi) , \tag{3.26}$$

$$p_{\mathrm{q}} \equiv -\frac{1}{2}g^{\sigma\nu}\partial_\sigma \Psi \partial_\nu \Psi - W(\Psi) = \frac{1}{2}\rho_{\mathrm{q}}\mu_{\mathrm{q}} - W(\Psi) . \tag{3.27}$$

One notices that a relationship

$$\mu_{\mathrm{q}} = \frac{\epsilon_{\mathrm{q}} + p_{\mathrm{q}}}{\rho_{\mathrm{q}}} , \tag{3.28}$$

between the specific enthalpy μ_{q}, the density ρ_{q}, the pressure p_{q}, and the energy density ϵ_{q}, of the scalar field "fluid" formally holds on the same form (3.5) as in the case of the barotropic ideal fluid.

After applying the above-given definitions in equation (3.21), it is formally reduced to the tensor of energy–momentum of an ideal fluid

$$T^{\mathrm{q}}_{\alpha\beta} = \left(\epsilon_{\mathrm{q}} + p_{\mathrm{q}}\right) v_\alpha v_\beta + p_{\mathrm{q}} g_{\alpha\beta} . \tag{3.29}$$

It is worth emphasizing that the analogy between the tensor of energy–momentum (3.29) of the scalar field "fluid" with that of the barotropic ideal fluid (3.18) is rather formal since the scalar field, in the most general case, does not satisfy *all* required thermodynamic equations because of the presence of the potential $W = W(\Psi)$ in the energy density ϵ_q, and pressure p_q of the scalar field. For example, equation of continuity (4.66) for scalar field differs from that for the ideal fluid (4.58) if the potential $W(\Psi) \neq 0$.

3.4 Lagrangian of a localized astronomical system

The Lagrangian $\mathcal{L}^p$ of matter of a localized astronomical system (a small-scale structure inhomogeneity) which perturbs the geometry of the background manifold of the FLRW metric, can be chosen arbitrary. We shall call the perturbation of the background manifold that is induced by $\mathcal{L}^p$, the *bare* perturbation. We assume that the matter of the bare perturbation is described by a (multicomponent) field variable, θ, which physical meaning depends on a specific problem we want to solve. The Lagrangian density of the bare perturbation is given by $\mathcal{L}^p = \sqrt{-g}L^p(\theta, g_{\alpha\beta})$. The tensor of energy–momentum of the matter of the bare perturbation, $\mathfrak{T}_{\alpha\beta}$, is obtained by taking a variational derivative

$$\mathfrak{T}_{\alpha\beta} = \frac{2}{\sqrt{-g}}\frac{\delta\mathcal{L}^p}{\delta g^{\alpha\beta}} \,. \tag{3.30}$$

Tensor $\mathfrak{T}_{\alpha\beta}$ is a source of the small-scale gravitational perturbation of the background manifold that is associated with a particular solution of the linearized Einstein equations which will be derived in the next sections.

4 Background manifold

4.1 Hubble flow

We shall consider the background universe as described by the Friedmann-Lemître–Robertson–Walker (FLRW) metric. The functional form of the metric depends on the coordinates introduced on the manifold. Because the FLRW metric describes homogeneous and isotropic spacetime there is a preferred class of coordinates which clearly reveal these properties of the background manifold. These coordinates materialize a special set of freely falling observers, called comoving observers. These observers are following with the flow of the expanding universe and have constant values of spatial coordinates. The proper distance between the comoving observers increases in proportion to the scale factor $R(T)$. In the preferred cosmological coordinates, the time coordinate of the FLRW metric is just the proper time as measured by the comoving observers. A particle moving relative to the local comoving observers has a peculiar velocity with respect to the Hubble flow. An observer with a nonzero peculiar velocity does not see the universe as isotropic.

For example, the peculiar velocity of the solar system implies the dipole anisotropy of cosmic microwave background (CMBR) radiation corresponding to $|\boldsymbol{v}_{\odot}| = 369.0 \pm 0.9\ \mathrm{km\,s^{-1}}$, toward a point with the galactic coordinates $(l, b) = (264^{\circ}, 48^{\circ})$ [52, 59]. Such a solar system's velocity implies a velocity $|\boldsymbol{v}_{LG}| = 627 \pm 22\,\mathrm{km\,s^{-1}}$ toward $(l, b) = (276^{\circ}, 30^{\circ})$ for our Galaxy and the Local Group of galaxies relative to the CMBR [43, 63]. The existence of the preferred frame in cosmology should not be understood as a violation of the Einstein principle of relativity. Indeed, any coordinate chart can be used in order to describe the FLRW metric. A preferred frame exists merely because the FLRW metric admits only six-parametric group (three spatial translations and three spatial rotations) as contrasted with the ten-parametric group of Minkowski (or de Sitter) spacetime which includes the time translation and three Lorentz boosts as well. The metric of FLRW does not remain invariant with respect to the time translation and the Lorentz transformations because its expansion makes different spacelike hypersurfaces nonequivalent. It may lead to some interesting observational predictions of cosmological effects within the solar system [69].

4.2 Friedmann–Lemître–Robertson–Walker metric

In what follows, we shall consider the problem of calculation of the post-Newtonian perturbations in the expanding universe described by the FLRW class of models. The FLRW metric is an exact solution of Einstein's field equations of general relativity that describes a homogeneous, isotropically expanding or contracting universe. The general form of the metric follows from the geometric properties of homogeneity and isotropy of the manifold [113, 114]. Einstein's equations are only needed to derive the scale factor of the universe as a function of time.

The most general form of the FLRW metric is given by

$$ds^2 = -dT^2 + R^2 \left[\frac{d\rho^2}{1 - k\rho^2} + \rho^2 \left(d^2\vartheta + \sin^2 \vartheta d^2 \upsilon \right) \right] , \tag{4.1}$$

where T is the coordinate time, $\{\rho, \vartheta, \upsilon\}$ are spherical coordinates, $R = R(T)$ is the scale factor depending on time and characterizing the size of the universe compared to the present value of $R = 1$. The time T has a physical meaning of the proper time of a comoving observer that is being at rest with respect to the cosmological frame of reference. The present epoch corresponds to the value of the time $T = T_0$. The constant k can take on three different values $k = \{-1, 0, +1\}$, where $k = -1$ corresponds to the spatial hyperbolic geometry, $k = 0$ does the spatially flat FLRW model, and $k = +1$ does the spatially closed world [89].

The Hubble parameter H characterizes the rate of the temporal evolution of the universe. It is defined by

$$H \equiv \frac{\dot{R}}{R} = \frac{1}{R}\frac{dR}{dT} . \tag{4.2}$$

For mathematical reasons, it is convenient to introduce a conformal time, η, via differential equation

$$d\eta = \frac{dT}{R(T)} \,. \tag{4.3}$$

If the time dependence of the scale factor is known, equation (4.3) can be solved; thus, yielding $T = T(\eta)$. It allows us to re-express the scale factor $R(T)$ in terms of the conformal time, $R(T(\eta)) \equiv a(\eta)$. The conformal Hubble parameter is, then, defined as

$$\mathcal{H} \equiv \frac{a'}{a} = \frac{1}{a}\frac{da}{d\eta} \,. \tag{4.4}$$

The two expressions for the Hubble parameters are related by means of equation

$$H = \frac{\mathcal{H}}{a} \,, \tag{4.5}$$

that allows us to link their time derivatives

$$a^2\dot{H} = \mathcal{H}' - \mathcal{H}^2 \,, \tag{4.6}$$

$$a^3\ddot{H} = \mathcal{H}'' - 4\mathcal{H}\mathcal{H}' + 2\mathcal{H}^3 \,, \tag{4.7}$$

and so on.

It is also convenient to introduce the isotropic Cartesian coordinates $X^i = \{X, Y, Z\}$, by transforming the radial coordinate

$$\rho = \frac{r}{1 + \frac{k}{4}r^2} \,, \tag{4.8}$$

and defining $r^2 = X^2 + Y^2 + Z^2 = \delta_{ij}X^iX^j$. In the isotropic coordinates the interval (4.1) takes on the following form:

$$ds^2 = G_{\alpha\beta}dX^\alpha dX^\beta \,, \tag{4.9}$$

where the coordinates $X^\alpha = \{X^0, X^1, X^2, X^3\} = \{\eta, X, Y, Z\}$, and the metric have a conformal form

$$G_{\alpha\beta} = a^2(\eta)g_{\alpha\beta} \tag{4.10}$$

$$g_{00} = -1 \,, \quad g_{0i} = 0 \,, \quad g_{ij} = \frac{\delta_{ij}}{\left(1 + \frac{k}{4}r^2\right)^2} \,. \tag{4.11}$$

Determinant of the metric $G_{\alpha\beta}$ is $G = \det(G_{\alpha\beta}) = a^8 g$, where $g = \det(g_{\alpha\beta}) = -(1 + kr^2/4)^{-6}$. The spacetime interval in the isotropic Cartesian coordinates reads

$$ds^2 = a^2(\eta)\left[-d\eta^2 + \frac{\delta_{ij}dX^idX^j}{\left(1 + \frac{k}{4}r^2\right)^2}\right] \,. \tag{4.12}$$

The distinctive property of the isotropic coordinates in the FLRW metric is that the radial coordinate r is defined in such a way that the three-dimensional space looks exactly Euclidean and null cones appear in it as round spheres irrespectively of the value of the space curvature k. The isotropic coordinates do not represent proper distances on the sphere, nor does the radial coordinate r represents a proper radial distance measured with the help of radar astronomy technique. The proper spatial distance in the isotropic coordinates is $(1+kr^2/4)^{-1}ar$ [113].

The FLRW metric presented in the conformal form by equation (4.12) singles out a preferred cosmological reference frame defined by the congruence of world lines of the fiducial test particles being at rest with respect to the spatial coordinates X^i. 4-velocity of a fiducial particle is denoted as $\overline{U}^\alpha = dX^\alpha/d\tau$, where $d\tau = -ds$ is the proper time on the world line of the particle. In the isotropic conformal coordinates, $\overline{U}^\alpha = (1/a, 0, 0, 0)$. The 4-velocity is a unit vector, $\overline{U}^\alpha\overline{U}_\alpha = G_{\alpha\beta}\overline{U}^\alpha\overline{U}^\beta = -1$. It implies that the covariant components of the 4-velocity are $\overline{U}_\alpha = (-a, 0, 0, 0)$. In the preferred frame the universe looks homogeneous and isotropic. The choice of the isotropic Cartesian coordinates reflects these fundamental properties explicitly in the symmetric form of the metric (4.10). However, the set of the fiducial particles is a mathematical idealization. In reality, any isolated astronomical systems (galaxy, binary star, the solar system, etc.) have a peculiar velocity with respect to the preferred cosmological frame formed by the Hubble flow. We have to introduce a locally inertial coordinate chart which is associated with the isolated system and moves along with it. Transformation from the preferred cosmological frame to the local chart must include the Lorentz boost and a geometric part due to the expansion and curvature of cosmological spacetime. It can take on multiple forms which originate from certain geometric and/or experimental requirements [24, 26, 54, 61].

We do not impose specific limitations on the choice of coordinates on the background manifold and keep the overall formalism of the post-Newtonian approximations, covariant. The arbitrary coordinates are denoted as $x^\alpha = (x^0, x^i)$ and they are related to the preferred isotropic coordinates $X^\alpha = (\eta, X^i)$ by the coordinate transformation $x^\alpha = x^\alpha(X^\beta)$. This transformation has inverse $X^\alpha = X^\alpha(x^\beta)$, at least in some local domain of the background manifold. In this domain, the matrices of the coordinate transformations

$$\Lambda^\alpha{}_\beta = \frac{\partial x^\alpha}{\partial X^\beta}\,, \qquad \mathrm{M}^\alpha{}_\beta = \frac{\partial X^\alpha}{\partial x^\beta}\,, \tag{4.13}$$

and they satisfy to the apparent equalities $\Lambda^\alpha{}_\gamma \mathrm{M}^\gamma{}_\beta = \delta^\alpha_\beta$ and $\mathrm{M}^\alpha{}_\gamma \Lambda^\gamma{}_\beta = \delta^\alpha_\beta$.

4-velocity of the Hubble observers written in the arbitrary coordinates has the following form:

$$\bar{u}^\alpha = \Lambda^\alpha{}_\beta \overline{U}^\beta = a^{-1}\Lambda^\alpha{}_0\,, \quad \bar{u}_\alpha = \mathrm{M}^\beta{}_\alpha U_\beta = -a\mathrm{M}^0{}_\alpha\,. \tag{4.14}$$

The background FLRW metric written down in the arbitrary coordinates, x^α, takes the following form:

$$\bar{g}_{\alpha\beta}(x^\alpha) = a^2 \bar{\mathfrak{f}}_{\alpha\beta}(x^\alpha) . \tag{4.15}$$

Here the scalar function $a(x^\alpha) \equiv a\left[\eta(x^\alpha)\right]$, and the conformal metric

$$\bar{\mathfrak{f}}_{\alpha\beta}(x^\alpha) = \mathrm{M}^\mu{}_\alpha \mathrm{M}^\nu{}_\beta \mathrm{g}_{\mu\nu}(X^i) . \tag{4.16}$$

Any metric admits 1+3 decomposition with respect to a congruence of a timelike vector field [89]. FLRW metric admits a privileged congruence formed by the 4-velocity $\bar{u}^\alpha$ of the Hubble observers which is a physically privileged vector field. The 1+3 decomposition of the FLRW metric is applied in arbitrary coordinates and has the following form:

$$\bar{g}_{\alpha\beta} = -\bar{u}_\alpha \bar{u}_\beta + \bar{P}_{\alpha\beta} , \tag{4.17}$$

where the tensor

$$\bar{P}_{\alpha\beta} = a^2 \mathrm{M}^i{}_\alpha \mathrm{M}^j{}_\beta \mathrm{g}_{ij} , \tag{4.18}$$

describes the metric on the spacelike hypersurface being everywhere orthogonal to the 4-velocity $\bar{u}^\alpha$ of the Hubble flow. Tensor $\bar{P}_{\alpha\beta}$ is the operator of projection on this hypersuface. It can be also interpreted as a metric on the hypersurace of orthogonality to the Hubble vector flow. Equation (4.17) can be used in order to prove that $\bar{P}_{\alpha\beta}$ satisfies the following relationship:

$$\bar{P}^{\beta\mu} \bar{P}_\beta{}^\nu = \bar{P}^{\mu\nu} , \tag{4.19}$$

which can be confirmed by inspection. The trace $\bar{P}^\alpha{}_\alpha = \bar{g}^{\alpha\beta} \bar{P}_{\alpha\beta} = \bar{P}^{\alpha\beta} \bar{P}_{\alpha\beta} = 3$.

Now, we consider how to express the partial derivatives of any scalar function $F = F(\eta)$, which depends only on the conformal time $\eta = \eta(x^\alpha)$, in terms of the 4-velocity $\bar{u}^\alpha$ of the Hubble flow. Taking into account that $\eta = x^0$ and applying equation (4.14), we obtain

$$F_{,\alpha} = \frac{\partial F}{\partial x^\alpha} = \frac{dF}{d\eta}\frac{\partial \eta}{\partial x^\alpha} = F' \mathrm{M}^0{}_\alpha = -\frac{F'}{a}\bar{u}_\alpha = -\dot{F}\bar{u}_\alpha . \tag{4.20}$$

In particular, the partial derivative from the scale factor, $a_{,\alpha} = -\dot{a}\bar{u}_\alpha = -\mathcal{H}\bar{u}_\alpha$, and the partial derivative from the Hubble parameter $\mathcal{H}_{,\alpha} = -\dot{\mathcal{H}}\bar{u}_\alpha$.

4.3 Christoffel symbols and covariant derivatives

In the following sections, we will need to calculate the covariant derivatives from various geometric objects on the background cosmological manifold covered by an arbitrary coordinate chart $x^\alpha = (x^0, x^i)$. The calculation engages the affine connection $\bar{\Gamma}^\alpha_{\beta\gamma}$ of the background manifold which is decomposed into an algebraic sum of two connections (the Christoffel symbols) because of the conformal structure of the FLRW metric [111]. By definition,

$$\overline{\Gamma}^{\alpha}{}_{\beta\gamma} = \frac{1}{2}\overline{g}^{\alpha\nu}\left(\overline{g}_{\nu\beta,\gamma} + \overline{g}_{\nu\gamma,\beta} - \overline{g}_{\beta\gamma,\nu}\right), \tag{4.21}$$

where

$$\overline{g}_{\alpha\beta,\gamma} = -2H\overline{g}_{\alpha\beta}\overline{u}_{\gamma} + a^2\overline{\mathfrak{f}}_{\alpha\beta,\gamma}\,. \tag{4.22}$$

Separating terms on the right-hand side of (4.21) yields

$$\overline{\Gamma}^{\alpha}{}_{\beta\gamma} = \overline{A}^{\alpha}{}_{\beta\gamma} + \overline{B}^{\alpha}{}_{\beta\gamma}\,, \tag{4.23}$$

where

$$\overline{A}^{\alpha}{}_{\beta\gamma} = -H\left(\delta^{\alpha}_{\beta}\overline{u}_{\gamma} + \delta^{\alpha}_{\gamma}\overline{u}_{\beta} - \overline{u}^{\alpha}\overline{g}_{\beta\gamma}\right), \tag{4.24}$$

and

$$\overline{B}^{\alpha}{}_{\beta\gamma} = \frac{1}{2}\overline{\mathfrak{f}}^{\alpha\mu}\left(\overline{\mathfrak{f}}_{\mu\beta,\gamma} + \overline{\mathfrak{f}}_{\mu\gamma,\beta} - \overline{\mathfrak{f}}_{\beta\gamma,\mu}\right). \tag{4.25}$$

The nonvanishing components of the connections are given in the isotropic Cartesian coordinates X^{α} by

$$\overline{A}^{\alpha}{}_{0\beta} = \mathcal{H}\delta^{\alpha}_{\beta}\,, \quad \overline{A}^{0}{}_{ij} = \mathcal{H}\mathrm{g}_{ij}\,, \quad \overline{B}^{i}{}_{pq} = -\frac{k}{2}\frac{\delta^{i}_{p}X_{q} + \delta^{i}_{q}X_{p} - \delta_{pq}X^{i}}{1 + \frac{k}{4}r^2}\,, \tag{4.26}$$

where $X_q \equiv \delta_{qj}X^j$, and all the other components of the connections vanish.

A covariant derivative of a geometric object (scalar, vector, etc.) on the background manifold is denoted in this chapter with a vertical bar. For example, the covariant derivative of a vector field F^{α} is

$$F^{\alpha}{}_{|\beta} = F^{\alpha}{}_{,\beta} + \overline{\Gamma}^{\alpha}{}_{\beta\gamma}F^{\gamma}\,, \tag{4.27}$$

where a comma in front of subindex β denotes a partial derivative with respect to coordinate x^{β}. Equation (4.27) can be brought to yet another form if we denote the covariant derivative of the affine connection $\overline{B}^{\alpha}{}_{\beta\gamma}$ with a semicolon. Making use of (4.23) in equation (4.27) transforms it to the following form:

$$F^{\alpha}{}_{|\beta} = F^{\alpha}{}_{;\beta} + \overline{A}^{\alpha}{}_{\beta\gamma}F^{\gamma}\,. \tag{4.28}$$

The covariant derivative of a covector F_{α} is defined in a similar way

$$F_{\alpha|\beta} = F_{\alpha,\beta} - \overline{\Gamma}^{\gamma}{}_{\alpha\beta}F_{\gamma} \tag{4.29}$$

which is equivalent to

$$F_{\alpha|\beta} = F_{\alpha;\beta} - \overline{A}^{\gamma}{}_{\alpha\beta}F_{\gamma}\,, \tag{4.30a}$$

$$F_{\alpha;\beta} = F_{\alpha,\beta} - \overline{B}^{\gamma}{}_{\alpha\beta}F_{\gamma}\,. \tag{4.30b}$$

Equations for tensors of higher rank can be presented in a similar way. Of course, the covariant derivative of a scalar field F always coincides with its covariant derivative by definition,

$$F_{|\alpha} = F_{;\alpha} = F_{,\alpha} . \tag{4.31}$$

We also provide an equation for the covariant derivative of the 4-velocity of the Hubble flow. Doing calculations in the isotropic coordinates X^α for the 4-velocity $\overline{U}^\alpha$, and applying the tensor law of transformation to arbitrary coordinates x^α, results in

$$\overline{u}_{\alpha|\beta} = H\overline{P}_{\alpha\beta} , \quad \overline{u}^\alpha{}_{|\beta} = H\left(\delta^\alpha_\beta + \overline{u}^\alpha\overline{u}_\beta\right) , \quad \overline{u}^{\alpha|\beta} = H\overline{P}^{\alpha\beta} , \tag{4.32}$$

where the tensor indices are raised and lowered with the metric $\overline{g}_{\alpha\beta}$.

4.4 Riemann tensor

The Riemann tensor is defined by

$$\overline{R}^\alpha{}_{\beta\mu\nu} = \overline{\Gamma}^\alpha{}_{\beta\nu,\mu} - \overline{\Gamma}^\alpha{}_{\beta\mu,\nu} + \overline{\Gamma}^\alpha{}_{\mu\gamma}\overline{\Gamma}^\gamma{}_{\beta\nu} - \overline{\Gamma}^\alpha{}_{\nu\gamma}\overline{\Gamma}^\gamma{}_{\beta\mu} \tag{4.33}$$

and can be calculated directly from this equation. We prefer a slightly different way by making use of the algebraic decomposition of the Riemann tensor into the irreducible parts

$$\overline{R}_{\alpha\beta\mu\nu} = \overline{C}_{\alpha\beta\mu\nu} + \frac{1}{2}\left(\overline{S}_{\alpha\mu}\overline{g}_{\beta\nu} + \overline{S}_{\beta\nu}\overline{g}_{\alpha\mu} - \overline{S}_{\alpha\nu}\overline{g}_{\beta\mu} - \overline{S}_{\beta\mu}\overline{g}_{\alpha\nu}\right) + \frac{\overline{R}}{12}\left(\overline{g}_{\alpha\mu}\overline{g}_{\beta\nu} - \overline{g}_{\alpha\nu}\overline{g}_{\beta\mu}\right) , \tag{4.34}$$

where $\overline{C}_{\alpha\beta\mu\nu}$ is the Weyl tensor,

$$\overline{S}_{\mu\nu} = \overline{R}_{\mu\nu} - \frac{1}{4}\overline{R}\overline{g}_{\mu\nu} , \tag{4.35}$$

$\overline{R}_{\mu\nu} = \overline{g}^{\alpha\beta}\overline{R}_{\alpha\mu\beta\nu}$ is the Ricci tensor, and $R = \overline{g}^{\alpha\beta}\overline{R}_{\alpha\beta}$ is the Ricci scalar. The Weyl tensor of a conformally flat spacetime vanishes identically,

$$\overline{C}_{\alpha\beta\mu\nu} \equiv 0 . \tag{4.36}$$

Therefore, FLRW cosmological metric (4.1) has a remarkable property – it can be always brought up to the conformally flat form by applying an appropriate coordinate transformation [55].

Direct evaluation of other tensors entering (4.34) by making use of the FLRW metric (4.10), (4.11) yields

$$\overline{R}_{\mu\nu} = \frac{1}{a^2}\left[\mathcal{H}'\left(\overline{g}_{\mu\nu} - 2\overline{u}_\mu\overline{u}_\nu\right) + 2\left(\mathcal{H}^2 + k\right)\left(\overline{g}_{\mu\nu} + \overline{u}_\mu\overline{u}_\nu\right)\right] , \tag{4.37}$$

$$\overline{S}_{\mu\nu} = \frac{2}{a^2}\left(-\mathcal{H}' + \mathcal{H}^2 + k\right)\left(\overline{u}_\mu\overline{u}_\nu + \frac{1}{4}\overline{g}_{\mu\nu}\right) , \tag{4.38}$$

$$\overline{R} = \frac{6}{a^2}\left(\mathcal{H}' + \mathcal{H}^2 + k\right) . \tag{4.39}$$

Making use of equations (4.36)–(4.39) in the decomposition (4.34) of the Riemann tensor, yields the following result:

$$\bar{R}_{\alpha\beta\mu\nu} = \frac{1}{a^2}\left[\mathcal{H}'\left(\bar{g}_{\alpha\mu}\bar{g}_{\beta\nu} - \bar{g}_{\alpha\nu}\bar{g}_{\beta\mu}\right) - \left(\mathcal{H}' - \mathcal{H}^2 - k\right)\left(\bar{P}_{\alpha\mu}\bar{P}_{\beta\nu} - \bar{P}_{\alpha\nu}\bar{P}_{\beta\mu}\right)\right], \tag{4.40}$$

where

$$\bar{P}_{\alpha\beta} = \bar{g}_{\alpha\beta} + \bar{u}_\alpha\bar{u}_\beta, \tag{4.41}$$

is the operator of projection that was introduced earlier in (4.18).

4.5 The Friedmann equations

The Einstein tensor $\bar{\mathcal{E}}_{\alpha\beta} \equiv \bar{R}_{\alpha\beta} - \bar{g}_{\alpha\beta}\bar{R}/2$ of the FLRW cosmological model is derived from equations (4.37) and (4.39). It reads

$$\bar{\mathcal{E}}_{\alpha\beta} = -\frac{1}{a^2}\left[2\left(\mathcal{H}' - \mathcal{H}^2 - k\right)\bar{P}_{\alpha\beta} + 3\left(\mathcal{H}^2 + k\right)\bar{g}_{\alpha\beta}\right]. \tag{4.42}$$

Einstein's field equations on the background spacetime takes the following form:

$$\bar{\mathcal{E}}_{\alpha\beta} = 8\pi\bar{T}_{\alpha\beta}, \tag{4.43}$$

where the tensor of energy–momentum of the background spacetime manifold includes the background matter and the scalar field

$$\bar{T}_{\alpha\beta} = \bar{T}^{\mathrm{m}}_{\alpha\beta} + \bar{T}^{\mathrm{q}}_{\alpha\beta}. \tag{4.44}$$

Here, tensors of energy–momentum on the right-hand side of Einstein's equations are derived from the Lagrangians (3.16) and (3.19), and represent an algebraic sum of tensors (3.18) and (3.22). Each tensor of energy–momentum, $\bar{T}^{\mathrm{m}}_{\alpha\beta}$ and $\bar{T}^{\mathrm{q}}_{\alpha\beta}$, is Lie invariant with respect to the group of symmetry of the background FLRW metric independently, and each of them has the form of the tensor of energy–momentum of the perfect (ideal) fluid. Hence, the tensor of energy–momentum $\bar{T}_{\alpha\beta}$ on the right-hand side of (4.43) has the form of a perfect fluid as well,

$$\bar{T}_{\alpha\beta} = (\bar{\epsilon} + \bar{p})\,\bar{u}_\alpha\bar{u}_\beta + \bar{p}\,\bar{g}_{\alpha\beta}. \tag{4.45}$$

It imposes a certain restriction on the effective energy density $\bar{\epsilon}$ and pressure $\bar{p}$ which must obey Dalton's law for a partial energy density and pressure of the background matter and the scalar field components [10]

$$\bar{\epsilon} = \bar{\epsilon}_{\mathrm{m}} + \bar{\epsilon}_{\mathrm{q}}, \tag{4.46}$$

$$\bar{p} = \bar{p}_{\mathrm{m}} + \bar{p}_{\mathrm{q}}. \tag{4.47}$$

Here, $\bar{\epsilon}_{\rm m}$ and $\bar{p}_{\rm m}$ are the energy density and pressure of the ideal fluid, and $\bar{\epsilon}_{\rm q}$ and $\bar{p}_{\rm q}$ are the energy density and pressure of the scalar field which are related to the time derivative $\dot{\bar{\Psi}}$ of the scalar field and its potential $\overline{W} = \overline{W}(\overline{\Psi})$ by equations (3.26) and (3.27). On the background spacetime these equations takes the following form:

$$\bar{\epsilon}_{\rm q} = \frac{1}{2}\bar{\rho}_{\rm q}\bar{\mu}_{\rm q} + \overline{W}\,, \tag{4.48}$$

$$\bar{p}_{\rm q} = \frac{1}{2}\bar{\rho}_{\rm q}\bar{\mu}_{\rm q} - \overline{W}\,, \tag{4.49}$$

where $\bar{\mu}_{\rm q}$ is the background specific enthalpy of the scalar field defined by (3.22), and $\bar{\rho}_{\rm q} = \bar{\mu}_{\rm q}$ is the background density of the scalar field "fluid." It is worthwhile to remind to the reader that due to the homogeneity and isotropy of the FLRW metric, all matter variables on the background manifold are functions of the conformal time η only when being expressed in the isotropic Cartesian coordinates.

Einstein's equations (4.43) can be projected on the direction of the background 4-velocity of matter and on the spatial hypersurface being orthogonal to it. It yields two Friedmann equations for the evolution of the scale factor a,

$$H^2 = \frac{8\pi}{3}\bar{\epsilon} - \frac{k}{a^2}\,, \tag{4.50}$$

$$2\dot{H} + 3H^2 = -8\pi\bar{p} - \frac{k}{a^2}\,, \tag{4.51}$$

where $\bar{\epsilon}$ and $\bar{p}$ are the effective energy density and pressure of the mixture of matter and scalar field as defined above.

A consequence of the Friedmann equations (4.50), (4.51) is an equation

$$\dot{H} = -4\pi\left(\bar{\epsilon} + \bar{p}\right) + \frac{k}{a^2}\,, \tag{4.52}$$

relating the time derivative of the Hubble parameter with the sum of the overall energy density and pressure, which can be expressed in terms of the density and specific enthalpy of the background components of matter,

$$\bar{\epsilon} + \bar{p} = \bar{\rho}_{\rm m}\bar{\mu}_{\rm m} + \bar{\rho}_{\rm q}\bar{\mu}_{\rm q}\,. \tag{4.53}$$

In order to solve the Friedmann equations (4.50) and (4.51) we have to employ the equation of state of matter. Customarily, it is assumed that each matter component obeys its own cosmological equation of state,

$$\bar{p}_{\rm m} = w_{\rm m}\bar{\epsilon}_{\rm m}\,, \quad \bar{p}_{\rm q} = w_{\rm q}\bar{\epsilon}_{\rm q}\,, \tag{4.54}$$

where $w_{\rm m}$ and $w_{\rm q}$ are parameters lying in the range from -1 to $+1$. In the most simple cosmological models, parameters $w_{\rm m}$ and $w_{\rm q}$ are fixed. More realistic models admit that the parameters of the equation of state may change in the course of the cosmological expansion, that is they may depend on time. The equation of state does not close the system of the Friedmann equations, which have to be complemented with the equations of motion of the scalar field and of the ideal fluid in order to make the system of differential equations for the gravitational and matter field variables complete.

4.6 Hydrodynamic equations of the ideal fluid

The background value of the Clebsch potential of the ideal fluid, $\overline{\Phi}$, depends only on the conformal time η of the FLRW metric. The partial derivative of the potential, taken in arbitrary coordinate chart on the background manifold, can be expressed by the following equation (3.14) in terms of the background 4-velocity $\overline{u}^{\alpha}$ as follows:

$$\overline{\Phi}_{|\alpha} = -\overline{\mu}_{\rm m}\overline{u}_{\alpha} , \tag{4.55}$$

where the background value of the specific enthalpy is

$$\overline{\mu}_{\rm m} = \sqrt{-\overline{g}^{\alpha\beta}\overline{\Phi}_{,\alpha}\overline{\Phi}_{,\beta}} \tag{4.56}$$

in accordance with definition (3.14). It allows us to write down the specific enthalpy of the ideal fluid in terms of a derivative from the Clebsch potential $\overline{\Phi}$. Multiplying both sides of (4.55) with $\overline{u}_{\alpha}$, and accounting for $\overline{u}^{\alpha}\overline{u}_{\alpha} = -1$, we obtain

$$\overline{\mu}_{\rm m} \equiv \overline{u}^{\alpha}\overline{\Phi}_{|\alpha} = \dot{\overline{\Phi}} . \tag{4.57}$$

The background equation of continuity for the rest mass density $\overline{\rho}_{\rm m}$ of the ideal fluid is

$$\left(\overline{\rho}_{\rm m}\overline{u}^{\alpha}\right)_{|\alpha} = 0 , \tag{4.58}$$

that is equivalent to

$$\overline{\rho}_{{\rm m}|\alpha} - 3H\overline{\rho}_{\rm m}\overline{u}_{\alpha} = 0 , \tag{4.59}$$

where we have used (4.32). The background equation of conservation of energy is

$$\overline{\epsilon}_{{\rm m}|\alpha} - 3H\left(\overline{\epsilon}_{\rm m} + \overline{p}_{\rm m}\right)\overline{u}_{\alpha} = 0 , \tag{4.60}$$

where we have employed definition of the energy (3.4), and equation (4.59) along with (3.6).

4.7 Scalar field equations

Background equation for the scalar field $\overline{\Psi}$ is derived from the action (3.1) by taking variational derivatives with respect to $\overline{\Psi}$. It yields

$$\overline{g}^{\alpha\beta}\overline{\Psi}_{|\alpha\beta} - \frac{\partial\overline{W}}{\partial\overline{\Psi}} = 0 . \tag{4.61}$$

In terms of the time derivatives with respect to the Hubble time T, equation (4.61) reads

$$\ddot{\overline{\Psi}} + 3H\dot{\overline{\Psi}} + \frac{\partial\overline{W}}{\partial\overline{\Psi}} = 0 . \tag{4.62}$$

Here, we have taken into account that the background value of the scalar field, $\overline{\Psi}$, depends only on time $T = T(\eta)$, and its derivative with respect to T (denoted with a dot) is proportional to the background 4-velocity

$$\overline{\Psi}_{|\alpha} = -\dot{\overline{\Psi}}\overline{u}_\alpha , \tag{4.63}$$

which follows directly (4.20). If we use the definition of the background enthalpy of the scalar field

$$\overline{\mu}_{\rm q} \equiv \overline{u}^\alpha\overline{\Psi}_{|\alpha} = \dot{\overline{\Psi}} , \tag{4.64}$$

and account for definition (3.26) of the specific energy $\epsilon_{\rm q}$ of the scalar field, equation (4.62) will become

$$\overline{\epsilon}_{{\rm q}|\alpha} - 3H\left(\overline{\epsilon}_{\rm q} + \overline{p}_{\rm q}\right)\overline{u}_\alpha = 0 . \tag{4.65}$$

that looks similar to the hydrodynamic equation (4.59) of conservation of energy of the ideal fluid. Because of this similarity, the second Friedmann equation (4.51) can be derived from the first Friedmann equation (4.50) by taking a time derivative and applying the energy conservation equations (4.60) and (4.65).

The background density $\overline{\rho}_{\rm q}$ of the scalar filed "fluid" is $\overline{\rho}_{\rm q} = \overline{\mu}_{\rm q}$ in accordance with (3.25). The equation of continuity for the density $\overline{\rho}_{\rm q}$ of the ideal fluid is obtained by differentiating definition of $\overline{\rho}_{\rm q}$, and making use of (4.62). It yields

$$\left(\overline{\rho}_{\rm q}\overline{u}^\alpha\right)_{|\alpha} = -\frac{\partial\overline{W}}{\partial\overline{\Psi}} , \tag{4.66}$$

or, equivalently,

$$\overline{\rho}_{{\rm q}|\alpha} - 3H\overline{\rho}_{\rm q}\overline{u}_\alpha = \frac{\partial\overline{W}}{\partial\overline{\Psi}}\overline{u}_\alpha , \tag{4.67}$$

which shows that the "density" $\overline{\rho}_{\rm q}$ of the scalar field "fluid" is not conserved in the most general case of an arbitrary potential function $\overline{W}(\overline{\Psi})$. We emphasize that there is no any violation of physical laws, since (4.67) is simply another way of writing equation (4.61), and the scalar field is not thermodynamically equivalent to the ideal fluid. Equation (4.67) is convenient in the calculations that follow in the next sections.

4.8 Equations of motion of matter of the localized astronomical system

Matter of the localized astronomical system is described by the tensor of energy–momentum $\mathfrak{T}^{\alpha\beta}$ defined in (3.30) in terms of the Lagrangian derivative. It can be given explicitly as a function of field variables after we chose a specific form of matter, for example, gas, liquid, solid, or something else. We do not restrict ourselves with a particular form of this tensor, and shall develop a more generic approach that is applicable to any kind of matter comprising the localized astronomical system..

Background equation of motion of matter of the astronomical system is given by the conservation law

$$\mathfrak{T}^{\alpha\beta}{}_{|\beta} = 0 \,. \tag{4.68}$$

It can be also written down in terms of a covariant derivative of the conformal metric

$$\left(\sqrt{-\bar{g}}\mathfrak{T}^{\alpha\beta}\right)_{;\beta} + \sqrt{-\bar{g}}\overline{A}^{\alpha}{}_{\beta\gamma}\mathfrak{T}^{\beta\gamma} = 0 \,, \tag{4.69}$$

where the connection $\overline{A}^{\alpha}{}_{\beta\gamma}$ is defined in (4.24). Equation (4.68) tells us that the matter of the small-scale perturbation follows geodesics of the background manifold. This is the starting point for doing the post-Newtonian approximations in cosmology. In the geodesic approximation, the matter of the isolated astronomical system has no self-interaction through its own gravitational field. The self-interaction appears at the next step of the post-Newtonian iteration procedure.

It is natural to write down equation (4.68) in 1+3 form by projecting it on the direction of 4-velocity of the Hubble flow, $\bar{u}^{\alpha}$, and on the hypersurface being orthogonal to it. This is achieved by introducing the following projections:

$$\sigma \equiv \bar{u}^{\mu}\bar{u}^{\nu}\mathfrak{T}_{\mu\nu} \,, \tag{4.70a}$$

$$\tau \equiv \overline{P}^{\mu\nu}\mathfrak{T}_{\mu\nu} \,, \tag{4.70b}$$

$$\tau_{\alpha} \equiv -\overline{P}_{\alpha}{}^{\mu}\bar{u}^{\nu}\mathfrak{T}_{\mu\nu} \,, \tag{4.70c}$$

$$\tau_{\alpha\beta} \equiv \overline{P}_{\alpha}{}^{\mu}\overline{P}_{\beta}{}^{\nu}\mathfrak{T}_{\mu\nu} \,, \tag{4.70d}$$

which corresponds to the kinemetric-invariant decomposition[4] of $\mathfrak{T}_{\mu\nu}$ introduced by Zelmanov [123, 124]. Quantity σ is the energy density of matter of the localized system, τ_{α} is a density of a linear momentum of the matter, and $\tau_{\alpha\beta}$ is the stress tensor of the matter.

Equations of motion (4.68) of the localized matter can be rewritten in terms of the chronometric quantities as follows:

$$\left(\sigma\bar{u}^{\alpha} + \tau^{\alpha}\right)_{|\alpha} = -H\tau \,, \tag{4.71a}$$

$$\left(\tau^{\alpha\beta} + \bar{u}^{\beta}\tau^{\alpha}\right)_{|\beta} = -H\left(\tau^{\alpha} - \bar{u}^{\alpha}\tau\right) \,, \tag{4.71b}$$

where $\tau^{\alpha} \equiv \bar{g}^{\alpha\beta}\tau_{\beta}$ and $\tau^{\alpha\beta} \equiv \bar{g}^{\alpha\mu}\bar{g}^{\beta\nu}\tau_{\mu\nu}$. Equation (4.71a) is equivalent to the law of conservation of energy of matter of the localized system. Equation (4.71b) is analog to the Euler equation of motion of fluid or the equation of the force balance in case of solids.

4 This decomposition is also known as a threading approach or 1 + 3 orthonormal frame approach [110].

5 Lagrangian perturbations of FLRW manifold

5.1 The concept of perturbations

In this chapter, FLRW background manifold is defined by the metric $\bar{g}_{\alpha\beta}$ which dynamics is governed by background matter fields – the Clebsch potential $\bar{\Phi}$ of the ideal fluid and the scalar field $\bar{\Psi}$. We assume that the background metric and the background values of the fields are perturbed by a localized astronomical system which is considered as a *bare* perturbation associated with a field variable Θ. Perturbations of the metric and the matter fields caused by the *bare* perturbation are considered to be small so that the perturbed metric and the matter fields can be split in their backgrounds values and the corresponding perturbations,

$$g_{\alpha\beta} = \bar{g}_{\alpha\beta} + \varkappa_{\alpha\beta}\,, \quad \Phi = \bar{\Phi} + \phi\,, \quad \Psi = \bar{\Psi} + \psi\,. \tag{5.1}$$

These equations are exact. We emphasize that all functions entering equation (5.1) are taken at one and the same point of the background manifold. The *bare* perturbation does not remain the same in the presence of the perturbations of the metric and the matter fields. Therefore, the field variable Θ corresponding to the *bare* perturbation, is also perturbed

$$\Theta = \bar{\Theta} + \theta\,. \tag{5.2}$$

We consider the perturbations of the metric – $\varkappa_{\alpha\beta}$, the Clebsch potential – ϕ, and the scalar field – ψ as being weak with respect to their corresponding background values $\bar{g}_{\alpha\beta}$, $\bar{\Phi}$, and $\bar{\Psi}$, which dynamics is governed by the background equations that have been explained in Section 4. Because the field variable Θ is the source of the *bare* perturbation, we postulate that its background value is equal to zero: $\bar{\Theta} = 0$. The perturbations $\varkappa_{\alpha\beta}$, ϕ, and ψ have the same order of magnitude as θ.

Perturbation of the contravariant component of the metric is determined from the condition $g_{\alpha\gamma}g^{\gamma\beta} = \bar{g}_{\alpha\gamma}\bar{g}^{\gamma\beta} = \delta^{\beta}_{\alpha}$, and is given by

$$g^{\alpha\beta} = \bar{g}^{\alpha\beta} - \varkappa^{\alpha\beta} + \varkappa^{\alpha}{}_{\gamma}\varkappa^{\gamma\beta} + \cdots\,, \tag{5.3}$$

where the ellipses denote terms of the higher order.

It turns out [50, 96] that a more convenient field variable of the gravitational field in the theory of Lagrangian perturbations of curved manifolds, is a contravariant (Gothic) metric

$$\mathfrak{g}^{\alpha\beta} = \sqrt{-g}g^{\alpha\beta}\,. \tag{5.4}$$

The convenience of the Gothic metric stems from the fact that it enters the de Donder (harmonic) gauge conditions which significantly simplifies the Einstein equations [75, 111]. The Gothic metric variable is also indispensable for concise and elegant formulation of dynamic field theories on curved manifolds [32]. Making use of the Gothic metric allows us to significantly reduce the amount of algebra in taking the first and second variational derivatives from the Hilbert Lagrangian and the Lagrangian of the background matter in FLRW metric as explains in the rest of this section.

The covariant Gothic metric $\mathfrak{g}_{\beta\gamma}$ is defined by means of equation

$$\mathfrak{g}^{\alpha\beta}\mathfrak{g}_{\beta\gamma} = \delta^{\alpha}_{\gamma} , \tag{5.5}$$

that yields $\mathfrak{g}_{\alpha\beta} = g_{\alpha\beta}/\sqrt{-g}$. We accept that $\mathfrak{g}^{\alpha\beta}$ is expanded around its background value, $\bar{\mathfrak{g}}^{\alpha\beta} = \sqrt{-\bar{g}}\bar{g}^{\alpha\beta}$, as follows:

$$\mathfrak{g}^{\alpha\beta} = \bar{\mathfrak{g}}^{\alpha\beta} + \mathfrak{h}^{\alpha\beta} , \tag{5.6}$$

which is an exact equation.

Further calculations prompt that it is more suitable to single out $\sqrt{-\bar{g}}$ from $\mathfrak{h}^{\alpha\beta}$, and operate with a variable

$$l^{\alpha\beta} \equiv \frac{\mathfrak{h}^{\alpha\beta}}{\sqrt{-\bar{g}}} . \tag{5.7}$$

This variable splits the dynamic degrees of freedom of the gravitational perturbations from the background manifold which evolves in according with the unperturbed Friedmann equations. Tensor indices of $l^{\alpha\beta}$ are raised and lowered with the help of the background metric, for example, $l_{\alpha\beta} \equiv \bar{g}_{\alpha\mu}\bar{g}_{\beta\nu}l^{\mu\nu}$. The field variable $l^{\alpha\beta}$ relates to the perturbation $\varkappa_{\alpha\beta}$ of the metric tensor. To establish this relationship, we start from (5.4), substitute equation (5.6) to its left-hand side, and expand its right-hand side in the Taylor series with respect to $\varkappa_{\alpha\beta}$. It results in

$$\mathfrak{h}^{\alpha\beta} = \frac{\partial\bar{\mathfrak{g}}^{\alpha\beta}}{\partial\bar{g}_{\mu\nu}}\varkappa_{\mu\nu} + \frac{1}{2}\frac{\partial^2\bar{\mathfrak{g}}^{\alpha\beta}}{\partial\bar{g}_{\mu\nu}\partial\bar{g}_{\rho\sigma}}\varkappa_{\mu\nu}\varkappa_{\rho\sigma} + \cdots . \tag{5.8}$$

where the partial derivatives are calculated by successive application of the following rules:

$$\frac{\partial\bar{\mathfrak{g}}^{\alpha\beta}}{\partial\bar{g}_{\mu\nu}} = -\frac{1}{2}\sqrt{-\bar{g}}\left(\bar{g}^{\alpha\mu}\bar{g}^{\beta\nu} + \bar{g}^{\alpha\nu}\bar{g}^{\beta\mu} - \bar{g}^{\alpha\beta}\bar{g}^{\mu\nu}\right) , \tag{5.9a}$$

$$\frac{\partial\bar{g}^{\alpha\beta}}{\partial\bar{g}_{\mu\nu}} = -\frac{1}{2}\left(\bar{g}^{\alpha\mu}\bar{g}^{\beta\nu} + \bar{g}^{\alpha\nu}\bar{g}^{\beta\mu}\right) , \tag{5.9b}$$

$$\frac{\partial\sqrt{-\bar{g}}}{\partial\bar{g}_{\mu\nu}} = +\frac{1}{2}\sqrt{-\bar{g}}\bar{g}^{\mu\nu} , \tag{5.9c}$$

which can be easily confirmed by inspection. Replacing the partial derivatives in (5.8) and making use of the definition (5.7), yields the relationship between $l^{\alpha\beta}$ and $\varkappa^{\alpha\beta}$ as follows:

$$l^{\alpha\beta} = -\varkappa^{\alpha\beta} + \frac{1}{2}\bar{g}^{\alpha\beta}\varkappa + \varkappa^{\mu(\alpha}\varkappa^{\beta)}{}_{\mu} - \frac{1}{2}\varkappa^{\alpha\beta}\varkappa - \frac{1}{4}\bar{g}^{\alpha\beta}\left(\varkappa^{\mu\nu}\varkappa_{\mu\nu} - \frac{1}{2}\varkappa^2\right) + \cdots , \tag{5.10}$$

where $\varkappa \equiv \varkappa^{\sigma}{}_{\sigma} = \bar{g}^{\rho\sigma}\varkappa_{\rho\sigma}$, and ellipses denote terms of the cubic and higher order in $\varkappa_{\alpha\beta}$.

Perturbations of 4-velocities, w^{α} and v^{α}, entering definitions of the energy–momentum tensors (3.18) and (3.29), are fully determined by the perturbations of the metric and the potentials of the matter fields. Indeed, according to definitions (3.13) and (3.25) the 4-velocities are defined by the following equations:

$$w_{\alpha} = -\frac{\Phi_{,\alpha}}{\mu_{\rm m}} , \qquad v_{\alpha} = -\frac{\Psi_{,\alpha}}{\mu_{\rm q}} . \tag{5.11}$$

where $\mu_{\rm m} = \sqrt{-g^{\alpha\beta}\Phi_{,\alpha}\Phi_{,\beta}}$ and $\mu_{\rm q} = \sqrt{-g^{\alpha\beta}\Psi_{,\alpha}\Psi_{,\beta}}$ in accordance with (3.14) and (3.22), respectively. We define perturbation of the covariant components of the 4-velocities as follows:

$$w_{\alpha} = \bar{u}_{\alpha} + \delta w_{\alpha} , \qquad v_{\alpha} = \bar{u}_{\alpha} + \delta v_{\alpha} , \tag{5.12}$$

where the unperturbed values of the 4-velocities coincide and are equal to the 4-velocity of the Hubble flow due to the requirement of the homogeneity and isotropy of the background FLRW metric. Substituting these expansions to the left-hand side of definitions (5.11), and expanding its right-hand side by making use of the expansions (5.1) and (5.3) of the scalar fields and the metric, yields

$$\delta w_{\alpha} = -\frac{1}{\bar{\mu}_{\rm m}}\bar{P}^{\beta}{}_{\alpha}\phi_{|\beta} - \frac{1}{2}\mathfrak{q}\bar{u}_{\alpha} , \qquad \delta v_{\alpha} = -\frac{1}{\bar{\mu}_{\rm q}}\bar{P}^{\beta}{}_{\alpha}\psi_{|\beta} - \frac{1}{2}\mathfrak{q}\bar{u}_{\alpha} , \tag{5.13}$$

where we have introduced a new notation

$$\mathfrak{q} \equiv -\bar{u}^{\alpha}\bar{u}^{\beta}\varkappa_{\alpha\beta} , \tag{5.14}$$

for the gravitational perturbation of the metric tensor projected on the background 4-velocity of the Hubble flow. Making use of $l_{\alpha\beta}$, the previous equation can be recast to

$$\mathfrak{q} \equiv \bar{u}^{\alpha}\bar{u}^{\beta}l_{\alpha\beta} + \frac{l}{2} , \tag{5.15}$$

where $l \equiv l^{\alpha}{}_{\alpha} = \bar{g}^{\alpha\beta}l_{\alpha\beta}$. Remembering that $\bar{g}^{\alpha\beta} = \bar{P}^{\alpha\beta} - \bar{u}^{\alpha}\bar{u}^{\beta}$, we can put equation (5.15) yet to another form

$$\mathfrak{q} \equiv \frac{1}{2}\left(\bar{u}^{\alpha}\bar{u}^{\beta} + \bar{P}^{\alpha\beta}\right)l_{\alpha\beta} , \tag{5.16}$$

which is useful in the calculations that follow.

5.2 The perturbative expansion of the Lagrangian

We have introduced the Lagrangian of the theory in Section 3. The Hilbert Lagrangian of the gravitational field is $\mathcal{L}^{\mathrm{g}} = -\sqrt{-g}R/16\pi$, where R is the Ricci scalar. The Lagrangian density of matter is $\mathcal{L}^{\mathrm{m}} = \sqrt{-g}L^{\mathrm{m}}(\Phi, g_{\alpha\beta})$, and the Lagrangian density of the scalar field $\mathcal{L}^{\mathrm{q}} = \sqrt{-g}L^{\mathrm{q}}(\Psi, g_{\alpha\beta})$. The matter, the scalar field, as well as the spacetime manifold are perturbed by a matter of an isolated astronomical system described by a set of field variables Θ with the Lagrangian density $\mathcal{L}^{\mathrm{p}} = \sqrt{-g}L^{\mathrm{p}}(\Theta, g_{\alpha\beta})$.

The action of the unperturbed FLRW metric is a functional

$$\overline{\mathcal{S}} = \int_{\mathcal{M}} d^4x\overline{\mathcal{L}}\,, \tag{5.17}$$

depending on the unperturbed Lagrangian

$$\overline{\mathcal{L}} = \overline{\mathcal{L}}^{\mathrm{g}} + \overline{\mathcal{L}}^{\mathrm{m}} + \overline{\mathcal{L}}^{\mathrm{q}}\,, \tag{5.18}$$

taken on the background values of the field variables $\overline{g}_{\alpha\beta}$, $\overline{\Phi}$, and $\overline{\Psi}$.

The presence of a localized astronomical system perturbs the spacetime manifold and the background values of the field variables. The perturbed Lagrangian becomes an algebraic sum of four terms

$$\mathcal{L} = \mathcal{L}^{\mathrm{g}} + \mathcal{L}^{\mathrm{m}} + \mathcal{L}^{\mathrm{q}} + \mathcal{L}^{\mathrm{p}}\,, \tag{5.19}$$

where the Lagrangian $\mathcal{L}^{\mathrm{p}}$ describes the *bare* perturbation, and $\mathcal{L}^{\mathrm{g}}$, $\mathcal{L}^{\mathrm{m}}$, $\mathcal{L}^{\mathrm{q}}$ are perturbed values of the Lagrangian of the FLRW metric.

The perturbed Lagrangian can be decomposed in a Taylor series with respect to the perturbed values of the field variables. It is achieved by substituting expansions (5.1) to the Lagrangian (5.19) and expanding it around the background values of the variables in an infinite Taylor series. It yields

$$\mathcal{L} = \mathcal{L}^{\mathrm{p}} + \sum_{n=0}^{\infty} \mathcal{L}_n\,, \tag{5.20}$$

where $\mathcal{L}^{\mathrm{p}}$ is the Lagrangian of the bare perturbation, $\mathcal{L}_0 \equiv \overline{\mathcal{L}}$ is the Lagrangian describing the dynamic properties of the background manifold, and for any $n \leq 1$,

$$\mathcal{L}_n = \frac{1}{n}\left(\mathfrak{h}^{\mu\nu}\frac{\delta\mathcal{L}_{n-1}}{\delta\overline{\mathfrak{g}}^{\mu\nu}} + \phi\frac{\delta\mathcal{L}_{n-1}}{\delta\overline{\Phi}} + \psi\frac{\delta\mathcal{L}_{n-1}}{\delta\overline{\Psi}}\right)\,, \tag{5.21}$$

is the Lagrangian perturbation defined iteratively by taking variational derivatives from the Lagrangian perturbations of the previous iteration. In particular,

$$\mathcal{L}_1 = \mathfrak{h}^{\mu\nu}\frac{\delta\overline{\mathcal{L}}}{\delta\overline{\mathfrak{g}}^{\mu\nu}} + \phi\frac{\delta\overline{\mathcal{L}}}{\delta\overline{\Phi}} + \psi\frac{\delta\overline{\mathcal{L}}}{\delta\overline{\Psi}}\,, \tag{5.22a}$$

$$\mathcal{L}_2 = \frac{1}{2}\left(\mathfrak{h}^{\mu\nu}\frac{\delta\mathcal{L}_1}{\delta\overline{\mathfrak{g}}^{\mu\nu}} + \phi\frac{\delta\mathcal{L}_1}{\delta\overline{\Phi}} + \psi\frac{\delta\mathcal{L}_1}{\delta\overline{\Psi}}\right)\,, \tag{5.22b}$$

and so on. Here, the variational derivatives from the Lagrangian density, $\overline{\mathcal{L}}$, depending on the field variables and their derivatives, are defined as follows (see Section 2.1 for more details):

$$\frac{\delta \mathcal{L}_n}{\delta \overline{\mathfrak{g}}^{\mu\nu}} \equiv \frac{\partial \mathcal{L}_n}{\partial \overline{\mathfrak{g}}^{\mu\nu}} - \frac{\partial}{\partial x^\alpha}\frac{\partial \mathcal{L}_n}{\partial \overline{\mathfrak{g}}^{\mu\nu}_{\ ,\alpha}} + \frac{\partial^2}{\partial x^\alpha \partial x^\beta}\frac{\partial \mathcal{L}_n}{\partial \overline{\mathfrak{g}}^{\mu\nu}_{\ ,\alpha\beta}} , \tag{5.23a}$$

$$\frac{\delta \mathcal{L}_n}{\delta \overline{\Phi}} \equiv \frac{\partial \mathcal{L}_n}{\partial \overline{\Phi}} - \frac{\partial}{\partial x^\alpha}\frac{\partial \mathcal{L}_n}{\partial \overline{\Phi}_{,\alpha}} + \frac{\partial^2}{\partial x^\alpha \partial x^\beta}\frac{\partial \mathcal{L}_n}{\partial \overline{\Phi}_{,\alpha\beta}} , \tag{5.23b}$$

$$\frac{\delta \mathcal{L}_n}{\delta \overline{\Psi}} \equiv \frac{\partial \mathcal{L}_n}{\partial \overline{\Psi}} - \frac{\partial}{\partial x^\alpha}\frac{\partial \mathcal{L}_n}{\partial \overline{\Psi}_{,\alpha}} + \frac{\partial^2}{\partial x^\alpha \partial x^\beta}\frac{\partial \mathcal{L}_n}{\partial \overline{\Psi}_{,\alpha\beta}} . \tag{5.23c}$$

The variational derivative with respect to the metric density $\overline{\mathfrak{g}}^{\mu\nu}$ relates to the derivative with respect to the metric $\overline{g}^{\mu\nu}$ by an algebraic operator

$$\frac{\delta}{\delta \overline{\mathfrak{g}}^{\mu\nu}} = \frac{\partial \overline{g}^{\alpha\beta}}{\partial \overline{\mathfrak{g}}^{\mu\nu}}\frac{\delta}{\delta \overline{g}^{\alpha\beta}} = \frac{1}{2\sqrt{-\overline{g}}}\left(\delta^\alpha_\mu \delta^\beta_\nu + \delta^\alpha_\nu \delta^\beta_\mu - \overline{g}_{\mu\nu}\overline{g}^{\alpha\beta}\right)\frac{\delta}{\delta \overline{g}^{\alpha\beta}} . \tag{5.24}$$

One has to notice that the expansion (5.20) is defined up to the divergence terms which are represented as a total covariant derivative from a vector density. The divergences can be important in the discussion of the boundary conditions but they do not enter equations of motion of fields which represent a system of the differential equations in partial derivatives for the perturbations of the dynamic (field) variables. Furthermore, it is straightforward to prove that any of the Lagrangian derivatives (5.23a)–(5.23c), applied to a partial derivative of a geometric object $F = F(\overline{\mathfrak{g}}^{\alpha\beta}; \overline{\Phi}; \overline{\Psi}; \overline{\mathfrak{g}}^{\alpha\beta}_{\ ,\gamma}; \overline{\Phi}_{,\gamma}; \overline{\Psi}_{,\gamma}; \ldots)$, vanishes [90] (see (2.10))

$$\frac{\delta}{\delta \overline{\mathfrak{g}}^{\alpha\beta}}\left(\frac{\partial F}{\partial x^\alpha}\right) = 0 , \quad \frac{\delta}{\delta \overline{\Phi}}\left(\frac{\partial F}{\partial x^\alpha}\right) = 0 , \quad \frac{\delta}{\delta \overline{\Psi}}\left(\frac{\partial F}{\partial x^\alpha}\right) = 0 . \tag{5.25}$$

Equations (5.25) do not hold for a covariant derivative [90]. We shall use equation (5.25) for simplification of the Lagrangian derivatives.

The field equations are obtained by taking the variational derivatives from the perturbed action with respect to various variables subject to the least action principle. In accordance with this principle, the variational derivatives from the perturbed Lagrangian must vanish,

$$\frac{\delta \mathcal{L}^{\mathrm{g}}}{\delta \mathfrak{g}_{\alpha\beta}} + \frac{\delta \mathcal{L}^{\mathrm{m}}}{\delta \mathfrak{g}_{\alpha\beta}} + \frac{\delta \mathcal{L}^{\mathrm{q}}}{\delta \mathfrak{g}_{\alpha\beta}} = -\frac{\delta \mathcal{L}^{\mathrm{p}}}{\delta \mathfrak{g}_{\alpha\beta}} , \tag{5.26a}$$

$$\frac{\delta \mathcal{L}^{\mathrm{m}}}{\delta \Phi} = 0 , \tag{5.26b}$$

$$\frac{\delta \mathcal{L}^{\mathrm{q}}}{\delta \Psi} = 0 . \tag{5.26c}$$

The post-Newtonian approximations of these equations are easily derived by the following procedure. We substitute the Taylor decomposition (5.20) of the Lagrangian to equations (5.26a) and separate the background value of the derivatives from their perturbed values. We assume that gravitational dynamics of the unperturbed FLRW metric obeys the background field equations shown below in Section 5.3. Then, the perturbed part of the equations represent a series of the post-Newtonian equations of the first, second, third, etc. order, which can be solved by successive iterations. In this chapter we restrict ourselves with the linearized approximation of the first order with respect to the perturbations. It generalizes the first post-Newtonian field equations in asymptotically flat spacetime to the case of the expanding universe.

5.3 The background field equations

Dynamics of the background universe is governed exclusively by the background matter. Hence, we have to take into account only the background values of functions entering variational Euler–Lagrange equations (5.26a)–(5.26c) and drop out the Lagrangian $\mathcal{L}^{\mathrm{p}}$ of the bare perturbation. We get

$$\frac{\delta\overline{\mathcal{L}}^{\mathrm{g}}}{\delta\bar{\mathfrak{g}}_{\alpha\beta}}+\frac{\delta\overline{\mathcal{L}}^{\mathrm{m}}}{\delta\bar{\mathfrak{g}}_{\alpha\beta}}+\frac{\delta\overline{\mathcal{L}}^{\mathrm{q}}}{\delta\bar{\mathfrak{g}}_{\alpha\beta}}=0\,, \tag{5.27a}$$

$$\frac{\delta\overline{\mathcal{L}}^{\mathrm{m}}}{\delta\overline{\Phi}}=0\,, \tag{5.27b}$$

$$\frac{\delta\overline{\mathcal{L}}^{\mathrm{q}}}{\delta\overline{\Psi}}=0\,. \tag{5.27c}$$

After performing the derivatives, equation (5.27a) becomes the Einstein equation (4.43), equation (5.27b) is reduced to equation of continuity (4.58) after taking into account (4.55), (4.56), and equation (5.27c) is equivalent to (4.61). These equations have been thoroughly discussed in Section 4. Solution of these equations depends on the equation of state of the background matter. We assume that the solution exists and that the time dependence of the FLRW metric $\bar{g}_{\alpha\beta}=\bar{g}_{\alpha\beta}(\eta)$, the Clebsch potential $\overline{\Phi}=\overline{\Phi}(\eta)$, and the scalar field $\overline{\Psi}=\overline{\Psi}(\eta)$ are explicitly known.

5.4 The Lagrangian equations for gravitational field perturbations

We substitute the Taylor decomposition of the Lagrangian in equation (5.26a) and account for the background field equations (5.27a)–(5.27c). After taking the variational derivatives, we find out [70] that the gravitational field perturbations obey the following (exact) differential equation:

$$F_{\mu\nu}=8\pi\left(\mathfrak{T}_{\mu\nu}+\mathfrak{I}_{\mu\nu}\right)\,, \tag{5.28}$$

which generalizes the Einstein field equations in asymptotically flat spacetime to the case of the expanding FLRW metric. Tensor $F_{\mu\nu}$ is an algebraic superposition

$$F_{\mu\nu} \equiv F^{\rm g}_{\mu\nu} + F^{\rm m}_{\mu\nu} + F^{\rm q}_{\mu\nu}\,, \tag{5.29}$$

where the linear operators on the right-hand side are defined through the Lagrangian derivatives as follows:

$$F^{\rm g}_{\mu\nu} \equiv -\frac{16\pi}{\sqrt{-\bar g}}\frac{\delta}{\delta\bar g^{\mu\nu}}\left(\mathfrak{h}^{\alpha\beta}\frac{\delta\overline{\mathcal{L}}^{\rm g}}{\delta\bar{\mathfrak{g}}^{\alpha\beta}}\right), \tag{5.30a}$$

$$F^{\rm m}_{\mu\nu} \equiv -\frac{16\pi}{\sqrt{-\bar g}}\frac{\delta}{\delta\bar g^{\mu\nu}}\left(\mathfrak{h}^{\alpha\beta}\frac{\delta\overline{\mathcal{L}}^{\rm m}}{\delta\bar{\mathfrak{g}}^{\alpha\beta}} + \phi\frac{\delta\overline{\mathcal{L}}^{\rm m}}{\delta\bar\Phi}\right), \tag{5.30b}$$

$$F^{\rm q}_{\mu\nu} \equiv -\frac{16\pi}{\sqrt{-\bar g}}\frac{\delta}{\delta\bar g^{\mu\nu}}\left(\mathfrak{h}^{\alpha\beta}\frac{\delta\overline{\mathcal{L}}^{\rm q}}{\delta\bar{\mathfrak{g}}^{\alpha\beta}} + \psi\frac{\delta\overline{\mathcal{L}}^{\rm q}}{\delta\bar\Psi}\right). \tag{5.30c}$$

The right-hand side of equation (5.28) contains the tensor of energy–momentum $\mathfrak{T}_{\mu\nu}$ of the *bare* gravitational perturbation which is generated by the matter of the localized astronomical system and can be calculated as the Lagrangian derivative (3.30). The right-hand side of (5.28) contains the nonlinear corrections of the second and higher order of magnitude. They are given by

$$\mathcal{T}_{\mu\nu} = \frac{2}{\sqrt{-\bar g}}\left(\frac{\delta\mathcal{L}_2}{\delta\bar g^{\mu\nu}} + \frac{\delta\mathcal{L}_3}{\delta\bar g^{\mu\nu}} + \cdots\right). \tag{5.31}$$

Tensor $\mathcal{T}_{\mu\nu}$ can be split in two algebraically independent parts

$$\mathcal{T}_{\mu\nu} = \mathfrak{t}_{\mu\nu} + \tau_{\mu\nu}\,, \tag{5.32}$$

where $\mathfrak{t}_{\mu\nu}$ is the stress–energy tensor of pure gravitational perturbations $\mathfrak{h}^{\mu\nu}$ while $\tau_{\mu\nu}$ is the stress–energy tensor characterizing gravitational coupling of the matter field ϕ^A with the gravitational perturbations $\mathfrak{h}^{\mu\nu}$. If we restrict ourselves only with the second-order nonlinear corrections, the corresponding stress–energy tensors are given by variational derivatives

$$\mathfrak{t}_{\mu\nu} = -\frac{1}{16\pi\sqrt{-\bar g}}\frac{\delta}{\delta\bar g^{\mu\nu}}\left(\mathfrak{h}^{\rho\sigma}F^{\rm g}_{\rho\sigma} - \frac{1}{2}\mathfrak{h}\bar g^{\rho\sigma}F^{\rm g}_{\rho\sigma}\right), \tag{5.33}$$

$$\begin{aligned}\tau_{\mu\nu} = &-\frac{1}{16\pi\sqrt{-\bar g}}\frac{\delta}{\delta\bar g^{\mu\nu}}\left(\mathfrak{h}^{\rho\sigma}F^{\rm m}_{\rho\sigma} - \frac{1}{2}\mathfrak{h}g^{\rho\sigma}F^{\rm m}_{\rho\sigma} + \sqrt{-\bar g}\phi F^{\rm m}_{\Phi}\right)\\ &-\frac{1}{16\pi\sqrt{-\bar g}}\frac{\delta}{\delta\bar g^{\mu\nu}}\left(\mathfrak{h}^{\rho\sigma}F^{\rm q}_{\rho\sigma} - \frac{1}{2}\mathfrak{h}g^{\rho\sigma}F^{\rm q}_{\rho\sigma} + \sqrt{-\bar g}\psi F^{\rm q}_{\Psi}\right),\end{aligned} \tag{5.34}$$

where $F^{\rm m}_{\Phi}$ and $F^{\rm q}_{\Psi}$ are defined below in (5.47) and (5.53).

It is worth noticing that the gravitational stress–energy tensor $\mathfrak{t}_{\mu\nu}$ can be derived in exact analytic form and reads [50]

$$\begin{aligned}\mathfrak{t}_{\mu\nu} &= \frac{1}{8\pi}\left(\delta^{\rho}_{\mu}\delta^{\sigma}_{\nu} - \frac{1}{2}\overline{g}_{\mu\nu}\overline{g}^{\rho\sigma}\right)\left(\mathfrak{G}^{\alpha}{}_{\rho\beta}\mathfrak{G}^{\beta}{}_{\sigma\alpha} - \mathfrak{G}^{\alpha}{}_{\rho\sigma}\mathfrak{G}^{\beta}{}_{\alpha\beta}\right) \\ &+ \frac{1}{8\pi}\Big[\frac{1}{2}\mathfrak{h}_{\mu\nu}\overline{g}^{\rho\beta}\mathfrak{G}^{\alpha}{}_{\alpha\beta} - \frac{1}{2}\overline{g}_{\mu\nu}\mathfrak{h}^{\alpha\beta}\mathfrak{G}^{\rho}{}_{\alpha\beta} - \mathfrak{h}^{\rho}{}_{(\mu}\mathfrak{G}^{\alpha}{}_{\nu)\alpha} \\ &+ \mathfrak{h}^{\rho}{}_{\beta}\overline{g}_{\alpha(\mu}\mathfrak{G}^{\alpha}{}_{\nu)\beta} + \mathfrak{h}^{\beta}{}_{(\mu}\mathfrak{G}^{\rho}{}_{\nu)\beta} - \mathfrak{h}^{\beta}{}_{(\mu}\overline{g}_{\nu)\alpha}\overline{g}^{\rho\sigma}\mathfrak{G}^{\alpha}{}_{\beta\sigma}\Big]_{|\rho}\,, \end{aligned} \tag{5.35}$$

where $\mathfrak{G}^{\alpha}{}_{\beta\gamma} \equiv \Gamma^{\alpha}{}_{\beta\gamma} - \bar{\Gamma}^{\alpha}{}_{\beta\gamma}$ is the difference between the Christoffel symbols of the perturbed and unperturbed (background) manifolds

$$\mathfrak{G}^{\alpha}{}_{\beta\gamma} = \frac{1}{2}g^{\alpha\rho}\left(\varkappa_{\rho\beta|\gamma} + \varkappa_{\rho\gamma|\beta} - \varkappa_{\beta\gamma|\rho}\right)\,, \tag{5.36}$$

where $\varkappa_{\mu\nu} \equiv g_{\mu\nu} - \overline{g}_{\mu\nu}$. We emphasize that the geometric object $\mathfrak{G}^{\alpha}{}_{\beta\gamma}$ is a tensor since it represents the difference between the two Christoffel symbols defined on one and the same background manifold possessing two metrics, $\overline{g}_{\mu\nu}$ and $g_{\mu\nu} = \overline{g}_{\mu\nu} + \varkappa_{\mu\nu}$. It does not mean, of course, that we employ a bi-metric theory of gravity being different from general theory of relativity. The background metric $\overline{g}_{\mu\nu}$ is simply the lowest (unperturbed) state of the full metric $g_{\mu\nu}$, and each metric obeys the Einstein field equations. As the background metric $\overline{g}_{\mu\nu}$, its perturbation $\varkappa_{\mu\nu}$, and the symbol $\mathfrak{G}^{\alpha}{}_{\beta\gamma}$ are tensors, $\mathfrak{t}_{\mu\nu}$ is also a tensor. It defines energy density, a linear momentum density, and other physical characteristics of the gravitational perturbations at each point of the background spacetime [95]. Contribution of $\mathfrak{T}_{\mu\nu}$ to the linearized field equations should be neglected as it is of the higher order as compared with other terms in (5.28).

The differential operator, $F^{\mathrm{g}}_{\mu\nu}$, represents a linearized perturbation of the Ricci tensor, and after calculation of the variational derivative in (5.30a), is given by

$$F^{\mathrm{g}}_{\mu\nu} = \frac{1}{2}\left(l_{\mu\nu}{}^{|\alpha}{}_{|\alpha} + \overline{g}_{\mu\nu}l^{\alpha\beta}{}_{|\alpha\beta} - l_{\alpha\mu|\nu}{}^{|\alpha} - l_{\alpha\nu|\mu}{}^{\alpha}\right)\,, \tag{5.37}$$

where each vertical bar denotes a covariant derivative with respect to the background metric $\overline{g}_{\mu\nu}$.

Operators $F^{\mathrm{m}}_{\mu\nu}$ and $F^{\mathrm{q}}_{\mu\nu}$ depend essentially on a particular choice of the Lagrangian of matter and scalar field, and take on different forms depending on the specific analytic dependence of $\mathcal{L}^{\mathrm{m}}$ and $\mathcal{L}^{\mathrm{q}}$ on the field variables. In the particular case of the ideal fluid, the term embraced in the round parentheses on the right-hand side of equation (5.30b) is

$$\mathfrak{h}^{\alpha\beta}\frac{\delta\overline{\mathcal{L}}^{\mathrm{m}}}{\delta\overline{\mathfrak{g}}^{\alpha\beta}} + \phi\,\frac{\delta\overline{\mathcal{L}}^{\mathrm{m}}}{\delta\overline{\Phi}} = \frac{1}{2}\mathfrak{h}^{\alpha\beta}\left(\overline{T}^{\mathrm{m}}_{\alpha\beta} - \frac{1}{2}\overline{g}_{\alpha\beta}\overline{T}^{\mathrm{m}}\right) + \phi\,\partial_{\alpha}\left(\overline{\rho}_{\mathrm{m}}\sqrt{-\overline{g}}\overline{u}^{\alpha}\right)\,, \tag{5.38}$$

where $\overline{u}^{\alpha} \equiv -\overline{g}^{\alpha\beta}\overline{\Phi}_{,\beta}/\overline{\mu}_{\mathrm{m}}$, and $\overline{T}^{\mathrm{m}}_{\alpha\beta}$ are given in (3.18). We emphasize that though the ideal fluid satisfies the equation of continuity (4.58), it should not be immediately implemented in (5.38) because this expression is to be further differentiated with respect to the metric tensor according to (5.30b).

For the scalar field, the term enclosed to the round parentheses on the right-hand side of (5.30c) is

$$\mathfrak{h}^{\alpha\beta}\frac{\delta\overline{\mathcal{L}}^{\mathrm{q}}}{\delta\bar{\mathfrak{g}}^{\alpha\beta}}+\psi\frac{\delta\overline{\mathcal{L}}^{\mathrm{q}}}{\delta\overline{\Psi}}=\frac{1}{2}\mathfrak{h}^{\alpha\beta}\left(\overline{T}^{\mathrm{q}}_{\alpha\beta}-\frac{1}{2}\bar{g}_{\alpha\beta}\overline{T}^{\mathrm{q}}\right)+\psi\left[\sqrt{-\bar{g}}\frac{\partial\overline{W}}{\partial\overline{\Psi}}+\partial_\alpha\left(\bar{\rho}_{\mathrm{q}}\sqrt{-\bar{g}}\bar{u}^\alpha\right)\right], \tag{5.39}$$

where $\bar{u}^\alpha \equiv -\bar{g}^{\alpha\beta}\overline{\Psi}_{,\beta}/\bar{\mu}_{\mathrm{q}}$, $\bar{\rho}_{\mathrm{q}}=\bar{\mu}_{\mathrm{q}}$, $\overline{T}^{\mathrm{q}}_{\alpha\beta}$ is given in (3.29), and the equation of continuity for the scalar field (4.66) should not be implemented until differentiation with respect to the metric tensor (5.30c) is completed.

Taking the variational derivatives with respect to $\bar{g}^{\mu\nu}$ from the expressions (5.38) and (5.39), and applying thermodynamic equations (3.11), allows us to write down the right-hand sides of equations (5.30b) and (5.30c) in a more explicit form as follows:

$$\begin{aligned}F^{\mathrm{m}}_{\mu\nu}&=-4\pi\left[(\bar{p}_{\mathrm{m}}-\bar{\epsilon}_{\mathrm{m}})l_{\mu\nu}+\left(1-\frac{c^2}{v_{\mathrm{s}}^2}\right)(\bar{\epsilon}_{\mathrm{m}}+\bar{p}_{\mathrm{m}})q\bar{u}_\mu\bar{u}_\nu\right] \\ &+8\pi\bar{p}_{\mathrm{m}}\left\{\bar{u}_\mu\phi_{,\nu}+\bar{u}_\nu\phi_{,\mu}+\left[\left(1-\frac{c^2}{v_{\mathrm{s}}^2}\right)\bar{u}_\mu\bar{u}_\nu-\bar{g}_{\mu\nu}\right]\bar{u}^\alpha\phi_{,\alpha}\right\},\end{aligned} \tag{5.40}$$

$$F^{\mathrm{q}}_{\mu\nu}=-4\pi\left[\left(p_{\mathrm{q}}-\epsilon_{\mathrm{q}}\right)l_{\mu\nu}-2\bar{g}_{\mu\nu}\frac{\partial\overline{W}}{\partial\overline{\Psi}}\psi\right]+8\pi\bar{\rho}_{\mathrm{q}}\left(\bar{u}_\mu\psi_{,\nu}+\bar{u}_\nu\psi_{,\mu}-\bar{g}_{\mu\nu}\bar{u}^\alpha\psi_{,\alpha}\right), \tag{5.41}$$

where $\bar{\rho}_{\mathrm{q}}\equiv\dot{\overline{\Psi}}=\bar{\mu}_{\mathrm{q}}$ in accordance with definition (3.25). The potential energy of the scalar field, $\overline{W}=\overline{W}(\overline{\Psi})$, is kept arbitrary.

It is important to emphasize that in the most general case the ratio v_{s}^2/c^2 of the speed of sound in fluid to the fundamental speed c, may be not equal to the parameter w_{m} of the equation of state (4.54), that is there are physical equations of state such that $w_{\mathrm{m}}\neq(v_{\mathrm{s}}/c)^2$. Indeed, the speed of sound is defined as a partial derivative of pressure p_{m} with respect to the energy density ϵ_{m} taken under the condition of a constant entropy s_{m},

$$\frac{v_{\mathrm{s}}^2}{c^2}=\left(\frac{\partial p_{\mathrm{m}}}{\partial\epsilon_{\mathrm{m}}}\right)_{s_{\mathrm{m}}=\mathrm{const.}}. \tag{5.42}$$

This equation is equivalent to the following relation:

$$\frac{v_{\mathrm{s}}^2}{c^2}=\frac{(\partial p_{\mathrm{m}}/\partial\mu_{\mathrm{m}})_{s_{\mathrm{m}}=\mathrm{const.}}}{(\partial\epsilon_{\mathrm{m}}/\partial\mu_{\mathrm{m}})_{s_{\mathrm{m}}=\mathrm{const.}}}, \tag{5.43}$$

which is a consequence of thermodynamic relations and a definition of the partial derivative. The ratio of the partial derivatives in (5.43) is not reduced to w_{m} in case when w_{m} depends on some other thermodynamic parameters which are functions of the specific enthalpy. For example, in case of an ideal gas the equation of state $p_{\mathrm{m}}=w_{\mathrm{m}}\epsilon_{\mathrm{m}}$, where $w_{\mathrm{m}}=kT/mc^2$, k is the Boltzmann constant, m is the mass of a particle of the ideal fluid, and T is the fluid temperature. The speed of sound $v_s^2 = c^2(\partial p_m/\partial\epsilon_m)_{s_m=\text{const.}} = \Gamma w_m > w_m = p_m/\epsilon_m$, where $\Gamma > 1$ is the ratio of the heat capacities of the gas taken for the constant pressure and the constant volume, respectively [74].

The scalar field with the potential function $W(\Psi) \neq 0$ does not bear all thermodynamic properties of an ideal fluid. Nevertheless, we can formally define the speed of "sound" c_s propagating in the scalar field "fluid," by equation being similar to (5.43). More specifically,

$$\frac{c_s^2}{c^2} = \frac{\left(\partial p_q/\partial\mu_q\right)_{\Psi=\text{const.}}}{\left(\partial\epsilon_q/\partial\mu_q\right)_{\Psi=\text{const.}}}\,. \tag{5.44}$$

Simple calculation reveals that the speed of "sound" for scalar field is always equal to the fundamental speed

$$c_s = c\,, \tag{5.45}$$

irrespectively of the value of the potential function $W(\Psi)$. It explains why the terms being proportional to the factor $1 - c^2/c_s^2$, do not appear in the expression (5.41) as contrasted with (5.40).

5.5 The Lagrangian equations for dark matter perturbations

The perturbed field equations for the dark matter which is modeled by the ideal fluid, are obtained by taking the variational derivatives with respect to the field Φ from the Lagrangian (5.19) – it corresponds to the middle equation in (5.26a). Taking into account the background equation (5.27b) yields the equation of sound waves propagating in the fluid as small perturbations of the potential ϕ,

$$F_\Phi^{\text{m}} = 8\pi\Sigma^{\text{m}}\,, \tag{5.46}$$

where the linear differential operator

$$F_\Phi^{\text{m}} \equiv -\frac{1}{\sqrt{-\bar{g}}}\frac{\delta}{\delta\bar{\Phi}}\left(\mathfrak{h}^{\mu\nu}\frac{\delta\bar{\mathcal{L}}^{\text{m}}}{\delta\bar{\mathfrak{g}}^{\mu\nu}} + \phi\frac{\delta\bar{\mathcal{L}}^{\text{m}}}{\delta\bar{\Phi}}\right), \tag{5.47}$$

and the source term

$$\Sigma^{\text{m}} \equiv \frac{1}{8\pi\sqrt{-\bar{g}}}\left(\frac{\delta\mathcal{L}_2^{\text{m}}}{\delta\bar{\Phi}} + \frac{\delta\mathcal{L}_3^{\text{m}}}{\delta\bar{\Phi}} + \ldots\right). \tag{5.48}$$

In the case of a single-component ideal fluid, the Lagrangian (3.16) depends merely on the derivative of the Clebsch potential Φ and on the metric tensor. Therefore, the explicit form of the linear operator F_Φ^{m} is reduced to a covariant divergence

$$F_\Phi^{\text{m}} = Y^\alpha{}_{|\alpha}\,, \tag{5.49}$$

where a vector field

$$Y^{\alpha} \equiv \frac{\partial}{\partial \overline{\Phi}_{,\alpha}} \left[\left(l^{\mu\nu} - \frac{1}{2} l \overline{g}^{\mu\nu} \right) \left(\frac{\partial \overline{L}^{\rm m}}{\partial \overline{g}^{\mu\nu}} - \frac{1}{2} \overline{g}_{\mu\nu} \overline{L}^{\rm m} \right) + \phi_{,\beta} \frac{\partial \overline{L}^{\rm m}}{\partial \overline{\Phi}_{,\beta}} \right] , \tag{5.50}$$

and the partial derivatives are taken from the Lagrangian $L^{\rm m}$. More specifically, calculations yield

$$Y^{\alpha} \equiv \frac{\overline{\rho}_{\rm m}}{\overline{\mu}_{\rm m}} \phi^{|\alpha} - \overline{\rho}_{\rm m} l^{\alpha\beta} \overline{u}_{\beta} + \left(1 - \frac{c^2}{v_{\rm s}^2} \right) \left(\frac{\overline{\rho}_{\rm m}}{\overline{\mu}_{\rm m}} \overline{u}^{\alpha} \overline{u}^{\beta} \phi_{|\beta} - \frac{1}{2} \overline{\rho}_{\rm m} \overline{u}^{\alpha} \mathfrak{q} \right) . \tag{5.51}$$

Similar expression was derived by Lukash [83] who used the variational method to analyze the production and quantization of sound waves in the early universe.

5.6 The Lagrangian equations for dark energy perturbations

Equations for the perturbations ψ of dark energy, which is modeled by a scalar field Ψ, are derived by taking the variational derivative from the Lagrangian (5.19) with respect to the field variable Ψ – see equation (5.26c). Subtracting the background equation (5.27c) from (5.26c) and making use of the Lagrangian decomposition in the Taylor (post-Newtonian) series leads to

$$F_{\Psi}^{\rm q} = 8\pi \Sigma^{\rm q} , \tag{5.52}$$

where the linear differential operator

$$F_{\Psi}^{\rm q} \equiv -\frac{1}{\sqrt{-\overline{g}}} \frac{\delta}{\delta \overline{\Psi}} \left(\mathfrak{h}^{\mu\nu} \frac{\delta \overline{\mathcal{L}}^{\rm q}}{\delta \overline{\mathfrak{g}}^{\mu\nu}} + \psi \frac{\delta \overline{\mathcal{L}}^{\rm q}}{\delta \overline{\Psi}} \right) , \tag{5.53}$$

and the source term

$$\Sigma^{\rm q} \equiv \frac{1}{8\pi \sqrt{-\overline{g}}} \left(\frac{\delta \mathcal{L}_2^{\rm q}}{\delta \overline{\Psi}} + \frac{\delta \mathcal{L}_3^{\rm q}}{\delta \overline{\Psi}} + \dots \right) . \tag{5.54}$$

According to equation (3.19), the Lagrangian density of the scalar field $\mathcal{L}^{\rm q} = \sqrt{-g} L^{\rm q}$ depends on both the field Ψ and its first derivative, $\Psi_{,\alpha}$. For this reason, the differential operator $F^{\rm q}$ is not reduced to the covariant derivative from a vector field as the partial derivative of the Lagrangian with respect to Ψ does not vanish. We have

$$F_{\Psi}^{\rm q} \equiv Z^{\alpha}{}_{|\alpha} - \frac{l}{2} \frac{\partial \overline{W}}{\partial \overline{\Psi}} - \psi \frac{\partial^2 \overline{W}}{\partial \overline{\Psi}^2} , \tag{5.55}$$

where $l \equiv \overline{g}^{\alpha\beta} l_{\alpha\beta}$, and vector field

$$Z^{\alpha} \equiv \frac{\partial}{\partial \overline{\Psi}_{,\alpha}} \left[\left(l^{\mu\nu} - \frac{1}{2} l \overline{g}^{\mu\nu} \right) \left(\frac{\partial \overline{L}^{\rm q}}{\partial \overline{g}^{\mu\nu}} - \frac{1}{2} \overline{g}_{\mu\nu} \overline{L}^{\rm q} \right) + \psi_{,\beta} \frac{\partial \overline{L}^{\rm q}}{\partial \overline{\Psi}_{,\beta}} \right] . \tag{5.56}$$

Performing the partial derivatives in equation (5.56), yields a rather simple expression

$$Z^{\alpha} \equiv \psi^{|\alpha} - \overline{p}_{\mathrm{q}} l^{\alpha\beta} \overline{u}_{\beta} \,, \tag{5.57}$$

where we have used equation $\overline{\Psi}_{|\alpha} = -\overline{u}^{\beta}\overline{\Psi}_{|\beta}\overline{u}_{\alpha} = -\overline{p}_{\mathrm{q}}\overline{u}_{\alpha}$. The reader is invited to compare equation (5.57) with (5.51) to observe the differences between the Lagrangian perturbations of the ideal fluid and the scalar field. One may observe that (5.51) becomes identical with (5.57) in the limit $v_{\mathrm{s}} \to c$, and $\overline{p}_{\mathrm{m}} \to \overline{\mu}_{\mathrm{m}}$. This corresponds to the case of an extremely stiff equation of state $w_{\mathrm{m}} = 1$ in equation (4.54). According to the discussion following equations (5.44) and (5.45) the speed of “sound” c_{s} in the scalar field “fluid” is always equal to c. However, it does not assume that the parameter w_{q} of the equation of state of the scalar field, $\overline{p}_{\mathrm{q}} = w_{\mathrm{q}}\overline{\epsilon}_{\mathrm{q}}$, in (4.54) is equal to unity. This is because the scalar field is not completely equivalent to the ideal fluid in the sense of thermodynamic [2].

5.7 Linearized post-Newtonian equations for field variables

Equations for the metric tensor perturbations

Linearized field equations for gravitational field variables, $l_{\mu\nu}$, are obtained from (5.28) after neglecting in its right-hand side the nonlinear source $\mathcal{T}_{\mu\nu}$, and making a series of transformations to sort out similar terms. First, let us make use of equations (5.40) and (5.41) to find out

$$\begin{aligned} F^{\mathrm{m}}_{\mu\nu} + F^{\mathrm{q}}_{\mu\nu} = {} & 4\pi\left(\overline{\epsilon} - \overline{p}\right) l_{\mu\nu} \\ & + 8\pi\overline{p}_{\mathrm{m}}\left[\overline{u}_{\mu}\phi_{,\nu} + \overline{u}_{\nu}\phi_{,\mu} - \overline{g}_{\mu\nu}u^{\alpha}\phi_{,\alpha} + \left(1 - \frac{c^2}{v_{\mathrm{s}}^2}\right)\left(\overline{u}^{\alpha}\phi_{,\alpha} - \frac{1}{2}\overline{\mu}_{\mathrm{m}}\mathrm{q}\right)\overline{u}_{\mu}\overline{u}_{\nu}\right] \\ & + 8\pi\overline{p}_{\mathrm{q}}\left[\overline{u}_{\mu}\psi_{,\nu} + \overline{u}_{\nu}\psi_{,\mu} - \overline{g}_{\mu\nu}u^{\alpha}\psi_{,\alpha} + \overline{g}_{\mu\nu}\frac{\partial \overline{W}(\overline{\Psi})}{\partial\overline{\Psi}}\frac{\psi}{\overline{\mu}_{\mathrm{q}}}\right] , \end{aligned} \tag{5.58}$$

where we used the superposition $\overline{\epsilon} = \overline{\epsilon}_{\mathrm{m}} + \overline{\epsilon}_{\mathrm{q}}$, $\overline{p} = \overline{p}_{\mathrm{m}} + \overline{p}_{\mathrm{q}}$. Second step is to transform the linear differential operator $F^{\mathrm{g}}_{\mu\nu}$ in (5.37) to a more convenient form that will allow us to single out the gauge-dependent vector denoted by

$$A^{\mu} \equiv l^{\mu\nu}{}_{|\nu} \,. \tag{5.59}$$

Changing the order of the covariant derivatives in (5.37) and taking into account that the commutator of the second covariant derivatives is proportional to the Riemann tensor, we recast (5.37) to the following form:

$$F^{\mathrm{g}}_{\mu\nu} \equiv \frac{1}{2}\left(l_{\mu\nu}{}^{|\alpha}{}_{|\alpha} + \overline{g}_{\mu\nu}A^{\alpha}{}_{|\alpha} - A_{\mu|\nu} - A_{\nu|\mu}\right) - \overline{R}^{\alpha}{}_{(\mu}l_{\nu)\alpha} - \overline{R}_{\mu\alpha\beta\nu}l^{\alpha\beta} \,, \tag{5.60}$$

where the round brackets around indices denote symmetrization. The terms with the Ricci and Riemann tensors can be expressed in terms of the total background energy and pressure of the ideal fluid and scalar field by making use of equations (4.34) and (4.36) and Einstein's equations (4.43). It yields

$$R^{\alpha}{}_{(\mu}l_{\nu)\alpha} + \overline{R}_{\mu\alpha\beta\nu}l^{\alpha\beta} = 4\pi\Big[\left(\frac{5\overline{\epsilon}}{3} - \overline{p}\right)l_{\mu\nu} + \frac{l}{2}\left(\overline{p} - \frac{\overline{\epsilon}}{3}\right)\overline{g}_{\mu\nu} + (\overline{\epsilon} + \overline{p})\left(2\overline{u}^{\alpha}\overline{u}_{\mu}l_{\nu\alpha} + 2\overline{u}^{\alpha}\overline{u}_{\nu}l_{\mu\alpha} - \overline{u}_{\mu}\overline{u}_{\nu}l - \overline{g}_{\mu\nu}\mathfrak{q}\right)\Big]. \tag{5.61}$$

Finally, substituting equations (5.58), (5.60), and (5.61) to (5.28) results in

$$\begin{aligned} & l_{\mu\nu}{}^{|\alpha}{}_{|\alpha} + \overline{g}_{\mu\nu}A^{\alpha}{}_{|\alpha} - A_{\mu|\nu} - A_{\nu|\mu} \\ & - 16\pi\left[\frac{\overline{\epsilon}}{3}l_{\mu\nu} + \frac{l}{4}\left(\overline{p} - \frac{\overline{\epsilon}}{3}\right)\overline{g}_{\mu\nu} + (\overline{\epsilon} + \overline{p})\left(\frac{1}{2}\overline{u}^{\alpha}\overline{u}_{(\mu}l_{\nu)\alpha} - \frac{1}{2}\overline{u}_{\mu}\overline{u}_{\nu}l - \frac{1}{2}\overline{g}_{\mu\nu}\mathfrak{q}\right)\right] \\ & + 16\pi\overline{\rho}_{\mathrm{m}}\left[\overline{u}_{\mu}\phi_{,\nu} + \overline{u}_{\nu}\phi_{,\mu} - \overline{g}_{\mu\nu}u^{\alpha}\phi_{,\alpha} + \left(1 - \frac{c^2}{v_s^2}\right)\left(\overline{u}^{\alpha}\phi_{,\alpha} - \frac{1}{2}\overline{\mu}_{\mathrm{m}}\mathfrak{q}\right)\overline{u}_{\mu}\overline{u}_{\nu}\right] \\ & + 16\pi\overline{\rho}_{\mathrm{q}}\left[\overline{u}_{\mu}\psi_{,\nu} + \overline{u}_{\nu}\psi_{,\mu} - \overline{g}_{\mu\nu}u^{\alpha}\psi_{,\alpha} + \overline{g}_{\mu\nu}\frac{\partial\overline{W}(\overline{\Psi})}{\partial\overline{\Psi}}\frac{\psi}{\overline{\mu}_{\mathrm{q}}}\right] = 16\pi\mathfrak{T}_{\mu\nu}, \end{aligned} \tag{5.62}$$

where the nonlinear term, $\mathcal{T}_{\mu\nu}$, was neglected on the right-hand side of (5.62).

The very first term in (5.62) is a tensor Laplace–Beltrami operator, $l_{\mu\nu}{}^{|\alpha}{}_{|\alpha} \equiv \overline{g}^{\alpha\beta}l_{\mu\nu|\alpha\beta}$, that is a rather complicated geometric object. Its explicit expression can be developed by making use of the Christoffel symbols given in (4.23). Tedious but straightforward calculation yields [70]

$$\begin{aligned} l_{\mu\nu}{}^{|\alpha}{}_{|\alpha} = {} & l_{\mu\nu}{}^{;\alpha}{}_{;\alpha} + 2H\overline{u}^{\alpha}l_{\mu\nu;\alpha} - 2\left(H\overline{u}^{\alpha}l_{\alpha\mu}\right)_{|\nu} - 2\left(H\overline{u}^{\alpha}l_{\alpha\nu}\right)_{|\mu} \\ & + 2H\left(\overline{u}_{\mu}A_{\nu} + \overline{u}_{\nu}A_{\mu}\right) + 2\dot{H}\left(l_{\mu\nu} - \overline{u}^{\alpha}\overline{u}_{\mu}l_{\nu\alpha} - \overline{u}^{\alpha}\overline{u}_{\nu}l_{\mu\alpha}\right) \\ & + 2H^2\left(2l_{\mu\nu} + 3\overline{u}_{\mu}\overline{u}^{\alpha}l_{\alpha\nu} + 3\overline{u}_{\nu}\overline{u}^{\alpha}l_{\alpha\mu} - \overline{g}_{\mu\nu}\overline{u}^{\alpha}\overline{u}^{\beta}l_{\alpha\beta} - \overline{u}_{\mu}\overline{u}_{\nu}l\right), \end{aligned} \tag{5.63}$$

where the semicolon denotes a covariant derivative that is calculated with the Christoffel symbols $B^{\alpha}{}_{\mu\nu}$ like in (4.30b), and the spatial Laplace–Beltrami operator, $l_{\mu\nu}{}^{;\alpha}{}_{;\alpha} \equiv \overline{g}^{\alpha\beta}l_{\mu\nu;\alpha\beta}$.

Further derivation of the differential post-Newtonian field equation for the linearized metric tensor perturbations can be significantly simplified if we redefine the gauge function, A^{α}, in the following form:

$$A^{\alpha} = -2Hl^{\alpha\beta}\overline{u}_{\beta} + 16\pi\left(\overline{\rho}_{\mathrm{m}}\phi + \overline{\rho}_{\mathrm{q}}\psi\right)\overline{u}^{\alpha} + B^{\alpha}, \tag{5.64}$$

where B^{α} is an arbitrary gauge vector function. This choice of the gauge function A^{α} allows us to eliminate two terms in equation (5.63) which depend on the first covariant derivatives with respect to the background metric $\overline{g}_{\alpha\beta}$. Moreover, it allows us to eliminate a number of terms depending on the first derivatives of the fields ϕ and ψ in equation (5.62). Since we keep the gauge function B^α arbitrary, equation (5.64) does not fix any gauge. The choice of the gauge is controlled by the gauge function B^α.

One substitutes the gauge function (5.64) to equations (5.63) and (5.62) and make use of the background Friedmann equations (4.50) and (4.51) to replace the background values of the energy density, $\bar{\epsilon}$, and pressure, $\bar{p}$, with the Hubble parameter H and its time derivative $\dot{H}$. It brings about equation (5.62) to the following form:

$$\begin{aligned} & l_{\mu\nu}{}^{;\alpha}{}_{;\alpha} + 2H\bar{u}^\alpha l_{\mu\nu;\alpha} + 2\left(\dot{H} + H^2\right)\left(l_{\mu\nu} + \bar{u}_\mu \bar{u}^\alpha l_{\alpha\nu} + \bar{u}_\nu \bar{u}^\alpha l_{\alpha\mu} - l\bar{u}_\mu \bar{u}_\nu\right) \\ & - \frac{2k}{a^2}\left[l_{\mu\nu} + 2\bar{u}_\mu \bar{u}^\alpha l_{\alpha\nu} + 2\bar{u}_\nu \bar{u}^\alpha l_{\alpha\mu} - l\bar{u}_\mu \bar{u}_\nu - \left(\mathfrak{q} + \frac{l}{2}\right)\bar{g}_{\mu\nu}\right] \\ & + 16\pi \bar{u}_\mu \bar{u}_\nu \left[\bar{\rho}_{\mathrm{m}}\left(1 - \frac{c^2}{v_{\mathrm{s}}^2}\right)\left(\bar{u}^\alpha \phi_{,\alpha} - \frac{1}{2}\bar{\mu}_{\mathrm{m}}\mathfrak{q}\right) - 2\frac{\partial \overline{W}}{\partial \overline{\Psi}}\psi - 4H\left(\bar{\rho}_{\mathrm{m}}\phi + \bar{\rho}_{\mathrm{q}}\psi\right)\right] \\ & + \bar{g}_{\mu\nu} B^\alpha{}_{|\alpha} - B_{\mu|\nu} - B_{\nu|\mu} + 2H\left(\bar{u}_\mu B_\nu + \bar{u}_\nu B_\mu - \bar{g}_{\mu\nu}\bar{u}^\alpha B_\alpha\right) = 16\pi \mathfrak{T}_{\mu\nu}\,. \end{aligned} \tag{5.65}$$

This equation is fully covariant and is valid in any gauge and/or coordinate chart.

Now, let us fix the gauge by selecting a specific gauge function B^α in (5.64). The task is to decouple the linearized field equations for l_{00}, l_{0i}, and l_{ij} components of the metric tensor perturbations. For this purpose, let us work in the isotropic coordinates associated with the Hubble flow, where $\bar{u}^\alpha = (1/a, 0, 0, 0)$ and choose the gauge condition, $B^\alpha = 0$. It brings equation (5.65) for different components of the metric perturbations to the form

$$\begin{aligned} & \Box q + 2\mathcal{H}q_{;0} + 4kq - 4\pi\left(1 - \frac{c^2}{v_{\mathrm{s}}^2}\right)\bar{\rho}_{\mathrm{m}}\bar{\mu}_{\mathrm{m}}\mathfrak{q} \\ & = 8\pi a^2\left(\mathfrak{T}_{00} + \mathfrak{T}_{kk}\right) - 8\pi a^3\left[\left(1 - \frac{c^2}{v_{\mathrm{s}}^2}\right)\bar{\rho}_{\mathrm{m}}\phi_{,0} - 2a\frac{\partial \overline{W}}{\partial \overline{\Psi}}\psi - \mathcal{H}\left(\bar{\rho}_{\mathrm{m}}\phi + \bar{\rho}_{\mathrm{q}}\psi\right)\right], \end{aligned} \tag{5.66a}$$

$$\Box l_{0i} + 2\mathcal{H}l_{0i;0} + 2kl_{0i} = 16\pi a^2 \mathfrak{T}_{0i}\,, \tag{5.66b}$$

$$\Box l_{\langle ij\rangle} + 2\mathcal{H}l_{\langle ij\rangle;0} + 2\left(\mathcal{H}' - k\right)l_{\langle ij\rangle} = 16\pi a^2 \mathfrak{T}_{\langle ij\rangle}\,, \tag{5.66c}$$

$$\Box l_{kk} + 2\mathcal{H}l_{kk;0} + 2\left(\mathcal{H}' + 2k\right)l_{kk} = 16\pi a^2 \mathfrak{T}_{kk}\,. \tag{5.66d}$$

Here, $\mathcal{H}$ is a conformal Hubble parameter (4.4), a prime means a derivative with respect to η, and we denote by $\Box l_{\mu\nu} \equiv \bar{\mathfrak{f}}^{\alpha\beta} l_{\mu\nu;\alpha\beta}$, $q \equiv (l_{00} + l_{kk})/2$, $l_{kk} \equiv l_{11} + l_{22} + l_{33}$, $l_{\langle ij\rangle} \equiv l_{ij} - (1/3)\delta_{ij}l_{kk}$, and the same index notations are applied to the tensor of energy–momentum $\mathfrak{T}_{ij}$ of the localized astronomical system, $\mathfrak{T}_{\langle ij\rangle} = \mathfrak{T}_{ij} - (1/3)\delta_{ij}\mathfrak{T}_{kk}$. These equations are clearly decoupled from one another, thus, demonstrating the advantage of the gauge condition $B^\alpha = 0$.

Equations (5.66b)–(5.66d) can be solved independently if the initial and boundary conditions are known, and the tensor of energy–momentum of the localized astronomical system is well defined. Equation (5.66a) for a scalar q demands besides knowledge of $\mathfrak{T}_{\alpha\beta}$, knowing the scalar field perturbations, ϕ and ψ, that contribute to the source of the field equation for q on the right-hand side of (5.66a). Equations for these perturbations are obtained below.

Equations for the dark matter perturbations

The dark matter perturbations, ϕ, evolve in accordance with the Lagrangian equation (5.46). In the linear approximation, we can neglect the nonlinear source term $\Sigma^{\rm m}$ in its right-hand side. The covariant derivative in the definition of the linear operator $F^{\rm m}$ given on the right-hand side of (5.49) can be explicitly performed, thus, yielding equation for the Clebsch potential

$$\phi^{|\alpha}{}_{|\alpha} - 2\bar{\mu}_{\rm m}H\mathfrak{q} + 16\pi\bar{\mu}_{\rm m}\left(\bar{\rho}_{\rm m}\phi + \bar{\rho}_{\rm q}\psi\right) + \left(1 - \frac{c^2}{v_{\rm s}^2}\right)\left(\bar{u}^\alpha\bar{u}^\beta\phi_{|\alpha\beta} - \frac{1}{2}\bar{\mu}_{\rm m}\bar{u}^\alpha\mathfrak{q}_{,\alpha}\right)$$
$$= \bar{\mu}_{\rm m}\bar{u}^\alpha B_\alpha\,, \tag{5.67}$$

where equation (5.64) has been used. The gauge B^α remains yet unspecified so that equation (5.67) is covariant and is valid in any coordinate chart. To make it compatible with equations (5.66a)–(5.66d) for the metric tensor perturbations, we have to choose $B^\alpha = 0$.

Equations for the dark energy perturbations

Linearized equation for the dark energy perturbations, ψ, is obtained from the Lagrangian equation (5.52) after neglecting the (nonlinear) source term $\Sigma^{\rm q}$. After performing the covariant differentiation in equation (5.55), we conclude that the dark energy perturbation obeys the following equation:

$$\psi^{|\alpha}{}_{|\alpha} - \left(2\bar{\mu}_{\rm q}H + \frac{\partial\bar{W}}{\partial\bar{\Psi}}\right)\mathfrak{q} + 16\pi\bar{\mu}_{\rm q}\left(\bar{\rho}_{\rm m}\phi + \bar{\rho}_{\rm q}\psi\right) - \frac{\partial^2\bar{W}}{\partial\bar{\Psi}^2}\psi = \bar{\mu}_{\rm q}\bar{u}^\alpha B_\alpha\,, \tag{5.68}$$

where equation (5.64) has been used along with the equality $\bar{\rho}_{\rm q} = \bar{\mu}_{\rm q}$. The gauge function B^α is kept unspecified so that equation (5.68) is covariant and is valid in any coordinates. To make it compatible with equations (5.66a)–(5.66d) for the metric tensor perturbations, we have to choose $B^\alpha = 0$.

6 Gauge-invariant scalars and field equations in 1+3 threading formalism

6.1 Threading decomposition of the metric perturbations

We have derived the system of coupled differential equations (5.65), (5.67), and (5.68) for the field variables $l_{\alpha\beta}$, ϕ, and ψ, describing perturbations of the gravitational field, dark matter and dark energy respectively. These equations are gauge invariant and written down in arbitrary coordinates on the background manifold. Nonethelees, they operate with the field variables which are not gauge-invariant in themselves. Therefore, solutions of equations (5.65), (5.67), and (5.68) that are found in a particular gauge has no direct physical interpretation and must be connected to physical observables to match theory with observations. Another way around is to find out some gauge-invariant geometric objects built out of $l_{\alpha\beta}$, ϕ, and ψ which will not depend on a particular choice of gauge and coordinates. This program was initiated by Bardeen [8] who proposed to split the perturbations of the metric tensor in scalar, vector, and tensor components by making use of 3+1 spacetime slicing ADM technique [5], and to build gauge-invariant cosmological variables out of these elements. Gauge-invariant scalars are the most important quantities in cosmology as they describe the structure formation in the universe. Ellis and Bruni [39] pointed out that Bardin's variables are not directly related to the density fluctuations but to it second derivatives which makes them less useful in relativistic calculations of structure formation. They proposed their own gauge-invariant variables that are build out of gradients of the geometric objects which vanish on the background manifold so that only their perturbations make physical sense.

In this section, we propose even more direct approach to the definition of the gauge-invariant scalars by making use of the scalar potentials Φ and Ψ for description of the dark matter and dark energy. In this way, we shall find out the gauge-invariant scalars that are equivalent to the matter density fluctuation itself but not to its gradient or a second-order derivative. We shall employ 1+3 threading approach to split four-dimensional tensors into scalar, vector, and three-dimensional tensors. The original idea was proposed by Zelmanov [121] who called the elements of the tensorial decomposition the chronometric invariants. Later on, the theory of chronometric invariants was reinvented by a number of researchers. The central ingredient of the theory is a congruence of world lines threading spacetime. In FLRW cosmology, this congruence is naturally associated with the Hubble flow and the Hubble velocity $\bar{u}^{\alpha}$. Threading (chronometric) decomposition is achieved with the invariant operator of projection $\bar{P}_{\alpha\beta}$ onto a hypersurface being orthogonal to the congruence of world lines of the Hubble flow,

$$\bar{P}_{\alpha\beta} = \bar{g}_{\alpha\beta} + \bar{u}_{\alpha}\bar{u}_{\beta} , \tag{6.1}$$

where $\bar{g}_{\alpha\beta}$ is FLRW background metric. The operator $\bar{P}_{\alpha\beta}$ can be considered as a metric on the spatial hypersurface of the background FLRW manifold.

The post-Newtonian theory under development admits four algebraically independent scalar perturbations. Two of them are the Clebsch potential of the ideal fluid ϕ and the scalar field ψ. The two other scalars characterize the scalar perturbations of the gravitational field. They can be chosen, for example, as a projection of the metric tensor perturbation on the direction of the background 4-velocity field, $\bar{u}^\alpha \bar{u}^\beta l_{\alpha\beta}$, and the trace of the metric tensor perturbation, $l = \bar{g}^{\alpha\beta} l_{\alpha\beta}$. However, it is more convenient to work with two other scalars, defined as their linear combinations,

$$\mathfrak{q} \equiv \frac{1}{2}\left(\bar{u}^\alpha \bar{u}^\beta + \bar{P}^{\alpha\beta}\right) l_{\alpha\beta} , \tag{6.2a}$$

$$\mathfrak{p} \equiv \bar{P}^{\alpha\beta} l_{\alpha\beta} , \tag{6.2b}$$

Notice that the scalar $\mathfrak{q}$ has been introduced earlier in (5.16). The scalar $\mathfrak{p}$ is, in fact, a projection of $l_{\alpha\beta}$ onto the space-like hypersurface being orthogonal everywhere to the world lines of Hubble observers.

Vectorial chronometric perturbations are defined by a spacial-temporal projection

$$\mathfrak{p}_\alpha \equiv -\bar{P}_\alpha{}^\beta \bar{u}^\gamma l_{\beta\gamma} , \tag{6.3}$$

where minus sign was taken for the sake of mathematical convenience. Due to its definition, vector $\mathfrak{p}^\alpha = \bar{g}^{\alpha\beta}\mathfrak{p}_\beta$ is orthogonal to the 4-velocity $\bar{u}^\alpha$, that is $\bar{u}^\alpha \mathfrak{p}_\alpha = 0$. Hence, it describes a space-like vector-like gravitational perturbations with three algebraically independent components.

Tensorial chronometric perturbations are associated with the projection

$$\mathfrak{p}^{\mathrm{T}}_{\alpha\beta} \equiv \mathfrak{p}_{\alpha\beta} - \frac{1}{3}\bar{P}_{\alpha\beta}\mathfrak{p} , \tag{6.4}$$

where

$$\mathfrak{p}_{\alpha\beta} \equiv \bar{P}_\alpha{}^\mu \bar{P}_\beta{}^\nu l_{\mu\nu} . \tag{6.5}$$

Here, the tensor $\mathfrak{p}_{\alpha\beta}$ is a double projection of $l_{\alpha\beta}$ onto space-like hypersurface being orthogonal to the world lines of Hubble observers. The trace of this tensor coincides with the scalar $\mathfrak{p}$. Indeed,

$$\bar{g}^{\alpha\beta}\mathfrak{p}_{\alpha\beta} = \bar{g}^{\alpha\beta}\bar{P}_\alpha{}^\mu \bar{P}_\beta{}^\nu l_{\mu\nu} = \bar{P}^{\beta\mu}\bar{P}_\beta{}^\nu l_{\mu\nu} = \bar{P}^{\mu\nu} l_{\mu\nu} = \mathfrak{p} , \tag{6.6}$$

where the property of the projection tensor $\bar{P}^{\beta\mu}\bar{P}_\beta{}^\nu = \bar{P}^{\mu\nu}$ has been used. Equation (6.6) makes it clear that tensor $\mathfrak{p}^{\mathrm{T}}_{\alpha\beta}$ is traceless, that is $\bar{g}^{\alpha\beta}\mathfrak{p}^{\mathrm{T}}_{\alpha\beta} = 0$. Because of this property, and four orthogonality conditions, $\bar{u}^\alpha \mathfrak{p}^{\mathrm{T}}_{\alpha\beta} = 0$, the symmetric tensor $\mathfrak{p}^{\mathrm{T}}_{\alpha\beta}$ has only five, algebraically independent components.

Gravitational perturbation $l_{\alpha\beta}$ can be decomposed into the algebraically irreducible scalar, vector, and tensor parts as follows:

$$l_{\alpha\beta} = \mathfrak{p}^{\mathrm{T}}_{\alpha\beta} + \bar{u}_\alpha \mathfrak{p}_\beta + \bar{u}_\beta \mathfrak{p}_\alpha + \left(\bar{u}_\alpha \bar{u}_\beta + \frac{1}{3}\bar{P}_{\alpha\beta}\right)\mathfrak{p} + 2\bar{u}_\alpha \bar{u}_\beta \left(\mathfrak{q} - \mathfrak{p}\right) . \tag{6.7}$$

One should not confuse the pure algebraic (threading) decomposition of the metric tensor perturbation with its functional (slicing) decomposition. The slicing (or kinemetric, according to Zelmanov [122]) decomposition was pioneered by Arnowitt et al. [5, 89]. It is commonly used in the research on the relativistic theory of formation of the large-scale structure in the universe. The ADM decomposition of the metric tensor perturbations is done by foliating spacetime [8, 62] with a set of spacelike hypersurfaces and making use of three-dimensional Helmholtz theorem [7] which singles out the longitudinal (L), transversal (T), and transverse-traceless (TT) parts of the perturbations. In other words, the slicing decomposition make vector $\mathfrak{p}_\alpha$ and tensor parts of the gravitational perturbation, $\mathfrak{p}^{\mathrm{T}}_{\alpha\beta}$, are further decomposed in the functionally irreducible components which include two more scalars, and two transverse spatial vectors each having only two (out of three) independent components. The remaining part of the tensor perturbations, $\mathfrak{p}^{\mathrm{T}}_{\alpha\beta}$, is transverse-trackless and has only two functionally independent components denoted as $\mathfrak{p}^{\mathrm{TT}}_{\alpha\beta}$. The ADM decomposition of the metric tensor is a powerful technique in the theory of gauge-invariant cosmological perturbations [9, 91]. However, it is not convenient in the development of the systematic post-Newtonian approximations and celestial dynamics of inhomogeneities in cosmology. Thus, we do not use it in this chapter.

Our next step is a to find the gauge-invariant scalars directly reproducing the density fluctuation and to derive the post-Newtonian field equations for the algebraically irreducible components of matter and gravitational field. We, first, discuss the gauge transformations of the corresponding field variables.

6.2 Gauge transformation of the field variables

Gauge invariance is a cornerstone of the modern theoretical physics with an interesting but somehow controversial history [58]. The gauge invariance should be distinguished from coordinate (diffeomorphism) invariance or general covariance because, by definition, gauge transformation changes merely the field variables of the theory under consideration but not coordinates on the underlying spacetime manifold. Discussing gauge-transformation and gauge-invariance requires introduction of a supplementary gauge field and associated with it geometric structures – an affine connection and a fiber bundle manifold describing the intrinsic degrees of freedom of corresponding field variables of the gauge field theory [33, 101].

This chapter discusses physical perturbations of tensor $l_{\alpha\beta}$, and scalars Φ, Ψ in the framework of general relativity where the affine connection is represented by the

Christoffel symbols of spacetime manifold while the gauge transformation is generated by a flow of an arbitrary vector (gauge) field ξ^α that maps the manifold into itself. Generic gauge transformation of the fields on a curved manifold is associated with their Lie transport along the vector flow ξ^α [75, 113] while an infinitesimal gauge transformation is a Lie derivative of the field taken at the value of the parameter on the curves of the vector flow equal to 1 [66, chapter 3.6].

Let us consider a mapping of spacetime manifold into itself induced by a vector flow, $\xi^\alpha = \xi^\alpha(x^\beta)$. It means that each point of the manifold with coordinates x^α is mapped to another point with coordinates

$$\hat{x}^\alpha = x^\alpha - \xi^\alpha(x)\,. \tag{6.8}$$

This mapping of the manifold into itself can be interpreted as a local diffeomorphism which transforms the field variables in accordance to their tensor properties. The transformed value of the field variable is pulled back to the point of the manifold having the original coordinates x^α, and is compared with the original value of the field at this point. The difference between the transformed and the original value of the field, generated by the diffeomorphism (6.8) is the gauge transformation of the field that is given by the Lie derivative taken along the vector flow ξ^α at the point of the manifold with coordinates x^α.

Let us denote the transformed values of the field variables with a hat. In the linearized perturbation theory of the cosmological manifold, the gauge transformations of the field variables (the metric tensor perturbation $\varkappa_{\alpha\beta}$, the scalar field ϕ, and ψ) are given by equations

$$\hat{\varkappa}_{\alpha\beta} = \varkappa_{\alpha\beta} + \xi_{\alpha|\beta} + \xi_{\beta|\alpha}\,, \tag{6.9a}$$

$$\hat{l}_{\alpha\beta} = l_{\alpha\beta} - \xi_{\alpha|\beta} - \xi_{\beta|\alpha} + \bar{g}_{\alpha\beta}\xi^\gamma{}_{|\gamma}\,, \tag{6.9b}$$

$$\hat{\phi} = \phi + \bar{\Phi}_{|\alpha}\xi^\alpha\,, \tag{6.9c}$$

$$\hat{\psi} = \psi + \bar{\Psi}_{|\alpha}\xi^\alpha\,, \tag{6.9d}$$

where the hat above each symbol denotes a new value of the field variable after applying the gauge transformation (6.8), and all functions are calculated at the same value of coordinates x^α. The gauge transformations of the field variables are expressed in terms of the covariant derivatives on the manifold and are coordinate independent. Equation (6.9b) is derived from the Lie transformation (6.9a) of the metric tensor perturbation, and the relation (5.10) connecting $\varkappa_{\alpha\beta}$ and $l_{\alpha\beta}$.

Gauge invariance of the Lagrangian perturbation theory of geometric manifolds means that the gauge transformations of the field variables cannot change the content of the theory. In other words, equations for the field variables must be invariant with respect to the gauge transformations (6.9a)–(6.9d). However, direct inspection of equations (5.65), (5.67), and (5.68) shows that they do depend on the choice of the gauge in the form of the gauge function B^α introduced in equation (5.64). To find out the gauge-invariant content of the theory one should search for the gauge-invariant field variables and to derive the gauge-invariant equations for them. This program has been completed by Bardeen [9] who used the functional 3+1 slicing decomposition of the metric tensor perturbations and the vector field ξ^α to build the gauge-invariant variables out of the various projections of the metric tensor components on space an time. Modifications of Bardeen's approach can be found in [21, 36, 39, 41, 83, 93] and in the book by Mukhanov [91]. We use algebraic 1+3 threading decomposition of the metric tensor perturbations (6.7) that allows us to build gauge-invariant scalars. Vector and tensor perturbations remain gauge dependent in the threading approach. In order to suppress the gauge degrees of freedom in these variables we impose a particular gauge condition $B^\alpha = 0$ in equation (5.64). This limits the freedom of the gauge field ξ^α by a particular set of differential equations which are discussed in Section 6.7.

6.3 Gauge-invariant scalars

The existence of the preferred 4-velocity, $\bar{u}^\alpha$, of the Hubble flow in the expanding universe provides a natural way of separating the perturbations of the field variables in scalar, vector, and tensor components. This section discusses how to build the gauge-invariant scalars. Vector and tensor perturbations are discussed afterward.

The gauge-invariant scalar perturbations can be build from the perturbation of the Clebsch potential, ϕ, the perturbation of the scalar field ψ, and a scalar q defined in (6.2a). To build the first gauge-invariant scalar, we introduce the scalar perturbations

$$\chi_{\mathrm{m}} \equiv \frac{\phi}{\bar{\mu}_{\mathrm{m}}}\,, \qquad \chi_{\mathrm{q}} \equiv \frac{\psi}{\bar{\mu}_{\mathrm{q}}}\,, \tag{6.10}$$

that normalize perturbations of the Clebsch potential ϕ and that of the scalar field ψ to the corresponding background values of the specific enthalpy, $\bar{\mu}_{\mathrm{m}}$ and $\bar{\mu}_{\mathrm{q}}$. The gauge transformations for the three scalars q, χ_{m}, and χ_{q} are obtained from (6.9b)–(6.9d), and read

$$\hat{\mathrm{q}} = \mathrm{q} - 2\bar{u}^\alpha \bar{u}^\beta \xi_{\alpha|\beta}\,, \tag{6.11a}$$

$$\hat{\chi}_{\mathrm{m}} = \chi_{\mathrm{m}} - \bar{u}_\alpha \xi^\alpha\,, \tag{6.11b}$$

$$\hat{\chi}_{\mathrm{q}} = \chi_{\mathrm{q}} - \bar{u}_\alpha \xi^\alpha\,, \tag{6.11c}$$

where we have used the definition of the background 4-velocity $\bar{u}_\alpha = -\bar{\Phi}_{|\alpha}/\bar{\mu}_{\mathrm{m}} = -\bar{\Psi}_{|\alpha}/\bar{\mu}_{\mathrm{q}}$ in terms of the partial derivatives of the background values of the scalar fields Φ and Ψ. Equations (6.11b) and (6.11c) immediately reveal that the linear combination

$$\chi \equiv \chi_{\mathrm{m}} - \chi_{\mathrm{q}}\,, \tag{6.12}$$

is gauge invariant, $\hat{\chi} = \chi$, that is the diffeomorphism (6.8) does not change the value of the scalar variable χ.

Two other gauge-invariant scalars are defined by the following equations:

$$V_{\rm m} \equiv \bar{u}^{\alpha}\chi_{{\rm m}|\alpha} - \frac{\rm q}{2}\,, \tag{6.13a}$$

$$V_{\rm q} \equiv \bar{u}^{\alpha}\chi_{{\rm q}|\alpha} - \frac{\rm q}{2}\,, \tag{6.13b}$$

or, more explicitly,

$$V_{\rm m} = \frac{1}{\bar{\mu}_{\rm m}}\bar{u}^{\alpha}\phi_{|\alpha} - \frac{\rm q}{2} + 3\frac{v_{\rm s}^2}{c^2}H\chi_{\rm m}\,, \tag{6.14a}$$

$$V_{\rm q} = \frac{1}{\bar{\mu}_{\rm q}}\bar{u}^{\alpha}\psi_{|\alpha} - \frac{\rm q}{2} + 3H\chi_{\rm q} + \frac{\chi_{\rm q}}{\mu_{\rm q}}\frac{\partial\overline{W}}{\partial\overline{\Psi}}\,, \tag{6.14b}$$

where the last terms on the right-hand side of these equations were obtained by making use of thermodynamic relationships (3.11), the equality $\bar{\rho}_{\rm q} = \bar{\mu}_{\rm q}$, and the equations of continuity (4.59) and (4.67) for the density of the ideal fluid, $\bar{\rho}_{\rm m}$, and that of the scalar field, $\bar{\rho}_{\rm q}$, respectively.

One can easily check that both scalars, $V_{\rm m}$ and $V_{\rm q}$ remain unchanged after making the infinitesimal coordinate transformation (6.8). Indeed, the gauge transformation of the derivatives

$$\hat{\chi}_{{\rm m}|\alpha} = \chi_{{\rm m}|\alpha} - H\bar{P}_{\alpha\beta}\xi^{\beta} - \bar{u}_{\beta}\xi^{\beta}{}_{|\alpha}\,, \tag{6.15a}$$

$$\hat{\chi}_{{\rm q}|\alpha} = \chi_{{\rm q}|\alpha} - H\bar{P}_{\alpha\beta}\xi^{\beta} - \bar{u}_{\beta}\xi^{\beta}{}_{|\alpha}\,, \tag{6.15b}$$

where $\bar{P}_{\alpha\beta} = \bar{g}_{\alpha\beta} + \bar{u}_{\alpha}\bar{u}_{\beta}$ is the operator of projection on the hypersurface being orthogonal to the Hubble flow of 4-velocity $\bar{u}^{\alpha}$. After performing the gauge transformation (6.8), and substituting the gauge transformations of functions q, $\chi_{\rm m}$ and $\chi_{\rm q}$ to the definitions of $V_{\rm m}$ and $V_{\rm q}$, we find out

$$\hat{V}_{\rm m} = V_{\rm m}\,, \quad \hat{V}_{\rm q} = V_{\rm q}\,, \tag{6.16}$$

that proves the gauge-invariant property of the scalars $V_{\rm m}$ and $V_{\rm q}$.

Physical meaning of the gauge-invariant quantity $V_{\rm m}$ can be understood as follows. We consider the perturbation of the specific enthalpy $\mu_{\rm m}$ defined in equation (3.14). Substituting the decomposition (5.1) of the field variables to equation (3.14) and expanding, we obtain

$$\mu_{\rm m} = \bar{\mu}_{\rm m} + \delta\mu_{\rm m}\,, \tag{6.17}$$

where the perturbation $\delta\mu_{\rm m}$ of the specific enthalpy is defined (in the linearized order) by

$$\delta\mu_{\rm m} = \bar{u}^{\alpha}\phi_{|\alpha} - \frac{1}{2}\bar{\mu}_{\rm m}{\rm q}\,. \tag{6.18}$$

It helps us to recognize that

$$V_{\rm m} = \frac{\delta\mu_{\rm m}}{\bar{\mu}_{\rm m}} + 3\frac{v_{\rm s}^2}{c^2}H\chi_{\rm m}\,. \tag{6.19}$$

Fractional perturbation of the specific enthalpy can be rewritten with the help of thermodynamic equations (3.11) in terms of the perturbation $\delta\epsilon_{\rm m}$ of the energy density of the ideal fluid,

$$\frac{\delta\mu_{\rm m}}{\bar{\mu}_{\rm m}} = \frac{v_s^2}{c^2}\frac{\delta\epsilon_{\rm m}}{\bar{\epsilon}_{\rm m} + \bar{p}_{\rm m}} , \tag{6.20}$$

or, by making use of equation (3.8), in terms of the perturbation $\delta\rho_{\rm m}$ of the density of the ideal fluid

$$\frac{\delta\mu_{\rm m}}{\bar{\mu}_{\rm m}} = \frac{v_s^2}{c^2}\frac{\delta\rho_{\rm m}}{\bar{\rho}_{\rm m}} . \tag{6.21}$$

This allows us to write down equation (6.19) as follows:

$$V_{\rm m} = \frac{v_s^2}{c^2}\left(\frac{\delta\rho_{\rm m}}{\bar{\rho}_{\rm m}} + 3H\chi_{\rm m}\right) , \tag{6.22}$$

which elucidates the relationship between the gauge-invariant variable $V_{\rm m}$ and the perturbation $\delta\rho_{\rm m}$ of the rest mass density of the dark matter. More specifically, $V_{\rm m}$ is an algebraic sum of two scalar functions, $\delta\rho_{\rm m}$ and $\chi_{\rm m}$ neither of each is gauge-invariant. The gauge transformation of the dark matter density perturbation is

$$\delta\hat{\rho}_{\rm m} = \delta\rho_{\rm m} - \bar{\rho}_{{\rm m}|\alpha}\xi^\alpha = \delta\rho_{\rm m} + 3H\bar{\rho}_{\rm m}\bar{u}_\alpha\xi^\alpha , \tag{6.23}$$

and the gauge transformation of the variable $\chi_{\rm m}$ is given by (6.11b). Their algebraic sum in equation (6.22) does not change under the diffeomorphism (6.8) showing that $V_{\rm m}$ is the gauge-invariant density fluctuation that does not depend on a particular choice of coordinates on spacetime manifold.

Similar considerations, applied to function $V_{\rm q}$ reveals that it can be represented as an algebraic sum of the perturbation, $\delta\rho_{\rm q}$, of the density of the dark energy, and function $\chi_{\rm q}$,

$$V_{\rm q} = \frac{\delta\rho_{\rm q}}{\bar{\rho}_{\rm q}} + 3H\chi_{\rm q} . \tag{6.24}$$

It is easy to check out that each term on the right-hand side of this equation taken separately, is not gauge invariant but their linear combination does. It is worth emphasizing that standard textbooks on cosmological theory (see, for example, [80, 94, 113, 114]) derive equations for the density perturbations $\delta\rho/\bar{\rho}$ but those equations are not gauge invariant and, hence, their solutions have no direct physical meaning and should be interpreted with care (see discussion in Section 8.1).

6.4 Field equations for the scalar perturbations

Equation for a scalar q

Function q was defined in (6.2a). In order to derive a differential equation for q, we apply the covariant Laplace–Beltrami operator to q, and make use of the covariant equations (5.62) and (5.64). Straightforward but fairly long calculation yields

$$\mathfrak{q}^{|\alpha}{}_{|\alpha} - 2\left(\dot{H} + H^2 - \frac{2k}{a^2}\right)\mathfrak{q} + 8\pi\overline{\rho}_{\rm m}\overline{\mu}_{\rm m}\left[\left(1 - \frac{c^2}{v_s^2}\right)V_{\rm m} - \left(1 + 3\frac{v_{\rm s}^2}{c^2}\right)H\chi_{\rm m}\right] \tag{6.25}$$
$$- 16\pi\overline{\rho}_{\rm q}\left(\frac{\partial\overline{W}}{\partial\overline{\Psi}} + 2H\overline{\mu}_{\rm q}\right)\chi_{\rm q} - 2\overline{u}^\alpha\overline{u}^\beta B_{\alpha|\beta} - 4H\overline{u}^\alpha B_\alpha = 8\pi\left(\sigma + \tau\right),$$

where the source density $\sigma + \tau$ for the field $\mathfrak{q}$ is

$$\sigma + \tau = \left(\overline{u}^\alpha\overline{u}^\beta + \overline{P}^{\alpha\beta}\right)\mathfrak{T}_{\alpha\beta}, \tag{6.26}$$

in accordance with the definitions introduced in (4.70a) and (4.70b). The reader should notice that equation (6.25) depends on the gauge function B^α which remains arbitrary so far.

Equation for a scalar $\mathfrak{p}$

Function $\mathfrak{p}$ was defined in (6.2b). In order to derive equation for $\mathfrak{p}$, we apply the covariant Laplace–Beltrami operator to the definition of $\mathfrak{p}$, and make use of the covariant equations (5.62) and (5.64). It results in a wave equation

$$\mathfrak{p}^{|\alpha}{}_{|\alpha} + \frac{4k}{a^2}\mathfrak{p} + B^\alpha{}_{|\alpha} - 2\overline{u}^\alpha\overline{u}^\beta B_{\alpha|\beta} - 6H\overline{u}^\alpha B_\alpha = 16\pi\tau, \tag{6.27}$$

where the source density τ has been defined in (4.70b). Equation (6.27) depends on the arbitrary gauge function B^α.

Equation for a scalar χ

Equation for the gauge-invariant scalar, $\chi = \chi_{\rm m} - \chi_{\rm q}$, is derived from the definitions (6.10) and the field equations (5.67) and (5.68). Replacing ϕ and ψ in those equations with $\chi_{\rm m}$ and $\chi_{\rm q}$, and making use of equations (4.52) and (4.53) for reshuffling some terms, yields

$$\chi_{\rm m}{}^{|\alpha}{}_{|\alpha} + 2H\overline{u}^\alpha\chi_{{\rm m}|\alpha} - \left(\dot{H} - \frac{4k}{a^2}\right)\chi_{\rm m} \tag{6.28a}$$
$$+ 4HV_{\rm m} + \left(1 - \frac{c^2}{v_{\rm s}^2}\right)\overline{u}^\alpha V_{{\rm m}|\alpha} - 16\pi\overline{\rho}_{\rm q}\overline{\mu}_{\rm q}\chi = \overline{u}^\alpha B_\alpha,$$

$$\chi_{\rm q}{}^{|\alpha}{}_{|\alpha} + 2H\overline{u}^\alpha\chi_{{\rm q}|\alpha} - \left(\dot{H} - \frac{4k}{a^2}\right)\chi_{\rm q} \tag{6.28b}$$
$$+ 4HV_{\rm q} + \frac{2}{\overline{\mu}_{\rm q}}\frac{\partial\overline{W}}{\partial\overline{\Psi}}V_{\rm q} + 16\pi\overline{\rho}_{\rm m}\overline{\mu}_{\rm m}\chi = \overline{u}^\alpha B_\alpha.$$

Subtracting (6.28b) from (6.28a) cancels the gauge-dependent term, $\bar{u}^\alpha B_\alpha$, and brings about the field equation for χ,

$$\chi^{|\alpha}{}_{|\alpha} + 6H\bar{u}^\alpha \chi_{|\alpha} + 3\dot{H}\chi = \frac{2}{\bar{\mu}_{\rm q}}\frac{\partial \overline{W}}{\partial \overline{\Psi}} V_{\rm q} - \left(1 - \frac{c^2}{v_{\rm s}^2}\right)\bar{u}^\alpha V_{{\rm m}|\alpha} \,. \tag{6.29}$$

This equation is apparently gauge invariant since any dependence on the arbitrary gauge function B^α disappeared. It is also covariant that is valid in any coordinates.

Equation (6.29) can be recast to the form of an inhomogeneous wave equation:

$$(\bar{\rho}_{\rm m}\chi)^{|\alpha}{}_{|\alpha} = 2\frac{\bar{\rho}_{\rm m}}{\bar{\rho}_{\rm q}}\frac{\partial \overline{W}}{\partial \overline{\Psi}} V_{\rm q} - \left(1 - \frac{c^2}{v_{\rm s}^2}\right)\bar{\rho}_{\rm m}\bar{u}^\alpha V_{{\rm m}|\alpha} \,. \tag{6.30}$$

Yet another form of equation (6.29) is obtained in terms of the variable $\psi = \bar{\rho}_{\rm q}\chi = \bar{\mu}_{\rm q}\chi$. By simple inspection we can check that equation (6.29) is transformed to

$$\psi^{|\alpha}{}_{|\alpha} - m_\psi^2 \psi = 2\frac{\partial \overline{W}}{\partial \overline{\Psi}} V_{\rm m} - \left(1 - \frac{c^2}{v_{\rm s}^2}\right)\bar{\rho}_{\rm q}\bar{u}^\alpha V_{{\rm m}|\alpha} \,, \tag{6.31}$$

where we introduced notation $m_\psi \equiv \sqrt{\partial^2 \overline{W}/\partial \overline{\Psi}^2}$. This is an inhomogeneous Klein–Gordon equation for the field ψ governed by $V_{\rm m}$. The "mass" m_ψ of the scalar field excitation, ψ, depends on the second derivative of the potential function $\overline{W}$ which defines the 'coefficient of elasticity' of the background scalar field $\overline{\Psi}$.

Inhomogeneous equations (6.29), (6.30), and (6.31) have the source terms that is determined by variables $V_{\rm m}$ and $V_{\rm q}$. We derive differential equations for these field variables in the next sections.

Equation for a scalar $V_{\rm m}$

Equation for the field variable $V_{\rm m}$ is derived from the equations for functions $\chi_{\rm m}$ and q that enter its definition (6.13a). By applying the Laplace–Beltrami operator to function $V_{\rm m}$ we get

$$\begin{aligned} V_{\rm m}{}^{|\alpha}{}_{|\alpha} &= \bar{u}^\beta \left(\chi_{\rm m}{}^{|\alpha}{}_{|\alpha}\right)_{|\beta} + 2H\chi_{\rm m}{}^{|\alpha}{}_{|\alpha} - \frac{1}{2}{\rm q}^{|\alpha}{}_{|\alpha} + \bar{u}^\beta \overline{R}^\alpha{}_\beta \chi_{{\rm m}|\alpha} \\ &+ 2H\bar{u}^\alpha \left(V_{\rm m} + \frac{1}{2}{\rm q}\right)_{|\alpha} + 3H^2 \left(V_{\rm m} + \frac{1}{2}{\rm q}\right) . \end{aligned} \tag{6.32}$$

The Laplace–Beltrami operator for function $\chi_{\rm m}$ is given in equation (6.28a) which is not gauge invariant. Taking the covariant derivative from this equation and contracting it with $\bar{u}^{\alpha}$ brings about the first term on the right-hand side of equation (6.32),

$$\bar{u}^{\beta}\left(\chi_{\mathrm{m}\,|\alpha}^{\,|\alpha}\right)_{|\beta} = -\left(1-\frac{c^2}{v_{\mathrm{s}}^2}\right)\bar{u}^{\alpha}\bar{u}^{\beta}V_{\mathrm{m}|\alpha\beta} - 6H\bar{u}^{\alpha}V_{\mathrm{m}|\alpha}$$
$$-\left(5\dot{H}+\frac{4k}{a^2}\right)V_{\mathrm{m}} - H\bar{u}^{\alpha}\mathfrak{q}_{|\alpha} - \left(\frac{1}{2}\dot{H}+\frac{2k}{a^2}\right)\mathfrak{q}$$
$$-3H\left[\left(1+\frac{v_{\mathrm{s}}^2}{c^2}\right)\dot{H}-\left(3+\frac{v_{\mathrm{s}}^2}{c^2}\right)\frac{k}{a^2}\right]\chi_{\mathrm{m}}$$
$$+8\pi\bar{\rho}_{\mathrm{q}}\frac{\partial\overline{W}}{\partial\overline{\Psi}}\left(4\chi_{\mathrm{q}}-3\chi_{\mathrm{m}}\right)$$
$$+16\pi\bar{\rho}_{\mathrm{q}}\bar{\mu}_{\mathrm{q}}\left[\bar{u}^{\alpha}\chi_{|\alpha}-6H\chi+\frac{3}{4}H\left(1-\frac{v_{\mathrm{s}}^2}{c^2}\right)\chi_{\mathrm{m}}\right]+\bar{u}^{\alpha}\bar{u}^{\beta}B_{\alpha|\beta}\,. \tag{6.33}$$

The Laplace–Beltrami operator for function $\mathfrak{q}$ has been derived in (6.25). Now, we make use of equations (6.25), (6.28a), and (6.33) in calculating the right-hand side of (6.32). After a significant amount of algebra, we find out that all terms explicitly depending on q and the gauge functions B^{α} cancel out, so that equation for V_{m} becomes

$$V_{\mathrm{m}\,|\alpha}^{\,|\alpha} + \left(1-\frac{c^2}{v_{\mathrm{s}}^2}\right)\bar{u}^{\alpha}\bar{u}^{\beta}V_{\mathrm{m}|\alpha\beta} + 2\left(3-\frac{c^2}{v_{\mathrm{s}}^2}\right)H\bar{u}^{\alpha}V_{\mathrm{m}|\alpha}$$
$$+\left[2\left(\dot{H}+3H^2+\frac{2k}{a^2}\right)-4\pi\bar{\rho}_{\mathrm{m}}\bar{\mu}_{\mathrm{m}}\left(1-\frac{c^2}{v_{s}^2}\right)\right]V_{\mathrm{m}}$$
$$-16\pi\bar{\rho}_{\mathrm{q}}\bar{\mu}_{\mathrm{q}}\left[\bar{u}^{\alpha}\chi_{|\alpha}-3\left(H+\frac{1}{2\bar{\mu}_{\mathrm{q}}}\frac{\partial\overline{W}}{\partial\overline{\Psi}}\right)\chi\right] = -4\pi\left(\sigma+\tau\right)\,. \tag{6.34}$$

Second-order covariant derivatives in this equation read

$$\left[\bar{g}^{\alpha\beta}+\left(1-\frac{c^2}{v_{\mathrm{s}}^2}\right)\bar{u}^{\alpha}\bar{u}^{\beta}\right]V_{\mathrm{m}|\alpha\beta} \equiv \left(-\frac{c^2}{v_{\mathrm{s}}^2}\bar{u}^{\alpha}\bar{u}^{\beta}+\bar{P}^{\alpha\beta}\right)V_{\mathrm{m}|\alpha\beta}\,, \tag{6.35}$$

and they form a hyperbolic-type operator describing propagation of sound waves in the expanding universe from the source of the sound waves towards the field point with the constant velocity v_{s}^2. Additional terms on the left-hand side of equation (6.34) depend on the Hubble parameter H, and take into account the expansion of the universe.

Equation (6.34) contains only gauge-invariant scalars, V_{m} and χ. Moreover, it does not depend on the choice of coordinates on the background manifold. It also becomes clear that the field variables V_{m} and χ are coupled through the differential equations (6.31) and (6.34), which should be solved simultaneously in order to determine these variables. Solution of the coupled system of differential equations is a very complicated task which cannot be rendered analytically in the most general case. Only in some simple cases, the equations can be decoupled. We discuss such cases in Section 8.

Equation for a scalar V_q

The field variable V_q is not independent since it relates to V_m and χ by a simple relationship

$$V_q = V_m - \bar{u}^\alpha \chi_{|\alpha} , \tag{6.36}$$

which is obtained after subtraction of equation (6.13a) from (6.13b). Equation for V_q is derived directly from (6.36) and equations (6.34) and (6.29) for V_m and χ, respectively. We obtain

$$\begin{aligned} & V_q{}^{|\alpha}{}_{|\alpha} + 4\left(H + \frac{1}{2\bar{\mu}_q}\frac{\partial \overline{W}}{\partial \overline{\Psi}}\right)\bar{u}^\alpha V_{q|\alpha} \\ & + \left[2\left(\dot{H} + 3H^2 + \frac{2k}{a^2}\right) - 4\pi\bar{\rho}_m\bar{\mu}_m\left(1 - \frac{c^2}{v_s^2}\right)\right. \\ & \left. + \frac{2}{\bar{\mu}_q}\left(5H + \frac{1}{\bar{\mu}_q}\frac{\partial \overline{W}}{\partial \overline{\Psi}}\right)\frac{\partial \overline{W}}{\partial \overline{\Psi}} + 2\frac{\partial^2 \overline{W}}{\partial \overline{\Psi}^2}\right] V_q \\ & + 4\pi\bar{\rho}_m\bar{\mu}_m\left(3 + \frac{c^2}{v_s^2}\right)\left(\bar{u}^\alpha \chi_{|\alpha} - 3\frac{v_s^2}{c^2}H\chi\right) = -4\pi\,(\sigma + \tau) . \end{aligned} \tag{6.37}$$

This equation can be also derived by the procedure being similar to that used in the previous subsection in deriving equation for V_m. We followed this procedure and confirm that it leads to (6.37) as expected. Equation (6.37) is clearly gauge invariant. Besides V_q it depends on variable χ and should be solved along with equation (6.29).

6.5 Field equations for vector perturbations

Vector perturbations of the ideal fluid and scalar field are gradients, $\phi_{|\alpha}$ and $\psi_{|\alpha}$. However, they are insufficient to build a gauge-invariant vector perturbation out of the vector perturbation of the metric tensor $\mathfrak{p}_\alpha$. Field equations for vector $\mathfrak{p}_\alpha$ can be derived by applying the covariant Laplace–Beltrami operator to both sides of definition (6.3) and making use of equation (5.65). After performing the covariant differentiation and a significant amount of algebra, we derive the field equation

$$\begin{aligned} & \mathfrak{p}_\alpha{}^{|\beta}{}_{|\beta} - 2H\bar{u}_\alpha \mathfrak{p}_\beta{}^{|\beta} - \left(2\dot{H} + 3H^2 - \frac{2k}{a^2}\right)\mathfrak{p}_\alpha \\ & + \overline{P}_\alpha{}^\beta \bar{u}^\gamma\left(B_{\beta|\gamma} + B_{\gamma|\beta} + 2H\bar{u}_\gamma B_\beta\right) = 16\pi\tau_\alpha , \end{aligned} \tag{6.38}$$

where the matter current τ_α is defined in (4.70c). This equation is apparently gauge dependent as shown by the appearance of the gauge function B^α. This equation reduces to a much simpler form

$$\mathfrak{p}_\alpha{}^{|\beta}{}_{|\beta} - 2H\bar{u}_\alpha \mathfrak{p}_\beta{}^{|\beta} - \left(2\dot{H} + 3H^2 - \frac{2k}{a^2}\right)\mathfrak{p}_\alpha = 16\pi\tau_\alpha , \tag{6.39}$$

in a special gauge B^α=0, which imposes a restriction on the divergence of the metric tensor perturbation in equation (5.64).

Equation (6.38) points out that the vector perturbations are generated by the current of matter τ_a existing in the localized astronomical system which physical origin may be a relict of the primordial perturbations. We do not discuss this interesting scenario in this chapter as it would require a nonconservation of entropy and nonisentropic background fluid – the case which we have intentionally excluded in order to focus on derivation of cosmological generalization of the post-Newtonian equations of relativistic celestial dynamics [66].

6.6 Field equations for tensor perturbations

Field equations for traceless tensor $\mathfrak{p}^{\mathsf{T}}_{\alpha\beta}$ can be derived by applying the covariant Laplace–Beltrami operator to the definition (6.4) and making use of equation (5.65) along with a tedious algebraic transformations. This yields the following equation:

$$\mathfrak{p}^{\mathsf{T}}_{\alpha\beta}{}^{|\gamma}{}_{|\gamma} - 2H\left(\bar{u}_\alpha \mathfrak{p}^{\mathsf{T}}_{\beta\gamma}{}^{|\gamma} + \bar{u}_\beta \mathfrak{p}^{\mathsf{T}}_{\alpha\gamma}{}^{|\gamma}\right) - 2\left(H^2 + \frac{k}{a^2}\right)\mathfrak{p}^{\mathsf{T}}_{\alpha\beta}$$
$$- \bar{P}_\alpha{}^\mu \bar{P}_\beta{}^\nu \left(B_{\mu|\nu} + B_{\nu|\mu}\right) + \frac{2}{3}\bar{P}_{\alpha\beta}\bar{P}^{\mu\nu}B_{\mu|\nu} = 16\pi\tau^{\mathsf{T}}_{\alpha\beta}\,. \tag{6.40}$$

Here the transverse and traceless tensor source of the tensor perturbations is

$$\tau^{\mathsf{T}}_{\alpha\beta} \equiv \tau_{\alpha\beta} - \frac{1}{3}\bar{P}_{\alpha\beta}\,\tau\,, \tag{6.41}$$

where $\tau_{\alpha\beta}$ has been introduced in (4.70d), and $\tau = \bar{P}^{\alpha\beta}\tau_{\alpha\beta}$ in accordance with equation (4.70b). Tensor $\tau^{\mathsf{T}}_{\alpha\beta}$ is traceless, that is $\bar{g}^{\alpha\beta}\tau^{\mathsf{T}}_{\alpha\beta} = \bar{P}^{\alpha\beta}\tau^{\mathsf{T}}_{\alpha\beta} = 0$.

Equation (6.40) is gauge dependent. The gauge freedom is significantly reduced by imposing the gauge condition $B^\alpha = 0$ which brings equation (6.40) to the following form:

$$\mathfrak{p}^{\mathsf{T}}_{\alpha\beta}{}^{|\gamma}{}_{|\gamma} - 2H\left(\bar{u}_\alpha \mathfrak{p}^{\mathsf{T}}_{\beta\gamma}{}^{|\gamma} + \bar{u}_\beta \mathfrak{p}^{\mathsf{T}}_{\alpha\gamma}{}^{|\gamma}\right) - 2\left(H^2 + \frac{k}{a^2}\right)\mathfrak{p}^{\mathsf{T}}_{\alpha\beta} = 16\pi\tau^{\mathsf{T}}_{\alpha\beta}\,. \tag{6.42}$$

6.7 Residual gauge freedom

The gauge freedom of the theory under discussion is associated with the gauge function B^α appearing in equation (5.64). The most favorable choice of the gauge condition is

$$B^\alpha = 0\,, \tag{6.43}$$

which drastically simplifies the above equations for vector and tensor gravitational perturbations. The gauge (6.43) is a generalization of the harmonic (de Donder) gauge condition used in the gravitational wave astronomy and in the post-Newtonian dynamics of extended bodies. This choice of the gauge establishes differential relationships between the algebraically independent metric tensor components introduced in Section 6.1. Indeed, substituting the algebraic decomposition (6.7) of the metric tensor perturbations to equation (5.64) and imposing the condition (6.43) yields

$$\begin{aligned} \mathfrak{p}^{\top}_{\alpha\beta}{}^{|\beta} + \bar{u}_\alpha \mathfrak{p}_\beta{}^{|\beta} + \bar{u}_\beta \mathfrak{p}_\alpha{}^{|\beta} - \left(\bar{u}_\alpha \bar{u}_\beta - \frac{1}{3}\bar{P}_{\alpha\beta}\right)\mathfrak{p}^{|\beta} + 2H\mathfrak{p}_\alpha \\ + 2\bar{u}_\alpha \bar{u}_\beta \mathfrak{q}^{|\beta} + 2H\mathfrak{q}\bar{u}_\alpha = 16\pi\left(\bar{\rho}_{\rm m}\bar{\mu}_{\rm m}\chi_{\rm m} + \bar{\rho}_{\rm q}\bar{\mu}_{\rm q}\chi_{\rm q}\right)\bar{u}_\alpha \,. \end{aligned} \tag{6.44}$$

Projecting this relationship on the direction of the background 4-velocity, $\bar{u}^\alpha$, and on the hypersurface being orthogonal to it, we derive two algebraically independent equations between the perturbations of metric tensor components and of the matter variables. They are

$$\mathfrak{p}_\beta{}^{|\beta} + \bar{u}_\beta\left(2\mathfrak{q} - \mathfrak{p}\right)^{|\beta} + 2H\mathfrak{q} = 16\pi\left(\bar{\rho}_{\rm m}\bar{\mu}_{\rm m}\chi_{\rm m} + \bar{\rho}_{\rm q}\bar{\mu}_{\rm q}\chi_{\rm q}\right), \tag{6.45a}$$

$$\mathfrak{p}^{\top}_{\alpha\beta}{}^{|\beta} + \bar{u}_\beta \mathfrak{p}_\alpha{}^{|\beta} + \frac{1}{3}\bar{P}_{\alpha\beta}\mathfrak{p}^{|\beta} + 2H\mathfrak{p}_\alpha = 0\,. \tag{6.45b}$$

The gauge (6.43) does not fix the gauge function ξ^α uniquely. The residual gauge freedom is described by the gauge transformations that preserve equations (6.45a) and (6.45b). Substituting the gauge transformation (6.9b) of the gravitational field perturbation $l_{\alpha\beta}$ to equation (5.64) and holding on the gauge condition (6.43), yields the differential equation for the vector function ξ^α

$$\begin{aligned} \xi^{\alpha|\beta}{}_{|\beta} + \bar{g}^{\alpha\gamma}\left(\xi^\beta{}_{|\gamma\beta} - \xi^\beta{}_{|\beta\gamma}\right) + 2H\left(\xi^{\alpha|\beta}\bar{u}_\beta + \xi^{\beta|\alpha}\bar{u}_\beta - \xi^\beta{}_{|\beta}\bar{u}^\alpha\right) \\ -16\pi\left(\bar{\rho}_{\rm m}\bar{\mu}_{\rm m} + \bar{\rho}_{\rm q}\bar{\mu}_{\rm q}\right)\xi^\beta \bar{u}_\beta \bar{u}^\alpha = 0\,, \end{aligned} \tag{6.46}$$

which can be further recast to

$$\begin{aligned} \xi^{\alpha|\beta}{}_{|\beta} + 2H\left(\xi^{\alpha|\beta}\bar{u}_\beta + \xi^{\beta|\alpha}\bar{u}_\beta - \xi^\beta{}_{|\beta}\bar{u}^\alpha\right) \\ +2\left(\dot{H} - \frac{k}{a^2}\right)\xi^\beta \bar{u}_\beta \bar{u}^\alpha + \left(\dot{H} + 3H^2 + \frac{2k}{a^2}\right)\xi^\alpha = 0\,. \end{aligned} \tag{6.47}$$

The gauge function ξ^α can be decomposed in time like, $\xi \equiv -\xi^\beta \bar{u}_\beta$, and space-like, $\zeta^\alpha \equiv \bar{P}^\alpha{}_\beta \xi^\beta$, components,

$$\xi^\alpha = \zeta^\alpha + \bar{u}^\alpha \xi\,. \tag{6.48}$$

Calculating covariant derivatives from ξ and ζ^α and making use of equation (6.47), yield equations

$$\xi^{|\beta}{}_{|\beta} + 2H\bar{u}^\beta \xi_{|\beta} - \left(\dot{H} - \frac{4k}{a^2}\right)\xi = 0\,, \tag{6.49a}$$

$$\zeta^{\alpha|\beta}{}_{|\beta} + 2H\left(\bar{u}^\beta \zeta^\alpha{}_{|\beta} - \bar{u}^\alpha \zeta^\beta{}_{|\beta}\right) + \left(\dot{H} + H^2 + \frac{2k}{a^2}\right)\zeta^\alpha = 0\,. \tag{6.49b}$$

These equations have nontrivial solutions that describe the residual gauge freedom in choosing the coordinates on the background manifold subject to the gauge condition (6.43). It is remarkable that equations (6.49a) and (6.49b) are decoupled and can be solved separately. It means that the residual gauge transformations along the world lines of the Hubble flow are functionally independent of those performed on the hypersurface being orthogonal to the Hubble flow. Equations (6.49a) and (6.49b) of the residual gauge freedom in the cosmological setting given in this subsection generalize equations of the residual gauge freedom in harmonic coordinates of asymptotically flat spacetime [13, 31].

7 Post-Newtonian field equations in a spatially flat universe

7.1 Cosmological parameters and scalar field potential

Linearized equations of the field perturbations given in the previous section are valid for a wide class of matter models of the FLRW metric. They neither specify the equation of state of dark matter, nor that of dark energy. We also keep the parameter of the space curvature k free. By choosing a specific model of matter and picking up a value of $k = -1, 0, +1$, we can solve, at least, in principle the field equations governing the time evolution of the background cosmological manifold. Realistic models of the cosmological dark matter and dark energy are rather sophisticated and, as a rule, include several components. It leads to the system of coupled field equations which can be solved only numerically [2]. However, the large scale structure of the universe is formed at rather late stages of the cosmological evolution being fairly close to the present epoch. Therefore, the study of the impact of cosmological expansion on the post-Newtonian dynamics of isolated astronomical systems is based on recent and present equation of state of matter in the universe.

Precise radiometric observations of the relic CMB radiation and photometry of type Ia supernova explosions reveal that at the present epoch the space curvature of the universe, $k = 0$, and the evolution of the universe is primarily governed by the dark energy and dark matter, which make up to 74% and 24% of the total energy density of the universe, respectively, while 4% of the energy density of the universe belongs to visible matter (baryons), and a tiny fraction of the energy density occupies by the CMBR radiation [43, 52, 59, 63]. It means that we can neglect the effects of the baryonic matter and CMB radiation field in consideration of the post-Newtonian dynamics of astronomical systems in the expanding universe.

We model dark matter by an ideal fluid and dark energy is represented by a scalar field with a potential function $\overline{W}$ which structure has not yet been specified. We also follow the discussion given in [2] by assuming that the spatial curvature $k = 0$, and the potential, $\overline{W}$, of the scalar field relates to its derivative by a simple equation

$$\frac{\partial \overline{W}}{\partial \overline{\Psi}} = -\sqrt{8\pi}\lambda \overline{W} , \tag{7.1}$$

where the time-dependent parameter, $\lambda = \lambda(\overline{\Psi})$, characterizes the slope of the field potential $\overline{W}$. The time evolution of the background universe can be described in terms of the parameter λ and two other parameters, $x_1 = x_1(\overline{\Psi})$ and $x_2 = x_2(\overline{\Psi})$, which are functions of the density, $\overline{\rho}_q = \overline{\mu}_q = \overline{\Psi}$, of the background scalar field, and the potential, $\overline{W}$, scaled to the Hubble parameter, H. These parameters are defined as follows:

$$\overline{\rho}_q^2 = \frac{3H^2}{4\pi} x_1 , \tag{7.2}$$

$$\overline{W} = \frac{3H^2}{8\pi} x_2 . \tag{7.3}$$

The energy density of the scalar field, $\overline{\epsilon}_q$, is expressed in terms of the parameters x_1 and x_2 and the parameter $\Omega_q \equiv 8\pi\overline{\epsilon}_q/3H^2$, by a simple relationship

$$\Omega_q = x_1 + x_2 . \tag{7.4}$$

Time evolution of the parameters x_1 and x_2 is given by a system of two ordinary differential equations which are obtained by differentiating definitions (7.2) and (7.3) and making use of equations (4.67) taken along with the Friedmann equation (4.52) with $k = 0$. It yields

$$\frac{dx_1}{d\omega} = -6x_1 + \lambda\sqrt{6x_1}x_2 + 3x_1\left[(1 - w_m)x_1 + (1 + w_m)(1 - x_2)\right] , \tag{7.5a}$$

$$\frac{dx_2}{d\omega} = -\lambda\sqrt{6x_1}x_2 + 3x_2\left[(1 - w_m)x_1 + (1 + w_m)(1 - x_2)\right] , \tag{7.5b}$$

where $\omega \equiv \ln a$ is the logarithmic scale factor characterizing the number of e-folding of the universe, w_m is the parameter entering the hydrodynamic equation of state (4.54), and the parameters x_1 and x_2 are restricted by the condition imposed by the Friedmann equation (4.50), that is $\Omega_q + \Omega_m = 1$, or

$$x_1 + x_2 = 1 - \Omega_m , \tag{7.6}$$

where $\Omega_m \equiv 8\pi\overline{\epsilon}_m/3H^2$.

The parameter λ obeys the following equation:

$$\frac{d\lambda}{d\omega} = -\sqrt{6x_1}\lambda^2\left(\Gamma_q - 1\right) , \tag{7.7}$$

where

$$\Gamma_q = \frac{\partial^2\overline{W}/\partial\overline{\Psi}^2}{\left(\partial\overline{W}/\partial\overline{\Psi}\right)^2}\overline{W} . \tag{7.8}$$

If $\Gamma_q = 1$, the parameter λ is constant, and equation (7.1) can be integrated yielding an exponential potential

$$\overline{W}(\overline{\Psi}) = \overline{W}_0 \exp(-\sqrt{8\pi}\lambda\overline{\Psi}) . \tag{7.9}$$

In this case, and under assumption that, w_m = const., the system of two differential equations (7.5a) and (7.5b) is closed. If $\Gamma_q \neq 1$, three equations (7.5a), (7.5b), and (7.7) must be solved together in order to describe temporal evolution of the background cosmological manifold.

In the general case, derivatives of the potential $\overline{W}$ are expressed in terms of the parameters under discussion. Namely,

$$\frac{\partial \overline{W}}{\partial \overline{\Psi}} = -\frac{3\lambda}{\sqrt{8\pi}} H^2 x_2 , \qquad \frac{\partial^2 \overline{W}}{\partial \overline{\Psi}^2} = 3\Gamma_q \lambda^2 H^2 x_2 . \tag{7.10}$$

It is also useful to express the products $\overline{\rho}_q\overline{\mu}_q$ and $\overline{\rho}_m\overline{\mu}_m$ in terms of the parameters x_1 and x_2. For $\overline{\mu}_q = \overline{\rho}_q$, one can use definition (7.2) to obtain

$$\overline{\rho}_q\overline{\mu}_q = \frac{3H^2}{4\pi} x_1 . \tag{7.11}$$

The product $\overline{\rho}_m\overline{\mu}_m = \overline{\epsilon}_m + \overline{p}_m$, so that making use of the matter equation of state, $\overline{p}_m = w_m\overline{\epsilon}_m$, and equation (7.6), we derive

$$\overline{\rho}_m\overline{\mu}_m = \frac{3H^2}{8\pi}(1 + w_m)\Omega_m , \tag{7.12}$$

where $\Omega_m = 1 - x_1 - x_2$. These equations allow us to recast equation (4.52) for the time derivative of the Hubble parameter to the following form:

$$\dot{H} = -\frac{3}{2}\left(1 + w_{\text{eff}}\right) H^2 , \tag{7.13}$$

where

$$w_{\text{eff}} \equiv w_m + (1 - w_m)x_1 - (1 + w_m)x_2 , \tag{7.14}$$

is the (time-dependent) parameter of the effective equation of state of the mixture of the ideal fluid and the scalar field.

7.2 Conformal cosmological perturbations

The FLRW metric (4.15) is a product of the scale factor a and a conformal metric $\overline{\mathfrak{f}}_{\alpha\beta}$. The conformal spacetime is comoving with the Hubble flow and is not globally expanding. In case of the flat spatial curvature, $k = 0$, the conformal spacetime becomes equivalent to the Minkowski spacetime which is used as a starting point in the standard theory of the post-Newtonian approximations [31]. Therefore, it is mathematically instructive to formulate the field equations for cosmological perturbations in the conformal spacetime. It also allows us to simplify the differential operators on the left-hand side of the equations for perturbations (see Section 7.3 below). Nonetheless, the reader must keep in mind that the conformal spacetime is unphysical and additional scale transformations of coordinates are required to convert mathematical results from the conformal spacetime to a real physical world [69].

Let us associate the cosmological perturbation, $h_{\alpha\beta}$, of gravitational field in the conformal spacetime with the background metric $\bar{\mathtt{f}}_{\alpha\beta}$ with physical perturbation $\varkappa_{\alpha\beta}$ of the metric as follows [65, 70]:

$$\varkappa_{\alpha\beta} = a^2(\eta) h_{\alpha\beta} , \tag{7.15}$$

where perturbation $\varkappa_{\alpha\beta}$ has been defined in (5.1) and $a(\eta)$ is the scale factor of the FLRW metric. Gravitational perturbation $l_{\alpha\beta}$ relates to $\varkappa_{\alpha\beta}$ by equation (5.10), and can be also represented in the conformal form

$$l_{\alpha\beta} = a^2(\eta)\gamma_{\alpha\beta} , \tag{7.16}$$

where

$$\gamma_{\alpha\beta} = -h_{\alpha\beta} + \frac{1}{2}\bar{\mathtt{f}}_{\alpha\beta} h , \tag{7.17}$$

with $h \equiv \bar{\mathtt{f}}^{\alpha\beta} h_{\alpha\beta}$. In what follows, tensor indices of geometric objects in the conformal spacetime are raised and lowered with the help of the conformal metric $\bar{\mathtt{f}}_{\alpha\beta}$.

We assume that the scale factor a of the universe remains unperturbed. This assumption is justified since we can always include the perturbation of the scale factor to the perturbation $h_{\alpha\beta}$ of the conformal metric. Thus, the perturbed physical spacetime interval, ds, of the FLRW metric relates to the perturbed conformal spacetime interval, $d\tilde{s}$, by the conformal transformation

$$ds^2 = a^2(\eta) d\tilde{s}^2 . \tag{7.18}$$

Here, the perturbed conformal spacetime interval reads

$$d\tilde{s}^2 = f_{\alpha\beta} dx^\alpha dx^\beta , \tag{7.19}$$

where

$$f_{\alpha\beta} = \bar{\mathtt{f}}_{\alpha\beta} + h_{\alpha\beta} , \tag{7.20}$$

is the perturbed conformal metric. Here, $\bar{\mathtt{f}}_{\alpha\beta}$ is the unperturbed conformal metric defined in (4.16), $h_{\alpha\beta}$ is the perturbation of the conformal metric, and $x^\alpha = (x^0, x^i)$ are arbitrary coordinates which are the same as in the physical spacetime manifold in correspondence with the definition of the conformal metric transformation [88].

It is worth emphasizing that in case of the space curvature $k = 0$, the background conformal metric, $\mathrm{g}_{\alpha\beta}(\eta, X^i)$, expressed in the isotropic Cartesian coordinates (η, X^i), is the diagonal Minkowski metric, $\mathrm{g}_{\alpha\beta}(\eta, X^i) = \eta_{\alpha\beta} = \mathrm{diag}(-1, 1, 1, 1)$. Therefore, in this case the background metric $\overline{\mathfrak{f}}_{\alpha\beta}$ remains the Minkowski metric with the components expressed in arbitrary coordinates by means of tensor transformation

$$\overline{\mathfrak{f}}_{\alpha\beta} = \mathrm{M}^{\mu}{}_{\alpha}\mathrm{M}^{\nu}{}_{\beta}\eta_{\mu\nu} , \tag{7.21}$$

where the matrix of transformation has been defined in (4.13). If the matrix of transformation, $\mathrm{M}^{\mu}{}_{\alpha}$, is the Lorentz boost, the conformal metric, $\overline{\mathfrak{f}}_{\alpha\beta}$, remains flat, $\overline{\mathfrak{f}}_{\alpha\beta} = \eta_{\alpha\beta}$. It is worth noticing that, in general, the unperturbed conformal metric can be chosen flat even in case of $k = -1, +1$ [55]. Hence, all equations given above will remain intact which means that, in fact, our formalism is applicable to FLRW metric with any space curvature. The only change will be in the conformal factor which, in the case of $k = \pm 1$, is not merely the scale factor $a(\eta)$ of the FLRW metric but a more complicated function, $a(\eta, x^a)$, of time and spatial coordinates [55]. Though it is not difficult to handle all three cases of $k = -1, 0, +1$ on the same footing but it burdens equations for the field perturbations with a number of terms being proportional to k. Moreover, consideration of the dark energy equations with $k = \pm 1$ given in the preceding section gets complicated [2]. For this reason, we restrict ourselves with the case of the spatially flat universe with $k = 0$ which is an excellent approximation in treating cosmological observations [60].

Similarly to (6.7) the conformal metric perturbation, $\gamma_{\alpha\beta}$, can be split in 1+3 algebraically irreducible components

$$\gamma_{\alpha\beta} = p^{\mathrm{T}}_{\alpha\beta} + \overline{\mathrm{v}}_{\alpha}p_{\beta} + \overline{\mathrm{v}}_{\beta}p_{\alpha} + \left(\overline{\mathrm{v}}_{\alpha}\overline{\mathrm{v}}_{\beta} + \frac{1}{3}\overline{\pi}_{\alpha\beta}\right)p + 2\overline{\mathrm{v}}_{\alpha}\overline{\mathrm{v}}_{\beta}\left(q - p\right) , \tag{7.22}$$

where the 4-velocity $\overline{\mathrm{v}}^{\alpha} = a\bar{u}^{\alpha}$, $\overline{\mathrm{v}}_{\alpha} = \overline{\mathfrak{f}}_{\alpha\beta}\overline{\mathrm{v}}^{\beta} = a^{-1}\bar{g}_{\alpha\beta}\bar{u}^{\beta} = a^{-1}\bar{u}_{\alpha}$, and

$$\overline{\pi}_{\alpha\beta} = \overline{\mathfrak{f}}_{\alpha\beta} + \overline{\mathrm{v}}_{\alpha}\overline{\mathrm{v}}_{\beta} , \tag{7.23}$$

is the operator of projection on the conformal space which represents a hypersurface being everywhere orthogonal to the congruence of world lines of 4-velocity $\overline{\mathrm{v}}^{\alpha}$. 4-velocity $\overline{\mathrm{v}}^{\alpha}$ is an analog of the Hubble flow in the conformal spacetime. We also notice that $\overline{P}_{\alpha\beta} = a^2\overline{\pi}_{\alpha\beta}$.

Different pieces of the conformal metric perturbation, $\gamma_{\alpha\beta}$, are related to those of the physical metric perturbation, $l_{\alpha\beta}$, by the powers of the scale factor,

$$\mathfrak{p}^{\mathrm{T}}_{\alpha\beta} = a^2 p^{\mathrm{T}}_{\alpha\beta} , \qquad \mathfrak{p}_{\alpha} = a p_{\alpha} , \qquad \mathfrak{p} = p , \qquad \mathfrak{q} = q . \tag{7.24}$$

More specifically,

$$q = \frac{1}{2}\left(\overline{\mathrm{v}}^{\mu}\overline{\mathrm{v}}^{\nu} + \overline{\pi}^{\mu\nu}\right)\gamma_{\mu\nu} , \tag{7.25a}$$

$$p = \overline{\pi}^{\mu\nu}\gamma_{\mu\nu} , \tag{7.25b}$$

$$p_{\alpha} = -\overline{\pi}_{\alpha}{}^{\beta}\overline{\mathrm{v}}^{\gamma}\gamma_{\beta\gamma} , \tag{7.25c}$$

$$p^{\mathrm{T}}_{\alpha\beta} = p_{\alpha\beta} - \frac{1}{3}\overline{\pi}_{\alpha\beta}p , \tag{7.25d}$$

where

$$p_{\alpha\beta} = \overline{\pi}_\alpha{}^\mu \overline{\pi}_\beta{}^\nu \gamma_{\mu\nu} \,. \tag{7.26}$$

The trace of the gravitational perturbation, $\gamma \equiv \overline{\mathrm{f}}^{\alpha\beta} \gamma_{\alpha\beta} = 2(p - q)$. The components $h_{\alpha\beta} = -\gamma_{\alpha\beta} + \overline{\mathrm{f}}_{\alpha\beta}\gamma/2$ are used in calculating dynamical behavior of particles and light in the conformal spacetime as well as for matching theory with observables. The components of $h_{\alpha\beta}$ are

$$h_{\alpha\beta} = -p^{\mathrm{T}}_{\alpha\beta} - \overline{\nabla}_\alpha p_\beta - \overline{\nabla}_\beta p_\alpha + \frac{2}{3}\overline{\pi}_{\alpha\beta} p - \left(\overline{\nabla}_\alpha \overline{\nabla}_\beta + \overline{\pi}_{\alpha\beta}\right) q \,, \tag{7.27}$$

and $h \equiv \overline{\mathrm{f}}^{\alpha\beta} h_{\alpha\beta} = 2(p - q) = \gamma$.

It turns out that the conformal Hubble parameter, $\mathcal{H} = a'/a$ is more convenient in the conformal spacetime than the "canonical" Hubble parameter, $H = \dot{R}/R = R^{-1}dR/dT$, where T is the cosmological time (see Section 4.2). Relations between $\mathcal{H}$ and H, and their derivatives are shown in equations (4.5)–(4.7). These relations are employed along with equations (4.6) and (7.13) in order to express the time derivative, $\mathcal{H}'$, of the conformal Hubble parameter in terms of $\mathcal{H}^2$ and the parameter w_{eff} of the effective equation of state

$$\mathcal{H}' = -\frac{1}{2}(1 + 3w_{\mathrm{eff}})\mathcal{H}^2 \,. \tag{7.28}$$

We shall use this expression in the calculations that follow.

7.3 Post-Newtonian field equations in conformal spacetime

The set of the post-Newtonian field equations in cosmology consists of equations for the perturbations of the background dark matter, dark energy, and gravitational field. Perturbations of dark matter and dark energy are described by four scalars, V_{m}, V_{q}, χ_{m} and χ_{q} but only three of them are functionally independent because of equality (6.36), that is

$$V_{\mathrm{m}} - V_{\mathrm{q}} = \overline{u}^\alpha \left(\chi_{\mathrm{m}} - \chi_{\mathrm{q}}\right)_{|\alpha} \,. \tag{7.29}$$

Depending on a particular situation, any of the three scalars can be taken as independent variables in description of scalar perturbations.

The gravitational field perturbations are q, p, p_α, and $p^{\mathrm{T}}_{\alpha\beta}$ but among them the scalar q is not independent and can be expressed in terms of χ_{m} and V_{m} in accordance with (6.13a),

$$q = -2(V_{\mathrm{m}} - \overline{u}^\alpha \chi_{\mathrm{m},\alpha}) \,, \tag{7.30}$$

where we have also used the equality $\mathrm{q} = q$ as follows from (7.24). The scalar q can be also expressed in terms of χ_{q} and V_{q} in accordance with (6.13b). Hence, as soon as the pairs, V_{m} and χ_{m} or V_{q} and χ_{q} are known, the scalar gravitational perturbation q can be easily calculated from (7.30). Functions p, p_α, $p^{\mathrm{T}}_{\alpha\beta}$ are independent and decouple both from each other and from the other perturbations. Thus, the most difficult part of the perturbation theory is to find out solutions of the scalar perturbations which are coupled one to another.

The post-Newtonian field equations in the conformal spacetime for variables χ_{m}, χ_{q}, V_{m}, and for p, p_α, $p^{\mathrm{T}}_{\alpha\beta}$ are derived from the equations of the previous section by transforming all functions and operators from physical to conformal spacetime. The important part of the transformation technique is based on formulas converting the covariant Laplace–Beltrami wave operators, defined on the background spacetime manifold, to their conformal spacetime counterparts.

Laplace–Beltrami operator in conformal spacetime

Let F be an arbitrary scalar, F_α – an arbitrary covector, and $F_{\alpha\beta}$ – an arbitrary covariant tensor of the second rank. We have three different types of the Laplace–Beltrami operators on the curved background manifold: a scalar – $F^{|\mu}{}_{|\mu}$, a vector – $F_\alpha{}^{|\mu}{}_{|\mu}$, and a tensor – $F_{\alpha\beta}{}^{|\mu}{}_{|\mu}$ where the covariant derivatives are taken with the help of the affine connection $\overline{\Gamma}^\alpha{}_{\beta\gamma}$ being compatible with the metric $\overline{g}_{\alpha\beta}$ as shown in (4.21). Covariant derivatives give the invariant description of differential equations of mathematical physics on curved manifolds. However, for handling a more pragmatic purpose of finding solution of a differential equation, the covariant operators must be expressed in terms of partial derivatives with respect to the coordinates chosen for solving the equation.

Transformation of the covariant Laplace–Beltrami operators to the partial derivatives is achieved after writing down the covariant derivatives for a scalar F, a vector f_α, and a tensor $F_{\alpha\beta}$ in explicit form by making use of the Christoffel symbols given in (4.23)–(4.25). Tedious but straightforward calculations of the covariant derivatives yield the scalar, vector and tensor Laplace–Beltrami operators in the following form [70]:

$$F^{|\mu}{}_{|\mu} = \frac{1}{a^2}\left[\Box F - 2\mathcal{H}\overline{\mathrm{v}}^\mu F_{;\mu}\right], \tag{7.31a}$$

$$\begin{aligned} F_\alpha{}^{|\mu}{}_{|\mu} = \frac{1}{a^2}\Big[&\Box F_\alpha - 2\mathcal{H}\overline{\mathrm{v}}^\mu F_{\mu;\alpha} + 2\mathcal{H}\overline{\mathrm{v}}_\alpha \overline{\mathrm{f}}^{\mu\nu} F_{\mu;\nu} \\ &+ \left(\mathcal{H}' + 2\mathcal{H}^2\right) F_\alpha - 2\mathcal{H}^2 \overline{\mathrm{v}}_\alpha \overline{\mathrm{v}}^\mu F_\mu\Big], \end{aligned} \tag{7.31b}$$

$$\begin{aligned} F_{\alpha\beta}{}^{|\mu}{}_{|\mu} = \frac{1}{a^2}\Big[&\Box F_{\alpha\beta} + 2\mathcal{H}\overline{\mathrm{v}}^\mu F_{\alpha\beta;\mu} - 2\mathcal{H}\overline{\mathrm{v}}^\mu F_{\mu\alpha;\beta} - 2\mathcal{H}\overline{\mathrm{v}}^\mu F_{\mu\beta;\alpha} \\ &+ 2\mathcal{H}\overline{\mathrm{f}}^{\mu\nu}\left(\overline{\mathrm{v}}_\alpha F_{\beta\mu;\nu} + \overline{\mathrm{v}}_\beta F_{\alpha\mu;\nu}\right) + 2\left(\mathcal{H}' + \mathcal{H}^2\right) F_{\alpha\beta} \\ &- 4\mathcal{H}^2\left(\overline{\mathrm{v}}^\mu \overline{\mathrm{v}}_\alpha F_{\beta\mu} + \overline{\mathrm{v}}^\mu \overline{\mathrm{v}}_\beta F_{\alpha\mu} - \frac{1}{2}\overline{\mathrm{v}}_\alpha \overline{\mathrm{v}}_\beta \overline{\mathrm{f}}^{\mu\nu} F_{\mu\nu} - \frac{1}{2}\overline{\mathrm{f}}_{\alpha\beta}\overline{\mathrm{v}}^\mu \overline{\mathrm{v}}^\nu F_{\mu\nu}\right)\Big], \end{aligned} \tag{7.31c}$$

where we have introduced notations

$$\Box F \equiv \overline{f}^{\mu\nu} F_{;\mu\nu} , \quad \Box F_\alpha \equiv \overline{f}^{\mu\nu} F_{\alpha;\mu\nu} , \quad \Box F_{\alpha\beta} \equiv \overline{f}^{\mu\nu} F_{\alpha\beta;\mu\nu} , \tag{7.32}$$

of the wave operators for the scalar, vector, and tensor fields in the conformal spacetime and in arbitrary coordinates. Notice that although the conformal spacetime coincides, in case of $k = 0$, with the Minkowski spacetime, the metric $\overline{f}_{\alpha\beta}$ is not the diagonal Minkowski metric $\eta_{\alpha\beta}$ unless the coordinates are Cartesian. Of course, the covariant derivative from a scalar must be understood as a partial derivative, that is $F_{;\alpha} = F_{,\alpha}$.

We will need several other equations to complete the transformation of the Laplace–Beltrami operators to the conformal spacetime since the wave operator $\Box$ acts on functions like those shown in (7.24), which are made of a product of the scale factor, $a = a(\eta)$ in some power n (may be not an integer), with a geometric object, $\mathcal{F} = \mathcal{F}(x^\alpha)$, which can be a scalar, a vector or a tensor of the second rank (we have suppressed the tensor indices of $\mathcal{F}$ since they do not interfere with the derivation of the equations which follow). These equations are

$$(a^n \mathcal{F})_{;\mu} = a^n \left(\mathcal{F}_{;\mu} - n\mathcal{H}\overline{v}_\mu \mathcal{F}\right) , \tag{7.33a}$$

$$(a^n \mathcal{F})_{;\mu\nu} = a^n \left[\mathcal{F}_{;\mu\nu} - n\mathcal{H}\left(\overline{v}_\mu \mathcal{F}_{;\nu} + \overline{v}_\nu \mathcal{F}_{;\mu}\right) + n\left(\mathcal{H}' + n\mathcal{H}^2\right)\overline{v}_\mu \overline{v}_\nu\right] , \tag{7.33b}$$

and they allow us to write down the wave operator from the product of a^n and $\mathcal{F}$ in the following form:

$$\Box\left(a^n \mathcal{F}\right) = a^n \left[\Box \mathcal{F} - 2n\mathcal{H}\overline{v}^\mu \mathcal{F}_{;\mu} - n\left(\mathcal{H}' + n\mathcal{H}^2\right)\mathcal{F}\right] . \tag{7.34}$$

It is easy to confirm that contraction of (7.33b) with the conformal 4-velocity, $\overline{v}^\alpha$, brings about another differential operator

$$\overline{v}^\mu \overline{v}^\nu \left(a^n \mathcal{F}\right)_{;\mu\nu} = a^n \left[\overline{v}^\mu \overline{v}^\nu \mathcal{F}_{;\mu\nu} + 2n\mathcal{H}\overline{v}^\mu \mathcal{F}_{;\mu} + n\left(\mathcal{H}' + n\mathcal{H}^2\right)\mathcal{F}\right] . \tag{7.35}$$

We remind that if the object $\mathcal{F}$ is a scalar, the covariant derivative $\mathcal{F}_{;\alpha} = \mathcal{F}_{,\alpha}$ is reduced to a partial derivative. In case, when $\mathcal{F}$ is either a vector or a tensor, the covariant derivative must be calculated with taking into account the affine connection $\overline{B}^\alpha{}_{\beta\gamma}$ defined in (4.25).

It is also interesting to notice that in the expanding universe the conformal Laplace operator, $\Delta \mathcal{F} \equiv \overline{\pi}^{\mu\nu} \mathcal{F}_{;\mu\nu}$ is the scale invariant in the sense that

$$\Delta\left(a^n \mathcal{F}\right) = a^n \Delta \mathcal{F} , \tag{7.36}$$

where $\mathcal{F}$ is a tensor of an arbitrary rank. Equation (7.36) can be proven by adding up (7.34) and (7.35), and accounting for definition (7.23) of the projection operator on the hypersurface being orthogonal to $\overline{v}^\alpha$.

Now, we are ready to formulate the field equations for cosmological perturbations in the conformal spacetime.

Equations for perturbations of dark matter and dark energy

Dark matter and dark energy are described by scalar fields Φ and Ψ. The fields themselves are not gauge invariant. Therefore, physical meaning have only the equations for the gauge-invariant perturbations of these fields which are $V_{\rm m}$, $V_{\rm q}$, and χ. We consider, first, equation (6.34) for the gauge-invariant scalar $V_{\rm m}$. We convert the covariant derivatives taken with respect to the background metric, $\bar{g}_{\alpha\beta}$, to the partial derivatives of the conformally flat metric, $\bar{\mathrm{f}}_{\alpha\beta}$ and use equation (7.31a) for the Laplace–Beltrami operator along with the expressions for various cosmological parameters given in Section 7.2. After arranging terms with respect to the powers of the Hubble parameter $\mathcal{H}$, we obtain the scalar equation for function $V_{\rm m}$ describing the perturbations of dark matter,

$$\begin{aligned}\Box V_{\rm m} &+ \left(1-\frac{c^2}{v_{\rm s}^2}\right)\bar{\rm v}^\alpha\bar{\rm v}^\beta V_{{\rm m};\alpha\beta} + \left(3-\frac{c^2}{v_{\rm s}^2}\right)\mathcal{H}\bar{\rm v}^\alpha V_{{\rm m},\alpha} \\ &+3\left[1-w_{\rm eff}-\frac{1}{2}(1+w_{\rm m})\left(1-\frac{c^2}{v_{\rm s}^2}\right)\Omega_{\rm m}\right]\mathcal{H}^2 V_{\rm m} \\ &+12\mathcal{H}^2\left[\bar{\rm v}^\alpha\chi_{,\alpha} - 3\left(1-\sqrt{\frac{3}{8{\rm x}_1}}\lambda{\rm x}_2\right)\mathcal{H}\chi\right]\frac{{\rm x}_1}{a} = -4\pi a^2(\sigma+\tau)\ .\end{aligned} \tag{7.37}$$

This is a wave equation with the speed of sound $v_{\rm s}$ which determines the speed of propagation of the scalar perturbations in the dark matter considered as an ideal fluid. These perturbations can be interpreted as acoustic or sound waves of different wavelengths propagating in spacetime. Solution of homogeneous equation (7.37) describes the propagation of primordial scalar perturbations of dark matter. A particular solution of the inhomogeneous equation (7.37) tells us how the perturbation of dark matter caused by the isolated astronomical system propagate.

Similar procedure is applied to equation (6.37) and leads to a wave equation for function $V_{\rm q}$ describing propagation of perturbations of dark energy considered as a scalar field,

$$\begin{aligned}\Box V_{\rm q} &+ 2\left(1-\sqrt{\frac{3}{2{\rm x}_1}}\lambda{\rm x}_2\right)\mathcal{H}\bar{\rm v}^\mu V_{{\rm q},\mu} \\ &+3\left[1-w_{\rm eff}-\frac{1}{2}(1+w_{\rm m})\left(1-\frac{c^2}{v_{\rm s}^2}\right)\Omega_{\rm m}\right]\mathcal{H}^2 V_{\rm q} \\ &+\lambda{\rm x}_2\left[3\lambda\left(2\Gamma_{\rm q}+\frac{{\rm x}_2}{{\rm x}_1}\right)-5\sqrt{\frac{6}{{\rm x}_1}}\right]\mathcal{H}^2 V_{\rm q} \\ &+\frac{3}{2}\mathcal{H}^2(1+w_{\rm m})\left(3+\frac{c^2}{v_{\rm s}^2}\right)\left[\bar{\rm v}^\mu\chi_{,\mu}-3\frac{v_{\rm s}^2}{c^2}\mathcal{H}\chi\right]\frac{\Omega_{\rm m}}{a} = -4\pi a^2(\sigma+\tau)\ .\end{aligned} \tag{7.38}$$

The speed of propagation of dark energy is naturally equal to the fundamental speed c as contrasted with dark matter. Dark matter has an intrinsic elasticity associated with the bulk modulus $K = \epsilon(dp/d\epsilon)$ that is proportional to pressure p, and where ϵ is the energy density of the fluid. The speed of sound $v_s = \sqrt{K/\epsilon} < c$ for a fluid because in this case $K < \epsilon$. However, in case of the scalar field $K = |\epsilon|$, and $v_s = c$.

Equations (7.37) and (7.38) depend on the scalar function χ which obeys equation (6.29). Making use of the same transformations as applied to derivation of (7.37) and (7.38), we can recast (6.29) to a wave equation for χ,

$$\Box\chi + 4\mathcal{H}\left(1 - \sqrt{\frac{3}{8x_1}}\lambda x_2\right)\bar{\nabla}^\alpha\chi_{,\alpha} - \frac{9}{2}(1 + w_{\text{eff}})\mathcal{H}^2\chi$$
$$= -a\left[\sqrt{\frac{6}{x_1}}\lambda x_2\mathcal{H}V_{\text{m}} + \left(1 - \frac{c^2}{v_s^2}\right)\bar{\nabla}^\alpha V_{\text{m},\alpha}\right]. \tag{7.39}$$

We can observe that the speed of propagation of the field χ is equal to the fundamental speed c. Moreover, (7.39) depends on V_{m} and should be solved simultaneously with equation (7.37) for V_{m} after imposing certain boundary conditions. As soon as the gauge-invariant scalar χ is known, the potential, V_{q}, can be determined either as a particular solution of the inhomogeneous equation (7.38) or, more simple, from algebraic relation (6.36).

We also need equations for the normalized Clebsch and scalar potentials, χ_{m} and χ_{q}. These potentials are required to determine the gravitational perturbation, q, with the help of (7.30) and/or to check on self-consistency of the solutions of the field equations in the matter sector of perturbation theory. Conformal-spacetime equations for χ_{m} and χ_{q} are derived from their definition (6.10) and the field equations (5.67) and (5.68). They are

$$\Box\chi_{\text{m}} + \frac{3}{2}(1 + w_{\text{eff}})\mathcal{H}^2\chi_{\text{m}} = 12\mathcal{H}^2 x_1\chi - a\left[4\mathcal{H}V_{\text{m}} + \left(1 - \frac{c^2}{v_s^2}\right)\bar{\nabla}^\alpha V_{\text{m},\alpha}\right], \tag{7.40}$$

$$\Box\chi_{\text{q}} + \frac{3}{2}(1 + w_{\text{eff}})\mathcal{H}^2\chi_{\text{q}} = -6\mathcal{H}^2(1 + w_{\text{m}})\Omega_{\text{m}}\chi - a\left(4 - \sqrt{\frac{6}{x_1}}\lambda x_2\right)\mathcal{H}V_{\text{q}}. \tag{7.41}$$

By subtracting one of these equations from another, we get back to equation (7.39). Notice that χ_{m} and χ_{q} are not gauge-invariant perturbations and, hence, the solutions of (7.40) and (7.41) should be interpreted with care.

Equations for the metric perturbations

Post-Newtonian equations for gravitational perturbations in physical spacetime are (6.25), (6.27), (6.38), and (6.40). These equations are gauge dependent. In order to fix the gauge we imposed the gauge conditions (5.64) and (6.43). In this gauge, equations for the conformal metric tensor perturbations become

$$\Box q - 2\mathcal{H}\bar{\mathrm{v}}^{\alpha} q_{,\alpha} + (1 + 3w_{\mathrm{eff}})\,\mathcal{H}^2 q \tag{7.42a}$$
$$= 8\pi a^2 (\sigma + \tau) - 24\mathcal{H}^2 \left[\sqrt{\frac{3\mathrm{x}_1}{8}}\lambda \mathrm{x}_2 - \mathcal{H}\mathrm{x}_1\right]\frac{\chi_{\mathrm{q}}}{a} - 3(1 + w_{\mathrm{eff}})\,\mathcal{H}^2 \Omega_{\mathrm{m}}$$
$$\times \left[\left(1 - \frac{c^2}{v_{\mathrm{s}}^2}\right) V_{\mathrm{m}} - \mathcal{H}\left(1 + 3\frac{v_{\mathrm{s}}^2}{c^2}\right)\frac{\chi_{\mathrm{m}}}{a}\right],$$

$$\Box p - 2\mathcal{H}\bar{\mathrm{v}}^{\alpha} p_{,\alpha} = 16\pi a^2 \tau\,, \tag{7.42b}$$

$$\Box p_{\alpha} - 2\mathcal{H}\bar{\mathrm{v}}^{\beta} p_{\alpha;\beta} + (1 + 3w_{\mathrm{eff}})\,\mathcal{H}^2 p_{\alpha} = 16\pi a \tau_{\alpha}\,, \tag{7.42c}$$

$$\Box p^{\mathrm{T}}_{\alpha\beta} - 2\mathcal{H}\bar{\mathrm{v}}^{\gamma} p_{\alpha\beta;\gamma} = 16\pi \tau^{\mathrm{T}}_{\alpha\beta}\,. \tag{7.42d}$$

The reader can observe that equations (7.42a)–(7.42d) for linearized metric perturbations are decoupled from each other. Moreover, equations (7.42b–(7.42d) are decoupled from the matter perturbations V_{m}, χ_{m}, etc. Only equation (7.42a) for q is coupled with the matter perturbations governed by equations (7.37), (7.40), and (7.41) so that these equations should be solved together. As we have mentioned above, function q is a linear combination of V_{m} and χ_{m} according to (7.30). Hence, in order to determine q it is, in fact, sufficient to solve (7.37), (7.38), and (7.40). Nevertheless, it is convenient to present the differential equation (7.42a) for q explicitly for the sake of mathematical completeness and rigor. It can be used for independent validation of the solution of the system of equations (7.37), (7.40), and (7.38). Unfortunately, these equations are strongly coupled and cannot be solved analytically in the most general situation of a multicomponent background universe governed by dark energy and dark matter. Solution of (7.37)–(7.41) would require a numerical integration.

It would be instrumental to get better insight to the post-Newtonian theory of cosmological perturbations by making some simplifying assumptions about the background model of the expanding universe in order to decouple the system of the post-Newtonian equations and to find their analytic solution explicitly. We discuss these assumptions and the corresponding system of the decoupled post-Newtonian equations in Section 8 below.

7.4 Residual gauge freedom in the conformal spacetime

The gauge conditions (5.64), and (6.43) in physical space are given by equations (6.45a) and (6.45b). After transforming to the conformal spacetime the equations for the gauge condition reads

$$p^{\beta}{}_{;\beta} + \bar{\mathrm{v}}^{\beta}(2q - p)_{,\beta} + 2\mathcal{H}q = 16\pi a\left(\bar{\rho}_{\mathrm{m}}\bar{\mu}_{\mathrm{m}}\chi_{\mathrm{m}} + \bar{\rho}_{\mathrm{q}}\bar{\mu}_{\mathrm{q}}\chi_{\mathrm{q}}\right), \tag{7.43a}$$

$$p^{\mathrm{T}\alpha\beta}{}_{;\beta} + \bar{\mathrm{v}}^{\beta} p^{\alpha}{}_{;\beta} + \frac{1}{3}\bar{\pi}^{\alpha\beta} p_{,\beta} + 2\mathcal{H}p^{\alpha} = 0\,. \tag{7.43b}$$

The residual gauge freedom in the conformal spacetime is described by two gauge functions, $\zeta \equiv \xi/a$ and ζ^α, where ξ and ζ^α have been defined in Section 6.7. Differential equations for ζ and ζ^α are obtained by making transformation of equations (6.49a) and (6.49b) to the conformal spacetime. The calculation is straightforward and results in

$$\Box\zeta - 2\mathcal{H}\overline{v}^\beta\zeta_{,\beta} + \left(1 + 3w_{\rm eff}\right)\mathcal{H}^2\zeta = 0\,, \tag{7.44a}$$

$$\Box\zeta^\alpha - 2\mathcal{H}\overline{v}^\beta\zeta^\alpha{}_{;\beta} = 0\,. \tag{7.44b}$$

Solutions of equations (7.42a)–(7.42d) are determined up to the gauge transformations

$$\hat{q} = q + 2\overline{v}^\alpha\zeta_{,\alpha} + 2\mathcal{H}\zeta\,, \tag{7.45a}$$

$$\hat{p} = p + \zeta^\alpha{}_{;\alpha} + 3\overline{v}^\alpha\zeta_{,\alpha} + 6\mathcal{H}\zeta\,, \tag{7.45b}$$

$$\hat{p}_\alpha = p_\alpha + \overline{\pi}_{\alpha\beta}\left(\overline{v}^\gamma\zeta^\beta{}_{;\gamma} - \zeta^{,\beta} + 2\mathcal{H}\zeta^\beta\right)\,, \tag{7.45c}$$

$$\hat{p}_{\alpha\beta} = p_{\alpha\beta} - \left(\overline{\pi}_{\mu\alpha}\overline{\pi}_\beta{}^\nu + \overline{\pi}_{\mu\beta}\overline{\pi}_\alpha{}^\nu\right)\zeta^\mu{}_{;\nu} + \overline{\pi}_{\alpha\beta}\left(\zeta^\alpha{}_{;\alpha} + \overline{v}^\alpha\zeta_{,\alpha} + 2\mathcal{H}\zeta\right)\,, \tag{7.45d}$$

where the gauge functions ζ, ζ^α are solutions of the differential equations (7.44a) and (7.44b).

8 Decoupled system of the post-Newtonian field equations

8.1 The universe governed by dark matter and cosmological constant

Case 1: Arbitrary equation of state of dark matter

Let us consider a special case of the background value of dark energy represented by cosmological constant $\Lambda = 8\pi\overline{W}$. In this case, the equation of state of the scalar field is $w_{\rm q} = -1$, and we have $\overline{\rho}_{\rm q}\overline{\mu}_{\rm q} = \overline{\epsilon}_{\rm q} + \overline{p}_{\rm q} = 0$. The parameter $x_1 = 0$, and $x_2 = \Lambda/(3H^2)$. It yields the parameter $\Omega_{\rm q} = x_2$, and $\Omega_{\rm m} = 1 - x_2$. Since the cosmological constant corresponds to a constant potential $\overline{W}$ of the scalar field, we get for its derivative $\partial\overline{W}/\partial\overline{\Psi} = 0$, and equation (7.1) points out that the parameter $\lambda = 0$.

In the universe governed by dark matter and cosmological constant, the parameter of the effective equation of state of the dark matter is

$$w_{\rm eff} = w_{\rm m} - (1 + w_{\rm m})\frac{\Lambda}{3H^2}\,. \tag{8.1}$$

Hence, the time derivative of the Hubble parameter defined in (7.13), is reduced to a more simple expression,

$$\dot{H} = \frac{1}{2}\left(1 + w_{\rm m}\right)\left(\Lambda - 3H^2\right)\,. \tag{8.2}$$

On the other hand, equation (4.52) tells us that in this model of the universe the time derivative of the Hubble parameter is

$$\dot{H} = -4\pi\bar{\rho}_{\rm m}\bar{\mu}_{\rm m} \, . \tag{8.3}$$

The field equation (7.37) for scalar $V_{\rm m}$ is reduced to that describing the time evolution of the perturbation of the ideal fluid density, $\delta\rho_{\rm m}$. Indeed, the scalar $V_{\rm m}$ defined by equation (6.13a), can be recast to the form given by equation (6.22), that is

$$V_{\rm m} = \frac{v_{\rm s}^2}{c^2}\delta_{\rm m} \, , \tag{8.4}$$

where the gauge-invariant scalar perturbation

$$\delta_{\rm m} \equiv \frac{\delta\rho_{\rm m}}{\bar{\rho}_{\rm m}} + 3H\chi_{\rm m} \, , \tag{8.5}$$

is a linear combination of the perturbation of the mass density of the dark matter and the normalized Clebsch potential $\chi_{\rm m}$. Replacing expression (8.4) in equation (7.37), yields an exact equation for $\delta_{\rm m}$ that is

$$\begin{aligned}\left(1 - \frac{v_{\rm s}^2}{c^2}\right)\bar{\rm v}^\alpha\bar{\rm v}^\beta\delta_{{\rm m};\alpha\beta} - \frac{v_{\rm s}^2}{c^2}\Box\delta_{\rm m} + \left(1 - 3\frac{v_{\rm s}^2}{c^2}\right)\mathcal{H}\bar{\rm v}^\alpha\delta_{{\rm m},\alpha} \\ -\frac{3}{2}\left[\left(1 - 3w_{\rm m}\right)\frac{v_{\rm s}^2}{c^2} + \left(1 + w_{\rm m}\right)\right]\mathcal{H}^2\delta_{\rm m} \\ +\frac{1}{2}\left(1 + w_{\rm m}\right)\left(1 - 3\frac{v_{\rm s}^2}{c^2}\right)a^2\Lambda\delta_{\rm m} = 4\pi a^2\left(\sigma + \tau\right) \, .\end{aligned} \tag{8.6}$$

This equation describes propagation of the density perturbation of dark matter, $\delta_{\rm m}$, in the form of sound waves with velocity $v_{\rm s}$. Equation (8.6) is decoupled from any other perturbation and can be solved separately after the boundary conditions are specified. For this reason, we call (8.6) *master equation.*

Equation (7.39) for potential χ makes no sense since the normalized perturbation $\chi_{\rm q} = \psi/\bar{\mu}_{\rm q}$ of dark energy in the form of cosmological constant diverges due to the condition $\mu_{\rm q} = \rho_{\rm q} = 0$. Equation for the perturbation of dark energy, ψ, itself is obtained from (5.68) and is reduced to a homogeneous wave equation

$$\Box\psi - 2\mathcal{H}\bar{\rm v}^\mu\psi_{,\mu} = 0 \, . \tag{8.7}$$

Equation for the normalized Clebsch potential, $\chi_{\rm m}$, is derived from equation (7.40) and, in the case of the universe under consideration, reads

$$\Box\chi_{\rm m} + \frac{1}{2}\left(1 + w_{\rm m}\right)\left(3\mathcal{H}^2 - a^2\Lambda\right)\chi_{\rm m} = \left(1 - \frac{v_{\rm s}^2}{c^2}\right)a\bar{\rm v}^\mu\delta_{{\rm m},\mu} - 4a\mathcal{H}\frac{v_{\rm s}^2}{c^2}\delta_{\rm m} \, . \tag{8.8}$$

This is an inhomogeneous equation that can be solved as soon as one knows δ_m from the master equation (8.6). The potential χ_m is necessary to determine the perturbation of the 4-velocity of dark matter. We also need it to find out the metric perturbation q.

Gravitational potential, q, can be determined directly from equation (7.30) after solving equations (8.6) and (8.8) or by solving equation (7.42a) which (in the dark matter+cosmological constant universe) takes on the following form:

$$\Box q - 2\mathcal{H}\bar{v}^{\mu} q_{,\mu} + \left[(1+3w_m)\mathcal{H}^2 - (1+w_m)a^2\Lambda\right] q \tag{8.9a}$$
$$= 8\pi a^2 \left\{\sigma + \tau + \bar{\rho}_m\bar{\mu}_m \left[\left(1 - \frac{v_s^2}{c^2}\right)\delta_m + H\left(1 + 3\frac{v_s^2}{c^2}\right)\chi_m\right]\right\} .$$

Equations for other components of the metric tensor perturbations are found from (7.42b)–(7.42d). In dark matter+cosmological constant universe they read

$$\Box p - 2\mathcal{H}\bar{v}^{\mu} p_{,\mu} = 16\pi a^2 \tau , \tag{8.9b}$$

$$\Box p_{\alpha} - 2\mathcal{H}\bar{v}^{\mu} p_{\alpha;\mu} + \left[(1+3w_m)\mathcal{H}^2 - (1+w_m)a^2\Lambda\right] p_{\alpha} = 16\pi a \tau_{\alpha} , \tag{8.9c}$$

$$\Box p^{\mathsf{T}}_{\alpha\beta} - 2\mathcal{H}\bar{v}^{\mu} p^{\mathsf{T}}_{\alpha\beta;\mu} = 16\pi \tau^{\mathsf{T}}_{\alpha\beta} . \tag{8.9d}$$

Equations given in this section are valid for arbitrary cosmological equation of state of dark matter, $\bar{p}_m = w_m\bar{\epsilon}_m$, that is physically reasonable and makes sense. The parameter w_m of the equation of state should not be replaced with the ratio of v_s^2/c^2 which characterizes the derivative of pressure, $\bar{p}_m$, with respect to the energy density, $\bar{\epsilon}$, of dark matter. This is because the parameter w_m can depend in the most general case on the other thermodynamic quantities (like enthropy, temperature, etc.) which may implicitly depend on $\bar{\epsilon}$. Equations (8.6)–(8.9d) are decoupled in the sense that all of them can be solved one after another starting from solving the master equation (8.6) for δ_m.

Case 2: Cold dark matter

Equations of the previous section can be further simplified for some particular equations of state of dark matter. For example, in the case of CDM we can think about it as being made out of collisionless dust. Background pressure of dust drops down to zero making the parameter of the CDM equation of state $w_m = 0$. Sound waves do not propagate in dust. Hence, the speed of sound $v_s = 0$. For this reason, all terms being proportional to v_s^2 and w_m vanish in equation (8.6). Moreover, dust has the specific enthalpy, $\mu_m = 1$ making the energy density of dust equal to its rest mass density $\bar{\epsilon}_m = \bar{\rho}_m$, and the normalized perturbation χ_m of the Clebsch potential of dust equal to the perturbation ϕ of the Clebsch potential itself, $\chi_m = \phi$. The Friedmann equation (4.50) (for $k = 0$) tells us that

$$\mathcal{H}^2 = \frac{a^2}{3}\left(8\pi\bar{\rho}_m + \Lambda\right) . \tag{8.10}$$

Accounting for this result in the master equation (8.6), and neglecting all terms being proportional to the speed of sound, v_s and w_m, we obtain

$$\overline{v}^{\alpha}\overline{v}^{\beta}\delta_{m;\alpha\beta} + \mathcal{H}\overline{v}^{\alpha}\delta_{m,\alpha} - 4\pi a^2\overline{\rho}_m\delta_m = 4\pi a^2(\sigma + \tau) , \tag{8.11}$$

where the terms depending on the cosmological constant, Λ, have canceled out. This equation is more familiar when is written down in the preferred FLRW frame, where $\overline{v}^{\alpha} = (1, 0, 0, 0)$. Equation (8.11) assumes the following form:

$$\delta''_m + \mathcal{H}\delta'_m - 4\pi a^2\overline{\rho}_m\delta_m = 4\pi a^2(\sigma + \tau) , \tag{8.12}$$

where the time derivatives (denoted with a prime) are taken with respect to the conformal time η. Converting the time derivatives in (8.12) from the conformal time η to the cosmic time T reduces it to a canonical form

$$\ddot{\delta}_m + 2H\dot{\delta}_m - 4\pi\delta_m = 4\pi(\sigma + \tau) \tag{8.13}$$

which can be found in many textbooks on cosmology [80, 91, 94, 113, 114]. All textbooks always dropped off the source of the bare perturbation on the right-hand side of (8.13) as they are concerned with the description of the formation of the large scale structure in the universe out of the primordial perturbations. However, omitting the bare perturbation on the right-hand side of (8.13) is equivalent to neglecting the contribution of the small-scale density fluctuations in the early universe to the formation of the large scale structures – the process which can be physically significant in the CDM scenario of galaxy formation [15, 16].

Equation (8.13) has been derived by previous researchers without resorting to the concept of the Clebsch potential of the ideal fluid. For this reason, the density contrast, δ_m, was interpreted as the ratio of the perturbation of the dust density to its background value, $\delta = \delta\rho_m/\overline{\rho}_m$, without taking into account the perturbation, ϕ, of the Clebsch potential. However, the quantity δ is not gauge invariant which was considered as a drawback. The scrutiny analysis of the underlying principles of hydrodynamics in the expanding universe given in this chapter, reveals that equation (8.12) is, in fact, valid for the gauge-invariant density perturbation δ_m defined above in (8.5). Another distinctive feature of equation (8.12) is the presence of the source of a *bare* perturbation in its right-hand side. The *bare* perturbation is caused by the effective density $\sigma + \tau$ of the matter which comprises the isolated astronomical system and initiates the growth of instability in the cosmological matter that, in its own turn, induces formation of the large scale structure of the universe [94, 114]. Standard approach to cosmological perturbation theory always set $\sigma + \tau = 0$ and operates with the spectrum of the primordial perturbation of the density $\delta\rho_m/\rho_m$ (but not with the spectrum for δ_m).

Equation (8.8) in case of dust reads

$$\Box\chi_m + \frac{1}{2}\left(3\mathcal{H}^2 - a^2\Lambda\right)\chi_m = a\overline{v}^{\alpha}\delta_{m,\alpha} , \tag{8.14}$$

where $\chi_m = \phi$ is reduced to the perturbation ϕ of the Clebsch potential Φ for the reason that in case of dust $\mu_m = 1$. If equations (8.12) and (8.14) are solved, the gravitational perturbations can be found from equations (8.9a)–(8.9d), which take the following form:

$$\Box q - 2\mathcal{H}\overline{\mathrm{v}}^{\alpha} q_{,\alpha} + \left(\mathcal{H}^2 - a^2\Lambda\right) q = 8\pi a^2 \left[\sigma + \tau + \overline{\rho}_m \left(\delta_m + H\chi_m\right)\right] , \tag{8.15a}$$

$$\Box p - 2\mathcal{H}\overline{\mathrm{v}}^{\alpha} p_{,\alpha} = 16\pi a^2 \tau , \tag{8.15b}$$

$$\Box p_{\alpha} - 2\mathcal{H}\overline{\mathrm{v}}^{\beta} p_{\alpha;\beta} + \left(\mathcal{H}^2 - a^2\Lambda\right) p_{\alpha} = 16\pi a \tau_{\alpha} , \tag{8.15c}$$

$$\Box p^{\mathrm{T}}_{\alpha\beta} - 2\mathcal{H}\overline{\mathrm{v}}^{\gamma} p^{\mathrm{T}}_{\alpha\beta;\gamma} = 16\pi \tau^{\mathrm{T}}_{\alpha\beta} . \tag{8.15d}$$

It is interesting to note that besides the *bare* density perturbation, $\sigma + \tau$, caused by an isolated astronomical system, the source for the scalar gravitational perturbation, q, contains on the right-hand side of equation (8.15a) also the *induced* density perturbation $\overline{\rho}_m \left(\delta_m + H\chi_m\right) = \delta\rho_m + H\overline{\rho}_m\phi$ of the background dark matter. This *induced* density perturbation depends on time and leads to a temporal change of the initial (bare) mass of the isolated astronomical system in the course of the Hubble expansion of the universe. Thus, our post-Newtonian approach to cosmology explains the origin of the time dependence of the central, point-like mass in the cosmological solution found by McVittie [87] (see also discussion in [24]).

Case 3: Hot dark matter

Hot dark matter (HDM) is a hypothetical form of dark matter which consists of ultra-relativistic particles that travel with velocities being very close to the fundamental speed c. A plausible candidate for the HDM is neutrino. Hot dark matter taken alone, cannot explain how individual galaxies were formed from the primordial perturbations. Therefore, HDM is discussed only as part of a mixed dark matter theory [11]. Nonetheless, the case of the HDM is interesting from mathematical point of view. Equation of state of the HDM is approximated by the radiative equation of state, $p_m = (1/3)\epsilon_m$ which yields the parameter $w_m = 1/3$. We assume that this parameter is constant and, hence, the speed of sound $v_s = \sqrt{1/3}c$. This value of v_s is comparable with the fundamental speed c which means that we have to keep the terms with the speed of sound in the master equation (8.6). The values of w_m and v_s for the HDM equation of state reduce the master equation for the gauge-invariant HDM density perturbation δ_m to the following form:

$$\Box_s \delta_m + 6\mathcal{H}^2 \delta_m = -12\pi a^2 (\sigma + \tau) , \tag{8.16}$$

where

$$\Box_s \equiv \left(-\frac{c^2}{v_s^2}\overline{\mathrm{v}}^{\alpha}\overline{\mathrm{v}}^{\beta} + \overline{\pi}^{\alpha\beta}\right)\partial_{\alpha\beta} , \tag{8.17}$$

is the D'Alembert wave operator with the speed of propagation of sound waves $v_s = c/\sqrt{3}$. Equation for the perturbation of the Clebsch potential of the HDM is derived from (8.8) and reads

$$\Box\chi_{\rm m} + 2\left(\mathcal{H}^2 - \frac{1}{3}a^2\Lambda\right)\chi_{\rm m} = \frac{2a}{3}\left(\bar{v}^\mu\delta_{{\rm m},\mu} - 2\mathcal{H}\delta_{\rm m}\right) . \qquad (8.18)$$

Equations for the gravitational perturbations are

$$\Box q - 2\mathcal{H}\bar{v}^\mu q_{,\mu} + 2\left(\mathcal{H}^2 - \frac{2}{3}a^2\Lambda\right)q = 8\pi a^2\left[\sigma + \tau + \frac{2}{3}\bar{\rho}_{\rm m}\bar{\mu}_{\rm m}\left(\delta_{\rm m} + 3H\chi_{\rm m}\right)\right] . \qquad (8.19a)$$

Equations for other components of the metric tensor perturbations are found from (7.42b)–(7.42d). In dark matter+cosmological constant universe they read

$$\Box p - 2\mathcal{H}\bar{v}^\mu p_{,\mu} = 16\pi a^2\tau , \qquad (8.19b)$$

$$\Box p_\alpha - 2\mathcal{H}\bar{v}^\mu p_{\alpha;\mu} + 2\left(\mathcal{H}^2 - \frac{2}{3}a^2\Lambda\right)p_\alpha = 16\pi a\tau_\alpha , \qquad (8.19c)$$

$$\Box p^{\rm T}_{\alpha\beta} - 2\mathcal{H}\bar{v}^\mu p^{\rm T}_{\alpha\beta;\mu} = 16\pi\tau^{\rm T}_{\alpha\beta} . \qquad (8.19d)$$

8.2 The universe governed by dark energy

In this section, we explore the case of the universe governed primarily by a dark energy (scalar field Ψ) with dark matter constituent being unimportant. In this case, the time evolution of the background universe is defined exceptionally by equations (7.5a) and (7.5b). The most general solution of (7.5a) and (7.5b) is complicated and can not be achieved analytically. Numerical analysis shows that the solution evolves in the phase space of the two variables $\{x_1, x_2\}$ from an unstable to a stable fixed point by passing through a saddle point [2]. The cosmic acceleration is realized by the stable point with the values of $x_1 = \lambda^2/6$ and $x_2 = 1 - \lambda^2/6$, which is equivalent to the equations of state (4.54) with the values of the parameters, $w_{\rm m} = 0$, and, $w_{\rm q} = -1 + \lambda^2/3$. It also requires the energy density of the background matter $\bar{\epsilon}_{\rm m} = 0$, that is $\Omega_{\rm m} = 0$. In such a universe the derivatives of the potential of the scalar field are

$$\frac{1}{\bar{\mu}_{\rm q}}\frac{\partial\overline{W}}{\partial\overline{\Psi}} = -\frac{3}{2}H\left(1 - w_{\rm q}\right) , \qquad \frac{\partial^2\overline{W}}{\partial\overline{\Psi}^2} = \frac{9}{2}H^2\left(1 - w_{\rm q}^2\right) . \qquad (8.20)$$

Moreover, because $\bar{\rho}_{\rm m}\bar{\mu}_{\rm m} = \bar{\epsilon}_{\rm m} + \bar{p}_{\rm m} = 0$, the time derivative of the Hubble parameter is

$$\dot{H} = -4\pi\bar{\rho}_{\rm q}\bar{\mu}_{\rm q} = -\frac{3}{2}H^2\left(1 + w_{\rm q}\right) . \qquad (8.21)$$

In the point of the attractor of the scalar field, perturbations of the dark matter are fully suppressed that is the normalized value of the perturbed Clebsch potential of the dark matter, $\chi_{\rm m} = 0$. It makes the function $V_{\rm m} = q/2$, that is reduced to the perturbation of the scalar component of the gravitational field only. Perturbations of the scalar field are described by the scalar field variable, χ_q. In particular, after substituting the derivatives (8.20) of the scalar field potential along with the derivative (8.21) of the Hubble parameter, to (6.37), we obtain the post-Newtonian equation for function V_q,

$$\Box V_q - \left(1 - 3w_q\right)\mathcal{H}\bar{v}^{\mu}V_{q,\mu} + \frac{3}{2}\mathcal{H}^2\left(1 - w_q\right)\left(1 + 3w_q\right)V_q = -4\pi a^2(\sigma + \tau) \,. \tag{8.22}$$

Field equation for the perturbation of the scalar field, χ_q, is reduced to

$$\Box\chi_q + \frac{1}{2}\left(1 + 3w_q\right)\mathcal{H}^2\chi_q = -\left(1 + 3w_q\right)a\mathcal{H}V_q \,. \tag{8.23}$$

Post-Newtonian equations for gravitational perturbations are (7.42a)–(7.42d). After substituting the values of the parameters x_1, x_2, w_{eff}, etc., corresponding to the model of the universe governed by the dark energy alone, the post-Newtonian equations for the metric perturbations become

$$\Box q - 2\mathcal{H}\bar{v}^{\mu}q_{,\mu} + \left(1 + 3w_q\right)\mathcal{H}^2 q = 8\pi a^2(\sigma + \tau) + \frac{3}{a}\left(1 + w_q\right)\left(1 + 3w_q\right)\mathcal{H}^3\chi_q \,, \tag{8.24a}$$

$$\Box p - 2\mathcal{H}\bar{v}^{\mu}p_{,\mu} = 16\pi a^2\tau \,, \tag{8.24b}$$

$$\Box p_\alpha - 2\mathcal{H}\bar{v}^{\mu}p_{\alpha;\mu} + \left(1 + 3w_q\right)\mathcal{H}^2 p_\alpha = 16\pi a\tau_\alpha \,, \tag{8.24c}$$

$$\Box p^{\mathrm{T}}_{\alpha\beta} - 2\mathcal{H}\bar{v}^{\mu}p^{\mathrm{T}}_{\alpha\beta;\mu} = 16\pi\tau^{\mathrm{T}}_{\alpha\beta} \,. \tag{8.24d}$$

One can see that the field equations for the perturbations of dark energy and gravitational field are decoupled, and can be solved separately starting from the master equation (8.22).

8.3 Post-Newtonian potentials in the linearized Hubble approximation

The metric tensor perturbations

The post-Newtonian equations for cosmological perturbations of gravitational and matter field variables crucially depend on the equation of state of the matter fields in the background universe. It determines the time evolution of the scale factor $a = a(\eta)$ and the Hubble parameter $\mathcal{H} = \mathcal{H}(\eta)$ which are described by the wide range of elementary and special functions of mathematical physics (see, for example, textbooks [2, 85, 106] and references therein). It is not the goal of this chapter to provide the reader with an exhaustive list of the mathematical solutions of the perturbed equations which requires theoretical development of cosmological Green's function (see, for example, [51, 78, 79, 97]). We notice that solving the field equations of the post-Newtonian approximations in cosmology is more complicated than in case of the post-Newtonian theory in asymptotically flat spacetime. The reason is twofold: (1) the system of the post-Newtonian equations on cosmological background involves, besides equations for the metric tensor perturbations, also the equations for the perturbations of the matter that curves the background manifold and governs its temporal evolution; and (2) the post-Newtonian field equations in cosmology depend on the time dependent Hubble parameter that makes finding the Green functions of the field equations pretty difficult task. If we are interested in finding the far zone (radiative) solution for the gravitational field of an isolated astronomical system, we have to fulfil this task exactly. This problem has not yet been solved though it is very important for doing precise cosmology with gravitational wave astronomy. On the other hand, we can employ the post-Friedmannian approximations by looking for the solution of the cosmological post-Newtonian equations as a series with respect to the Hubble parameter. In this section, we shall limit ourselves with the linearized Friedmann approximation. In other words, we shall take into account only the terms which are proportional to the Hubble parameter $\mathcal{H}$, and shall systematically neglect all terms which are quadratic, cubic, and higher orders with respect to $\mathcal{H}$.

As we shall see, in the linearized Friedmann approximation the post-Newtonian equations for the field perturbations have identical mathematical structure so that they are not only decoupled from one another, but also their generic solution can be found irrespectively of the equation of state governing the background universe. Indeed, if we neglect all quadratic with respect to $\mathcal{H}$ terms, the field equations for the conformal metric perturbations are reduced to the following set:

$$\Box q - 2\mathcal{H}\bar{v}^{\alpha} q_{,\alpha} = 8\pi a^2 (\sigma + \tau) \,, \tag{8.25a}$$
$$\Box p - 2\mathcal{H}\bar{v}^{\alpha} p_{,\alpha} = 16\pi a^2 \tau \,, \tag{8.25b}$$
$$\Box p_{\alpha} - 2\mathcal{H}\bar{v}^{\beta} p_{\alpha;\beta} = 16\pi a \tau_{\alpha} \,, \tag{8.25c}$$
$$\Box p^{\top}_{\alpha\beta} - 2\mathcal{H}\bar{v}^{\gamma} p^{\top}_{\alpha\beta;\gamma} = 16\pi \tau^{\top}_{\alpha\beta} \,, \tag{8.25d}$$

where the wave operator $\Box$ has been defined in (7.32), and the source of the *bare* perturbation is the tensor of energy–momentum of a localized astronomical system having a bounded matter support in space – see Section 4.8. The differential structure of the left-hand side of equations (8.25a)–(8.25d) is the same for all functions. The equations differ from each other only in terms of the order of $\mathcal{H}^2$ which have been omitted.

In order to bring equations (8.25a)–(8.25d) to a solvable form, we resort to relation (7.34) which reveals that in the linearized Friedmann approximation, the post-Newtonian equations for metric perturbations can be reduced to the form of a wave equation

$$\Box(aq) = 8\pi a^3 (\sigma + \tau) \,, \tag{8.26a}$$
$$\Box(ap) = 16\pi a^3 \tau \,, \tag{8.26b}$$

$$\Box\left(ap_\alpha\right) = 16\pi a^2 \tau_\alpha\,, \tag{8.26c}$$

$$\Box\left(ap^{\mathrm{T}}_{\alpha\beta}\right) = 16\pi a \tau^{\mathrm{T}}_{\alpha\beta}\,. \tag{8.26d}$$

So far, we did not impose any limitations on the curvature of space that can take three values: $k = \{-1, 0, +1\}$. Solution of wave equations (8.26a)–(8.26d) can be given in terms of special functions in case of the Riemann ($k = +1$) or the Lobachevsky ($k = -1$) geometry [78, 79]. The case of the spatial Euclidean geometry ($k = 0$) is more manageable, and will be discussed below.

If the FLRW metric is spatially flat universe, $k = 0$, and we chose the Cartesian coordinates x^α related to the isotropic coordinates X^α of the FLRW metric by a Lorentz transformation, $X^\alpha = L^\alpha{}_\beta x^\beta$, where $L^\alpha{}_\beta$ is the matrix of the Lorentz boost. In these coordinates the operator $\Box$ becomes a wave operator in the Minkowski spacetime,

$$\Box = \eta^{\mu\nu}\partial_{\mu\nu}\,. \tag{8.27}$$

and equations (8.26a)–(8.26d) are reduced to the inhomogeneous wave equations which solution depends essentially on the boundary conditions imposed on the metric tensor perturbations at conformal past-null infinity $\mathfrak{J}^-$ of the cosmological manifold [89]. We shall assume a no-incoming radiation condition also known as Fock–Sommerfeld's condition [31, 44]

$$\lim_{\substack{r\to+\infty \\ t+r=\mathrm{const.}}} n^\gamma \partial_\gamma \left[a(\eta) r l_{\alpha\beta}(x^\gamma)\right] = 0\,, \tag{8.28}$$

where $x^\gamma = (x^0, x^i)$, $\eta \equiv X^0$ is the conformal time in isotropic coordinates connected to the coordinates x^α by a Lorentz boost $\eta = \eta(x^\gamma) = L^0{}_\beta x^\beta$, the null vector $n^\alpha = \{1, x^i/r\}$, and $r = \delta_{ij}x^i x^j$ is the radial distance. This condition ensures that there is no infalling gravitational radiation arriving to the localized astronomical system from the future null infinity $\mathfrak{J}^+$. Effectively, it singles out the retarded solution of the wave equation. Whether the boundary condition (8.28) is valid or not, we do not know for sure because our knowledge of the universe is limited by the existence of the cosmological (also known as light or particle) horizon [81] that represents the boundary between the observable and the unobservable regions of the universe. Nonetheless, in case of spatially flat ($k = 0$) universe, the condition (8.28) seems to be highly plausible.

A particular solution of the wave equations satisfying condition (8.28), is the retarded integral [75]

$$q(t, \boldsymbol{x}) = -\frac{2}{a\left[\eta(t, \boldsymbol{x})\right]} \int\limits_{\mathcal{V}} \frac{a^3\left[\eta\left(s, \boldsymbol{x}'\right)\right]\left[\sigma\left(s, \boldsymbol{x}'\right) + \tau\left(s, \boldsymbol{x}'\right)\right] d^3x'}{|\boldsymbol{x} - \boldsymbol{x}'|}\,, \tag{8.29a}$$

$$p(t, \boldsymbol{x}) = -\frac{4}{a\left[\eta(t, \boldsymbol{x})\right]} \int\limits_{\mathcal{V}} \frac{a^3\left[\eta(s, \boldsymbol{x}')\right] \tau\left(s, \boldsymbol{x}'\right) d^3x'}{|\boldsymbol{x} - \boldsymbol{x}'|}\,, \tag{8.29b}$$

$$p_\alpha(t,\boldsymbol{x}) = -\frac{4}{a\left[\eta(t,\boldsymbol{x})\right]}\int\limits_{\mathcal{V}}\frac{a^2\left[\eta(s,\boldsymbol{x}')\right]\tau_\alpha(s,\boldsymbol{x}')\,d^3x'}{|\boldsymbol{x}-\boldsymbol{x}'|}\,, \tag{8.29c}$$

$$p^{\mathrm{T}}_{\alpha\beta}(t,\boldsymbol{x}) = -\frac{4}{a\left[\eta(t,\boldsymbol{x})\right]}\int\limits_{\mathcal{V}}\frac{a\left[\eta(s,\boldsymbol{x}')\right]\tau^{\mathrm{T}}_{\alpha\beta}(s,\boldsymbol{x}')\,d^3x'}{|\boldsymbol{x}-\boldsymbol{x}'|}\,, \tag{8.29d}$$

where the scale factor a in front of the integrals depends on the coordinates of the field point $a \equiv a\left[\eta(t,\boldsymbol{x})\right]$, and the functions in the integrand depend on the retarded time

$$s = t - |\boldsymbol{x}-\boldsymbol{x}'|\,, \tag{8.30}$$

because gravity propagates with finite speed. Equation (8.30) describes characteristic of the null cone in the conformal Minkowski spacetime that determines the causal nature of the gravitational field in the expanding universe with $k = 0$. Solutions (8.29a)–(8.29d) are Lorentz invariant as shown by calculations in Section 8.4.

Integration in (8.29a)–(8.29d) is performed over the finite volume, $\mathcal{V}$, occupied by the matter of the localized astronomical system. In case of the system comprised of N massive bodies which are separated by distances being much larger than their characteristic size, the matter occupies the volumes of the bodies. In this case the integration in equations (8.29a)–(8.29d) is practically performed over the volumes of the bodies. It means that each post-Newtonian potential q, p, p_α, $p^{\mathrm{T}}_{\alpha\beta}$ is split in the algebraic sum of N pieces

$$q = \sum_{A=1}^{N} q_A\,, \quad p = \sum_{A=1}^{N} p_A\,, \quad p_\alpha = \sum_{A=1}^{N} p_{A\alpha}\,, \quad p^{\mathrm{T}}_{\alpha\beta} = \sum_{A=1}^{N} p^{\mathrm{T}}_{A\alpha\beta}\,, \tag{8.31}$$

where each function with subindex A has the same form as one of the corresponding equations (8.29a)–(8.29d) with the integration performed over the volume, $\mathcal{V}_A$, of the body A. This confirms the principle of superposition in the linearized Friedmann approximation.

The gauge functions

The residual gauge freedom describes arbitrariness in adding a solution of homogeneous wave equations (8.29a)–(8.29d). It is described by two functions, $\zeta \equiv \xi/a$ and ζ^α as discussed in Section 7.4. Since we neglected the terms being quadratic with respect to the Hubble parameter, equations (7.44a) and (7.44b) get simpler, and read

$$\Box\zeta - 2\mathcal{H}\bar{\mathrm{v}}^\beta\zeta_{,\beta} = 0\,, \tag{8.32a}$$

$$\Box\zeta^\alpha - 2\mathcal{H}\bar{\mathrm{v}}^\beta\zeta^\alpha{}_{;\beta} = 0\,. \tag{8.32b}$$

They are equivalent to the homogeneous wave equations in the conformal flat spacetime

$$\square\,(a\zeta) = 0\,, \quad \square\,(a\zeta^{\alpha}) = 0\,, \tag{8.33}$$

which point out that (in the approximation under consideration) the products, $a\zeta$ and $a\zeta^{\alpha}$, are the harmonic functions.

Potentials q, p, p_{α}, $p^{\scriptscriptstyle\mathsf{T}}_{\alpha\beta}$ must satisfy the gauge conditions (7.43a) and (7.43b). Neglecting terms being quadratic with respect to the Hubble parameter, the gauge conditions (7.43a) and (7.43b) can be written down as follows:

$$(ap^{\alpha})_{,\alpha} + \overline{v}^{\alpha}\left(2aq - ap\right)_{,\alpha} + \mathcal{H}ap = 0\,, \tag{8.34a}$$

$$\left(ap^{\scriptscriptstyle\mathsf{T}\alpha\beta}\right)_{,\beta} + \overline{v}^{\beta}\left(ap^{\alpha}\right)_{,\beta} + \frac{1}{3}\overline{\pi}^{\alpha\beta}(ap)_{,\beta} + \mathcal{H}ap^{\alpha} = 0\,, \tag{8.34b}$$

where we have taken into account $a_{,\alpha} = -a\mathcal{H}\overline{v}_{\alpha}$, and $p^{\alpha}\overline{v}_{\alpha} = 0$, $p^{\scriptscriptstyle\mathsf{T}\alpha\beta}\overline{v}_{\beta} = 0$. The potentials p^{α} and $p^{\scriptscriptstyle\mathsf{T}\alpha\beta}$ are obtained from p_{α} and $p^{\scriptscriptstyle\mathsf{T}}_{\alpha\beta}$ by rising the indices with the Minkowski metric and taking into account that the indices of τ_{α} and $\tau^{\scriptscriptstyle\mathsf{T}}_{\alpha\beta}$ in the integrands of (8.29c) and (8.29d) should be raised with the full background metric $\overline{g}^{\alpha\beta} = a^{-2}\eta^{\alpha\beta}$ taken at the point of integration. This is because by convention having been adopted in Section 4.8, the notations $\tau^{\alpha} \equiv \overline{g}^{\alpha\beta}\tau_{\beta}$ and $\tau^{\alpha\beta} \equiv \overline{g}^{\alpha\mu}\overline{g}^{\beta\nu}\tau_{\mu\nu}$. It yields

$$p^{\alpha}(t,\boldsymbol{x}) = -\frac{4}{a\left[\eta(t,\boldsymbol{x})\right]}\int\limits_{\mathcal{V}}\frac{a^{4}\left[\eta(s,\boldsymbol{x}')\right]\tau^{\alpha}\left(s,\boldsymbol{x}'\right)d^{3}x'}{|\boldsymbol{x}-\boldsymbol{x}'|}\,, \tag{8.35a}$$

$$p^{\scriptscriptstyle\mathsf{T}\alpha\beta}(t,\boldsymbol{x}) = -\frac{4}{a\left[\eta(t,\boldsymbol{x})\right]}\int\limits_{\mathcal{V}}\frac{a^{5}\left[\eta(s,\boldsymbol{x}')\right]\tau^{\scriptscriptstyle\mathsf{T}\alpha\beta}\left(s,\boldsymbol{x}'\right)d^{3}x'}{|\boldsymbol{x}-\boldsymbol{x}'|}\,. \tag{8.35b}$$

It is instrumental to write down solutions for the products of the potentials p and $p^{\alpha} = \eta^{\alpha\beta}p_{\beta}$ with the Hubble parameter. Multiplying both sides of equations (8.26b) and (8.26c) with the Hubble parameter $\mathcal{H}$, and neglecting the quadratic with respect to $\mathcal{H}$ terms, we obtain

$$\square(a\mathcal{H}p) = 16\pi a^{3}\mathcal{H}\tau\,, \quad \square(a\mathcal{H}p^{\alpha}) = 16\pi a^{4}\mathcal{H}\tau^{\alpha}\,, \tag{8.36}$$

which solutions are the retarded potential

$$a\mathcal{H}p(t,\boldsymbol{x}) = -4\int\limits_{\mathcal{V}}\frac{a^{3}\left[\eta(s,\boldsymbol{x}')\right]\mathcal{H}\left[\eta(s,\boldsymbol{x}')\right]\tau\left(s,\boldsymbol{x}'\right)d^{3}x'}{|\boldsymbol{x}-\boldsymbol{x}'|}\,, \tag{8.37a}$$

$$a\mathcal{H}p^{\alpha}(t,\boldsymbol{x}) = -4\int\limits_{\mathcal{V}}\frac{a^{4}\left[\eta(s,\boldsymbol{x}')\right]\mathcal{H}\left[\eta(s,\boldsymbol{x}')\right]\tau^{\alpha}\left(s,\boldsymbol{x}'\right)d^{3}x'}{|\boldsymbol{x}-\boldsymbol{x}'|}\,. \tag{8.37b}$$

Substituting functions q, p, p^α, $p^{\mathrm{T}\alpha\beta}$ and $a\mathcal{H}p$, $a\mathcal{H}p^\alpha$ to the gauge equations (8.34a) and (8.34b), bring about the following integral equations:

$$\int\limits_{\mathcal{V}} \left[\left(a^4\tau^\alpha + \bar{v}^\alpha a^3\sigma\right)_{,\alpha} + a^3\mathcal{H}\tau\right] \frac{d^3x'}{|\boldsymbol{x}-\boldsymbol{x}'|} = 0\,, \tag{8.38a}$$

$$\int\limits_{\mathcal{V}} \left[\left(a^5\tau^{\mathrm{T}\alpha\beta} + a^4\bar{v}^\beta\tau^\alpha + \frac{1}{3}\bar{\pi}^{\alpha\beta}a^3\tau\right)_{,\beta} + a^4\mathcal{H}\tau^\alpha\right] \frac{d^3x'}{|\boldsymbol{x}-\boldsymbol{x}'|} = 0\,, \tag{8.38b}$$

where all functions in the integrands are taken at the retarded time s and at the point $\boldsymbol{x}'$, for example, $a \equiv a[\eta(s,\boldsymbol{x}')]$, $\mathcal{H} \equiv \mathcal{H}[\eta(s,\boldsymbol{x}')]$, $\sigma \equiv \sigma[(s,\boldsymbol{x}')]$, and so on. These equations are satisfied by the equations of motion (4.71a) and (4.71b) of the localized matter distribution. Indeed, divergences of any vector F^α and a symmetric tensor $F^{\alpha\beta}$ obey the following equalities:

$$F^\alpha{}_{|\alpha} = \frac{1}{\sqrt{-\bar{g}}}\left(\sqrt{-\bar{g}}F^\alpha\right)_{,\alpha}\,, \tag{8.39}$$

$$F^{\alpha\beta}{}_{|\beta} = \frac{1}{\sqrt{-\bar{g}}}\left(\sqrt{-\bar{g}}F^{\alpha\beta}\right)_{,\beta} + \bar{\Gamma}^\alpha_{\beta\gamma}F^{\beta\gamma}\,. \tag{8.40}$$

Moreover, the root square of the determinant of the background metric tensor is expressed in terms of the scale factor, $\sqrt{-\bar{g}} = a^4$, while the 4-velocity $\bar{u}^\alpha = \bar{v}^\alpha/a$. Employing these expressions along with equations (8.39) and (8.40) in equations of motion (4.71a) and (4.71b), transform them to

$$\left(a^4\tau^\alpha + \bar{v}^\alpha a^3\sigma\right)_{,\alpha} + a^3\mathcal{H}\tau = 0\,, \tag{8.41a}$$

$$\left(a^4\tau^{\alpha\beta} + a^3\bar{v}^\beta\tau^\alpha\right)_{,\beta} + 2a^3\mathcal{H}\tau^\alpha = 0\,. \tag{8.41b}$$

Equation (8.41a) proves that the integral equation (8.38a) is valid. In order to prove the second integral equation (8.38b), we multiply equation (8.41b) with the scale factor a, and reshuffle its terms. It brings (8.41b) to the following form:

$$\left(a^5\tau^{\alpha\beta} + a^4\bar{v}^\beta\tau^\alpha\right)_{,\beta} + a^4\mathcal{H}\tau^\alpha = 0\,. \tag{8.42}$$

Substituting, $\tau^{\alpha\beta} = \tau^{\mathrm{T}\alpha\beta} + (1/3a^2)\bar{\pi}^{\alpha\beta}\tau$, to (8.42) and comparing with the integrand in (8.38b) makes it clear that (8.38b) is valid. We conclude that the retarded integrals (8.29a)–(8.29d) yield the complete solution of the linearised wave equations (8.26a)–(8.26d) in the sense that there is no residual gauge freedom since the gauge functions $\zeta = \zeta^\alpha = 0$.

Perturbations of dark matter and dark energy

What remains is to find out solutions for the scalar functions V_m and V_q and χ_m and χ_q. In the linearized Friedmann approximation equation for V_m is obtained from (7.37) by discarding all terms of the order of $\mathcal{H}^2$. It yields

$$\Box V_{\rm m} + \left(1 - \frac{c^2}{v_{\rm s}^2}\right)\overline{\rm v}^{\alpha}\overline{\rm v}^{\beta} V_{{\rm m},\alpha\beta} + \left(3 - \frac{c^2}{v_{\rm s}^2}\right)\mathcal{H}\overline{\rm v}^{\alpha} V_{{\rm m},\alpha} = -4\pi a^2 (\sigma + \tau) \,. \tag{8.43}$$

Applying relations (7.34) and (7.35) in equation (8.43) allows us to recast it to

$$\frac{1}{a^n}\left[\Box\left(a^n V_{\rm m}\right) + \left(1 - \frac{c^2}{v_{\rm s}^2}\right)\overline{\rm v}^{\alpha}\overline{\rm v}^{\beta}\left(a^n V_{\rm m}\right)_{,\alpha\beta}\right] \tag{8.44}$$
$$+\left[3 + (2n-1)\frac{c^2}{v_{\rm s}^2}\right]\mathcal{H}\overline{\rm v}^{\alpha} V_{{\rm m},\alpha} = -4\pi a^2 (\sigma + \tau) \,,$$

where n is yet undetermined real number. Now, we postulate that the speed of sound $v_{\rm s}$ is constant. Then, choosing, $n \equiv n_{\rm s}$, with

$$n_{\rm s} = \frac{1}{2}\left(1 - 3\frac{v_{\rm s}^2}{c^2}\right) \,, \tag{8.45}$$

annihilates the term being proportional to $\mathcal{H}$ on the left-hand side of (8.44) and reduces it to

$$\Box\left(a^{n_{\rm s}} V_{\rm m}\right) + \left(1 - \frac{c^2}{v_{\rm s}^2}\right)\overline{\rm v}^{\alpha}\overline{\rm v}^{\beta}\left(a^{n_{\rm s}} V_{\rm m}\right)_{,\alpha\beta} = -4\pi a^{2+n_{\rm s}} (\sigma + \tau) \,. \tag{8.46}$$

This equation describes propagation of perturbation $V_{\rm m}$ with the speed of sound $v_{\rm s}$. Indeed, let us introduce the sound-wave Laplace–Beltrami operator (8.17). Then, equation (8.46) reads

$$\Box_{\rm s}\left(a^{n_{\rm s}} V_{\rm m}\right) = -4\pi a^{2+n_{\rm s}} (\sigma + \tau) \,. \tag{8.47}$$

This equation has a well-defined Green function with characteristics propagating with the speed of sound $v_{\rm s}$. We discard the advanced Green function because we assume that at infinity the function $V_{\rm m}$ and its first derivatives vanish. Solution of (8.47) is explained below in Section 8.5, and has the following form:

$$V_{\rm m}(t, \boldsymbol{x}) = \frac{1}{a^{n_{\rm s}}(t, \boldsymbol{x})}\int\limits_{\mathcal{V}} \frac{a^{2+n_{\rm s}}(\varsigma, \boldsymbol{x}')\left[\sigma(\varsigma, \boldsymbol{x}') + \tau(\varsigma, \boldsymbol{x}')\right]}{\sqrt{1 + \gamma^2\left(1 - \frac{c^2}{v_{\rm s}^2}\right)(\boldsymbol{\beta} \times \boldsymbol{n})^2}}\frac{d^3x'}{|\boldsymbol{x} - \boldsymbol{x}'|} \,, \tag{8.48}$$

where the retarded time ς is given by equation (8.96), $\boldsymbol{\beta} = \beta^i = \overline{\rm v}^i/c$, $\gamma = 1/\sqrt{1 - \boldsymbol{\beta}^2}$ is the Lorentz factor, and the unit vector $\boldsymbol{n} = (\boldsymbol{x} - \boldsymbol{x}')/|\boldsymbol{x} - \boldsymbol{x}'|$. The retardation in the solution (8.48) is due to the finite speed of propagation of acoustic (sound) waves in the ideal fluid that represents the dark matter.

Equation for $V_{\rm q}$ is obtained in the linearized Friedmann approximation from (7.38) after discarding all terms being proportional to $\mathcal{H}^2$. It yields

$$\Box V_{\rm q} + 2\left(1 - \sqrt{\frac{3}{2{\rm x}_1}}\lambda {\rm x}_2\right)\mathcal{H}\overline{\rm v}^{\mu} V_{{\rm q},\mu} = -4\pi a^2 (\sigma + \tau) \,. \tag{8.49}$$

Applying relation (7.34) in (8.49) allows us to recast it to

$$\frac{1}{a^n}\Box\left(a^n V_q\right) + 2\left(n + 1 - \sqrt{\frac{3}{2x_1}}\lambda x_2\right)\mathcal{H}\overline{v}^\alpha V_{m,\alpha} = -4\pi a^2\left(\sigma + \tau\right) . \tag{8.50}$$

If, and only if, the ratio $\lambda x_2/\sqrt{x_1}$ is constant, we can choose, $n \equiv n_q = -1 + \sqrt{3/(2x_1)}\lambda x_2$, in order to eradicate the second term on the left-hand side of (8.50). In those models of the universe where this condition is satisfied, the resulting equation for V_q is simplified and reads

$$\Box\left(a^{n_q} V_q\right) = -4\pi a^{2+n_q}\left(\sigma + \tau\right) . \tag{8.51}$$

This is the wave equation in flat spacetime. We pick up the retarded solution as the most physical one,

$$V_q = \frac{1}{a^{n_q}(t, \boldsymbol{x})}\int\limits_{\mathcal{V}} \frac{a^{2+n_q}(s, \boldsymbol{x}')\left[\sigma(s, \boldsymbol{x}') + \tau(s, \boldsymbol{x}')\right] d^3x'}{|\boldsymbol{x} - \boldsymbol{x}'|} , \tag{8.52}$$

where the retarded time s has been defined in (8.30).

Perturbations χ_m and χ_q can be found by integrating equations (6.13a) and (6.13b) that can be written as

$$\overline{v}^\alpha \chi_{m,\alpha} = a\left(V_m + \frac{q}{2}\right) , \quad \overline{v}^\alpha \chi_{q,\alpha} = a\left(V_q + \frac{q}{2}\right) . \tag{8.53}$$

These are the ordinary differential equations of the first order. Their solutions are

$$\chi_m = \int\limits_{t_0}^{t} a[t, \boldsymbol{x}(t)]\{V_m[t, \boldsymbol{x}(t)] + \frac{1}{2}q[t, \boldsymbol{x}(t)]\}dt , \tag{8.54a}$$

$$\chi_q = \int\limits_{t_0}^{t} a[t, \boldsymbol{x}(t)]\{V_q[t, \boldsymbol{x}(t)] + \frac{1}{2}q[t, \boldsymbol{x}(t)]\}dt , \tag{8.54b}$$

where t_0 is an initial epoch of integration, and the integration is performed along the Hubble flow of the background universe

$$\frac{dx^i}{dt} = \overline{v}^i(t, \boldsymbol{x}) . \tag{8.55}$$

Therefore, the most simple way to integrate equations (8.53) would be to work in the preferred coordinate frame $X^\alpha = (\eta, X^i)$ where the velocity $\overline{v}^i = 0$, and the spatial coordinates $X^i = \text{const}$. After the calculation in the rest frame of the Hubble flow is finished, the transformation to a moving frame of observer can be done with the help of the coordinate transformation between the two frames.

8.4 Lorentz invariance of retarded potentials

We use a prime in the appendices exclusively as a notation for time and spatial coordinates which are used as variables of integration in volume integrals (see, for example, equations (8.57) and (8.58), and so on). It should not be confused with the time derivative with respect to the conformal time used in the main text of this chapter.

Let us consider an inhomogeneous wave equation for a scalar field, $V = V(\eta, \boldsymbol{X})$, written down in a coordinate chart $X^\alpha = (X^0, X^i) = (\eta, \boldsymbol{X})$,

$$\Box V = -4\pi\sigma_X \,, \tag{8.56}$$

where $\Box \equiv \eta^{\alpha\beta}\partial_{\alpha\beta}$, $\partial_\alpha = \partial/\partial X^\alpha$, and $\sigma_X = \sigma_X(\eta, \boldsymbol{X})$ is the source (a scalar function) of the field V with a compact support (bounded by a finite volume in space). Equation (8.56) has a solution given as a linear combination of advanced and retarded potentials. Let us focus only on the retarded potential which is more common in physical applications. Advanced potential can be treated similarly.

We assume the field, V, and its first derivatives vanish at past null infinity. Then, the retarded solution (retarded potential) of (8.56) is given by an integral

$$V(\eta, \boldsymbol{X}) = \int\limits_{\mathcal{V}} \frac{\sigma_X(\zeta, \boldsymbol{X}')\, d^3X'}{|\boldsymbol{X} - \boldsymbol{X}'|} \,, \tag{8.57}$$

where

$$\zeta = \eta - |\boldsymbol{X} - \boldsymbol{X}'| \,, \tag{8.58}$$

is the retarded time, and we assume the fundamental speed $c = 1$. Physical meaning of the retardation is that the field V propagates in spacetime with the fundamental speed c from the source σ_X, to the point with coordinates $X^\alpha = (\eta, \boldsymbol{X})$ where the field V is measured in correspondence with equation (8.57). Left side of equation (8.56) is Lorentz invariant. Hence, we expect that solution (8.58) must be Lorentz invariant as well. As a rule, textbooks prove this statement for a particular case of the retarded (Liénard–Wiechert) potential of a moving point-like source but not for the retarded potential given in the form of the integral (8.57). This appendix fulfils this gap.

Lorentz transformation to coordinates, $x^\alpha = (t, \boldsymbol{x})$ linearly transforms the isotropic coordinates $X^\alpha = (\eta, \boldsymbol{X})$ of the FLRW metric as follows

$$x^\alpha = \Lambda^\alpha{}_\beta X^\beta \,, \tag{8.59}$$

where the matrix of the Lorentz boost [89]

$$\Lambda^0{}_0 = \gamma \,, \quad \Lambda^i{}_0 = \Lambda^0{}_i = -\gamma\beta^i \,, \quad \Lambda^i{}_j = \delta^{ij} + \frac{\gamma - 1}{\beta^2}\beta^i\beta^j \,, \tag{8.60}$$

the boost 4-velocity $u^\alpha = \{u^0, u^i\} = u^0\{1, \beta^i\}$ is constant, and

$$\gamma = u^0 = \frac{1}{\sqrt{1 - \beta^2}} \,, \tag{8.61}$$

is the constant Lorentz factor.

The inverse Lorentz transformation is given explicitly as follows:

$$\eta = \gamma(t + \boldsymbol{\beta}\cdot\boldsymbol{x}) , \tag{8.62}$$

$$\boldsymbol{X} = \boldsymbol{r} + \frac{\gamma^2}{1+\gamma}(\boldsymbol{\beta}\cdot\boldsymbol{r})\boldsymbol{b} , \tag{8.63}$$

where

$$\boldsymbol{r} = \boldsymbol{x} + \boldsymbol{\beta}t , \tag{8.64}$$

and the boost three velocity, $\boldsymbol{\beta} = \{\beta^i\} = \{u^i/u^0\}$.

Let us reiterate (8.57) by introducing a one-dimensional Dirac's delta function and integration with respect to time η,

$$V(\eta, \boldsymbol{X}) = \int_{-\infty}^{\infty}\int_{\mathcal{V}} \frac{\sigma_X(\eta', \boldsymbol{X}')\delta(\eta' - \zeta)\, d\eta' d^3X'}{|\boldsymbol{X} - \boldsymbol{X}'|} , \tag{8.65}$$

where ζ is the retarded time given by (8.58). Then, we transform coordinates $X'^{\alpha} = (\eta', \boldsymbol{X}')$ to $x'^{\alpha} = (t', \boldsymbol{x}')$ with the Lorentz boost (8.59). The Lorentz transformation changes functions entering the integrand of (8.65) as follows:

$$\sigma(\eta', \boldsymbol{X}') = \sigma_x(t', \boldsymbol{x}') , \tag{8.66}$$

$$|\boldsymbol{X} - \boldsymbol{X}'| = \sqrt{|\boldsymbol{r} - \boldsymbol{r}'|^2 + \gamma^2[\boldsymbol{\beta}\cdot(\boldsymbol{r} - \boldsymbol{r}')]^2} , \tag{8.67}$$

where the coordinate difference

$$\boldsymbol{r} - \boldsymbol{r}' = \boldsymbol{x} - \boldsymbol{x}' + \boldsymbol{\beta}(t - t') . \tag{8.68}$$

The coordinate volume of integration remains Lorentz invariant

$$d\eta' d^3X' = dt' d^3x' . \tag{8.69}$$

Let us denote $F_\eta(\eta') \equiv \eta' - \zeta$, where ζ is given by (8.58). After making the Lorentz transformation this function changes to

$$\begin{aligned} F_\eta(\eta') = F_t(t') = \gamma\left[t' - t - \boldsymbol{\beta}\cdot\left(\boldsymbol{x} - \boldsymbol{x}'\right)\right] \\ + \sqrt{|\boldsymbol{x} - \boldsymbol{x}'|^2 - (t' - t)^2 + \gamma^2[\boldsymbol{\beta}\cdot(\boldsymbol{x} - \boldsymbol{x}') - (t' - t)]^2} , \end{aligned} \tag{8.70}$$

where we have used equations (8.62), (8.63), and (8.67) and relationship $\gamma^2\beta^2 = \gamma^2 - 1$, to perform the transformation. Integral (8.65) in coordinates x^α becomes

$$V(t, \boldsymbol{x}) = \int_{-\infty}^{\infty}\int_{\mathcal{V}} \frac{\sigma_x(t', \boldsymbol{x}')\delta(F_t(t'))\, dt' d^3x'}{\sqrt{|\boldsymbol{r} - \boldsymbol{r}'|^2 + \gamma^2[\boldsymbol{\beta}\cdot(\boldsymbol{r} - \boldsymbol{r}')]^2}} , \tag{8.71}$$

The delta function has a complicated argument $F_t(t')$ in coordinates x^α. It can be simplified with a well-known formula

$$\delta\left[F_t(t')\right] = \frac{\delta(t'-s)}{\dot{F}_t(s)}\,, \tag{8.72}$$

where $\dot{F}_t(s) \equiv \left[dF_t(t')/dt'\right]_{t'=s}$, and s is one of the roots of equation $F_t(t') = 0$ that is associated with the retarded interaction. It is straightforward to confirm by inspection that the root is given by formula

$$s = t - |\boldsymbol{x} - \boldsymbol{x}'|\,. \tag{8.73}$$

The time derivative of function $F_t(t')$ is

$$\dot{F}_t(t') = \gamma + \gamma^2 \frac{\beta^2(t'-t) - \boldsymbol{\beta}\cdot(\boldsymbol{x}-\boldsymbol{x}')}{\sqrt{|\boldsymbol{x}-\boldsymbol{x}'|^2 - (t'-t)^2 + \gamma^2[\boldsymbol{\beta}\cdot(\boldsymbol{x}-\boldsymbol{x}') - (t'-t)]^2}}\,. \tag{8.74}$$

After substituting $t' = s$, with s taken from equation (8.73), we obtain

$$\dot{F}_t(s) = \frac{1}{\gamma}\frac{|\boldsymbol{x}-\boldsymbol{x}'|}{|\boldsymbol{x}-\boldsymbol{x}'| + \boldsymbol{\beta}\cdot(\boldsymbol{x}-\boldsymbol{x}')}\,. \tag{8.75}$$

Performing now integration with respect to t' in equation (8.71) with the help of the delta function, we arrive to

$$V(t,\boldsymbol{x}) = \int\limits_{\mathcal{V}} \frac{\sigma_x(s,\boldsymbol{x}')\,d^3x'}{\dot{F}_t(s)|\boldsymbol{X}-\boldsymbol{X}'|_{t'=s}}\,, \tag{8.76}$$

where $|\boldsymbol{X} - \boldsymbol{X}'|_{t'=s}$ must be calculated from (8.67) with $t' = s$ where s is taken from (8.73). It yields

$$\dot{F}_t(s)|\boldsymbol{X}-\boldsymbol{X}'|_{t'=s} = |\boldsymbol{x}-\boldsymbol{x}'|\,, \tag{8.77}$$

and proves that the retarded potential (8.57) is Lorentz invariant

$$\int\limits_{\mathcal{V}} \frac{\sigma_X(\zeta,\boldsymbol{X}')\,d^3X'}{|\boldsymbol{X}-\boldsymbol{X}'|} = \int\limits_{\mathcal{V}} \frac{\sigma_x(s,\boldsymbol{x}')\,d^3x'}{|\boldsymbol{x}-\boldsymbol{x}'|}\,. \tag{8.78}$$

We have verified the Lorentz invariance for the *scalar* retarded potential. However, it is not difficult to check that it is valid in case of a source $\sigma_{\alpha_1\alpha_2\ldots\alpha_l}$ that is a tensor field of rank l. Indeed, the Lorentz transformation of the source leads to $\Lambda^{\beta_1}{}_{\alpha_1}\Lambda^{\beta_2}{}_{\alpha_2}\ldots\Lambda^{\beta_l}{}_{\alpha_l}\sigma_{\beta_1\beta_2\ldots\beta_l}$ but the matrix $\Lambda^\alpha{}_\beta$ is constant, and can be taken out of the sign of the retarded integral. Because of this property, all mathematical operations given in this appendix for a scalar retarded potential, remain the same for the tensor of any rank. Hence, the Lorentz invariance of the retarded integral is a general property of the wave operator in Minkowski spacetime.

8.5 Retarded solution of the sound-wave equation

Let us consider an inhomogeneous sound-wave equation for a scalar function $\mathcal{U} = \mathcal{U}(\eta, \boldsymbol{X})$ describing a perturbation propagation in a medium. This equation written down in the isotropic coordinates $X^\alpha = (\eta, \boldsymbol{X})$, reads

$$\square_s \mathcal{U} = -4\pi\tau_X \,, \tag{8.79}$$

where $\tau_X = \tau_X(\eta, \boldsymbol{X})$ is the source of $\mathcal{U}$ having a compact support, and the sound-wave differential operator $\square_s$ was defined in (8.17). It is Lorentz invariant and reads

$$\square_s = \square + \left(1 - \frac{c^2}{c_s^2}\right)\overline{\mathrm{v}}^\alpha \overline{\mathrm{v}}^\beta \partial_{\alpha\beta} \,, \tag{8.80}$$

where $\overline{\mathrm{v}}^\alpha$ is 4-velocity of motion of the medium with respect to the coordinate chart, v_s is the constant speed of sound in the medium, and we keep the fundamental speed c in the definition of the operator for dimensional purposes. We assume that $v_s < c$. The case of $v_s = c$ is treated in Section 8.4, and the case of $v_s \geq c$ makes a formal mathematical sense in discussion of the speed of propagation of gravity in alternative theories of gravity since the equation describing propagation of gravitational potential $\mathcal{U}$ has the same structure as (8.79) after formal replacement of v_s with the speed of gravity c_g [68, 118]. In particular, in the Newtonian theory the speed of gravity $c_g = \infty$, and the operator (8.80) is reduced to the Laplace operator

$$\Delta = \square + \mathrm{v}^\alpha \mathrm{v}^\beta \partial_\alpha \partial_\beta = \overline{\pi}^{\alpha\beta} \partial_{\alpha\beta} \,, \tag{8.81}$$

where the constant projection operator, $\overline{\pi}^{\alpha\beta}$, has been defined in (7.23).

We are looking for the solution of (8.79) in the Cartesian coordinates $x^\alpha = (t, \boldsymbol{x})$ moving with respect to the isotropic coordinates X^α with constant velocity β^i. Transformation from X^α to x^α is given by the Lorentz transformation (8.59). In coordinates X^α the 4-velocity $\overline{\mathrm{v}}^\alpha = (1, 0, 0, 0)$. Therefore, in these coordinates, equation (8.79) is just a wave equation for the field $\mathcal{U}$ propagating with speed v_s. It has a well-known retarded solution,

$$\mathcal{U}(\eta, \boldsymbol{X}) = \int\limits_{\mathcal{V}} \frac{\tau_X(\eta_s, \boldsymbol{X}')\, d^3X'}{|\boldsymbol{X} - \boldsymbol{X}'|} \,, \tag{8.82}$$

where

$$\eta_s = \eta - \frac{c}{v_s}|\boldsymbol{X} - \boldsymbol{X}'| \,, \tag{8.83}$$

is the retarded time.

Equation (8.79) is Lorentz invariant. Hence, its solution must be Lorentz invariant as well. Our goal is to prove this statement. To this end, we take solution (8.82) and perform the Lorentz transformation (8.62) and (8.63). We recast the retarded integral

(8.82) to another form with the help of one-dimensional delta function

$$\mathcal{U}(\eta, \mathbf{X}) = \int\limits_{\infty}^{\infty}\int\limits_{\mathcal{V}} \frac{\tau_X(\eta', \mathbf{X}')\delta(\eta' - \eta_s)\, d\eta' d^3X'}{|\mathbf{X} - \mathbf{X}'|} . \tag{8.84}$$

It looks similar to (8.57) but one has to remember that the retarded time η_s differs from ζ that was defined in (8.58) on the characteristics of the null cone defined by the fundamental speed c. Transformation of functions entering integrand in (8.84) is similar to what we did in Section 8.4 but, because $v_s \neq c$, calculations become more involved. It turns out more preferable to handle the calculations in tensor notations, making transition to the coordinate language only at the end of the transformation procedure.

Let us consider two events with the isotropic coordinates $X^\alpha = (\eta, \mathbf{X})$ and $X'^\alpha = (\eta', \mathbf{X}')$. We postulate that in the coordinate chart, x^α, these two events have coordinates, $x^\alpha = (t, \boldsymbol{x})$, and, $x'^\alpha = (t', \boldsymbol{x}')$, respectively. We define the components of a four-vector, $r^\alpha = (t' - t, \boldsymbol{x} - \boldsymbol{x}')$ which is convenient for doing mathematical manipulations with the Lorentz transformations. For instance, the Lorentz transformation of the Euclidean distance between the spatial coordinates of the two events, is given by a

$$|\mathbf{X} - \mathbf{X}'| = \sqrt{\overline{\pi}_{\alpha\beta} r^\alpha r^\beta} , \tag{8.85}$$

where $\overline{\pi}^{\alpha\beta}$ is the operator of projection on the hyperplane being orthogonal to $\overline{\mathrm{v}}^\alpha$ (the same operator as in (8.81)). Equation (8.85) is a Lorentz-invariant analog of expression (8.67) and matches it exactly. Transformation of the source, $\tau_X(X^\alpha) = \tau_x(x^\alpha)$ is fully equivalent to that of σ_X as given by equation (8.66). Coordinate volume of integration transforms in accordance with (8.69). We need to transform the argument, $\eta' - \eta_s$, of delta function which we shall denote in coordinates X^α as $f_\eta(\eta') \equiv \eta' - \eta_s$. The argument is a scalar function which is transformed as $f_\eta(\eta') = f_t(t')$ where,

$$f_t(t') = -\overline{\mathrm{v}}_\alpha r^\alpha + \frac{c}{v_s}\sqrt{\overline{\pi}_{\alpha\beta} r^\alpha r^\beta} . \tag{8.86}$$

Transformation of the delta function in the integrand of integral (8.84) is

$$\delta\left[f_t(t')\right] = \frac{\delta(t' - \varsigma)}{\dot{f}_t(\varsigma)} , \tag{8.87}$$

where $\dot{f}_t(\varsigma) \equiv \left[df_t(t')/dt'\right]_{t'=\varsigma}$, and ς is one of the roots of equation $f_t(t') = 0$ that is associated with the retarded interaction. Eventually, after accounting for transformation of all functions and performing integration with respect to time, integral (8.84) assumes the following form:

$$\mathcal{U}(t, \boldsymbol{x}) = \int\limits_{\mathcal{V}} \frac{\tau_x(\varsigma, \boldsymbol{x}')d^3x'}{\dot{f}_t(\varsigma)|\mathbf{X} - \mathbf{X}'|_{t'=\varsigma}} , \tag{8.88}$$

where $|\boldsymbol{X}-\boldsymbol{X}'|_{t'=\varsigma}$ denotes the expression (8.85) taken at the value of $t' = \varsigma$. What remains is to calculate the instant of time, ς, and the value of functions entering denominator of the integrand in (8.88).

Calculation of ς is performed by solving equation $f_t(\varsigma) = 0$, that defines the characteristic cone of the sound waves, and has the following explicit form:

$$\left[\eta_{\alpha\beta} + \left(1 - \frac{v_s^2}{c^2}\right)\overline{v}_\alpha \overline{v}_\beta\right] r^\alpha r^\beta = 0\,, \tag{8.89}$$

which is derived from (8.86). This is a quadratic algebraic equation with respect to the time variable $r^0 = \varsigma - t$. It reads

$$A(\varsigma - t)^2 + 2B(\varsigma - t) + C = 0\,, \tag{8.90}$$

where the coefficients A, B, C of the quadratic form are

$$A = -1 + \left(1 - \frac{v_s^2}{c^2}\right)\gamma^2\,, \tag{8.91}$$

$$B = -\left(1 - \frac{v_s^2}{c^2}\right)\gamma^2 \boldsymbol{\beta}\cdot(\boldsymbol{x}-\boldsymbol{x}')\,, \tag{8.92}$$

$$C = |\boldsymbol{x}-\boldsymbol{x}'|^2 + \left(1 - \frac{v_s^2}{c^2}\right)\gamma^2 \left[\boldsymbol{\beta}\cdot(\boldsymbol{x}-\boldsymbol{x}')\right]^2\,, \tag{8.93}$$

and $\gamma = 1/\sqrt{1-\beta^2}$ is the Lorentz factor. Equation (8.90) has two roots corresponding to the advanced and retarded times. The root corresponding to the retarded-time solution of (8.90) is

$$\varsigma = t - \frac{1}{A}\left(B - \sqrt{B^2 - AC}\right)\,, \tag{8.94}$$

or, more explicitly,

$$\varsigma = t - |\boldsymbol{x}-\boldsymbol{x}'| \frac{\left(1 - \frac{v_s^2}{c^2}\right)\gamma^2(\boldsymbol{\beta}\cdot\boldsymbol{n}) + \sqrt{1 - \left(1 - \frac{v_s^2}{c^2}\right)\gamma^2\left[1 - (\boldsymbol{\beta}\cdot\boldsymbol{n})^2\right]}}{1 - \left(1 - \frac{v_s^2}{c^2}\right)\gamma^2}\,, \tag{8.95}$$

where the unit vector $\boldsymbol{n} = (\boldsymbol{x}-\boldsymbol{x}')/|\boldsymbol{x}-\boldsymbol{x}'|$. After some algebra equation (8.95) can be simplified to

$$\varsigma = t - \frac{\alpha_s}{v_s}|\boldsymbol{x}-\boldsymbol{x}'|\,, \tag{8.96}$$

where

$$\alpha_s = \frac{1-\beta^2}{1 - \frac{\beta^2}{v_s^2}}\left[\sqrt{1 + \left(1 - \frac{c^2}{v_s^2}\right)\gamma^2(\boldsymbol{\beta}\times\boldsymbol{n})^2} - \left(1 - \frac{c^2}{v_s^2}\right)\gamma^2(\boldsymbol{\beta}\cdot\boldsymbol{n})\right]. \tag{8.97}$$

Coefficient α_s defines the speed of propagation of the sound waves, $v_s \equiv v_s/\alpha_s$, as measured by observer moving with speed β^i with respect to the Hubble flow. Thus, the value of the speed of sound, v_s, depends crucially on the motion of observer.

Derivative of the function, $\dot{f}_t(\varsigma)$, is given by

$$\dot{f}_t(\varsigma) = \frac{\partial f_t}{\partial r^\alpha}\frac{\partial r^\alpha}{\partial \varsigma}\,, \tag{8.98}$$

where the partial derivative $\partial r^\alpha/\partial\varsigma = \delta_0^\alpha = (1, 0, 0, 0)$. Making use of (8.86), the partial derivative

$$\frac{\partial f_x}{\partial r^\alpha} = -\overline{\mathrm{v}}_\alpha + \frac{c}{v_s}\frac{\overline{\pi}_{\alpha\beta}r^\beta}{\sqrt{\overline{\pi}_{\alpha\beta}r^\alpha r^\beta}}\,, \tag{8.99}$$

which has to be calculated at the instant of time, $t' = \varsigma$, where ς is given by (8.96).

In order to calculate the denominator in the integrand in (8.88), we account for (8.85) and (8.89) and combine (8.98) and (8.99) together. We get

$$|\boldsymbol{X} - \boldsymbol{X}'|\dot{f}_x(\varsigma) = \frac{c}{v_s}\left[r_\alpha + \left(1 - \frac{v_s^2}{c^2}\right)\overline{\mathrm{v}}_\alpha\overline{\mathrm{v}}_\beta r^\beta\right]\delta_0^\alpha\,. \tag{8.100}$$

It is straightforward to check that after using (8.94) the above equation is reduced to $|\boldsymbol{X} - \boldsymbol{X}'|\dot{f}_x(\varsigma) = (c/v_s)\sqrt{B^2 - AC}$, or more explicitly,

$$|\boldsymbol{X} - \boldsymbol{X}'|\dot{f}_x(\varsigma) = |\boldsymbol{x} - \boldsymbol{x}'|\sqrt{1 + \left(1 - \frac{c^2}{v_s^2}\right)\gamma^2(\boldsymbol{\beta}\times\boldsymbol{n})^2}\,, \tag{8.101}$$

Finally, the retarded Lorentz-invariant solution of (8.79) is

$$\mathcal{U}(t, \boldsymbol{x}) = \int\limits_{\mathcal{V}} \frac{\tau_x(\varsigma, \boldsymbol{x}')}{\sqrt{1 + \gamma^2\left(1 - \frac{c^2}{v_s^2}\right)(\boldsymbol{\beta}\times\boldsymbol{n})^2}}\frac{d^3x'}{|\boldsymbol{x} - \boldsymbol{x}'|}\,, \tag{8.102}$$

with the retarded time ς calculated in accordance with (8.96). This solution reduces to the retarded potential (8.78) in the limit of $v_s \to c$.

References

[1] L. R. W. Abramo, R. H. Brandenberger and V. F. Mukhanov, Energy-momentum tensor for cosmological perturbations, *Physical Review D* **56** (1997), 3248–3257.

[2] L. Amendola and S. Tsujikawa, *Dark Energy: Theory and Observations*, Cambridge: Cambridge University Press, 2010.

[3] J. D. Anderson, P. A. Laing, E. L. Lau, A. S. Liu, M. M. Nieto and S. G. Turyshev, Study of the anomalous acceleration of Pioneer 10 and 11, *Physical Review D* **65** (2002), 082004.

[4] J. D. Anderson and M. M. Nieto, Astrometric solar-system anomalies, in: *IAU Symposium* (S. A. Klioner, P. K. Seidelmann and M. H. Soffel, eds.), IAU Symposium 261, pp. 189–197, 2010.
[5] R. Arnowitt, S. Deser and C. W. Misner, Dynamical Structure and Definition of Energy in General Relativity, *Physical Review* **116** (1959), 1322–1330.
[6] S. V. Babak and L. P. Grishchuk, Energy-momentum tensor for the gravitational field, *Physical Review D* **61** (2000), 024038.
[7] R. Baierlein, Representing a vector field: Helmholtz's theorem derived from a Fourier identity, *American Journal of Physics* **63** (1995), 180–182.
[8] J. M. Bardeen, Gauge-invariant cosmological perturbations, *Physical Review D* **22** (1980), 1882–1905.
[9] J. M. Bardeen, Gauge-Invariant Cosmological Perturbations, in: *The Early Universe: Reprints* (E. W. Kolb and M. S. Turner, eds.), Frontiers in Physics, Reading: Addison-Wesley, p. 563, 1988.
[10] A. N. Beris and B. J. Edwards, *Thermodynamics of Flowing Systems with internal microstructure*, Oxford: Oxford University Press, 1994.
[11] G. Bertone, *Particle Dark Matter: Observations, Models and Searches*, Cambridge: Cambridge University Press, 2010.
[12] S. Bildhauer and T. Futamase, The age problem in inhomogeneous universes, *General Relativity and Gravitation* **23** (1991), 1251–1264.
[13] L. Blanchet, Gravitational Radiation from Post-Newtonian Sources and Inspiralling Compact Binaries, *Living Reviews in Relativity* **9** (2006), 4.
[14] L. Blanchet and T. Damour, The post-Minkowskian iteration method and the structure of radiative gravitational fields, *Academie des Sciences Paris Comptes Rendus Serie B Sciences Physiques* **298** (1984), 431–434.
[15] G. R. Blumenthal, S. M. Faber, J. R. Primack and M. J. Rees, Formation of galaxies and large-scale structure with cold dark matter, *Nature* **311** (1984), 517–525.
[16] G. R. Blumenthal, S. M. Faber, J. R. Primack and M. J. Rees, Erratum: Formation of galaxies and large-scale structure with cold dark matter, *Nature* **313** (1985), 72.
[17] W. B. Bonnor, A generalization of the Einstein-Straus vacuole, *Classical and Quantum Gravity* **17** (2000), 2739–2748.
[18] V. A. Brumberg, *Relativistic Celestial Mechanics*, Nauka: Moscow, USSR, 1972.
[19] V. A. Brumberg, *Essential Relativistic Celestial Mechanics*, Bristol: Adam Hilger, 1991.
[20] V. A. Brumberg, Relativistic Celestial Mechanics on the verge of its 100 year anniversary. (Brouwer Award lecture), *Celestial Mechanics and Dynamical Astronomy* **106** (2010), 209–234.
[21] M. Bruni, G. F. R. Ellis and P. K. S. Dunsby, Gauge-invariant perturbations in a scalar field dominated universe, *Classical and Quantum Gravity* **9** (1992), 921–945.
[22] T. Buchert and J. Ehlers, Lagrangian theory of gravitational instability of Friedman-Lemaitre cosmologies – second-order approach: an improved model for non-linear clustering, *Monthly Notices of the Royal Astronomical Society* **264** (1993), 375–387.
[23] T. Buchert and S. Räsänen, Backreaction in Late-Time Cosmology, *Annual Review of Nuclear and Particle Science* **62** (2012), 57–79.
[24] M. Carrera and D. Giulini, Influence of global cosmological expansion on local dynamics and kinematics, *Reviews of Modern Physics* **82** (2010), 169–208.
[25] S. Chandrasekhar and F. P. Esposito, The $2\frac{1}{2}$-Post-Newtonian Equations of Hydrodynamics and Radiation Reaction in General Relativity, *The Astrophysical Journal* **160** (1970), 153.
[26] C. Chicone and B. Mashhoon, Explicit Fermi coordinates and tidal dynamics in de Sitter and Gödel spacetimes, *Physical Review D* **74** (2006), 064019.
[27] I. Ciufolini and J. A. Wheeler, *Gravitation and Inertia*, 1995.

[28] C. Clarkson and B. Osano, Locally extracting scalar, vector and tensor modes in cosmological perturbation theory, *Classical and Quantum Gravity* **28** (2011), 225002.

[29] C. Clarkson and B. Osano, Erratum: Locally extracting scalar, vector and tensor modes in cosmological perturbation theory, *Classical and Quantum Gravity* **29** (2012), 079601.

[30] J. Crowder and N. J. Cornish, Beyond LISA: Exploring future gravitational wave missions, *Physical Review D* **72** (2005), 083005.

[31] T. Damour, The problem of motion in Newtonian and Einsteinian gravity, in: *Three Hundred Years of Gravitation* (S. W. Hawking and W. Israel, eds.), pp. 128–198, 1987.

[32] S. Deser, Self-interaction and gauge invariance, *General Relativity and Gravitation* **1** (1970), 9–18.

[33] B. Dubrovin, A. Fomenko and S. Novikov, *Modern Geometry – Methods and Applications. Part II. The Geometry and Topology of Manifolds*, Graduate texts in mathematics, Springer: New York, 1985.

[34] R. Durrer, Some applications of the 3+1 formalism of general relativity, *Helvetica Physica Acta* **61** (1988), 1027–1062.

[35] R. Durrer, Gauge-invariant cosmological perturbation theory for collisionless matter – Numerical results, *Astronomy and Astrophysics* **208** (1989), 1–13.

[36] J. Ehlers and T. Buchert, Newtonian Cosmology in Lagrangian Formulation: Foundations and Perturbation Theory, *General Relativity and Gravitation* **29** (1997), 733–764.

[37] A. Einstein and E. G. Straus, The Influence of the Expansion of Space on the Gravitation Fields Surrounding the Individual Stars, *Reviews of Modern Physics* **17** (1945), 120–124.

[38] A. Einstein and E. G. Straus, Corrections and Additional Remarks to our Paper: The Influence of the Expansion of Space on the Gravitation Fields Surrounding the Individual Stars, *Reviews of Modern Physics* **18** (1946), 148–149.

[39] G. F. R. Ellis and M. Bruni, Covariant and gauge-invariant approach to cosmological density fluctuations, *Physical Review D* **40** (1989), 1804–1818.

[40] G. F. R. Ellis, M. Bruni and J. Hwang, Density-gradient-vorticity relation in perfect-fluid Robertson-Walker perturbations, *Physical Review D* **42** (1990), 1035–1046.

[41] G. F. R. Ellis, J. Hwang and M. Bruni, Covariant and gauge-independent perfect-fluid Robertson-Walker perturbations, *Physical Review D* **40** (1989), 1819–1826.

[42] G. F. R. Ellis and J.-P. Uzan, c is the speed of light, isn't it?, *American Journal of Physics* **73** (2005), 240–247.

[43] D. J. Fixsen, E. S. Cheng, J. M. Gales, J. C. Mather, R. A. Shafer and E. L. Wright, The Cosmic Microwave Background Spectrum from the Full COBE FIRAS Data Set, *The Astrophysical Journal* **473** (1996), 576.

[44] V. A. Fock, *The Theory of Space, Time and Gravitation*, Oxford: Pergamon Press, 1964.

[45] T. Futamase, Approximation scheme for constructing a clumpy universe in general relativity, *Physical Review Letters* **61** (1988), 2175–2178.

[46] T. Futamase, An approximation scheme for constructing inhomogeneous universes in general relativity, *Monthly Notices of the Royal Astronomical Society* **237** (1989), 187–200.

[47] G. Goldhaber, The Acceleration of the Expansion of the Universe: A Brief Early History of the Supernova Cosmology Project (SCP), in: *American Institute of Physics Conference Series* (D. B. Cline, ed.), American Institute of Physics Conference Series 1166, pp. 53–72, September 2009.

[48] W. Greiner and J. Reinhardt, *Field quantization*, Springer-Verlag, Berlin, 1996, Translated from the German, with a foreword by D. A. Bromley.

[49] L. P. Grishchuk, Amplification of gravitational waves in an isotropic universe, *Soviet Journal of Experimental and Theoretical Physics* **40** (1974), 409.

[50] L. P. Grishchuk, A. N. Petrov and A. D. Popova, Exact theory of the (Einstein) gravitational field in an arbitrary background space-time, *Communications in Mathematical Physics* **94** (1984), 379–396.
[51] R. Haas and E. Poisson, Mass change and motion of a scalar charge in cosmological space-times, *Classical and Quantum Gravity* **22** (2005), 739.
[52] G. Hinshaw, J. L. Weiland, R. S. Hill, N. Odegard, D. Larson, C. L. Bennett, J. Dunkley, B. Gold, M. R. Greason, N. Jarosik, E. Komatsu, M. R. Nolta, L. Page, D. N. Spergel, E. Wollack, M. Halpern, A. Kogut, M. Limon, S. S. Meyer, G. S. Tucker and E. L. Wright, Five-Year Wilkinson Microwave Anisotropy Probe Observations: Data Processing, Sky Maps, and Basic Results, *The Astrophysical Journal Supplement* **180** (2009), 225–245.
[53] S. Hojman, Problem of the identical vanishing of Euler-Lagrange derivatives in field theory, *Phys. Rev. D* **27** (1983), 451–453.
[54] L. Hongya, Cosmological models in globally geodesic coordinates. I. Metric, *Journal of Mathematical Physics* **28** (1987), 1920–1923.
[55] M. Ibison, On the conformal forms of the Robertson-Walker metric, *Journal of Mathematical Physics* **48** (2007), 122501.
[56] R. A. Isaacson, Gravitational Radiation in the Limit of High Frequency. I. The Linear Approximation and Geometrical Optics, *Physical Review* **166** (1968), 1263–1271.
[57] R. A. Isaacson, Gravitational Radiation in the Limit of High Frequency. II. Nonlinear Terms and the Effective Stress Tensor, *Physical Review* **166** (1968), 1272–1279.
[58] J. D. Jackson and L. B. Okun, Historical roots of gauge invariance, *Reviews of Modern Physics* **73** (2001), 663–680.
[59] N. Jarosik, C. L. Bennett, J. Dunkley, B. Gold, M. R. Greason, M. Halpern, R. S. Hill, G. Hinshaw, A. Kogut, E. Komatsu, D. Larson, M. Limon, S. S. Meyer, M. R. Nolta, N. Odegard, L. Page, K. M. Smith, D. N. Spergel, G. S. Tucker, J. L. Weiland, E. Wollack and E. L. Wright, Seven-year Wilkinson Microwave Anisotropy Probe (WMAP) Observations: Sky Maps, Systematic Errors, and Basic Results, *The Astrophysical Journal Supplement Series* **192** (2011), 14.
[60] C. L. Bennett, D. Larson, J. L. Weiland, N. Jarosik, G. Hinshaw, N. Odegard, K. M. Smith, R. S. Hill, B. Gold, M. Halpern, E. Komatsu, M. R. Nolta, L. Page, D. N. Spergel, E. Wollack, J. Dunkley, A. Kogut, M. Limon, S. S. Meyer, G. S. Tucker, E. L. Wright, Nine-year Wilkinson Microwave Anisotropy Probe (WMAP) Observations: Final Maps and Results, *The Astrophysical Journal Supplement Series* **208** (2013), 20.
[61] S. A. Klioner and M. H. Soffel, Refining the Relativistic Model for Gaia: Cosmological Effects in the BCRS, in: *The Three-Dimensional Universe with Gaia* (C. Turon, K. S. O'Flaherty and M. A. C. Perryman, eds.), ESA Special Publication 576, pp. 305–308, January 2005.
[62] H. Kodama and M. Sasaki, Cosmological Perturbation Theory, *Progress of Theoretical Physics Supplement* **78** (1984), 1.
[63] A. Kogut, C. Lineweaver, G. F. Smoot, C. L. Bennett, A. Banday, N. W. Boggess, E. S. Cheng, G. de Amici, D. J. Fixsen, G. Hinshaw, P. D. Jackson, M. Janssen, P. Keegstra, K. Loewenstein, P. Lubin, J. C. Mather, L. Tenorio, R. Weiss, D. T. Wilkinson and E. L. Wright, Dipole Anisotropy in the COBE Differential Microwave Radiometers First-Year Sky Maps, *The Astrophysical Journal* **419** (1993), 1.
[64] K. D. Kokkotas, Gravitational Wave Astronomy (With 2 Figures), in: *Reviews in Modern Astronomy* (S. Röser, ed.), Reviews in Modern Astronomy 20, p. 140, 2008.
[65] E. W. Kolb, V. Marra and S. Matarrese, Description of our cosmological spacetime as a perturbed conformal Newtonian metric and implications for the backreaction proposal for the accelerating universe, *Physical Review D* **78** (2008), 103002.
[66] S. Kopeikin, M. Efroimsky and G. Kaplan, *Relativistic Celestial Mechanics of the Solar System*, Weinheim: Wiley-VCH, 2011.

[67] S. Kopeikin and I. Vlasov, Parametrized post-Newtonian theory of reference frames, multipolar expansions and equations of motion in the N-body problem, *Physics Reports* **400** (2004), 209–318.
[68] S. M. Kopeikin, The speed of gravity in general relativity and theoretical interpretation of the Jovian deflection experiment, *Classical and Quantum Gravity* **21** (2004), 3251–3286.
[69] S. M. Kopeikin, Celestial ephemerides in an expanding universe, *Physical Review D* **86** (2012), 064004.
[70] S. M. Kopeikin and A. N. Petrov, Post-Newtonian celestial dynamics in cosmology: Field equations, *Physical Review D* **87** (2013), 044029.
[71] G. A. Krasinsky and V. A. Brumberg, Secular increase of astronomical unit from analysis of the major planet motions, and its interpretation, *Celestial Mechanics and Dynamical Astronomy* **90** (2004), 267–288.
[72] A. A. Kurskov and L. M. Ozernoi, Evolution of cosmological turbulence. IV – Gravitational growth of inhomogeneities with isolation of galaxies and galaxy systems, *Soviet Astronomy* **19** (1976), 569–576.
[73] L. D. Landau and E. M. Lifshitz, *Fluid mechanics*, Oxford: Pergamon Press, 1959.
[74] L. D. Landau and E. M. Lifshitz, *Statistical physics. Part 1*, Oxford: Pergamon Press, 1969.
[75] L. D. Landau and E. M. Lifshitz, *The classical theory of fields*, Oxford: Pergamon Press, 1975.
[76] N. Li and D. J. Schwarz, Onset of cosmological backreaction, *Physical Review D* **76** (2007), 083011.
[77] N. Li and D. J. Schwarz, Scale dependence of cosmological backreaction, *Physical Review D* **78** (2008), 083531.
[78] E. Lifshitz, On the gravitational stability of the expanding universe, *JETP* **10** (1946), 116–129, http://www.citebase.org/abstract?id=oai%3AarXiv.org%3A1102.5039.
[79] E. M. Lifshitz and I. M. Khalatnikov, Investigations in relativistic cosmology, *Advances in Physics* **12** (1963), 185–249.
[80] E. V. Linder, *First Principles of Cosmology*, Prentice Hall, 1997.
[81] B. Ryden, *Introduction to Cosmology*, Addison-Wesley, 2003.
[82] D. R. Lorimer and M. Kramer, *Handbook of Pulsar Astronomy*, Cambridge: Cambridge University Press, 2005.
[83] V. N. Lukash, Production of sound waves in the early universe, *Soviet Journal of Experimental and Theoretical Physics Letters* **31** (1980), 596.
[84] C.-P. Ma and E. Bertschinger, Cosmological Perturbation Theory in the Synchronous and Conformal Newtonian Gauges, *The Astrophysical Journal* **455** (1995), 7.
[85] A. Macías, J. L. Cervantes-Cota and C. Lämmerzahl (eds.), *Exact solutions and scalar fields in gravity: recent developments*, 2001.
[86] B. Mashhoon, N. Mobed and D. Singh, Tidal dynamics in cosmological spacetimes, *Classical and Quantum Gravity* **24** (2007), 5031–5046.
[87] G. C. McVittie, The mass-particle in an expanding universe, *Monthly Notices of the Royal Astronomical Society* **93** (1933), 325–339.
[88] E. W. Mielke, Conformal changes of metrics and the initial-value problem of general relativity, *General Relativity and Gravitation* **8** (1977), 321–345.
[89] C. W. Misner, K. S. Thorne and J. A. Wheeler, *Gravitation*, San Francisco: W. H. Freeman and Co., 1973.
[90] N. V. Mitskevich, *Physical Fields in General Relativity*, Nauka: Moscow, USSR, 1969.
[91] V. Mukhanov, *Physical Foundations of Cosmology*, Cambridge: Cambridge University Press, November 2005.
[92] V. F. Mukhanov, L. R. W. Abramo and R. H. Brandenberger, Backreaction Problem for Cosmological Perturbations, *Physical Review Letters* **78** (1997), 1624–1627.

[93] V. F. Mukhanov, H. A. Feldman and R. H. Brandenberger, Theory of cosmological perturbations, *Physics Reports* **215** (1992), 203–333.

[94] P. J. E. Peebles, *The Large-scale Structure of the Universe*, Princeton: Princeton University Press, 1980.

[95] A. N. Petrov and J. Katz, Conserved currents, superpotentials and cosmological perturbations, *Royal Society of London Proceedings Series A* **458** (2002), 319.

[96] A. D. Popova and A. N. Petrov, Exact Dynamic Theories on a Given Background in Gravitation, *International Journal of Modern Physics A* **3** (1988), 2651–2679.

[97] J. Ramírez and S. Kopeikin, A decoupled system of hyperbolic equations for linearized cosmological perturbations, *Physics Letters B* **532** (2002), 1–7.

[98] S. Räsänen, Applicability of the linearly perturbed FRW metric and Newtonian cosmology, *Physical Review D* **81** (2010), 103512.

[99] G. Schäfer, Post-Newtonian Methods: Analytic Results on the Binary Problem, in: *Mass and Motion in General Relativity* (L. Blanchet, A. Spallicci and B. Whiting, eds.), pp. 167–210, 2011.

[100] E. Schücking, Das Schwarzschildsche Linienelement und die Expansion des Weltalls, *Zeitschrift für Physik* **137** (1954), 595–603.

[101] B. Schutz, *Geometrical Methods of Mathematical Physics*, Cambridge, UK: Cambridge University Press, 1980.

[102] B. F. Schutz, Perfect Fluids in General Relativity: Velocity Potentials and a Variational Principle, *Physical Review D* **2** (1970), 2762–2773.

[103] M. Soffel, S. A. Klioner, G. Petit, P. Wolf, S. M. Kopeikin, P. Bretagnon, V. A. Brumberg, N. Capitaine, T. Damour, T. Fukushima, B. Guinot, T.-Y. Huang, L. Lindegren, C. Ma, K. Nordtvedt, J. C. Ries, P. K. Seidelmann, D. Vokrouhlický, C. M. Will and C. Xu, The IAU 2000 Resolutions for Astrometry, Celestial Mechanics, and Metrology in the Relativistic Framework: Explanatory Supplement, *The Astronomical Journal* **126** (2003), 2687–2706.

[104] M. H. Soffel, *Relativity in Astrometry, Celestial Mechanics and Geodesy*, Springer-Verlag: Berlin, 1989.

[105] I. H. Stairs, Binary pulsars and tests of general relativity, in: *IAU Symposium* (S. A. Klioner, P. K. Seidelmann and M. H. Soffel, eds.), IAU Symposium 261, pp. 218–227, 2010.

[106] H. Stephani, D. Kramer, M. MacCallum, C. Hoenselaers and E. Herlt, *Exact solutions of Einstein's field equations*, Cambridge: Cambridge University Press, 2003.

[107] M. Takada and T. Futamase, A post-Newtonian Lagrangian perturbation approach to large-scale structure formation, *Monthly Notices of the Royal Astronomical Society* **306** (1999), 64–88.

[108] A. H. Taub, General Relativistic Variational Principle for Perfect Fluids, *Physical Review* **94** (1954), 1468–1470.

[109] M. Tegmark, Measuring the metric: A parametrized post-Friedmannian approach to the cosmic dark energy problem, *Physical Review D* **66** (2002), 103507.

[110] H. van Elst and C. Uggla, General relativistic 1+3 orthonormal frame approach, *Classical and Quantum Gravity* **14** (1997), 2673–2695.

[111] R. M. Wald, *General relativity*, Chicago: University of Chicago Press, 1984.

[112] A. Liddle, *An Introduction to Modern Cosmology*, New York: John Wiley & Sons, Inc., 2003.

[113] S. Weinberg, *Gravitation and Cosmology: Principles and Applications of the General Theory of Relativity*, New York: John Wiley & Sons, Inc., 1972.

[114] S. Weinberg, *Cosmology*, Oxford University Press, 2008.

[115] J. M. Weisberg, D. J. Nice and J. H. Taylor, Timing Measurements of the Relativistic Binary Pulsar PSR B1913+16, *The Astrophysical Journal* **722** (2010), 1030–1034.

[116] A. Wiegand and D. J. Schwarz, Inhomogeneity-induced variance of cosmological parameters, *Astronomy & Astrophysics* **538** (2012), A147.

[117] C. M. Will, *Theory and Experiment in Gravitational Physics*, Cambridge: Cambridge University Press, 1993.

[118] C. M. Will, The Confrontation between General Relativity and Experiment, *Living Reviews in Relativity* **9** (2006), 3.

[119] C. M. Will, Inaugural Article: On the unreasonable effectiveness of the post-Newtonian approximation in gravitational physics, *Proceedings of the National Academy of Science* **108** (2011), 5938–5945.

[120] R. Zalaletdinov, The Averaging Problem in Cosmology and Macroscopic Gravity, *International Journal of Modern Physics A* **23** (2008), 1173–1181.

[121] A. Zelmanov, *Chronometric Invariants: On Deformations and the Curvature of Accompanying Space*, Ph.D. Thesis, 1944. American Research Press: Rehoboth, NM USA, 2006, Translated from the Russian by Dmitri Rabounski and Stephen J. Crothers. Formulae corrected by Larissa Borissova.

[122] A. L. Zelmanov, Chronometric Invariants and Frames of Reference in the General Theory of Relativity, *Soviet Physics Doklady* **1** (1956), 227.

[123] A. L. Zelmanov, Kinemetric invarants and their relation to chronometric invariants in Einstein's gravitation theory, *Soviet Physics Doklady* **18** (1973), 231.

[124] A. L. Zelmanov and V. G. Agakov, *Elements of the general theory of relativity*, 1989.

Index

De Gruyter Studies in Mathematical Physics

Volume 20
Valery A. Slaev, Anna G. Chunovkina, Leonid A. Mironovsky
Metrology and Theory of Measurement, 2013
ISBN 978-3-11-028473-7, e-ISBN 978-3-11-028482-9, Set-ISBN 978-3-11-028483-6

Volume 19
Edward A. Bormashenko
Wetting of Real Surfaces, 2013
ISBN 978-3-11-025853-0, e-ISBN 978-3-11-025879-0, Set-ISBN 978-3-11-220375-0

Volume 18
Michael V. Sadovskii
Statistical Physics, 2012
ISBN 978-3-11-027031-0, e-ISBN 978-3-11-027037-2, Set-ISBN 978-3-11-027038-9

Volume 17
Michael V. Sadovskii
Quantum Field Theory, 2013
ISBN 978-3-11-027029-7, e-ISBN 978-3-11-027035-8, Set-ISBN 978-3-11-027036-5

Volume 16
Alexander S. Holevo
Quantum Systems, Channels, Information: A Mathematical Introduction, 2012
ISBN 978-3-11-027325-0, e-ISBN 978-3-11-027340-3, Set-ISBN 978-3-11-027341-0

Volume 15
Vladimir N. Kukudzhanov
Numerical Continuum Mechanics, 2012
ISBN 978-3-11-027322-9, e-ISBN 978-3-11-027338-0, Set-ISBN 978-3-11-027339-7

Volume 14
Oleg N. Kirillov
Non-conservative Stability Problems of Modern Physics, 2013
ISBN 978-3-11-027034-1, e-ISBN 978-3-11-027043-3, Set-ISBN 978-3-11-027044-0